Biological
Data Mining

Chapman & Hall/CRC
Data Mining and Knowledge Discovery Series

SERIES EDITOR
Vipin Kumar
University of Minnesota
Department of Computer Science and Engineering
Minneapolis, Minnesota, U.S.A

AIMS AND SCOPE

This series aims to capture new developments and applications in data mining and knowledge discovery, while summarizing the computational tools and techniques useful in data analysis. This series encourages the integration of mathematical, statistical, and computational methods and techniques through the publication of a broad range of textbooks, reference works, and handbooks. The inclusion of concrete examples and applications is highly encouraged. The scope of the series includes, but is not limited to, titles in the areas of data mining and knowledge discovery methods and applications, modeling, algorithms, theory and foundations, data and knowledge visualization, data mining systems and tools, and privacy and security issues.

PUBLISHED TITLES

UNDERSTANDING COMPLEX DATASETS: Data Mining with Matrix Decompositions
David Skillicorn

COMPUTATIONAL METHODS OF FEATURE SELECTION
Huan Liu and Hiroshi Motoda

CONSTRAINED CLUSTERING: Advances in Algorithms, Theory, and Applications
Sugato Basu, Ian Davidson, and Kiri L. Wagstaff

KNOWLEDGE DISCOVERY FOR COUNTERTERRORISM AND LAW ENFORCEMENT
David Skillicorn

MULTIMEDIA DATA MINING: A Systematic Introduction to Concepts and Theory
Zhongfei Zhang and Ruofei Zhang

NEXT GENERATION OF DATA MINING
Hillol Kargupta, Jiawei Han, Philip S. Yu, Rajeev Motwani, and Vipin Kumar

DATA MINING FOR DESIGN AND MARKETING
Yukio Ohsawa and Katsutoshi Yada

THE TOP TEN ALGORITHMS IN DATA MINING
Xindong Wu and Vipin Kumar

GEOGRAPHIC DATA MINING AND KNOWLEDGE DISCOVERY, Second Edition
Harvey J. Miller and Jiawei Han

TEXT MINING: CLASSIFICATION, CLUSTERING, AND APPLICATIONS
Ashok N. Srivastava and Mehran Sahami

BIOLOGICAL DATA MINING
Jake Y. Chen and Stefano Lonardi

Chapman & Hall/CRC
Data Mining and Knowledge Discovery Series

Biological Data Mining

Edited by
Jake Y. Chen
Stefano Lonardi

CRC Press
Taylor & Francis Group
Boca Raton London New York

CRC Press is an imprint of the
Taylor & Francis Group, an **informa** business

A CHAPMAN & HALL BOOK

Chapman & Hall/CRC
Taylor & Francis Group
6000 Broken Sound Parkway NW, Suite 300
Boca Raton, FL 33487-2742

First issued in paperback 2017

© 2010 by Taylor and Francis Group, LLC
Chapman & Hall/CRC is an imprint of Taylor & Francis Group, an Informa business

No claim to original U.S. Government works

ISBN-13: 978-1-4200-8684-3 (hbk)
ISBN-13: 978-1-138-11658-0 (pbk)

Library of Congress Cataloging-in-Publication Data

Biological data mining / editors, Jake Y. Chen, Stefano Lonardi.
    p. cm. -- (Data mining and knowledge discovery series)
  Includes bibliographical references and index.
   ISBN 978-1-4200-8684-3 (hardcover : alk. paper)
   1. Bioinformatics. 2. Data mining. 3. Computational biology. I. Chen, Jake. II. Lonardi, Stefano. III. Title. IV. Series.

QH324.2.B578 2010
570.285--dc22   2009028067

Visit the Taylor & Francis Web site at
http://www.taylorandfrancis.com

and the CRC Press Web site at
http://www.crcpress.com

Contents

Preface

Modern biology has become an information science. Since the invention of a DNA sequencing method by Sanger in the late seventies, public repositories of genomic sequences have been growing exponentially, doubling in size every 16 months—a rate often compared to the growth of semiconductor transistor densities in CPUs known as Moore's Law. In the nineties, the public–private race to sequence the human genome further intensified the fervor to generate high-throughput biomolecular data from highly parallel and miniaturized instruments. Today, sequencing data from thousands of genomes, including plants, mammals, and microbial genomes, are accumulating at an unprecedented rate. The advent of second-generation DNA sequencing instruments, high-density cDNA microarrays, tandem mass spectrometers, and high-power NMRs have fueled the growth of molecular biology into a wide spectrum of disciplines such as personalized genomics, functional genomics, proteomics, metabolomics, and structural genomics. Few experiments in molecular biology and genetics performed today can afford to ignore the vast amount of biological information publicly accessible. Suddenly, molecular biology and genetics have become data rich.

Biological data mining is a *data-guzzling turbo engine* for postgenomic biology, driving the competitive race toward unprecedented biological discovery opportunities in the twenty-first century. Classical bioinformatics emerged from the study of macromolecules in molecular biology, biochemistry, and biophysics. Analysis, comparison, and classification of DNA and protein sequences were the dominant themes of bioinformatics in the early nineties. Machine learning mainly focused on predicting genes and proteins functions from their sequences and structures. The understanding of cellular functions and processes underlying complex diseases were out of reach. Bioinformatics scientists were a rare breed, and their contribution to molecular biology and genetics was considered marginal, because the computational tools available then for biomolecular data analysis were far more primitive than the array of experimental techniques and assays that were available to life scientists. Today, we are now witnessing the reversal of these past trends. Diverse sets of data types that cover a broad spectrum of genotypes and phenotypes, particularly those related to human health and diseases, have become available. Many interdisciplinary researchers, including applied computer scientists, applied mathematicians, biostatisticians, biomedical researchers, clinical scientists, and biopharmaceutical professionals, have discovered in biology a *gold mine* of knowledge leading to many exciting possibilities: the unraveling of the tree of life, harnessing the power of microbial organisms for renewable energy, finding new ways to diagnose disease early, and developing new therapeutic compounds that save lives. Much of the experimental high-throughput biology data are generated and analyzed "in haste," therefore leaving plenty of opportunities for knowledge discovery even after the original data are released. Most of the bets on the race to *separate the wheat from the chaff* have been placed on biological data mining techniques. After all, when easy, straightforward, first-pass data analysis has not yielded novel biological insights, data mining techniques must be able to help—or, many presumed so.

In reality, biological data mining is still much of an "art," successfully practiced by a few bioinformatics research groups that occupy themselves with solving real-world biological problems. Unlikely data mining in business, where the major concerns are often related to the bottom line—profit—the goals of biological data mining can be as diverse as the spectrum of biological questions that exist. In the business domain, association rules discovered between sales items are immediately actionable; in biology, any unorthodox hypothesis produced by computational models has to be first red-flagged and is lucky to be validated experimentally. In the Internet business domain, classification, clustering, and visualization of blogs, network traffic patterns, and news feeds add significant values to regular Internet users who are unaware of high-level patterns that may exist in the data set; in molecular biology and genetics, any clustering or classification of the data presented to biologists may promptly elicit questions like "great, but how and why did it happen?" or "how can you explain these results in the context of the biology I know?" The majority of general-purpose data mining techniques do not take into consideration the prior knowledge domain of the biological problem, leading them to often underperform hypothesis-driven biological investigative techniques. The high level of variability of measurements inherent in many types of biological experiments or samples, the general unavailability of experimental replicates, the large number of hidden variables in the data, and the high correlation of biomolecular expression measurements also constitute significant challenges in the application of classical data mining methods in biology. Many biological data mining projects are attempted and then abandoned, even by experienced data mining scientists. In the extreme cases, large-scale biological data mining efforts are jokingly labeled as *fishing expeditions* and dispelled, in national grant proposal review panels.

This book represents a culmination of our past research efforts in biological data mining. Throughout this book, we wanted to showcase a small, but noteworthy sample of successful projects involving data mining and molecular biology. Each chapter of the book is authored by a distinguished team of bioinformatics scientists whom we invited to offer the readers the widest possible range of application domains. To ensure high-quality standards, each contributed chapter went through standard peer reviews and a round of revisions. The contributed chapters have been grouped into five major sections.

The first section, entitled *Sequence, Structure, and Function*, collects contributions on data mining techniques designed to analyze biological sequences and structures with the objective of discovering novel functional knowledge. The second section, on *Genomics, Transcriptomics, and Proteomics*, contains studies addressing emerging large-scale data mining challenges in analyzing high-throughput "omics" data. The chapters in the third section, entitled *Functional and Molecular Interaction Networks*, address emerging system-scale molecular properties and their relevance to cellular functions. The fourth section is about *Literature, Ontology, and Knowledge Integrations*, and it collects chapters related to knowledge representation, information retrieval, and data integration for structured and unstructured biological data. The contributed works in the fifth and last section, entitled *Genome Medicine Applications*, address emerging biological data mining applications in medicine.

We believe this book can serve as a valuable guide to the field for graduate students, researchers, and practitioners. We hope that the wide range of topics covered will allow readers to appreciate the extent of the impact of data mining in molecular biology and genetics. For us, research in data mining and its applications to biology and genetics is fascinating and rewarding. It may even help to save human lives one day. This field offers great opportunities and rewards if one is prepared to learn molecular biology and genetics, design user-friendly software tools under the proper biological assumptions, and validate all discovered hypotheses rigorously using appropriate models.

In closing, we would like to thank all the authors that contributed a chapter in the book. We are also indebted to Randi Cohen, our outstanding publishing editor. Randi efficiently managed timelines and deadlines, gracefully handled the communication with the authors and the reviewers, and took care of every little detail associated with this project. This book could not have been possible without her. Our thanks also go to our families for their support throughout the book project.

Jake Y. Chen
Indianapolis, Indiana

Stefano Lonardi
Riverside, California

Editors

Jake Chen is an assistant professor of informatics at Indiana University School of Informatics and assistant professor of computer science at Purdue School of Science, Indiana. He is the founding director of the Indiana Center for Systems Biology and Personalized Medicine—the first research center in the region to promote the development of systems biology tools towards solving future personalized medicine problems. He is an IEEE senior member and a member of several other interdisciplinary Indiana research centers, including: Center for Computational Biology and Bioinformatics, Center for Bio-computing, Indiana University Cancer Center, and Indiana Center for Environmental Health. He was a scientific co-founder and chief informatics officer (2006–2008) of Predictive Physiology and Medicine, Inc. and the founder of Medeolinx, LLC-Indiana biotech startups developing businesses in emerging personalized medicine and translational bioinformatics markets.

Dr. Chen received PhD and MS degrees in computer science from the University of Minnesota at Twin Cities and a BS in molecular biology and biochemistry from Peking University in China. He has extensive industrial research and management experience (1998–2003), including developing commercial GeneChip microarrays at Affymetrix, Inc. and mapping the first human protein interactome at Myriad Proteomics. After rejoining academia in 2004, he concentrated his research on "translational bioinformatics," studies aiming to bridge the gaps between bioinformatics research and human health applications. He has over 60 publications in the areas of biological data management, biological data mining, network biology, systems biology, and various disease-related omics applications.

Stefano Lonardi is associate professor of computer science and engineering at the University of California, Riverside. He is also a faculty member of the graduate program in genetics, genomics and bioinformatics, the Center for Plant Cell Biology, the Institute for Integrative Genome Biology, and the graduate program in cell, molecular and developmental biology.

Dr. Lonardi received his "Laurea cum laude" from the University of Pisa in 1994 and his PhD, in the summer of 2001, from the Department of Computer Sciences, Purdue University, West Lafayette, IN. He also holds a PhD in electrical and information engineering from the University of Padua (1999). During the summer of 1999, he was an intern at Celera Genomics, Department of Informatics Research, Rockville, MD.

Dr. Lonardi's recent research interests include designing of algorithms, computational molecular biology, data compression, and data mining. He has published more than 30 papers in major theoretical computer science and computational biology journals and has about 45 publications in refereed international conferences. In 2005, he received the CAREER award from the National Science Foundation.

Contributors

Muhammad Abulaish
Department of Computer Science
Jamia Millia Islamia
New Delhi, India

Alberto Apostolico
College of Computing
Georgia Institute of Technology
Atlanta, Georgia

Simon Beaulah
InforSense, Ltd.
London, United Kingdom

Paola Bertolazzi
Istituto di Analisi dei Sistemi ed
    Informatica Antonio Ruberti
Consiglio Nazionale delle Ricerche
Rome, Italy

Paul E. Blower
Department of Pharmacology
Ohio State University
Columbus, Ohio

Charles Buck
Bindley Bioscience Center
Purdue University
West Lafayette, Indiana

Jennifer Cai
Department of Pathology
University of Texas Southwestern
    Medical Center
Dallas, Texas

Dongsheng Che
Department of Computer Science
East Stroudsburg University
East Stroudsburg, Pennsylvania

Yi-Ping Phoebe Chen
School of Information Technology
Deakin University
Melbourne, Australia

Hyeyoung Cho
Bindley Bioscience Center
Purdue University
West Lafayette, Indiana
and
Department of Bio and Brain
    Engineering
KAIST
Daejeon, South Korea

Jeong-Hyeon Choi
Center for Genomics and
    Bioinformatics and
    School of Informatics
Indiana University
Bloomington, Indiana

Giovanni Ciriello
Department of Information
    Engineering
University of Padova
Padova, Italy

Matteo Comin
Department of Information
    Engineering
University of Padua
Padova, Italy

Mick Correll
InforSense, LLC
Cambridge, Massachusetts

John Crispino
Hematology Oncology
Northwestern University
Chicago, Illinois

Carl Dahlke
Health Information Systems
Northrop Grumman, Inc.
Rockville, Maryland

Mehmet Dalkilic
School of Informatics
Indiana University
Bloomington, Indiana

Lipika Dey
Innovation Labs
Tata Consultancy Services
New Delhi, India

Patrick Dunn
Health Information Systems
Northrop Grumman, Inc.
Rockville, Maryland

Giovanni Felici
Istituto di Analisi dei Sistemi ed
    Informatica Antonio Ruberti
Consiglio Nazionale delle Ricerche
Rome, Italy

Raffaele Giancarlo
Dipartimento di Matematica ed
    Applicazioni
University of Palermo
Palermo, Italy

Concettina Guerra
College of Computing
Georgia Institute of Technology
Atlanta, Georgia and
Department of Information
    Engineering
University of Padua
Padova, Italy

Yike Guo
InforSense, Ltd.
London, United Kingdom

Herb Hagler
Department of Pathology
University of Texas Southwestern
    Medical Center
Dallas, Texas

Jing-Dong J. Han
Key Laboratory of Molecular
    Developmental Biology
Center for Molecular Systems
    Biology
Institute of Genetics and
    Developmental Biology
Chinese Academy of Sciences
Beijing, People's Republic of China

Christine E. Heitsch
School of Mathematics
Georgia Institute of Technology
Atlanta, Georgia

Hai Hu
Windber Research Institute
Windber, Pennsylvania

Yang Huang
National Institutes of Health
Bethesda, Maryland

Zan Huang
Hematology Oncology
Northwestern University
Chicago, Illinois

Hongmei Jiang
Department of Statistics
Northwestern University
Evanston, Illinois

David Karp
Division of Rheumatology
University of Texas Southwestern
    Medical Center
Dallas, Texas

George Karypis
Deparment of Computer Science
University of Minnesota
Minneapolis, Minnesota

Weimao Ke
University of North Carolina
Chapel Hill, North Carolina

Daisuke Kihara
Department of Biological Sciences
    and Department of Computer
    Science
Markey Center for Structural Biology
College of Science
Purdue University
West Lafayette, Indiana

Sun Kim
Center for Genomics and
    Bioinformatics and School of
    Informatics
Indiana University
Bloomington, Indiana

Megan Kong
Department of Pathology
University of Texas Southwestern
    Medical Center
Dallas, Texas

Yazhene Krishnaraj
Wayne State University
Detroit, Michigan

Giuseppe Lancia
Dipartimento di Matematica e
    Informatica
University of Udine
Udine, Italy

Chia-Ju Lee
Biomedical Informatics Center
Northwestern University
Chicago, Illinois

Jamie Lee
Department of Pathology
University of Texas Southwestern
    Medical Center
Dallas, Texas

Gang Li
School of Information Technology
Deakin University
Melbourne, Australia

Guojun Li
Department of Biochemistry and
    Molecular Biology and Institute of
    Bioinformatics
University of Georgia
Athens, Georgia
and
School of Mathematics and System
    Sciences
Shandong University
Jinan, People's Republic of China

Li Liao
Computer and Information Sciences
University of Delaware
Newark, Delaware

Simon Lin
Biomedical Informatics Center
Northwestern University
Chicago, Illinois

Elizabeth McClellan
Division of Biomedical Informatics
University of Texas Southwestern
    Medical Center
Dallas, Texas
and
Department of Statistical Science
Southern Methodist University
Dallas, Texas

Monnie McGee
Department of Statistical Science
Southern Methodist University
Dallas, Texas

Yehia Mechref
National Center for Glycomics and
    Glycoproteomics
Department of Chemistry
Indiana University
Bloomington, Indiana

Tijana Milenković
Department of Computer Science
University of California
Irvine, California

Jason H. Moore
Computational Genetics Laboratory
Norris-Cotton Cancer Center
Departments of Genetics and
    Community and Family Medicine
Dartmouth Medical School
Lebanon, New Hampshire

and

Department of Computer Science
University of New Hampshire
Durham, New Hampshire

and

Department of Computer Science
University of Vermont
Burlington, Vermont

and

Translational Genomics Research
    Institute
Phoenix, Arizona

Javed Mostafa
University of North Carolina
Chapel Hill, North Carolina

Robin Munro
InforSense, Ltd.
London, United Kingdom

Glenn J. Myatt
Myatt & Johnson, Inc.
Jasper, Georgia

Chris J. Needham
School of Computing
University of Leeds
Leeds, United Kingdom

Cheolhwan Oh
Bindley Bioscience Center
Purdue University
West Lafayette, Indiana

Mihai Pop
Center for Bioinformatics and
　Computational Biology
University of Maryland
College Park, Maryland

Teresa M. Przytycka
National Institutes of Health
Bethesda, Maryland

Nataša Pržulj
Department of Computer Science
University of California
Irvine, California

Yu Qian
Department of Pathology
University of Texas Southwestern
　Medical Center
Dallas, Texas

Naren Ramakrishnan
Department of Computer Science
Virginia Tech
Blacksburg, Virginia

Huzefa Rangwala
Department of Computer Science
George Mason University
Fairfax, Virginia

Chandan Reddy
Wayne State University
Detroit, Michigan

Catherine P. Riley
Bindley Bioscience Center
Purdue University
West Lafayette, Indiana

Jia Rong
School of Information Technology
Deakin University
Melbourne, Australia

Lee Sael
Department of Computer Science
Purdue University
West Lafayette, Indiana

Davide Scaturro
Dipartimento di Matematica
　ed Applicazioni
University of Palermo
Palermo, Italy

Richard H. Scheuermann
Department of Pathology
Division of Biomedical Informatics
University of Texas Southwestern
　Medical Center
Dallas, Texas

Kazuhiro Seki
Organization of Advanced Science
　and Technology
Kobe University
Kobe, Japan

Jonathan Sheldon
InforSense Ltd.
London, United Kingdom

Barry Smith
Department of Philosophy
University at Buffalo
Buffalo, New York

Junilda Spirollari
New Jersey Institute of
    Technology
Newark, New Jersey

Burke Squires
Department of Pathology
University of Texas
    Southwestern Medical Center
Dallas, Texas

Haixu Tang
School of Informatics
National Center for Glycomics
    and Glycoproteomics
Indiana University
Bloomington, Indiana

Jahiruddin
Department of Computer Science
Jamia Millia Islamia
New Delhi, India

Filippo Utro
Dipartimento di Matematica
    ed Applicazioni
University of Palermo
Palermo, Italy

Jason T. L. Wang
New Jersey Institute of Technology
Newark, New Jersey

Jeff Wiser
Health Information Systems
Northrop Grumman, Inc.
Rockville, Maryland

Mohammed Zaki
Department of Computer Science
Rensselaer Polytechnic Institute
Troy, New York

Giuseppe Zanotti
Department of Biological Chemistry
University of Padua
Padova, Italy

Xiang Zhang
Department of Chemistry
Center of Regulatory and
    Environmental Analytical
    Metabolomics
University of Louisville
Louisville, Kentucky

Jie Zheng
National Institutes of Health
Bethesda, Maryland

Part I

Sequence, Structure, and Function

Chapter 1

Consensus Structure Prediction for RNA Alignments

Junilda Spirollari and Jason T. L. Wang

New Jersey Institute of Technology

1.1 Introduction

RNA secondary structure prediction has been studied for quite awhile. Many minimum free energy (MFE) methods have been developed for predicting the secondary structures of single RNA sequences, such as mfold [1], RNAfold [2], MPGAfold [3], as well as recent tools presented in the literature [4, 5]. However, the accuracy of predicted structures is far from perfect. As evaluated by Gardner and Giegerich [6], the accuracy of the MFE methods for single sequences is 73% when averaged over many different RNAs.

Recently, a new concept of energy density for predicting the secondary structures of single RNA sequences was introduced [7]. The normalized free energy, or energy density, of an RNA substructure is the free energy of that substructure divided by the length of its underlying sequence. A dynamic programming algorithm, called Densityfold, was developed, which delocalizes the thermodynamic cost of computing RNA substructures and improves on secondary structure prediction via energy density minimization [7]. Here, we extend the concept used in Densityfold and present a tool, called RSpredict, for RNA secondary structure prediction. RSpredict computes the RNA structure with minimum energy density based on the loop decomposition scheme used in the nearest neighbor energy model [8]. RSpredict focuses on the loops in an RNA secondary structure, whereas Densityfold considers RNA substructures where a substructure may contain several loops.

While the energy density model creates a foundation for RNA secondary structure prediction, there are many limitations in Densityfold, just like in all other single sequence-based MFE methods. Optimal structures predicted by these methods do not necessarily represent real structures [9]. This happens due to several reasons. The thermodynamic model may not be accurate. The bases of structural RNAs may be chemically modified and these processes are not included in the prediction model. Finally, some functional RNAs may not have stable secondary structures [6]. Thus, a more reliable approach is to use comparative analysis to compute consensus secondary structures from multiple related RNA sequences [9].

In general, there are three strategies with the comparative approach. The first strategy is to predict the secondary structures of individual RNA sequences separately and then align the structures. Tools such as RNAshapes [10,11], MARNA [12], STRUCTURELAB [13], and RADAR [14,15] are based on this strategy. RNA Sampler [9] and comRNA [16] compare and find stems conserved across multiple sequences and then assemble conserved stem blocks to form consensus structures, in which pseudoknots are allowed.

The second strategy predicts common secondary structures of two or more RNA sequences through simultaneous alignment and consensus structure inference. Tools based on this strategy include RNAscf [17], Foldalign [18], Dynalign [19], stemloc [20], PMcomp [21], MASTR [22], and CARNAC [23]. These tools utilize either folding free energy change parameters or stochastic context-free grammars (SCFGs) and are considered derivations of Sankoff's method [24].

The third strategy is to fold multiple sequence alignments. RNAalifold [25, 26] uses a dynamic programming algorithm to compute the consensus secondary structure with MFE by taking into account thermodynamic stability, sequence covariation together with RIBOSUM-like scoring matrices [27]. Pfold [28] is a SCFG algorithm that produces a prior probability distribution of RNA structures. A maximum likelihood approach is used to estimate a phylogenetic tree for predicting the most likely structure for input sequences. A limitation of Pfold is that it does not run on alignments of more than 40 sequences and in some cases produces no structures due to under-flow errors [6]. Maximum weighted matching (MWM), based on a graph-theoretical approach and developed by Cary and Stormo [29] and Tabaska et al. [30], is able to predict common secondary structures allowing pseudo-knots. KNetFold [31] is a recently published machine learning method, implemented using a hierarchical network of k-nearest neighbor classifiers that analyzes the base pairings of alignment columns in the input sequences through their mutual information, Watson–Crick base pairing rules and thermodynamic base pair propensity derived from RNAfold [2]. The method presented in this chapter, RSpredict, joins the many tools using the third strategy; it accepts a multiple alignment of RNA sequences as input data and predicts the consensus secondary structure for the input sequences via energy density minimization and covariance score calculation.

We also considered two variants of RSpredict, referred to as RSefold and RSdfold respectively. Both RSefold and RSdfold use the same covariance score calculation as in RSpredict. The differences among the three approaches lie in the folding algorithms they adopt. Rse-fold predicts the consensus secondary structure for the input sequences via free energy minimization, as opposed to energy density minimization used in RSpredict. RSdfold does the prediction via energy density minimization, though its energy density is calculated based on RNA substructures as in Densityfold, rather than based on the loops used in RSpredict.

The rest of the chapter is organized as follows. We first describe the implementation and algorithms used by RSpredict, and analyze the time complexity of the algorithms (see Section 1.2). We then present experimental results of running the RSpredict tool as well as comparison with the existing tools (see Section 1.3). The experiments were performed on a variety of datasets. Finally we discuss some properties of RSpredict, possible ways to improve the tool and point out some directions for future research (see Section 1.4).

1.2 Algorithms

RSpredict, which can be freely downloaded from http://datalab.njit.edu/ biology/RSpredict, was implemented in the Java programming language. The program accepts, as input data, a multiple sequence alignment in the FASTA or ClustalW format and outputs the consensus secondary structure of the input sequences in both the Vienna style dot bracket format [26] and the connectivity table format [32]. Below, we describe the energy density model adopted by RSpredict. We then present a dynamic programming algorithm for folding a single RNA sequence via energy density minimization. Next, we describe techniques for calculating covariance scores based on the input alignment. Finally we summarize the algorithms used by RSpredict, combining both the folding technique and the covariance scores obtained from the input alignment, and show its time complexity.

1.2.1   Folding of a single RNA sequence

1.2.1.1   Preliminaries

We represent an RNA secondary structure as a fully decomposed set of loops. In general, a loop L can be one of the following (see Figure 1.1):

i. A hairpin loop (which is a loop enclosed by only one base pair; the smallest possible hairpin loop consists of three nucleotides enclosed by a base pair)

ii. A stack, composed of two consecutive base pairs

iii. A bulge loop, if two base pairs are separated only on one side by one or more unpaired bases

iv. An internal loop, if two base pairs are separated by one or more unpaired bases on both sides

v. A multibranched loop, if more than two base pairs are separated by zero or more unpaired bases in the loop

We now introduce some terms and definitions. Let S be an RNA sequence consisting of nucleotides or bases A, U, C, G. $S[i]$ denotes the base at position i of the sequence S and $S[i, j]$ is the subsequence starting at position i and ending at position j in S. A base pair between nucleotides at positions i and j is denoted as (i, j) or $(S[i], S[j])$, and its enclosed sequence is $S[i, j]$. Given a loop L in the secondary structure R of sequence S, the base pair (i^*, j^*) in L is called the *exterior pair* of L if $S[i^*]$($S[j^*]$, respectively) is closest to the $5'$ ($3'$, respectively) end of R among all nucleotides in L. All other nonexterior base pairs in L are called *interior pairs* of L. The length of a loop L is the number of nucleotides in L. Note that two loops may overlap on a base pair. For example, the interior pair of a stack may be the exterior pair of another stack, or the exterior pair of a hairpin loop. Also note that a bulge or an internal loop has exactly one exterior pair and one interior pair.

We use the energy density concept as follows. Given a secondary structure R, every base pair (i, j) in R is the exterior pair of some loop L. We assign (i, j) and L an energy density, which is the free energy of the loop L divided by the length of L. The set of free energy parameters for nonmultibranched loops used in our algorithm is acquired from [33]. The free energy of a multibranched loop is computed based on the approach adopted by mfold [1], which is a linear function of the number of unpaired bases and the number of base pairs inside the loop, namely $a + b \times n_1 + c \times n_2$, where a, b, c are constants, n_1 is the number of unpaired bases and n_2 is the number of base pairs inside the multibranched loop. We adopt the loop decomposition scheme used in the nearest neighbor energy model developed by Turner et al. [8]. The secondary structure R contains multiple loop components and the energy densities of

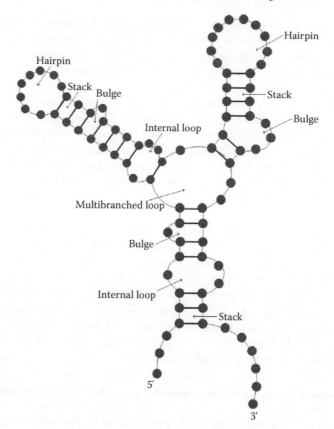

FIGURE 1.1: Illustration of the loops in an RNA secondary structure. Each loop has at least one base pair. A stem consists of two or more consecutive stacks shown in the figure.

the loop components are additive. Our folding algorithm computes the total energy density of R by taking the sum of the energy densities of the loop components in R. Thus, the RNA folding problem can be formalized as follows. Given an RNA sequence S, find the set of base pairs (i, j) and loops with (i, j) as exterior pairs, such that the total energy density of the loops (or equivalently, the exterior pairs) is minimized. The set of base pairs constitutes the optimal secondary structure of S.

When generalizing the folding of a single sequence to the prediction of the consensus structure of a multiple sequence alignment, we introduce the notion of refined alignments. At times, an input alignment may have some columns each of which contains more than 75% gaps. Some tools including RSpredict delete these columns to get a refined alignment [28]; some tools simply use the original input alignment as the refined alignment. Suppose the original input alignment A_o has N sequences and n_o columns, and the refined alignment A has N sequences and n columns, $n \leq n_o$. Formally, the consensus structure of the refined alignment A is a secondary structure R together with its sequence S such that each base pair $(S[i], S[j])$, $1 \leq i < j \leq n$, in R corresponds to the pair of columns i, j in the alignment A, and each base $S[i]$, $1 \leq i \leq n$, is the representative base of the ith column in the alignment A. There are several ways to choose the representative base. For example, $S[i]$ could be the most frequently occurring nucleotide, excluding gaps, in the ith column of the alignment A. Furthermore, there is an energy measure value associated with each base pair $(S[i], S[j])$ or more precisely its corresponding column pair (i, j), such that the total energy measure value of all the base pairs in R is minimized.

The consensus secondary structure of the original input alignment A_o is defined as the structure R_o, obtained from R, as follows: (i) the base (base pair, respectively) for column C_o (column pair (C_o1, C_o2), respectively) in A_o is identical to the base (base pair, respectively) for the corresponding column C (column pair $(C1, C2)$, respectively) in A if C_o $((C_o1, C_o2)$, respectively) is not deleted when getting A from A_o; (ii) unpaired gaps are inserted into R, such that each gap corresponds to a column that is deleted when getting A from A_o (see Figure 1.2). In Figure 1.2, the RSpredict algorithm transforms the original input alignment A_o to a refined alignment A by deleting the fourth column (the column in red) of A_o. The algorithm predicts the consensus structure of the refined alignment A. Then the algorithm generates the consensus structure of A_o by inserting an unpaired gap to the fourth position of the consensus structure of A. The numbers inside parentheses in the refined alignment A represent the original column numbers in A_o.

In what follows, we first present an algorithm for folding a single RNA sequence based on the energy density concept described here. We then generalize the algorithm to predict the consensus secondary structure for a set of aligned RNA sequences.

1.2.1.2   Algorithm

The functions and parameters used in our algorithm are defined below where $S[i, j]$ is a subsequence of S and $R[i, j]$ is the optimal secondary structure of $S[i, j]$.

  i. $NE(i, j)$ is the total energy density of all loops in $R[i, j]$, where nucleotides at positions i, j may or may not form a base pair.

 ii. $NE_p(i, j)$ is the total energy density of all loops in $R[i, j]$ if nucleotides at positions i, j form a base pair.

iii. $e_H(i, j)(E_H(i, j)$, respectively) is the free energy (energy density, respectively) of the hairpin with exterior pair (i, j).

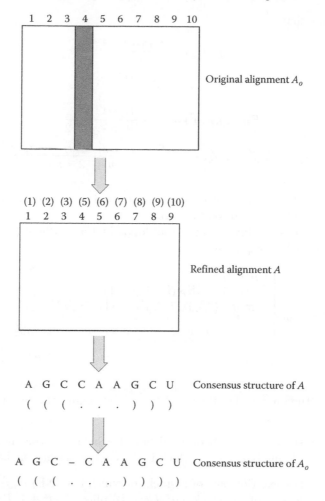

FIGURE 1.2: Illustration of the consensus structure definition used by RSpredict.

iv. $e_S(i,j)(E_S(i,j)$, respectively) is the free energy (energy density, respectively) of the stack with exterior pair (i,j) and interior pair $(i+1,j-1)$.

v. $e_B(i,j,i',j')$, $(E_B(i,j,i',j')$, respectively) is the free energy (energy density, respectively) of the bulge or internal loop with exterior pair (i,j) and interior pair (i',j').

vi. $e_J(i,j,i'_1,j'_1,i'_2,j'_2,\ldots,i'_k,j'_k)$ $E_J(i,j,i'_1,j'_1,i'_2,j'_2,\ldots,i'_k,j'_k)$ respectively, is the free energy (energy density, respectively) of the multibranched loop with exterior pair (i,j) and interior pairs (i'_1,j'_1), (i'_2,j'_2),$\ldots$, (i'_k,j'_k).

It is clear that

$$E_H(i,j) = \frac{e_H(i,j)}{j-i+1} \tag{1.1}$$

$$E_S(i,j) = \frac{e_S(i,j)}{4} \tag{1.2}$$

$$E_B(i,j,i',j') = \frac{e_B(i,j,i',j')}{i'-i+j-j'+2} \tag{1.3}$$

$$E_J(i,j,i'_1,j'_1,i'_2,j'_2,\ldots,i'_k,j'_k) = \frac{e_J(i,j,i'_1,j'_1,i'_2,j'_2,\ldots,i'_k,j'_k)}{n_1+2\times n_2} \tag{1.4}$$

Here n_1 is the number of unpaired bases and n_2 is the number of base pairs in the multibranched loop in (vi).

Thus, the total energy density of all loops in $R[i,j]$ where (i,j) is a base pair is computed by Equation 1.5:

$$\mathrm{NE}_P(i,j) = \min \begin{cases} E_H(i,j) \\ E_S(i,j) + \mathrm{NE}_P(i+1,j-1) \\ \displaystyle\min_{i<i'<j'<j}\{E_B(i,j,i',j') + \mathrm{NE}_P(i',j')\} \\ \displaystyle\min_{i<i'_1<j'_1<i'_2<j'_2<\cdots<i'_k<j'_k<j}\{E_J(i,j,i'_1,j'_1,i'_2,j'_2,\ldots,i'_k,j'_k) \\ \qquad\qquad\qquad + \sum_{r=1}^{k}\mathrm{NE}_P(i'_r,j'_r)\} \end{cases} \tag{1.5}$$

That is, the energy density is calculated by taking the minimum of the following four cases:

i. (i,j) is the exterior pair of a hairpin, in which case the energy density $\mathrm{NE}_P(i,j)$ equals $E_H(i,j)$, which is the energy density of the hairpin

ii. (i,j) is the exterior pair of a stack, in which case $\mathrm{NE}_P(i,j)$ equals the energy density of the stack, i.e., $E_S(i,j)$, plus $\mathrm{NE}_P(i+1,j-1)$

iii. (i,j) is the exterior pair of a bulge or an internal loop, in which case $\mathrm{NE}_P(i,j)$ equals the minimum of the energy density of the bulge or internal loop $E_B(i,j,i',j')$ plus $\mathrm{NE}_P(i',j')$ for all $i<i'<j'<j$

iv. (i,j) is the exterior pair of a multibranched loop, in which case $\mathrm{NE}_P(i,j)$ equals the minimum of the energy density of the multibranched loop $E_j(i,j,i'_1,j'_1,i'_2,j'_2,\ldots,i'_k,j'_k)$ plus $\sum_{r=1}^{k}\mathrm{NE}_P(i'_r,j'_r)$, for all $i<i'_1<j'_1<i'_2<j'_2<\cdots<i'_k<j'_k<j$

Equation 1.6 below shows the recurrence formula for calculating $\mathrm{NE}(i,j)$:

$$\mathrm{NE}(i,j) = \min \begin{cases} \mathrm{NE}(i,j-1) \\ \mathrm{NE}(i+1,j) \\ \mathrm{NE}_P(i,j) \\ \min_{i<h<j}\{\mathrm{NE}(i,h-1) + \mathrm{NE}(h,j)\} \end{cases} \tag{1.6}$$

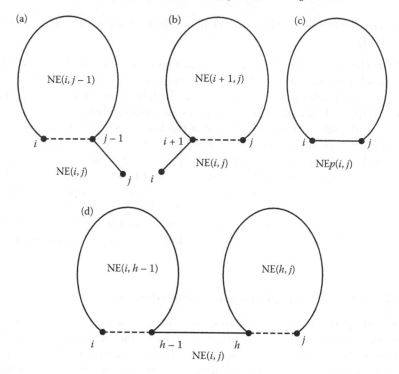

FIGURE 1.3: Illustration of the cases in Equation 1.6. a) the total normalized energy of all loops in the optimal secondary structure $R[i, j-1]$ of subsequence $S[i, j-1]$; b) the total normalized energy of all loops in the optimal secondary structure $R[i+1, j]$ of subsequence $S[i+1, j]$; c) the total normalized energy of all loops in the optimal secondary structure $R[i, j]$ of subsequence $S[i, j]$, where $S[i]$ and $S[j]$ form a base pair; d) the minimum of $NE(i, k-1)$ plus $NE(k, j)$ for all $i < k < j$; The dashed line between two nucleotides means that the two nucleotides may or may not form a base pair. The solid line between two nucleotides means that the two nucleotides form a base pair.

That is, the energy density is computed by taking the minimum of the following four cases:

i. The total energy density of all loops in the optimal secondary structure $R[i, j-1]$ of subsequence $S[i, j-1]$ (Figure 1.3a)

ii. The total energy density of all loops in the optimal secondary structure $R[i+1, j]$ of subsequence $S[i+1, j]$ (Figure 1.3b)

iii. The total energy density of all loops in the optimal secondary structure $R[i, j]$ of subsequence $S[i, j]$, where $S[i]$ and $S[j]$ form a base pair (Figure 1.3c)

iv. The minimum of NE$(i, h-1)$ plus NE(h, j) for all $i < h < j$ (Figure 1.3d)

Note that case (iii) of Equation 1.6 is not considered when the nucleotides at positions i, j are forbidden to form a base pair, i.e., $(S[i], S[j])$ is a nonstandard base pair. A standard base pair is any of the following: (A,U), (U,A), (G,C), (C,G), (G,U), (U,G); all other base pairs are nonstandard.

In calculating the time complexity of the folding algorithm, there is a need to check for finding the optimal i', j' where $i < i' < j' < j$ in case (iii) (the optimal $i'_1, j'_1, i'_2, j'_2, \ldots, i'_k, j'_k$ where $i < i'_1 < j'_1 < i'_2 < j'_2 < \cdots < i'_k < j'_k < j$ in case (iv), respectively) of Equation 1.5. It can be shown that it takes linear time to compute NE$_P(i, j)$ in Equation 1.5. Hence, the time complexity of the folding algorithm is $O(n^3)$ since we need to calculate NE$_P(i, j)$ for all $1 \le i < j \le n$, where n is the number of nucleotides in the given sequence S. The energy density of the optimal secondary structure R for the sequence S equals NE$(1, n)$.

1.2.2   Calculation of covariance scores

When applying the above folding algorithm to a multiple sequence alignment A_o, we take into consideration the correlation between columns of the alignment. In many cases, the sequences in the alignment may have highly varying lengths. We refine the alignment A_o by deleting columns containing more than 75% gaps to get a refined alignment A [28]. We will use this refined alignment throughout the rest of this subsection.

1.2.2.1   Covariance score

We use the covariance score introduced by RNAalifold [25, 26, 34] to quantify the relationship between two columns in the refined alignment. Let $f_{ij}(XY)$ be the frequency of finding both base X in column i and base Y in column j, where X, Y are in the same row of the refined alignment. We exclude the occurrences of gaps in column i or column j when calculating $f_{ij}(XY)$. The covariation measure for columns i, j, denoted C_{ij}, is calculated by Equation 1.7:

$$C_{ij} = \frac{\sum XY, X'Y' f_{ij}(XY) D_{ij}(XY, X'Y') f_{ij}(X'Y')}{2} \qquad (1.7)$$

Here, $D_{ij}(XY, X'Y')$ is the Hamming distance between the two base pairs (X, Y) and (X', Y') if both of the base pairs are standard base pairs, or 0 otherwise. The Hamming distance between (X, Y) and (X', Y') is calculated as follows:

$$D_{ij}(XY, X'Y') = 2 - \delta(X, X') - \delta(Y, Y') \qquad (1.8)$$

where

$$\delta(X, X') = \begin{cases} 1 & \text{if } X = X' \\ 0 & \text{otherwise} \end{cases} \qquad (1.9)$$

Observe that the information acquired from the two base pairs (X, Y) and (X', Y') is the same as that from (X', Y') and (X, Y). Thus, we divide the numerator in Equation 1.7 by two so as to obtain the non-redundant information between column i and column j in the refined alignment.

For every pair of columns i, j in the refined alignment, the covariance score of the two columns i and j, denoted Cov_{ij}, is calculated in Equation 1.10:

$$\text{Cov}_{ij} = C_{ij} + c_1 \times \text{NF}_{ij} \qquad (1.10)$$

Here, C_{ij} is as defined in Equation 1.7, c_1 is a user-defined coefficient (in the study presented here, c_1 has a value of -1), and

$$\text{NF}_{ij} = \frac{\text{NC}_{ij}}{N} \qquad (1.11)$$

where N is the total number of sequences and NC_{ij} is the total number of conflicting sequences in the refined alignment. A conflicting sequence is one that has a gap in column i or column j, or has a nonstandard base pair in the columns i, j of the refined alignment. A sequence with gaps in both columns i, j is not conflicting.

1.2.2.2 Pairing threshold

We say that column i and column j in the refined alignment can possibly form a base pair if their covariance score is greater than or equal to a pairing threshold; otherwise, column i and column j are forbidden to form a base pair. The pairing threshold, η, used in RSpredict is calculated as follows.

It is known that, on average, 54% of the nucleotides in an RNA sequence S are involved in the base pairs of its secondary structure [35]. We use this information to calculate an alignment-dependent pairing threshold, observing that the base pairs in the consensus secondary structure of a sequence alignment represent the column pairs with the highest covariance scores. Given that different structures contain different numbers of base pairs, we consider two different percentages of columns, namely, 30% and 65%, in the sequence alignment. For each percentage p, there are at most T_p possible base pairs, where

$$T_p = \frac{(p \times n) \times (p \times n - 1)}{2} \qquad (1.12)$$

and n is the number of columns in the sequence alignment.

Now, we calculate the covariance scores of all pairs of columns in the given refined alignment, and sort the covariance scores in descending order. We then select the top T_p largest covariance scores and store the covariance scores in the set ST_p. Thus, the set $\text{ST}_{0.65}$ contains the top largest covariance scores that involve 65% of the columns in the refined alignment; the set $\text{ST}_{0.30}$ contains the top largest covariance scores that involve 30% of the columns in the refined alignment; and $\text{ST}_{0.65} \backslash \text{ST}_{0.30}$ is the set difference that contains covariance scores in $\text{ST}_{0.65}$ but not in $\text{ST}_{0.30}$ (see Figure 1.4). The pairing

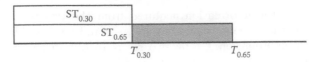

FIGURE 1.4: Illustration of the pairing threshold computation. The pairing threshold used in RSpredict is computed as the average of the covariance scores inside the shaded area.

threshold η used in RSpredict is calculated as the average of the covariance scores in $\mathrm{ST}_{0.65}\backslash \mathrm{ST}_{0.30}$, as shown in Equation 1.13:

$$\eta = \frac{\sum \mathrm{Cov}_{ij} \in \mathrm{ST}_{0.65}\backslash \mathrm{ST}_{0.30} \mathrm{Cov}_{ij}}{|\mathrm{ST}_{0.65}\backslash \mathrm{ST}_{0.30}|} \tag{1.13}$$

where the denominator is the cardinality of the set difference $\mathrm{ST}_{0.65}\backslash \mathrm{ST}_{0.30}$.

If the covariance score of columns i and j is greater than or equal to η, then column i and column j can possibly form a base pair, and we refer to (i, j) as a pairing column. If the covariance score of the columns i and j is less than η, we will check the covariance scores of the immediate neighboring column pairs of i, j to see if they are above a user-defined threshold [31] (in the study presented here, this threshold is set to 0). The immediate neighboring column pairs of i, j are $i + 1, j - 1$ and $i - 1, j + 1$. If the covariance scores of both of the immediate neighboring column pairs of i, j are greater than or equal to $\max\{\eta, 0\}$, then (i, j) is still considered as a paring column.

1.2.3   Algorithms for RSpredict

Given a refined multiple sequence alignment A with N sequences, let (i, j) be a pairing column in A. Let X_i^S (Y_j^S, respectively) be the nucleotide at position i (j, respectively) of the sequence S in the alignment A. $\left(X_i^S, Y_j^S\right)$ must be the exterior pair of some loop L in S. We use $e\left(X_i^S, Y_j^S\right)$ to represent the free energy of that loop L. If $\left(X_i^S, Y_j^S\right)$ is a nonstandard base pair, $e\left(X_i^S, Y_j^S\right) = 0$. We assign the pairing column (i, j) a pseudo-energy e_{ij} where

$$e_{ij} = \frac{1}{N} \sum_{S \in A} e\left(X_i^S, Y_j^S\right) + c_2 \times \mathrm{Cov}_{ij} \tag{1.14}$$

Here, c_2 is a user-defined coefficient (in the study presented here, $c_2 = -1$). Thus, every pairing column in the refined alignment A has a pseudo-energy. We then apply the minimum energy density folding algorithm described in the beginning of this section to the refined alignment A, treating each pairing column in A as a possible base pair considered in the folding algorithm.

Notice that when calculating the energy density for the loop L, the sequence S is in the refined alignment A, which may have fewer columns than the original input alignment A_o (cf. Figure 1.2). RSpredict computes all energy densities based on the refined alignment, and the program uses loop lengths from the refined alignment A rather than the original input alignment A_o. Let R be the consensus secondary structure, computed by RSpredict, for the refined alignment A. We obtain the consensus structure R_o of the original input alignment A_o by inserting unpaired gaps to the positions in R whose corresponding columns are deleted when getting A from A_o (cf. Figure 1.2). The following summarizes the algorithms for RSpredict:

1. Input an alignment A_o in the FASTA or ClustalW format.

2. Delete the columns with more than 75% gaps from A_o to obtain a refined alignment A.

3. Compute the pseudo-energy e_{ij} for every pairing column (i, j) in A as in Equation 1.14.

4. Run the minimum energy density folding algorithm on A, using the pseudo-energy values obtained from step (3) to produce the consensus secondary structure R of the refined alignment A. The base at position i of the consensus secondary structure R is the most frequently occurring nucleotide, excluding gaps, in the ith column of the refined alignment A.

5. Map the consensus structure R back to the original alignment A_o by inserting unpaired gaps to the positions of R whose corresponding columns are deleted in Step (2).

Notice that Equation 1.6 is used to compute the NE values only. To generate the optimal structure R in Step (4), we maintain a stack of pointers that point to the substructures of loops with minimum energy density as we compute the NE values. Once all the NE values are calculated and the energy density of the optimal secondary structure R is obtained, we pop up the pointers from the stack to extract the optimal predicted structure. In step (5), we map the bases (base pairs, respectively) for the columns (column pairs, respectively) in A to their corresponding columns (column pairs, respectively) in A_o. For example, consider Figure 1.2 again. In the figure, the refined alignment A is obtained by deleting column 4 from the original input alignment A_o. The bases for columns 1, 2, 3, 4 in A are mapped to columns 1, 2, 3, 5 in A_o. The base pair between column 1 and column 9 in A becomes the base pair between column 1 and column 10 in A_o; the base pair between column 2 and column 8 in A becomes the base pair between column 2 and column 9 in A_o. An unpaired gap is inserted to the position corresponding to the deleted column 4 in A_o.

Let N be the number of sequences and n_o be the number of columns in the input alignment A_o. Step (2) takes $O(Nn_o)$ time. Step (3) takes $O\left(n_o^2\right)$ time. Step (4) takes $O\left(n_o^3\right)$ time. Step (5) takes $O(n_o)$ time. Therefore, the time complexity of RSpredict is $O\left(Nn_o + n_o^3\right)$, which is approximately $O\left(n_o^3\right)$ as N is usually much smaller than n_o.

1.3   Results

We conducted a series of experiments to evaluate the performance of RSpredict and compared it with five related tools including KNetFold, Pfold, RNAalifold, RSefold, and RSdfold. We tested these tools on Rfam [36] sequence alignments with different similarities. The Rfam sequence alignments come with consensus structures. For evaluation purposes, we used the Rfam consensus structures as reference structures and compared them against the consensus structures predicted by the six tools. The similarity of a sequence alignment is determined by the average pairwise sequence identity (APSI) of that alignment [6]. In the study presented here, a sequence alignment is of high similarity if its APSI value is greater than 75%, is of medium similarity if its APSI value is between 55% and 75%, or is of low similarity if its APSI value is less than 55%. The data sets used in testing included 20 Rfam sequence alignments of high similarity and 36 Rfam sequence alignments of low and medium similarity. These data sets were chosen to form a collection of sequence alignments with different (low, medium and high) APSI values, different numbers of sequences, as well as different sequence alignment lengths. More specifically, the data sets contained sequence alignments that ranged in size from 2 to 160 sequences, in length from 33 to 262 nucleotides and had APSI values ranging from 42% to 99%.

The performance measures used in our study include sensitivity (SN) and selectivity (SL) [6], where

$$SN = \frac{TP}{TP + FN} \tag{1.15}$$

$$SL = \frac{TP}{TP + (FP - \xi)}. \tag{1.16}$$

Here, TP is the number of correctly predicted base pairs ("true positives"), FN is the number of base pairs in a reference structure that were not predicted ("false negatives") and FP is the number of incorrectly predicted base pairs ("false positives"). False positives are classified as inconsistent, contradicting or compatible [6]. When predicting the consensus secondary structure for a multiple sequence alignment, a predicted base pair (i, j) is inconsistent if column i in the alignment is paired with column $q, q \neq j$, or column j is paired with column $p, p \neq i$, and p, q form a base pair in the reference structure of the alignment. A base pair (i, j) is contradicting if there exists a base pair (p, q) in the reference structure of the alignment, such that $i < p < j < q$. A base pair (i, j) is compatible if it is a false positive but is neither inconsistent nor contradicting. The ξ in SL represents the number of compatible base pairs, which are considered neutral with respect to algorithmic accuracy. Therefore ξ is subtracted from FP. Finally, we used the Matthews correlation coefficient (MCC) to combine the sensitivity and selectivity, where MCC is approximated to the geometric mean of the two measures, i.e., MCC $\approx \sqrt{SN \times SL}$ [18]. The larger MCC, SN, SL values a tool has, the better performance that tool achieves and the more accurate that tool is.

1.3.1 Performance evaluation on Rfam alignments of high similarity

The first data set consisted of seed alignments of high similarity taken from 20 families in Rfam. The APSI values of these seed alignments ranged from 77% to 99%. The alignments ranged in size from 2 to 160 sequences and in length from 33 to 159 nucleotides. Table 1.1 presents the accession number, description, number of sequences, and length of the seed alignment of each of the 20 Rfam families used in the experiment. The seed alignments of the 20 families are of high similarity; their APSI values are shown in the last column of the table. The families are sorted, from top to bottom, in ascending order on the APSI values. All six tools including RSpredict, KNetFold, RNAalifold, Pfold, RSefold and RSdfold were tested on this data set.

The graphs in Figure 1.5 show the trend of the MCC, SN, and SL, which are sorted in descending order for each tool under analysis. The X-axis shows, therefore, the rank of the MCC (SN and SL, respectively) from highest to lowest. For example, number 1 in the X-axis corresponds to the highest score achieved by each tool. The Y-axis represents the MCC, SN, and SL, respectively.

It can be seen from Figure 1.5 that RSpredict performed the best while RSdfold performed the worst among the six tools. The Pfold tool had good performance in selectivity but did not perform well in sensitivity and as a result in MCC. It also suffered from a size limitation (the Pfold web server can accept a multiple alignment of up to 40 sequences). Only 17 out of the 20 sequence alignments used in the experiment were accepted by the Pfold server; the other three alignments (RF00386, RF00041, and RF00389) had more than 40 sequences and therefore could not be run on the Pfold server. RSpredict had stable performance with the best mean 0.85 (standard deviation 0.16, respectively) in MCC, while the other methods' MCC values varied a lot and had means (standard deviations, respectively) ranging from 0.37 to 0.82 (0.24 to 0.34, respectively).

1.3.2 Performance evaluation on Rfam alignments of medium and low similarity

In the second experiment, we compared RSpredict with the other five methods on multiple sequence alignments of low and medium similarity. The test dataset included seed alignments of 36 families taken from Rfam [36]. The APSI values of the seed alignments ranged from 42 to 75%, the number of sequences in the alignments ranged from 3 to 114, and the alignment lengths ranged from 43 to 262 nucleotides. Table 1.2 presents the accession number,

TABLE 1.1:   Rfam alignments of high similarity.

Accession	Description	Number of sequences	Length	APSI
RF00460	U1A polyadenylation inhibition element (PIE)	8	75	77%
RF00326	Small nucleolar RNA Z155	8	81	79%
RF00560	Small nucleolar RNA SNORA17	38	132	82%
RF00453	Cardiovirus cis-acting replication element (CRE)	12	33	82%
RF00386	Enterovirus 5′ cloverleaf cis-acting replication element	160	91	83%
RF00421	Small nucleolar RNA SNORA32	9	122	84%
RF00302	Small nucleolar RNA SNORA65	8	130	84%
RF00465	Japanese encephalitis virus (JEV) hairpin structure	20	60	86%
RF00501	Rotavirus cis-acting replication element (CRE)	14	68	87%
RF00041	Enteroviral 3′ UTR element	60	123	87%
RF00575	Small nucleolar RNA SNORD70	4	88	89%
RF00362	Pospiviroid RY motif stem loop	16	79	92%
RF00105	Small nucleolar RNA SNORD115	23	82	92%
RF00467	Rous sarcoma virus (RSV) primer binding site (PBS)	23	75	93%
RF00389	Bamboo mosaic virus satellite RNA cis-regulatory element	42	159	93%
RF00384	Poxvirus AX element late mRNA cis-regulatory element	7	62	93%
RF00098	Snake H/ACA box small nucleolar RNA	22	150	93%
RF00607	Small nucleolar RNA SNORD98	2	67	98%
RF00320	Small nucleolar RNA Z185	2	86	98%
RF00318	Small nucleolar RNA Z175	3	81	99%

description, number of sequences, and length of the seed alignment of each of the 36 Rfam families used in the experiment. The seed alignments of the 36 families are of low and medium similarity; their APSI values are shown in the last column of the table. The families are sorted, from top to bottom, in ascending order on the APSI values.

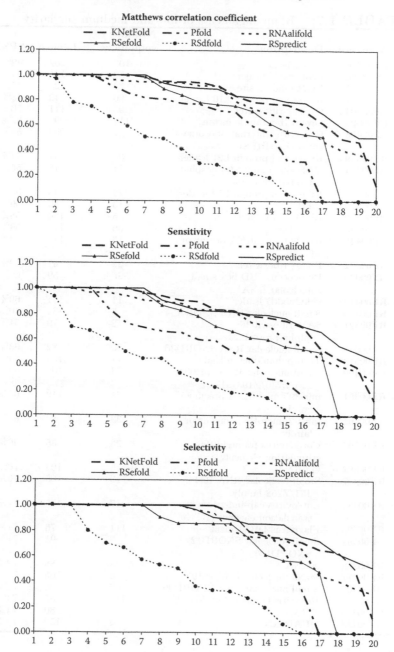

FIGURE 1.5: Comparison of the MCC, SN, and SL values of the six tools under analysis on the seed alignments of high similarity taken from the 20 families listed in Table 1.1.

TABLE 1.2:   Rfam alignments of low and medium similarity.

Accession	Description	Number of sequences	Length	APSI
RF00230	T-box leader	103	262	42%
RF00080	yybP-ykoY leader	50	131	44%
RF00515	PyrR binding site	72	125	47%
RF00557	Ribosomal protein L10 leader	66	149	48%
RF00504	Glycine riboswitch	93	111	50%
RF00029	Group II catalytic intron	114	94	52%
RF00458	Cripavirus internal ribosome entry site (IRES)	7	203	54%
RF00559	Ribosomal protein L21 leader	33	81	54%
RF00234	glmS glucosamine-6-phosphate activated ribozyme	11	218	55%
RF00556	Ribosomal protein L19 leader	24	43	55%
RF00519	suhB	13	80	56%
RF00379	ydaO/yuaA leader	25	150	58%
RF00380	ykoK leader	36	172	59%
RF00445	mir-399 microRNA precursor family	13	119	59%
RF00522	PreQ1 riboswitch	22	47	59%
RF00095	Pyrococcus C/D box small nucleolar RNA	25	59	60%
RF00442	ykkC-yxkD leader	11	111	60%
RF00430	Small nucleolar RNA SNORA54	5	134	60%
RF00521	SAM riboswitch (alpha-proteobacteria)	12	79	61%
RF00049	Small nucleolar RNA SNORD36	20	82	63%
RF00513	Tryptophan operon leader	11	100	63%
RF00309	Small nucleolar RNA snR60/ Z15/Z230/Z193/J17	23	106	63%
RF00451	mir-395 microRNA precursor family	21	112	64%
RF00464	mir-92 microRNA precursor family	33	80	64%
RF00507	Coronavirus frameshifting stimulation element	23	85	66%
RF00388	Qa RNA	5	103	70%
RF00357	Small nucleolar RNA R44/ J54/Z268 family	19	105	70%
RF00434	Luteovirus cap-independent translation element (BTE)	17	108	71%
RF00525	Flavivirus DB element	111	76	71%
RF00581	Small nucleolar SNORD12/ SNORD106	8	91	71%
RF00238	ctRNA	48	88	72%
RF00477	Small nucleolar RNA snR66	5	105	72%
RF00608	Small nucleolar RNA SNORD99	3	80	72%
RF00468	Heaptitis C virus stem-loop VII	110	66	74%
RF00489	ctRNA	14	80	74%
RF00113	QUAD RNA	14	150	75%

The MCC, SN, and SL values are sorted in descending order for each tool under analysis and placed in the graphs in Figure 1.6. The X-axis shows, therefore, the rank of the MCC (SN and SL, respectively) from highest to lowest. For example, number 1 in the X-axis corresponds to the highest score achieved by each tool. The Y-axis represents the MCC, SN, and SL, respectively.

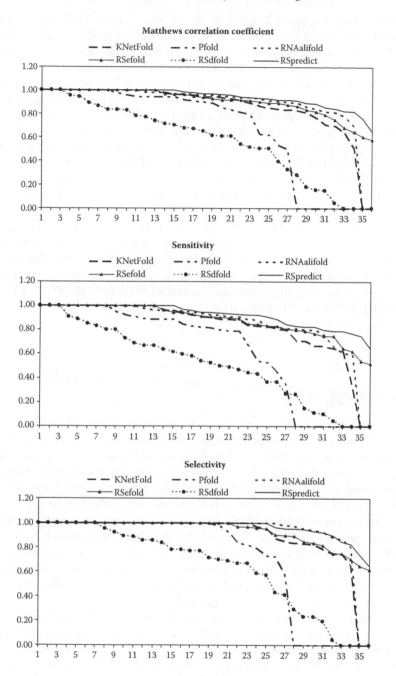

FIGURE 1.6: Comparison of the MCC, SN, and SL values of the six tools under analysis on the seed alignments of low and medium similarity taken from the 36 families listed in Table 1.2.

Comparing Figures 1.5 and 1.6, we see that the methods under analysis generally performed better on sequence alignments of medium and low similarity than on sequence alignments of high similarity. Like what was observed in the previous experiment, RSdfold performed the worst (cf. Figure 1.5). The structures predicted by RSdfold tend to be stem-like structures; therefore, many structures, particularly those containing multibranched loops, were mispredicted. For this reason, RSdfold yielded very low MCC, SN and SL values.

RSpredict outperformed the other five methods based on the three performance measures used in the experiment. The tool achieved a high mean value of 0.94 in MCC, better than those of KNetFold (0.86), Pfold (0.88) and RNAalifold (0.89). Similar results were observed for sensitivity and selectivity values. Furthermore, RSpredict exhibited stable performance across all the families tested in the experiment. The tool had an MCC, SN and SL standard deviation of 0.08, 0.09 and 0.08, respectively. These numbers were better than the standard deviation values obtained from the other five methods, which ranged from 0.11 to 0.34. Pfold suffered from a size limitation; it could not generate a structure for the large seed alignments with more than 40 sequences in 9 families, including RF00230, RF00080, RF00515, RF00557, RF00504, RF00029, RF00525, RF00238 and RF00468.

1.4 Conclusions

In this chapter we presented a software tool, called RSpredict, capable of predicting the consensus secondary structure for a set of aligned RNA sequences via energy density minimization and covariance score calculation. Our experimental results showed that RSpredict is competitive with some widely used tools including RNAalifold and Pfold on tested datasets, suggesting that RSpredict can be a choice when biologists need to predict RNA secondary structures of multiple sequence alignments, especially those with low and medium similarity. Notice that RSpredict differs from KNetFold [31] in that KNetFold is a machine learning method that relies on precompiled training data derived from existing RNA secondary structures. RSpredict, on the other hand, is based on a dynamic programming algorithm for folding sequences and does not utilize training data.

Given a multiple sequence alignment A_o, our work is focused on predicting the consensus structure of the aligned sequences in A_o, rather than folding each individual sequence in A_o. Our approach is to first transform A_o to a refined alignment A by deleting columns with more than 75% gaps from A_o, then predict the consensus structure for A, and finally extend the consensus structure by inserting gaps to the positions corresponding to the deleted columns in A_o (cf. Figure 1.2). The predicted structure may not correspond exactly to any individual sequence in the original alignment A_o. As an example, assume for simplicity that A_o is the same as A, i.e., no columns are deleted when getting A from A_o. Consider a particular sequence S in A_o. Assume that the position (column) i of S has a gap due to the alignment with the other sequences in A_o. On the other hand, the position i in the consensus structure of A_o has the most frequently occurring nucleotide in column i of A_o, which cannot be a gap. As a result, the consensus structure of A_o, which is at least one nucleotide longer than S, cannot be mapped exactly back onto S. In future work we plan to look into ways for improving on consensus structure prediction. Possible ways include the utilization of evolutionary information [37], more sophisticated models of covariance scoring, and training data for more accurate pairing thresholds.

References

[1] Zuker, M. 2003. Mfold web server for nucleic acid folding and hybridization prediction. *Nucleic Acids Res.* 31:3406–3415.

[2] Hofacker, I.L. 2003. Vienna RNA secondary structure server. *Nucleic Acids Res.* 31:3429–3431.

[3] Shapiro, B.A., Kasprzak, W., Grunewald, C., Aman, J. 2006. Graphical exploratory data analysis of RNA secondary structure dynamics predicted by the massively parallel genetic algorithm. *J. Mol. Graph. Model.* 25:514–531.

[4] Bellamy-Royds, A.B., Turcotte, M. 2007. Can Clustal-style progressive pairwise alignment of multiple sequences be used in RNA secondary structure prediction? *BMC Bioinformatics* 8:190.

[5] Horesh, Y., Doniger, T., Michaeli, S., Unger, R. RNAspa: a shortest path approach for comparative prediction of the secondary structure of ncRNA molecules. *BMC Bioinformatics* 8:366.

[6] Gardner, P.P., Giegerich, R. 2004. A comprehensive comparison of comparative RNA structure prediction approaches. *BMC Bioinformatics* 5:140.

[7] Alkan, C., Karakoc, E., Sahinalp, S.C., Unrau, P., Alexander, E., Zhang, K., Buhler, J. 2006. RNA secondary structure prediction via energy density minimization. In *Proceedings of the Research in Computational Molecular Biology (RECOMB)*, Springer Berlin/Heidelberg, Venice, Italy, 130–142.

[8] Xia, T., SantaLucia, J., Burkard, M.E., Kierzek, R., Schroeder, S.J., Jiao, X., Cox, C., Turner, D.H. 1998. Thermodynamic parameters for an

expanded nearest-neighbor model for formation of RNA duplexes with Watson-Crick base pairs. *Biochemistry* 37:14719–14735.

[9] Xu, X., Yongmei, J., Stormo, G.D. 2007. RNA Sampler: a new sampling based algorithm for common RNA secondary structure prediction and structural alignment. *Bioinformatics* 23:1883–1891.

[10] Giegerich, R., Voss, B., Rehmsmeier, M. 2007. Abstract shapes of RNA. *Nucleic Acids Res.* 32:4843–4851.

[11] Steffen, P., Voss, B., Rehmsmeier, M., Reeder, J., Giegerich, R. 2006. RNAshapes: an integrated RNA analysis package based on abstract shapes. *Bioinformatics* 22:500–503.

[12] Siebert, S., Backofen, R. 2005. MARNA: multiple alignment and consensus structure prediction of RNAs based on sequence structure comparisons. *Bioinformatics* 21:3352–3359.

[13] Shapiro, B.A., Bengali, D., Kasprzak, W., Wu, J.C. 2001. RNA folding pathway functional intermediates: their prediction and analysis. *J. Mol. Biol.* 312:27–44.

[14] Khaladkar, M., Bellofatto, V., Wang, J.T.L., Tian, B., Shapiro, B.A. 2007. RADAR: a web server for RNA data analysis and research. *Nucleic Acids Res.* 35:W300–W304.

[15] Liu, J., Wang, J.T.L., Hu, J., Tian, B. 2005. A method for aligning RNA secondary structures and its application to RNA motif detection. *BMC Bioinformatics* 6:89.

[16] Ji, Y., Xu, X., Stormo, G.D. 2004. A graph theoretical approach for predicting common RNA secondary structure motifs including pseudoknots in unaligned sequences. *Bioinformatics* 20:1591–1602.

[17] Bafna, V., Tang, H., Zhang, S. 2006. Consensus folding of unaligned RNA sequences revisited. *J. Comput. Biol.* 13:283–295.

[18] Gorodkin, J., Stricklin, S.L., Stormo, G.D. 2001. Discovering common stem-loop motifs in unaligned RNA sequences. *Nucleic Acids Res.* 29:2135–2144.

[19] Mathews, D.H., Turner, D.H. 2002. Dynalign: an algorithm for finding the secondary structure common to two RNA sequences. *J. Mol. Biol.* 317:191–203.

[20] Holmes, I., Rubin, G.M. 2002. Pairwise RNA structure comparison with stochastic context-free grammars. In *Proceedings of the Pacific Symposium Biocomputing*, Lihue, Hawaii, 163–174.

[21] Hofacker, I.L., Bernhart, S.H.F., Stadler, P.F. 2004. Alignment of RNA base pairing probability matrices. *Bioinformatics* 20:2222–2227.

[22] Lindgreen, S., Gardner, P.P., Krogh, A. 2007. MASTR: multiple alignment and structure prediction of non-coding RNAs using simulated annealing. *Bioinformatics* 23:3304–3311.

[23] Touzet, H., Perriquet, O. 2004. CARNAC: folding families of related RNAs. *Nucleic Acids Res.* 32:W142–W145.

[24] Sankoff, D. 1985. Simultaneous solution of the RNA folding, alignment and protosequence problems. *SIAM J. Appl. Math.* 45:810–825.

[25] Hofacker, I.L., Fekete, M., Stadler, P.F. 2002. Secondary structure prediction for aligned RNA sequences. *J. Mol. Biol.* 319:1059–1066.

[26] Bernhart, S.H., Hofacker, I.L., Will, S., Gruber, A.R., Stadler, P.F. 2008. RNAalifold: improved consensus structure prediction for RNA alignments. *BMC Bioinformatics* 9:474.

[27] Klein, R.J., Eddy, S.R. 2003. RSEARCH: finding homologs of single structured RNA sequences. *BMC Bioinformatics* 4:44.

[28] Knudsen, B., Hein, J. 2003. Pfold: RNA secondary structure prediction using stochastic context-free grammars. *Nucleic Acids Res.* 31:3423–3428.

[29] Cary, R.B., Stormo, G.D. 1995. Graph-theoretic approach to RNA modeling using comparative data. In *Proceedings of the Third International Conference on Intelligent Systems for Molecular Biology*, AAAI Press, Menlo Park, CA, 75–80.

[30] Tabaska, J.E., Cary, R.B., Gabow, H.N., Stormo, G.D. 1998. An RNA folding method capable of identifying pseudoknots and base triples. *Bioinformatics* 14:691–699.

[31] Bindewald, E., Shapiro, B.A. 2006. RNA secondary structure prediction from sequence alignments using a network of k-nearest neighbor classifiers. *RNA* 12:342–352.

[32] Mathews, D.H., Disney, M.D., Childs, J.L., Schroeder, S.J., Zuker, M., Turner, D.H. 2004. Incorporating chemical modification constraints into a dynamic programming algorithm for prediction of RNA secondary structure. *Proc. Natl. Acad. Sci. USA.* 101:7287–7292.

[33] Mathews, D.H., Sabina, J., Zuker, M., Turner, D.H. 1999. Expanded sequence dependence of thermodynamic parameters provides robust prediction of RNA secondary structure. *J. Mol. Biol.* 288:911–940.

[34] Lindgreen, S., Gardner, P.P., Krogh, A. 2006. Measuring covariation in RNA alignments: physical realism improves information measures. *Bioinformatics* 22:2988–2995.

[35] Mathews, D.H., Banerjee, A.R., Luan, D.D., Eickbush, T.H., Turner, D.H. 1997. Secondary structure model of the RNA recognized by the reverse transcriptase from the R2 retrotransposable element. *RNA* 3:1–16.

[36] Griffiths-Jones, S., Bateman, A., Marshall, M., Khanna, A., Eddy, S.R. 2003. Rfam: an RNA family database. *Nucleic Acids Res.* 31:439–441.

[37] Seemann, S.E., Gorodkin, J., Backofen, R. 2008. Unifying evolutionary and thermodynamic information for RNA folding of multiple alignments. *Nucleic Acids Res.* 36:6355–6362.

Chapter 2

Invariant Geometric Properties of Secondary Structure Elements in Proteins

Matteo Comin

University of Padua

Concettina Guerra

Georgia Institute of Technology and University of Padua

Giuseppe Zanotti

University of Padua

2.1 Introduction

2.1.1 The dilemma of protein folding

Proteins and nucleic acids represent the two major classes of biological macromolecules present in living organisms. They both are necessary to a cell to perform most of its functions, but their role is profoundly different: whilst in nucleic acids the information content is kept in the form of a string, i.e., it resides in the linear sequence of the four bases, the most important aspect of a protein (at least of the globular ones) is its three-dimensional (3D) architecture. Using the 20 different amino acids that can constitute a protein (we are neglecting here posttranslational modifications, which can be physiologically very important, but are not relevant for the problem of folding), it is in principle possible to build an impressive number of different sequences:* considering, for example, a polypetide chain of only 100 amino acids, this number is 20^{100}. Only a very small fraction of these sequences is actually present in a cell. For example, the genome of a simple gram-negative bacterium, like *Escherichia coli*, codes for less than 2000 genes, whilst the genome of a complex organism, like a man, contains many more genes (according to different estimates, between 20,000 and 30,000 genes) and consequently many more proteins. The previous numbers drastically decrease if we consider tertiary structures. It is in fact well known that the 3D structure of a protein is much more conserved than its amino acid sequence, and proteins with different primary structure can display the same fold.† Quite often the same fold corresponds to the same function, and this is one of the reasons why it is necessary to know the 3D structure of a protein and not simply its amino acid sequence; but there are also common protein folds that correspond to totally different functions. We will not discuss here if the latter phenomenon has to be ascribed to convergent or divergent evolution, but the practical consequence of this fact is the relatively limited number of different protein folds present in nature. If we consider the Protein Data Bank (PDB, http://www.rcsb.org), the database that collects all the 3D structures of biological macromolecules till now experimentally determined, either through X-ray or electron diffraction or NMR, there are at present about 47,000 structures of proteins deposited. They correspond to about 1,050 different folds according to SCOP (Murzin et al., 1995) or to 850 according to CATH (Orengo and Thornton, 2005). We do not know yet if they can be considered representative of all the possible folds present in living

---

* The amino acid sequence is also called the primary structure. The level of organization of a protein include three other levels: the secondary structure considers how the polypeptide chain folds on itself, forming pieces of repeated conformation; the tertiary structure describes how secondary structure elements (SSEs) organize in 3D space; the quaternary structure (which is not present in all the proteins) describes the organization of more than one polypeptide chain.

† The term "fold" is used to indicate the way SSEs are arranged in space and is roughly a synonym of "tertiary structure."

organisms: until some years ago it was estimated that the possible folds could have been about 1000; since completely new folds have not been discovered in the last four years, it is quite reasonable to assume that the number of folds we know is probably quite close to the total number of the existent ones. If so, this means that in nature a limited number of 3D architectures have been developed, and those are used to perform all the necessary functions of cells and organisms. Interestingly, similar 3D folds can be present in proteins that bear small or even undetectable sequence homology, but at present we are not yet able, given an amino acid sequence unrelated with that of previously known 3D structures, to predict with sufficient reliability which folding that particular sequence will assume.

2.1.2   Protein classification and the discovery of hidden rules

The concept of fold similarity is not exempt from ambiguities. Do protein families really exist, or is it more likely that there is a sort of "continuous" of similarities? The idea of grouping proteins into "families" according to their fold similarity possibly derives from our needs of classification and categorization (Gibrat et al., 1996). Whilst in some cases two proteins clearly share the same fold, in others this similarity is questionable, and, in fact, different programs estimate a different numbers of total folds and classify some proteins as belonging to the same family or not (Figure 2.1). This need of categorization has, however, a great practical relevance, both in structure prediction and in function assignment. The experimental determination of the 3D structure of a protein, either by X-ray or NMR, takes nowadays months or, in difficult cases, years, while the sequences of entire genomes, and consequently of the proteins coded by them, are determined at a very high rate.* In this respect, the ability of predicting the 3D structure of a protein is of paramount importance. At the same time, the recognition of structural similarities in proteins that present limited or nonexistent sequence similarity can sometimes be used to assign a biological role to a protein of unknown function, when its 3D structure has been determined.

Sometimes similarity does not involve entire structures, but only a portion of them: it is limited to a single domain, i.e., to a substructure that can be defined as an independent structural unit inside a larger protein. In order to detect similarities, at least in all cases that are not self-evident, the parameters and the algorithm used become relevant and can strongly influence the final results. Different algorithms have been devised to compare and superimpose protein structures, but none of them is completely free of failures. Some impose the constraint of continuity of the matched atoms along the primary sequence, in other words preserve the sequential order of the matched atoms; other methods try to minimize the so-called "soap-bubble area" between two structures, or involve other techniques, like lattice fitting (surveys

---

* At the time of writing of this chapter 680 genomes of bacteria (http://www. ebi.ac.uk/genomes/bacteria.html) and 33 of eukaryotes (http://www.ebi.ac.uk/2can/ genomes/eukaryotes.html) are available, and many others are in progress.

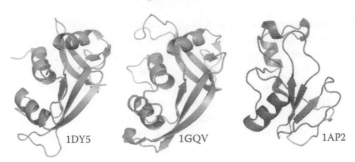

FIGURE 2.1: Cartoon representation of three proteins belonging to the $\alpha+\beta$ class of proteins according to the SCOP server (Murzin et al., 1995). 1DY5 is the bovine pancreatic ribonuclease A, 1 GQV is the human enosophil-derived neurotoxin, 1A2P the ribonuclease from *Bacillus amyloliquefaciens*. The first two are clearly superimposable, the third displays a similar motif (a β-sheet flanked by two or three α-helices), but with significant differences. According to SCOP, they belong to the same class, but the first two are classified in the same subclass and the third to a different one.

of the different methods of proteins comparison can be found in Gerstein and Hegyi (1998) and Kolodny et al. (2005)).

The method illustrated in this chapter integrates different techniques, namely geometric hashing and dynamic programming, to achieve good performance. Advantages (and disadvantages) of the method will be discussed.

2.2   The Use of Geometric Invariants and Hashing for a Simplified Representation of Secondary Structure Elements (SSEs)

2.2.1   Simplified representations of three-dimensional (3D) structures

The structure of a protein composed of n atoms (we consider here only "heavy" atoms, i.e., we neglect hydrogen atoms, that in general are too small to be observed in a diffraction experiment) is fully described by $3n$ numbers, with a triplet of numbers associated to each atom and corresponding to the three coordinates of the atomic center with respect to an arbitrary reference system.* In order to describe folds, to compare structures and detect

---

* In general, a fourth number characterizes the thermal motion of the atom around its equilibrium position, but this can be totally neglected for fold classification and structure comparison. In the case of X-ray diffraction data, the reference system is usually related to the crystal cell.

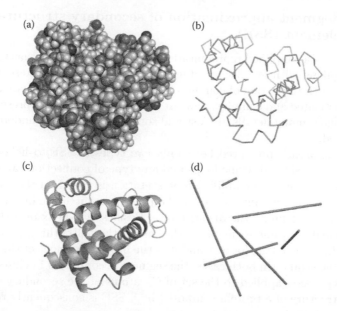

FIGURE 2.2: Simplified representations of protein folds. The structure of sperm whale Met-myoglobin (ODB:ID 1L2K) is shown has: (a) all atoms represented as spheres; (b) Cα atoms chain trace; (c) ribbon diagram. Helical ribbons represent α-helices; (d) helices are substituted by segments, each of them corresponding to the axis of the helix. To further simplify the representation, connections among SSEs have been neglected.

similarities, a simplified representation is desirable and necessary. One of the reasons is that amino acids side chains differ in similar proteins; moreover, the protein fold is defined by the torsion angles of the polypeptide main chain, making side chains useless. The most common simplification is represented by the use of a single atom for each amino acid, in general the α-carbon atom (Cα). This reduces the protein structure to a series of line segments, which is more manageable than a structure composed by points-atoms (Figure 2.2). An alternative is represented by the "distance or contact matrix," a square matrix where the a_{ij} element assumes the value 1 if amino acids i and j are in contact (i.e., the distance between their atoms is less than a given value), 0 otherwise. This latter representation can equivalently be expressed by a graph. It is needless to say that two similar proteins are characterized by similar distance matrices. A further reduction is represented by the use of the secondary structure element (SSE), either connected or not (Figure 2.2). The latter is the most simplified representation, since a medium size protein (or a protein domain) is on average composed of around 12 SSEs. The definition of the type (helix or strand) and orientation of secondary structure elements (SSEs) is sufficient to completely define a protein fold.

2.2.2 Segment approximation of secondary structure element (SSE)

The use of a simplified representation of a protein structure has several advantages (along with few disadvantages). A very simplified representation, in fact, enables the detection of common motifs or regularities that are concealed by a sophisticated and complex view. At the same time, if the representation is too simplified, many details are lost and some aspects can be underestimated or neglected.

In the approach presented here, only two types of SSE, α-helices and β-strands, are considered, while loops and any type of connections among them are neglected. This oversimplified representation is based on the assumption that SSEs are the building blocks of the protein architecture or, at least, are fundamental in defining the 3D structure. Besides, a α-helix or a β-strand is approximated by a line segment: in the case of the helix, this segment passes through the helical axis, for a strand it is the best-fit segment of the set of C_α atoms of the strand. In both cases, the segment is computed in closed form by a linear regression applied to the set of C_α atoms of the secondary structure. The 3D structure of a protein composed by n SSE is consequently reduced to a set of n oriented segments.

2.2.3 Building of the hash table for triplets of secondary structure element (SSE)

The technique of *geometric hashing* was originally developed within the field of pattern recognition and computer vision and then applied to problems in structural bioinformatics (Dror et al., 2003; Shatsky et al., 2004, 2006). It consists of representing a set of 3D structures by storing in a hash table redundant transformation-invariant information extracted from the structures. The hash table is used at recognition time, when similar invariants are extracted from a query structure and hashed into the table to retrieve the structures most similar to the query. Geometric hashing has generally been applied to point sets, either in 2D or 3D space, undergoing rigid transformations or the more general affine transformations. For matching 3D point sets, quadruples of points are used to define reference frames or bases in which the coordinates of all other points are computed. Such coordinates remain invariant for the class of affine transformations. Models are stored into the table by considering all possible combinations of quadruples of points as bases and using the invariant coordinates of the remaining points to index the table. At recognition time, if the correct quadruple of points is chosen from the query structure, the candidate matches are efficiently retrieved from the hash table. Geometric hashing suffers from sensitivity to noise and excessive memory requirements. Several heuristics have been proposed to solve either one or both: coarse quantization of the hash bin, the selection of few relevant bases, the detection of "seed matches" and their clustering, locality sensitive hashing. In our application, with a protein reduced to a set of n linear segments, the hashing is based on invariant properties of groups of SSEs, such as angles and distances, rather than on properties of atoms. If we consider a triplet of SSEs, the three angles associated to the three pairs of segments of the triplet are used to index a 3D hash table. A table entry stores information about all triplets that hashed into it. In addition, for each triplet the distance between pairs of segments is inserted into the table.

It is important to notice that SSEs are considered independently from their position in the polypeptide chain and thus all combinations of segments are considered. The number of triplets in a protein composed by n SSEs is $O(n^3)$. A major point must be stressed here: two triplets (or n-ets) that share the same angular values are not necessarily super-imposable, and may even correspond to totally different arrangements of SSEs. One reason is that the dihedral angle between two linear segments is defined as the angle formed by the two planes perpendicular to the straight lines that include the segments. Consequently, two different segments belonging to the same straight line form the same angle with any another segment belonging to a second straight line (Figure 2.3).

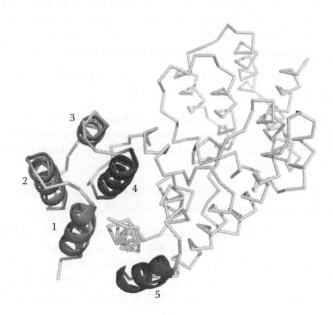

FIGURE 2.3: Cα chain trace of the influenza virus matrix protein (PDB:ID 1AA7). The two triplets of α-helices (1, 2, 3) and (3, 4, 5) belong to the same cell of the hash table, i.e., they form the same dihedral angles between pairs of helices. Nevertheless, they present a different arrangement in space and are not superimposable.

2.2.4 Building the hash table

Let P be a protein and $(p_1, \ldots, p_n)$ the best-fit line segments associated to its SSEs, listed according to their order along the polypeptide chain. Triplets of segments (p_u, p_v, p_z) of P are ordered in such a way that $u \leq v \leq z$; a triplet is characterized by three dihedral angles $(\alpha_{uv}, \alpha_{vz}, \alpha_{uz})$ and three distances between the mid-points of the segment (d_{uv}, d_{vz}, d_{uz}). A 4D hash table is built with the following index structure: the quantized angle values of a triplet of segments constitute the first three indices, the fourth index is a number that characterizes the composition of the triplet in terms of helices and strands. The latter index is fundamental in order to distinguish a segment representing a helix with that representing a strand. After various tests, the cell size of the hash table was empirically chosen equal to $18°$.

2.3 The Use of Geometric Invariants for Three-Dimensional (3D) Structures Comparison

Once the triplets hash table containing all the data for the entire PDB (or a representative subset) has been built, it can be used to efficiently find the proteins of the PDB that have high structural similarity with a query protein or domain. There are three main steps in our search, where each step refines the results of the previous step using a more computation-intensive procedure.

Step 1. Access the hash table to find a list of proteins that are good candidates for similarity with the query.

Step 2. For each candidate protein, perform a pair-wise structure alignment with the query protein. Rank the candidate proteins based on the score of the alignments and remove from the list the candidates with a score below a given threshold.

Step 3. Supcrimpose the query protein with each candidate protein and compute the root mean square deviation (RMSD).

Step 1 selects from the approximately 47,000 proteins of the PDB a list of candidates typically of size less than 1000. The exact number depends on the threshold used by the algorithm. This is the main advantage of hashing that avoids individual comparisons of the query with all 47,000 proteins, drastically reducing the number of proteins for which a pair-wise comparison is needed.

2.3.1 Retrieving similarity from the table

Step 1 accesses the hash table looking for triplets of SSEs of stored proteins equivalent to triplets of the query protein P. Consider triplets (p_u, p_v, p_z) of protein P and triplet (q_r, q_s, q_t), $r \leq s \leq t$, of protein Q. The latter has angles

$(\varphi_{rs}, \varphi_{st}, \varphi_{rt})$ and distances (h_{rs}, h_{st}, h_{rt}). The two triplets are considered equivalent if:

$$|\alpha_{uv} - \varphi_{rs}| < TA, \quad |\alpha_{vz} - \varphi_{st}| < TA, \quad |\alpha_{uz} - \varphi_{rt}| < TA$$

and

$$|d_{uv} - h_{rs}| < TD, \quad |d_{vz} - h_{st}| < TD, \quad |d_{ux} - h_{rt}| < TD,$$

with TA and TD given thresholds. Several threshold values were tested and $TA = 18°$ and $TD = 8\,\text{Å}$ were selected.

For every triplet of SSEs belonging to protein P, the hash table cell indexed by the triplet is accessed and a vote is cast to the records of the cell with similar distance values, corresponding to equivalent triplets. In the voting process, each triplet of P may cast a vote to a protein T only once, even of several triplets of the protein T are present in the same cell. Due to the quantization of the angles, equivalent triplets may not lie in the same cell, but in a neighboring one, thus votes are cast also for the adjacent cells.

Possible matches of the query protein with the proteins stored in the table are ranked according to the number of votes they accumulated. In order to avoid false positives, hypotheses of similarity are verified by pair-wise comparisons of the query protein with the candidate proteins, as described below.

2.3.2 Pair-wise alignment of secondary structures

The input to step 2 consists of the query protein P and the candidate protein Q represented by the segments $(p_1, \ldots, p_n)$ and $(q_1, \ldots, q_m)$, respectively associated to their secondary structures. The task is to align the SSEs of the two structures by finding the longest sequence of well matched pairs.

A solution to the alignment problem is obtained by the dynamic programming technique.

The scoring function used in the algorithm compares pairs of segments from each of the two proteins; the algorithm maximizes a score that represents the degree of similarity between these segments. Specifically, for a given scoring function $d(p_i, q_j)$ defined on all pairs (p_i, q_j) of SSEs, the DP algorithm maximizes the sum D of the scores over all aligned pairs of secondary structures, i.e.,

$$D = \max \sum_{(ph, qk) \in A} d(p_h, q_k)$$

The alignment preserves the order along the backbone chains, i.e., for all pairs in A, if (p_i, q_j) and (p_h, q_k) are two aligned pairs of A and $i < h$, then it must also be $j < k$. The alignment can introduce gaps that correspond to missing or inserted SSEs in one structure.

In the literature, different scoring functions for SSEs have been considered for the alignment of two proteins. They can roughly be classified as either orientation-independent or orientation-dependent. Orientation-independent scores are based on the comparison of internal geometric properties between pairs of segments selected from the two proteins, while orientation-dependent scores compare individual segment orientations and origins from the two proteins. For example, if one chooses to consider the angle of two segments, an orientation-independent score would compare the angle between a pair of segments of protein P to the angle between a pair of segments of Q. On the other hand, the orientation-dependent score would compare the orientation or origin of a segment from protein P to that of a segment from protein B.

In Singh and Brutlag (1997), both scores have been used in an iterative procedure based on DP. Initially, the scores between segments are orientation-independent; following each DP iteration, the new results are used to derive orientation-dependent scores for pairs of secondary structures. Every subsequent DP iteration involves orientation-independent as well as the newly computed orientation-dependent scores. In our approach (Comin et al., 2004) the score $d(p_i, q_j)$ is orientation-independent and is derived from the information stored in the hash table. The value $d(p_i, q_j)$ is given by the number of times the pair (p_i, q_j) occurs in two equivalent triplets of segments of secondary structures of P and Q. Recall that two equivalent triplets produce the same indexes into the hash table and consequently are stored in the same hash table cell. The reuse of data already computed allows a time-efficient implementation of the method.

2.3.3 Ranking candidate proteins

The DP algorithm is run on all candidate proteins. Based on the scores of the obtained alignments, the candidate proteins are re-ranked to produce the top n (a user specified parameter) solutions to the search. Specifically for each candidate protein Q a new similarity measure S with query protein P is evaluated as follows:

$$S = D(SS_{al}/SS_{tot})^2$$

where D is the value maximized by DP, SS_{al} is the number of aligned SSEs and SS_{tot} is the total number of SSEs of protein Q. The term $(SS_{al}/SS_{tot})^2$ is introduced to account for differences in protein size and penalize the occurrences of a small substructure in large proteins.

2.3.4 Atomic superposition

In the final step 3, we use the results of the highest scoring alignment of P and Q to find matches for the atoms of the two structures and obtain the rigid transformation that minimizes the RMSD between the matched atoms. This is done trough a number of iterations, with each iteration extending the matches to new atoms and refining the transformation. In the first iteration we use the method in (Horn, 1987) to compute the initial transformation T that optimally superimposes the atoms of the starting and ending residues of all aligned SSEs. We apply T to the entire candidate protein and extend the matches to other atoms of the SSEs while preserving the alignment of matched atoms. Specifically, if (a_i, b_j) is a pair of already aligned atoms of SSE a and b of P and Q, respectively, for each atom a_{i+1} of a we determine (if it exists) the unmatched atom $b_h (h > j)$ of b such that the distance $|T(b_h) - a_i|$ is minimum. The pair (a_{i+1}, b_h) is a new match. For the new set of matched atoms we derive the transformation that minimizes the RMSD. For the next iterations, the two steps, i.e., (1) refinement of the transformation from the new matches, (2) application of the transformation to derive new matches, are repeated. The iterations terminate when no more matches can be added. Finally, the match is extended to the atoms outside the aligned secondary structures, while maintaining the order along the primary structure. More precisely, we examine all atoms belonging to loop regions in between corresponding secondary structures of P and Q. For each atom of a loop in P we consider a spherical neighbor of a_i of radius ε. If there is an atom b_j of the corresponding loop of Q such that $T(b)$ falls in the neighbor of a_i, i.e., the distance $|T(b) - a_i|$ is less than ε, and if such atom preserves the sequential order along the backbone, then the pair (a_i, b_j) is a new match. By doing so, typically a considerable increase in the number of matched atoms is obtained.

2.3.5   Benchmark applications

We benchmarked our method on different query proteins or chains. A query protein or chain is compared with the entire set of proteins or chains of the PDB to retrieve those with high score when matched with the query. We used the SCOP classification (Murzin et al., 1995) as standard-of-truth. In Table 2.1, we show the results obtained using as a query one of the structures of the sperm whale myoglobin (PDB:ID 110M), an all-α protein fold. The structures listed in the table are some of the top 615 obtained using the score S defined in the previous section. Of them, the top 606 are all classified by SCOP in the same fold as 110 m. The first protein with a different fold appears at position 607. As another example, we have considered as a query protein the triose phosphate isomerase from chicken muscle (PDB:ID 1TIM); more than 400 chains are correctly recognized before a protein with a different fold (according to the SCOP classification) is found.

To assess the quality of our results we use the measure of accuracy defined as the percentage of correctly classified proteins in the top n items outputs, for various values of n. An output is considered correct, or true positive, if the protein is classified by SCOP in the same fold as the query protein. Extensive experimentation on a large set of query proteins has demonstrated that the approach has a good performance in terms of accuracy (Comin et al., 2004). Our results are typically consistent with SCOP, with few false positives intermixed with true positives at quite low positions in the output list.

Other experiments were done to benchmark the alignment and superposition procedures. Our results were compared to those of two existing methods, namely CE (Shindyalov and Bourne, 1998) and DALI (Holm and Sander,

TABLE 2.1:   Output obtained with query protein 110 M.

Rank	Protein	*n* Aligned SSE	Score *S*	Name	Na	RMSD
1	1 abs	8	172	OXYGEN STORAGE	154	0.3
2	2 mgf	8	170	OXYGEN STORAGE	154	0.4
3	2 mga	8	170	OXYGEN STORAGE	154	0.3
4	1 mtk	8	170	OXYGEN STORAGE	154	0.1
5	1 mti	8	170	OXYGEN STORAGE	154	0.2
6	1 mls	8	170	OXYGEN STORAGE	154	0.3
7	1 mlr	8	170	OXYGEN STORAGE	154	0.2
...						
...						
...						
604	1 f5p	7	8	OXYGEN STORAGE TRANSPORT	128	2.7
605	1 lfl	5	8	OXYGEN STORAGE TRANSPORT	130	3.1
606	1 gcw	5	8	OXYGEN STORAGE TRANSPORT	133	2.6
607	1 ny9	5	8	TRANSCRIPTION	31	2.9
608	1 aoi	4	8	NUCLEOSOME CORE/DNA	49	2.9
609	1 hbr	6	7	OXYGEN STORAGE/ TRANSPORT	139	2.9
610	1°0y	6	7	OXYGEN TRANSPORT	126	3

Notes: The proteins are ranked according to the score *S* (fourth column). The third column gives the number of aligned SSEs, Na is the number of aligned atoms, and RMSD is computed for the corresponding atoms.

1996), in terms of quality of the alignment and execution time. The quality of the alignment was expressed in terms of both the number of aligned atoms and the RMSD. On "easy" cases all three methods performed quite well and no significant variations could be found in their solutions. On some "difficult cases" identified in (Shindyalov and Bourne, 1998), our approach gives

TABLE 2.2: Comparison of three methods for structure alignment, PROuST, CE, and DALI for some "difficult" cases.

Prot 1 (size)	Prot 2 (size)	PROuST		CE		DALI	
		Na/RMSD	Time (seconds)	Na/RMSD	Time (seconds)	Na/RMSD	Time (seconds)
1 clc (639)	1 hoe (74)	61/3.1	0.6	66/3.1	10.5	66/3.4	24.4
3 hla B (99)	2 rhe (299)	71/3.1	0.4	85/3.5	1.3	75/3.0	15.8
2 aza A (276)	1 paz (120)	74/2.8	0.5	85/2.9	1.6	81/2.5	13.9
2 sim (799)	1 nsb A (466)	263/3.3	1.3	276/2.9	50.7	292/3.3	388.2
1 tie (272)	4 fgf (217)	101/3.1	0.4	115/2.8	2.3	114/3.1	29.7

Notes: Na is the number of aligned atoms and the execution time is in seconds. The size of a protein is expressed as the numbers of residues.

reasonable results. On the other hand, its execution time is typically much lower. Table 2.2 shows the details of the comparison of the three methods.

In conclusion, the experiments on protein structure comparison have demonstrated the flexibility of our method, which considers only spatial relationships among secondary structural elements, neglecting the connections among them, and is able to detect a partial similarity between two proteins of totally different size. Furthermore, the method does not need any a-priori assumptions about the regions to compare.

2.4 Statistical Analysis of Triplets and Quartets of Secondary Structure Element (SSE)

Occurrences of triplets and quartets of SSEs will be discussed mainly with the purposes of protein classification and categorization. In this section, we study the distribution of triplets of angles stored in the hash table and then extend the analysis to quartets. The goal is to find angular patterns of triplets and quartets that are over-represented in a dataset of protein structures, in other words patterns that occur in proteins more frequently than expected in sets of random segments. Obviously, angles do not capture all the structural information of a configuration of SSEs, since they are independent of the actual location of the segments on the supporting lines. For this reason, in the following we refer to segments as vectors and consider them applied in the origin of the reference system, as in Figure 2.4.

Angles represent concise geometric descriptors of the spatial arrangements of SSEs and have attracted the attention of researchers from the early days of structural bioinformatics. Initially, the interest focused mostly on the distribution of angles of contacting or packing helices. Based on the observed peaks in the distribution, interesting models were proposed that provided an

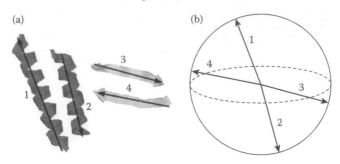

FIGURE 2.4: (a) An example of vector representation for a quartet of SSEs.
(b) The unit vectors translated to the origin (into the unit sphere).

explanation for the preference of packing helices for such angles. The angular preference remained, even though attenuated, when the analysis was
corrected for the statistical bias, which could be observed also in random
segments (Bowie, 1997). More recently, the attention has shifted toward the
analysis of global interaction of multiple SSEs. The distribution of the angles of triplets of SSEs, either interacting or noninteracting, was analyzed in
Guerra et al. (2002) and Platt et al. (2003), where it was observed that certain
angular patterns are over-represented when compared to sets of uniformly distributed vectors. Specifically, planar configurations of α-helices and β-strands
are highly preferred; in a planar configuration, one of the three angles is equal
to the sum of the other two, while in a non planar configuration each angle
must be less than or equal to the sum of the other two.

When the analysis moves from triplets to quartets of segments, the number
of angles that they form grows from three to six and the study becomes much
more complex. One reason is the combinatorial explosion of the data that
makes the enumeration of quartets computationally expensive, combined with
the difficulty of visualizing and interpreting 6D data.

2.4.1 Methodology for the analysis of angular patterns

We now describe an efficient a priori-based algorithm for finding over-
represented arrangements of quartets of segments without exhaustive enumeration (Comin et al., 2008). The original a priori algorithm was introduced
by Agrawal et al. (1993) to solve the problem of finding association rules in
large datasets. The simple idea of the method is that if a given set of k items
is frequent, then all its subsets of $k - 1$ items must also be frequent; this is
called the anti-monotone property. The general implementation of this idea is
that we can search for frequent item-sets of any size k by iteratively merging
frequent item-sets of size $k - 1$. In our scenario we want to implement this
notion to find over-represented quartets of segments.

A quartet of segments $S = (s_{i1}, s_{i2}, s_{i3}, s_{i4})$ is described by six angles $Q = (\alpha_{12}, \alpha_{13}, \alpha_{23}, \alpha_{24}, \alpha_{34}, \alpha_{14})$. The key idea of the algorithm is that if an angular 6D pattern of angles is frequent, then some specific triplets of angles extracted from the sextuple must be frequent. Thus, for a given set of proteins structures, the pattern discovery algorithm proceeds by first identifying the frequent (over-represented) triplets of angles and then joining them to generate candidate frequent sextuples and finally in verifying those candidates by determining their actual frequency. This method yields a significant reduction in the execution time with respect to the complete enumeration: from three days to 20 minutes for a set of 300 representative structures. An important aspect of the method is that it can be extended from quartets to any given number of SSEs.

Although our algorithm is inspired by a priori, it differs from its applications to basket analysis and bioinformatics. In particular, the basic a priori algorithm has the requirement that all subsets of items of an itemset must be frequent, in order for the itemset itself to be frequent. Unfortunately the anti-monotone property, that is the basis for every a priori algorithm, does not hold in the case of sextuple of angles, and more generally for the angles formed by a set of segments. There are triplets of angles that cannot be generated from three of the four segments under examination: these are the cases when the three angles involve all four segments. However, there are only four triplets of angles that are subsets of a frequent sextuple Q that must be frequent. These are $(\alpha_{12}, \alpha_{13}, \alpha_{23})$, $(\alpha_{23}, \alpha_{24}, \alpha_{34})$, $(\alpha_{13}, \alpha_{14}, \alpha_{34})$ and $(\alpha_{12}, \alpha_{14}, \alpha_{24})$. Indeed, the four triplets are obtained by the four different ways of choosing three segments out of four.

A sketch of the approach is presented here, for a detailed description see Comin et al. (2008). Given a representative set of protein structures, we assume that the hash table indexed by the discretized angles of the triplets has already been built as described in the previous sections. To determine if a triplet is over-represented, we need to remove the statistical bias that is present also in the distribution of random vectors, which is far from being uniform. For a given data set of n proteins, we generate n sets of random vectors; each set contains the same number of vectors as the SSEs of each protein of the data set. Then we construct the hash table for such sets and measure the deviation of the hash table of the real structures from that of the random ones; finally, we choose as over-represented the entries or triplets with the highest deviations from the random sets.

Once the over-represented triplets have been identified, the last two steps of the mining procedure, i.e., the joining of triplets into the candidate sextuples and the verification of such candidates, are performed as follows. The operation *join* merges four frequent triplets $(\alpha_{12}; \alpha_{13}, \alpha_{23})$, $(\alpha_{23}; \alpha_{24}; \alpha_{34})$, $(\alpha_{13}, \alpha_{14}; \alpha_{34})$ and $(\alpha_{12}; \alpha_{14}; \alpha_{24})$ into the candidate sextuple $(\alpha_{12}; \alpha_{13}; \alpha_{23}; \alpha_{24}; \alpha_{34}; \alpha_{14})$. In other words, four triplets are merged if the last angle of the first triplet is the same as the first angle of the second; the second element of the first triplet is the same as the first element of the third triplet, and so on.

The requirement that the triplets must be frequent does not guarantee that the sextuple is frequent. In fact this requirement provides a necessary, but not sufficient, condition. The obtained sextuple is only a candidate solution that needs to be verified by determining its actual frequency and its statistical significance.

2.4.2   Results of the statistical analysis

We selected a set of 300 nonredundant proteins from different families and computed all triplets of SSEs and their associated linear segments. We recall that all triplets of SSEs are considered without any distinction between interacting and non-interacting elements. The distribution of triplets of SSEs of the selected proteins and of random vectors are shown in Table 2.3 by displaying the frequencies of the entries of the two hash tables. Notice that not all the cells of the hash tables are populated by triplets of segments owing to the triangular inequality of the angles. By construction, the two hash tables contain the same total number of triplets and are computed using sets of segments with the same composition; consequently they can be directly compared. The hash table contains 520 nonempty cells (containing a total of 398,853 triplets of vectors), of which 242 were selected as frequent (corresponding to 189,270 triplets). The triplets of angles selected as frequent are marked in the last column of Table 2.3.

Starting from the frequent triplets, a set of over-represented arrangements of four SSEs was built, using the procedure described in the previous paragraph. It has to be considered that each arrangement of four vectors is described by six ordered angles, where an angle corresponds to a specific pair of SSEs, which is identified by the sequential order of SSEs along the primary structure. In this way, two arrangements forming the same six angles, but in a different order, correspond to two different angular patterns in a protein. On the contrary, if we neglect the physical connections among SSEs, they can be considered geometrically equivalent and, for this reason, they were merged together, ignoring the relative order of angles.

By merging the patterns, the discovery procedure selects a set of 785 over-represented patterns, formed by 485,021 quartets of segments, out of 2262 possible patterns and more than 3,000,000 quartets obtained by the exhaustive search. The top ten angular patterns ranked by their frequency are presented in Table 2.4. The overall discovery procedure is relatively fast: it takes approximately 20 minutes on a standard PC (AMD Athon 2.6 GHz). On the same machine, the exhaustive generation of all possible quartets of SSEs takes more than three days. On the other hand, the results of the faster procedure are comparable to those of the enumeration. The top angular patterns of proteins obtained by the exhaustive enumeration are shown in Table 2.5, ranked by their frequency (Table 2.5a) and by the difference of their frequency from that of random quartets (Table 2.5b). As can be seen from the tables, the top

TABLE 2.3:   Distributions of angles of triplets of SSEs ranked by the frequency of real triplets.

α_1	α_2	α_3	Expected Frequency	Real Frequency	Deviation	Over-represented
8	3	7	1030.58	1656	+625.42	*
7	3	8	1030.33	1589	+558.67	*
8	2	8	755.01	1567	+811.99	*
2	6	8	1031.26	1546	+514.74	*
8	5	5	1279.32	1542	+262.68	*
8	6	2	1032.71	1534	+501.29	*
3	5	8	1214	1505	+291	*
8	7	3	1033.88	1503	+469.12	*
8	5	3	1208.6	1495	+286.4	*
8	3	5	1219.6	1494	+274.4	*
8	4	6	1223.51	1456	+232.49	*
...	...	...	...	...	...	...
6	3	3	1561.27	1201	−360.27	
8	3	6	923.87	1200	+276.13	*
6	4	2	1477.28	1199	−278.28	
3	4	1	1216.74	1198	−18.74	
1	4	3	1207.47	1195	−12.47	
1	5	4	1275.53	1192	−83.53	
4	3	7	1474.58	1192	−282.58	
7	3	4	1477.07	1191	−286.07	
1	2	3	1031.96	1185	+153.04	*
5	2	7	1329.16	1185	−144.16	
6	6	6	1569.37	1183	−386.37	

TABLE 2.4:   The ten top over-represented patterns of angles.

α_0	α_1	α_2	α_3	α_4	α_5	Frequency
1	2	3	7	8	9	6,439
1	2	7	8	8	9	5,780
1	2	3	7	8	8	5,586
1	3	6	7	8	9	5,100
1	1	2	7	8	9	4,657
1	2	6	7	8	9	4,437
1	2	3	6	8	9	4,085
2	3	6	7	8	9	3,884
1	3	7	7	8	8	3,831
1	2	3	7	7	8	3,728

patterns in Table 2.5b are in almost all cases the same patterns detected as over-represented by comparison with the random sets.

Over-represented patterns tend to be arranged into specific spatial conformations that can be described in terms of groups of parallel and anti-parallel

TABLE 2.5: The top angular patterns of the exhaustive enumeration of quartets ranked by their frequency (a) and by the difference of their frequency from that of random quartets (b).

α_0	α_1	α_2	α_3	α_4	α_5	Frequency Rand	Frequency	Difference
(a)								
3	4	5	6	7	8	16119	17224	−1105
2	3	5	6	7	8	13447	12467	+979
2	4	5	6	7	8	12912	13274	−362
3	3	5	6	7	8	11054	10525	+528
2	4	5	6	7	7	10236	12575	−2339
3	3	4	6	7	8	9795	9613	+181
1	4	5	6	7	8	9503	8582	+921
2	3	4	6	7	8	9496	9089	+406
3	4	4	6	7	8	9355	10361	−1006
2	3	4	5	6	8	8627	10043	−1416
(b)								
1	2	3	7	8	9	7439	3267	+4171
1	2	7	8	8	9	5861	2087	+3773
1	2	3	7	8	8	6517	3258	+3258
1	3	6	7	8	9	5939	2807	+3131
1	2	3	6	8	9	5861	2813	+3047
1	1	2	7	8	9	4658	1831	+2826
1	3	5	6	8	9	6146	3373	+2772
2	3	6	7	8	9	5764	3022	+2741
1	2	6	7	8	9	5076	2378	+2697
1	3	4	6	8	9	5915	3364	+2550

SSEs. For instance, we observe that the top pattern (1,2,7,8,8,9) is compatible with a known motif of interacting strands. On the other hand, such quartets of interacting strands contribute less than 450 to the total frequency of 5780. Thus, although the number of interacting quartets of SSEs is very small compared to all possible quartets, still their angular patterns appear among the top over-represented arrangements.

2.4.3 Selection of subsets containing secondary structure element (SSE) in close contact

The over-represented patterns considered so far have included the SSEs of the selected set of proteins, regardless of their distances. Moreover the over-represented patterns are extracted by using only angular information; we now investigate whether such patterns are also close in space. We now consider only homogenous patterns of SSEs (i.e., composed by all strands or all helices) from the top ranked configuration (1,2,7,8,8,9) and we define two SSEs to be in contact if the distance between the mid-points of their associated vectors is less than a given threshold (18 Å in our analysis). Notice that, following our

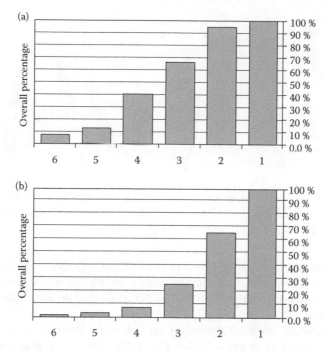

FIGURE 2.5: Cumulative number of pairs of SSEs in contact: (a) quartets of strands, (b) quartets of helices. The number of strands considered in contact (cutoff distance 18 Å) is reported in abscissa.

definition, in a sextuple not all pairs of SSEs must be in contact; it is sufficient that every SSE is in contact with at least another SSE. In the case of four SSEs, a sextuple should have at least three pairs of SSEs in contact. Figure 2.5 shows the cumulative number of SSEs in contact for the quartets of segments within the top angular pattern. It is interesting to observe that in all cases at least one pair of vectors are in contact, and very often this happens for three or more vectors. Notice that the use of the same threshold penalizes helices, owing to their bigger steric hindrance as compared to β-strands. Nevertheless, more than 65% of the helical quartets have at least two SSEs in contact. Better results are obtained when considering quartets formed only by β-strands: in this case 65% of the elements have at least three SSEs in contact. Again, we can observe that purely angular patterns that are over-represented are in most cases close in space.

In Figure 2.6 we present different examples of four strands, with angles (1,2,7,8,8,9). In all these examples the four strands are in contact. Although they display different arrangements, their pairwise angles are similar, thus they fall into the same cell of the hash table. These patterns of angles include SSEs from the same β-sheet (Figure 2.6c), as well as from different β-sheets

(a)   (b)   (c)

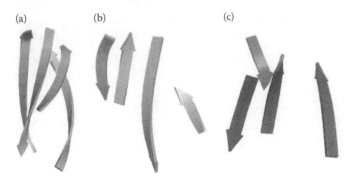

FIGURE 2.6: Three examples of the pattern of angles (1,2,7,8,8,9) composed by strands only: (a) Protein 1HPL, SSE: 16-17-18-20; (b) Protein 1ACC, SSE: 0-1-2-3; (c) Protein 1AOR, SSE: 4-6-8-12.

(Figure 2.6a and b). The fact that most, but not all, SSEs are close in space consolidates the idea that arrangements of angles are influenced by atomic interactions, either directly or through other SSEs that do not explicitly belong to the quartet.

In conclusion, we have described an efficient algorithm to extract over-represented quartets of SSEs, which avoids the exhaustive generation of patterns. We have shown that a careful analysis of the angular bias of random vectors is essential in the determination of over-represented arrangements of secondary structures. This study provides a generalized framework that can be easily extended to patterns composed by more than four SSEs.

2.5 Conclusions

In this chapter, we have introduced an over-simplified representation of the protein structures that can be efficiently used either for their comparison or for statistical analysis or discovery of common motifs. This representation, based on invariant geometric features of proteins' secondary structures, neglects any information about their connection along the polypeptide chain. In doing so, many relevant details are lost, but the remaining information is the essential one, and it can be efficiently used in pattern discovery. We have shown that this representation is sufficient, for example, to detect similarities among protein structures in a large database with a very high efficiency and without any *a priori* assumption, comparable to that used by other algorithms that require some human intervention. Given the over-simplified representation, part of the information can be missed in our method, however it can be efficiently used to detect overall similarities or to individuate concealed features. The results can then be used as the starting point for a more accurate analysis making use of an atomic representation.

Finally, the building of a hash table of an entire protein database have been used to compute over-represented patterns of triplets or quartets of SSEs, but the method is more general and could be used for discovering recurrent or hidden motifs in protein architectures. The study of the distribution of these geometric invariants could bring new insights to guide the engineering of stable protein modules. Other applications could be designed by replacing the null distribution with that of a specific family of proteins.

References

Agrawal, R., Imielinski, T., and Swami, A.N., 1993. Mining association rules between sets of items in large databases. *Proc. of the ACM SIGMOD Intl. Conference on Management of Data*, 207–216.

Bowie, J.U. 1997. Helix packing angle preferences. *Nature Structural and Molecular Biology*, 4:915–917.

Comin, M., Guerra, C., and Zanotti, G. 2008. Mining over-represented 3D patterns of secondary structures in proteins. *Journal of Bioinformatics and Computational Biology*, 6(6):1067–1087.

Comin, M., Guerra, C., and Zanotti, G. 2004. PROUST: A comparison method of three-dimensional structures of proteins using indexing techniques. *Journal of Computational Biology*, 11(6):1061–1072.

Dror, O., Benyamini, H., Nussinov, R., and Wolfson, H. 2003. Multiple structural alignment by secondary structures: algorithm and applications. *Protein Science*, 12:2492–2507.

Gerstein, M., and Hegyi, H. 1998. Comparing genomes in terms of protein structures: surveys of a finite parts list. *FEMS Microbiology*, 22:277–304.

Gibrat, J.-F., Madej, T., and Bryant, S.H. 1996. Surprising similarities in structure comparison. *Current Opinion in Structural Biology*, 6:377–385.

Guerra, C., Lonardi, S., and Zanotti, G. 2002. Analysis of secondary structures using indexing techniques. *IEEE Proc. First Int. Symposium on 3D Data Processing Visualization and Transmission*, 812–821.

Holm, L., and Sander, C. 1996. Mapping the protein universe. *Science*, 273:595–602.

Horn, B.K.P. 1987. Closed-form solution of absolute orientation using unit quaternions. *Journal of the Optical Society of America*, 4(4):629–642.

Kolodny, R., Koehl, P., and Levitt, M. 2005. Comprehensive evaluation of protein structure alignment methods: scoring by geometric measures. *Journal of Molecular Biology*, 346:1173–1188.

Murzin, A., Brenner, S.E., Hubbard, T., and Chotia, C. 1995. SCOP: a structural classification of proteins for the investigation of sequences and structures. *Journal of Molecular Biology*, 247:536–540.

Orengo, C.A., and Thornton, J.M. 2005. Protein families and their evolution—A structural perspective. *Annual Review of Biochemistry*, 74:867–900.

Platt, D.E., Guerra, C., Zanotti, G., and Rigotsous, I. 2003. Global secondary structure packing angle bias in proteins. *Proteins: Structure, Functions, and Genetics*, 53:252–261.

Shatsky, M., Nussinov, R., and Wolfson, H.J. 2004. A method for simultaneous alignment of multiple protein structures. *Proteins*, 156(1):143–156.

Shatsky, M., Nussinov, R., and Wolfson, H.J. 2006. Optimization of multiple-sequence alignment based on multiple-structure alignment. *Proteins: Structure, Functions Bioinformatics*, 62:209–217.

Shindyalov, I.N., and Bourne, P.E. 1998. Protein structure alignment by incremental combinatorial extension (CE) of the optimal path. *Protein Engineering*, 11(9):739–747.

Singh, A.P., and Brutlag, D.L. 1997. Hierarchical protein structure superposition using both secondary structures and atomic representations. *Proc. 5th Int. Conf. Intell. Sys. Mol. Biology*, 284–293.

Wang, X., Jason, T.L., Shasha D., Shapiro, B.A., Rigoutsos, I., and Zhang, K. 2002. Finding patterns in three-dimensional graphs: algorithms and applications to scientific data mining. *IEEE Transactions on Knowledge and Data Mining*, 14(4):731–749.

Chapter 3

Discovering 3D Motifs in RNA

Alberto Apostolico
Georgia Institute of Technology

Giovanni Ciriello
University of Padova

Concettina Guerra
Georgia Institute of Technology and University of Padua

Christine E. Heitsch
Georgia Institute of Technology

3.1 Introduction

RNA molecules are known to perform a variety of different biological functions, including roles as messenger (mRNA), transfer (tRNA), and ribosomal (rRNA) molecules in protein synthesis.

Each of these different types of RNA pose a common challenge: to understand the folding of the RNA sequence into a structure that performs biochemical functions. Of particular interest is the structure and function of ribosomal RNA molecules, both because of the size and complexity of the structures as well as their essential functionality in protein biosynthesis. Ribosomes are a complex of several RNA sequences, which comprise about 2/3 of the ribosome, and various small proteins. For instance, a bacterial (70S) ribosome consists of two subunits: a large (50S) and small (30S). These subunits in turn contain a 5S RNA and 23S RNA sequences in the large and a 16S RNA sequence in the small. (The "S" is a unit of time which roughly corresponds to molecular size.) The identification of small structural motifs and their organization into larger subassemblies is of fundamental interest in the prediction and design of 3D structures of large RNAs such as the ribosome. This problem has been studied only sparsely, as most of the existing work is limited to the characterization and discovery of motifs in RNA secondary structures.

This chapter reviews approaches to the characterization and identification of structural motifs in 3D rRNA molecules that find possible use in tasks such as: (i) alignment and homology searching of macromolecules by sequence and nonsequence criteria, (ii) structure prediction of RNA based on homology modeling, (iii) crossgenre alignment and homology searching, (iv) motif searching (cross-genre). In addition these techniques can help us dissect fundamental processes such as RNA folding, RNA-small molecule interaction and RNA-protein interaction. Our treatment specifically tackles the efficient automatic recognition of known 3D motifs, such as tetraloops, E-loops, kink-turns and others. On the other hand, it also suggests new ways of characterizing complex 3D motifs, notably junctions, that have been defined and identified in the secondary structure but have not been analyzed and classified in three dimensions.

3.1.1   The ribonucleic acid (RNA) structure

Ribonucleic acid (RNA) is an oriented (5′ to 3′) biochemical chain resulting from the concatenation of the four nucleotide units. The RNA nucleotides, like those of DNA, consist of a nitrogenous base (adenine, cytosine, guanine and uracil/thymine), a (deoxy)ribose sugar, and a phosphate. The sugars and phosphates form linkages creating the backbone of the RNA sequences, which progresses from the 5′ position on one ribose through the phosphodiester bond to the 3′ position on the next ribose. Similarly, RNA bases form hydrogen bonds leading to the Watson–Crick pairings of C, G, and A, U as well as the "wobble" G, U pairing. Unlike the DNA double helix, though, most RNA molecules are naturally single-stranded which, though intra-sequence base pairing and other molecular interactions, fold into 3D structures.

RNA folding is hierarchical (see Figure 3.1). The intermediate states, differentiated by the extent and type of hydrogen bonding and base stacking interactions, are characterized by their dimensionality. The 1D structure is, of course, the nucleotide sequence. The folding of this sequence is a process from initial to intermediate to final states where the initial unfolded state is conformationally polymorphic, while the final folded state is essentially unique. Advances in nucleotide sequencing technology, coupled with algorithms such as Needlesman–Wundsch, Smith–Waterman, and extensions

(1) Primary structure

AUGCCAGCUGGUGGAUUGCUCGGCUCAGGCGCUGAUGAAGGACGUGCCAAGCUGCGAUAAGCCAUGGGGAGCCG
CGAAUGAGAAUCUCUACAAUUGCUUCGC GCAAUGAGGAACCCCGAGAACUGAAACAUCUCAGUAUCGGGAGGAA
UAACCGCGAGUGAACGCGAUACAGCCCAAACCG .... UAACGAGACGU

(2) Secondary structure

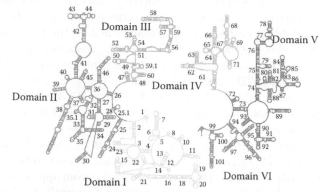

(3) Tertiary structure

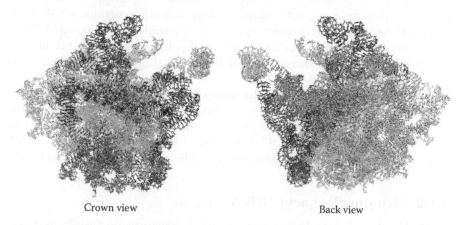

Crown view   Back view

FIGURE 3.1: RNA folding of the 23S RNA of *Haloarcula marismortui* large ribosomal subunit (PDBid: 1JJ2–Chain 0).

thereof, have been responsible for the explosive growth of ribosomal information at the 1D level. For instance, the Ribosomal Database Project II Release 9.55 (http://rdp.cme.msu.edu/) consists of 440,891 aligned and annotated 16S rRNA sequences.

Crucially, however, this abundance of sequence data does not automatically extend to information about higher dimensional structures. To illustrate, secondary structures determined through comparative sequence analysis of many phylogenetically related sequences are the "gold-standard" of establishing the 2D structure of an RNA molecule. The process remains very challenging, though, and the Gutell Lab CRW (http://www.rna.ccbb.utexas.edu/)

currently contains rRNA secondary structures for 668 of the 16S sequences, 209 of the 5S sequences, and a mere 86 of the 23S sequences. Structurally, a RNA secondary structure is a 2D map of local stems and loops, connected by single stranded regions. Local stems are base-paired (double-stranded) regions, with the two strands linked by a relatively short segment called loop. The secondary structure, stabilized by nonspecific metal ions, is thought to form early in the folding pathway, and is maintained in the final folded 3D structure, which is also stabilized by tertiary interactions and specific metal ions. Tertiary interactions are between nucleotides that are remote in the primary sequence. A folded 3D structure is given by the x, y, z coordinates of each RNA atom, along with localized ions. For large RNA molecules, x-ray diffraction is the only available experimental method for 3D structure determination. So far the structures of only four ribosomes are available.

Prediction of 3D structure is dependent on, and is significantly more difficult than, prediction of 2D structure. Folding of small RNAs can be simulated via all atom molecular dynamics. However, errors arise from approximations in force-fields and from limitations in simulation times. Currently, the best paths to 3D structure prediction uses RNA secondary structures, thermodynamics and phylogeny along with manual manipulation, constraint satisfaction, molecular mechanics, dynamics and structural homology. For example Burks and co-workers (Burks et al., 2005) converted paired regions of 2D structures to helices, and other regions to RNA motifs, using the structural classification provided by the SCOR database (Tamura et al., 2004). Models obtained thus far are at the level of folding architecture and relationships between folded elements, and do not provide precise atomic positions. Further details on the 3D modelling of RNA structures are given in Shapiro et al. (2007). In sum, the goal of producing reliable 3D structures, at atomic resolution, using RNA sequence as input data is a fundamental problem that has not been solved.

3.1.2 Ribonucleic acid (RNA) motifs

Besides base-paired helical regions, RNA structures are typically interspersed by loop regions with characteristic folding and, for some of them, sequence patterns. To distinguish different classes of loop motifs in three dimensions, we adopt the terminology which differentiates types of loops in nested secondary structures. RNA structures in two dimensions are composed of base-paired segments known as *helices* interspersed with regions called *loops*. Although loops are considered "single-stranded" in RNA secondary structures, we recognize that nearly all of these nonhelical nucleotides form some type of base-base interactions in the 3D structure. In 2D, however, these loops are distinguished by the number of Waston–Crick or *wobble* base pairs in the loop which is the same as the number of nonhelical strands within the loop. In the following we introduce some known characteristic loops.

A *1-loop* or *hairpin* includes, as the common name suggests, the contiguous fragment of nonhelical RNA which terminates a single helix. An extensively studied 1-loop motif is the tetraloop. A *tetraloop* is defined as a small loop composed of four nucleotides, connecting the two anti-parallel chains of an RNA helix. Tetraloops show characteristic conformations and stability, and consensus sequence patterns given by GNRA, UNCG, and CUUG, where N can be any nucleotide and R can be either G or A. In Hsiao et al. (2006) three main types of tetraloops are identified: standard, with deletion (where one nucleotide is omitted, e.g., the fourth one in GAA-) and with insertion (where one nucleotide is added, e.g., the fourth in CAG(Λ)Λ). A standard tetraloop is shown in Figure 3.2a.

Bulges and *internal* loops are known as *2-loops*, which join two helices and thus contain a closing base pair, an enclosed base pair, and the two single-stranded segments connecting those two base pairs. Well-known 2-loop motifs are the *E-loop* and the *kink-turn*. An E-loop is generally defined as an asymmetrical internal loop, where the elements of the loop are usually cross-strand paired (Leontis and Westhof, 1998). In our analysis we specifically refer to E-loop as defined in Leontis and Westhof (1998), see Figure 3.2b, sometimes also referred to as the *core* of the E-loop (Leontis et al., 2006). The E-loop bulged strand presents the consensus pattern AUGA.

The kink-turn consists of approximately 15 nucleotides from two distinct segments which base pair to form two helices and an internal loop. A representation of the kink-turn is given in Figure 3.2c. This is an important structural motif since it mediates RNA tertiary structure interactions making significant interactions with bound proteins. Six kink-turns in HM 23S were revealed in Klein et al. (2001).

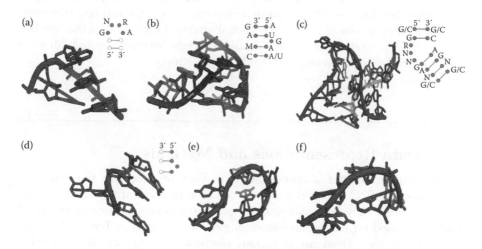

FIGURE 3.2: Secondary and tertiary structure of notable RNA motifs: (a) the standard tetraloop; (b) the kink-turn; (c) the E-loop; (d) the S2-motif. Only the tertiary structures of the π-turn (e) and Ω-turn (f) are reported, since these motifs do not share conserved characteristics at the secondary structure level.

Some special cases of 2-loops were introduced in Duarte et al. (2003) and Wadley and Pyle (2004) called *S2-motif*, *π-turn*, and *Ω-turn*. These motifs differ from the typical 2-loop motifs since they present only one characteristic fragment coupled with a fragment that has neither fixed length nor characteristic shape. We will refer to these motifs as *single-strand 2-loop motifs*, while to kink-turns and E-loops as *double-strand 2-loop motifs*. For these motifs we performed a single-strand search as done before for the 1-loops. The S2-motif is an internal loop, similar to the E-loop, composed by a characteristic strand with the base of the bulged nucleotide completely outside the loop usually involved in tertiary interaction (see Figure 3.2d). The π-turn is defined as a sequence of five consecutive nucleotides forming a pinched strand such that the backbone folds with an angle of ∼120°. Similarly, the Ω-turn is a strand composed by five consecutive nucleotides that changes direction twice: first between the second and the third nucleotide (∼180°), then between the fourth and the fifth (∼90°). The resulting bend is of ∼90°, generating the Ω shaped conformation. Examples of π-turn and of Ω-turn are given in Figure 3.2e and 3.2f, respectively.

Finally, *k-loops* for $k \geq 3$ are commonly called *junctions*, *multibranch* or *branching* loops. Again, as the names suggest, junctions are the loop regions where an RNA structure branches and which involves $k \geq 3$ helices and k distinct single-stranded segments. Precisely, in a k-junction we consider the ensemble of fragments that connect the terminal base-pairs of the helices, including these pairs and ignore, for the purpose of this analysis, the flanking helices. The connecting fragments have variable length and shape, thus they seem to lack common features. By contrast, the whole junction region at a lower level of resolution reveals recurrent 3D conformations. Moreover, junctions remain one of the least well-characterized aspects of minimum free energy RNA secondary structures, although recent thermodynamic studies are likely to lead to improvements in prediction accuracy.

3.2 Data Representations and Methods

Recently, a number of approaches have been proposed for the structural comparison of RNA molecules and for the identification of recurrent 3D motifs. Typically, these approaches use a geometric representation of the RNA molecule based on either inter-atom angles or distances or both. These basic measures have given rise to various geometric descriptors. We review the current literature focusing on methods that have been tested on ribosomes, specifically on the *Haloarcula marismortui* large ribosomal subunit (HM 23S–PDBid: 1JJ2) (Ban et al., 2000).

The study of the different approaches reveals an interesting trend: *when the search for a 3D motif is performed on a complex and variable structure, such as the ribosome, a coarse level of resolution is generally preferable to a fine one.* Inter-atom angles provide a reduced set of internal coordinates that, unlike Cartesian coordinates (from the PDB), are invariant under rigid transformations. At a fine level of resolution each nucleotide of an RNA molecule can be characterized by eight torsion angles: α, β, γ, δ, ε, and ζ formed by the backbone atoms, the glycosidic angle χ, and the ribose pseudo-rotation phase angle P. In Hershkovitz et al. (2003), fragments of contiguous nucleotides are described by sequences of the eight discretized torsion angles values; these sequences are then compared for possible similarities by well established pattern recognition techniques involving torsion matching and binning. The interesting feature of the approach is that substructure similarities can be established without the need for explicitly deriving a rigid transformation that best superimposes the structures. Although accurate, the use of all torsion angles provides a fine description that may not be adequate because often the structure of the motifs is not completely conserved but exhibits large variability in conformation. In fact, even a simplified application of the method, using only four out of seven torsion angles, identified only 25 out of observed 43 tetraloops of HM 23S.

A coarser level of resolution is proposed by the approach PRIMOS (Duarte et al., 2003), in which each nucleotide is represented by two pseudo-torsion angles: given three consecutive nucleotides $(i-1, i, i+1)$, the nucleotide i is described by two dihedral angles, η and θ, defined as $\eta(C4'_{i-1}-P_i-C4'_i-P_{i+1})$ and $\theta(P_i - C4'_i - P_{i+1} - C4'_{i+1})$. A sequence of η and θ values corresponding to contiguous nucleotides forms the so-called *worm*. A typical application of PRIMOS is the search within a set of RNA structures for the occurrence of a known motif or query, represented as a worm. The program PRIMOS matches the query against all possible fragments of the same length by comparing the angles η and θ of the corresponding nucleotides. An extension of PRIMOS is COMPADRES (Wadley and Pyle, 2004), a completely automated method that can identify previously undetected motifs. It is based on the all-to-all comparison of the worms extracted from the available RNA structures and on the clustering of the worms into groups of structurally similar fragments (within some user-defined tolerance). The method led to the discovery of two novel recurrent substructures in the ribosome, namely the already mentioned π-turns and Ω-turns. Although in general accurate, the method led to few misclassifications as will be described later (see Section 3.4.1).

More recently, an even coarser multiscale description of bio-macromolecules and, specifically, of motifs has been introduced in Hsiao et al. (2006), where a new space has been developed, called PBR space (P indicates phosphate, B indicates base, and R indicates ribose). This method decreases the level of resolution by determining the centers of mass and relative orientations of bases, riboses, and phosphates. Sequences of consecutive nucleotides are represented by patterns of angles and distances between the centers of mass in the PBR space. RNA motifs are then detectable by empirically established fingerprints incorporating information on molecular interactions. This method focused mainly on the tetraloop motifs in HM 23S and was successful in detecting all 43 instances. This result compares favourably with the one obtained with a finer resolution that detected 25 out of 43 tetraloops. Furthermore, the authors provided a classification of tetraloops into three main groups: standard tetraloops, tetraloops with insertion, tetraloops with deletion. However, the method in Hsiao et al. (2006) is not completely automated since it requires the setting of some parameters by visual inspection.

A similar trend from a fine to a coarse level of resolution can be noticed in approaches to motif recognition based on inter-atom distances. In Huang et al. (2005) the authors made use of the set of distances between 15 corresponding atoms of two fragments to evaluate the root mean square deviation (RMSD) after superimposition of the fragments; the RMSD distance is then used to cluster nucleotides sequences into groups of similar structures. By contrast, NASSAM (Harrison et al., 2003) considers only the distances between the start and end points of two vectors, that are chosen to represent the base of each nucleotide. Even though NASSAM looks at the structure at a lower level of resolution, it still requires for the match strong structural similarity, thus it fails to correctly identify some motifs, such as tetraloops, due to their not unique folding. In a more recent work (Sarver et al., 2006), an effective method for motif search was introduced, called FR3D, that combines geometric, symbolic and sequence information. This simplified model is base-centered: a fragment is represented by the set of distances between its bases, each defined as a single center point. Also FR3D enables the search of composite motifs, i.e., motif instances that are composed by discontinuous stretches of the sequence. Furthermore, pair-wise interaction constraints are introduced to reduce the size of the search.

Some methods for RNA structural alignment, although not explicitly designed for motif detection, can be adapted to this task. The method ARTS (Dror et al., 2005) compares two structures to identify the largest common substructures; it uses a simplified representation of an RNA molecule given by the 3D coordinates of the P atoms only. In each structure, all sets of four P atoms corresponding to two consecutive WC base pairs are considered. Two sets, one in each structure, are matched if the six internal distances between the four atoms in one structure are approximately the same as the distances in the second substructure. From two matched sets the geometric transformation that best superimposes the four P atoms of the sets is derived. Finally, the rigid transformations derived from all possible sets of two WC pairs are clustered and extended to other WC base pairs to obtain the transformation that superimposes the largest number of WC base pairs in the two structures.

Another method for the structural alignment of RNA molecules is DIAL (Ferre' et al., 2007). DIAL combines torsion angles information together with nucleotide sequence similarity and base-pairing similarity to compute a score function that is optimized by a dynamic programming algorithm. Actually DIAL put together methods for both global alignment (Needleman–Wunsch) and local alignment (Smith–Waterman), to develop a set of tools for RNA structural alignment and 3D motifs detection. The search for 3D motifs is performed by a semiglobal alignment method and applies both to 1-loops, i.e., contiguous fragments, and to 2-loops, i.e., noncontiguous fragments. Even though the overall alignment of RNA molecules seems to be effective and efficient, the method often produces mismatches when used for motif search and fails to correctly identify motifs, especially tetraloops. This may be due to the fact the results are heavily conditioned by the nucleotide sequence similarity constraint that perhaps should not be considered in motif characterization, and by the gap penalties imposed on the target structure where the search is performed.

The already mentioned SCOR database provides a classification of 3D structures and tertiary interactions for many of the currently known structural motifs. An extensive survey on RNA motifs characterization and discovery approaches can be found in Leontis et al. (2006).

RNA branching loops or junctions are the least well characterized loop regions of the RNA molecules. To the best of our knowledge, relatively few attempts have been made so far to characterize and then search for these loops in three dimensions. A database called RNA Junction (Bindewald et al., 2007) has been built that contains structure and sequence information for RNA helical junctions including all inter-helical angles. A simple classification of three junctions has been proposed in Lescoute and Westof (2006), where junctions are classified based on the length of the connecting fragments. We showed (Apostolico et al., 2009) that shape histograms together with inter-fragments angles capture the shape of the junction in the 3D space, providing a simple but effective structural classification of these regions. We believe that a structural characterization of the junctions may give good insights into the overall folding of the molecule.

3.3   Motif Search with Shape Histograms

In this section, we describe a method we recently developed to search for structural motifs on RNA molecules based on inter-atom distances. Specifically, our method uses the distribution of the Euclidean distances of the atoms from a fixed point to represent a motif. Such distribution is represented through a shape histogram. In the following sections, we will demonstrate that this yields a motif characterization which is distinctive and robust, yet also easy to compute. This approach has the further advantage that the difficult problem of identifying similar 3D motifs reduces to the comparison of two shape histograms. Thus, once we have computed the shape histogram for two 3D structural fragments, the similarity is evaluated based on a distance measure between histograms. In this way, our method addresses the two stage process of motif finding, first of characterizing a putative motif and then of searching for other such motifs. We find that shape histograms yield both a simple characterization of even complex structures, such as RNA junctions, and efficient methods for identifying other similar structures.

3.3.1 Computing the shape histograms

Shape histograms are computed from a fragment or a set of fragments evaluating the distribution of the Euclidean distances of the atoms composing the fragments from a fixed point, such as the centroid of the atoms. Precisely we compute the centroid C with respect to the phosphate atoms of the fragments, then we measure the geometric distances of all the backbone atoms from C. The obtained distances are then quantized with a step size equal to 1 Å. This results in a histogram vector $\mathbf{h} = [h_1, \ldots, h_k]$, where the component h_i is the frequency of the distance value d_i, denoted $h_i = f(d_i)$, i.e., the number of points/atoms at distance d_i from the centroid C. Even though shape histograms work in Cartesian space, they are invariant under rigid geometric transformations so fragments in arbitrary orientations can be matched without explicitly taking rotations into account. It should be noted that we consider only the backbone information. The reason behind this choice is that bases are more flexible and often involved in tertiary interaction within the RNA or with proteins, while the backbone has a more conserved 3D structure. This allowed us to find motif instances that otherwise we would have missed.

3.3.2 Comparing the shape histograms

Once the shape histograms have been computed for the 3D fragments of two substructures, we have to address the issue of how to compare them in order to produce a similarity measure. Given two histograms $\mathbf{h}$ and $\mathbf{g}$ we treated them as vectors and we experimented with four similarity measures between vectors: the L_2 norm (i.e., the Euclidean distance), a normalized version of the L_2 norm, the Pearson correlation coefficient and the cosine of the angle formed by $\mathbf{h}$ and $\mathbf{g}$ in the plane they define. Note that both the Pearson coefficient and the cosine measure are scale independent, which is a desirable property if the number of points taken into account when the histograms are computed is different. This, actually, will happen for example only if the base atoms too are taken into account (C and U have eight base atoms, while A has 10 and G 11). When the shape histogram is computed for a set of strands, distance values tend to vary over a wide range. This happens for instance when one loop has two strands close to each other in the 3D space, and the other has strands that are far apart. Thus the nonzero components in two histograms may have little or no overlap at all, furthermore a vector may include many components equal to zero. Zero components affect the value of the Pearson coefficient since two corresponding components with 0 value will positively contribute to the correlation, even if for the purpose of our analysis these components should not be considered in the computation. With respect to the Pearson coefficient, the cosine measure has the additional advantage of taking into account only the nonzero components. Generally we observed that the cosine is the best measure to highly rank the correct motifs instances. Given two histograms **h** and **g** the cosine is computed by:

$$\text{Cos}(h, g) = \frac{\sum_i h_i g_i}{\sqrt{\sum_i h_i^2} \sqrt{\sum_i g_i^2}}.$$

3.4 A Case Study: The *Haloarcula marismortui* Large Ribosomal Subunit

The crystal structure of the *Haloarcula marismortui* large ribosomal subunit (HM 23S-PDBid:1JJ2) have been determined at 2.4 Å resolution (Ban et al., 2000). HM 23S with more 2700 nucleotides represents the best testset, although our method can be applied to any RNA structure once its atomic positions have been determined.

3.4.1 Ribonucleic acid (RNA) motif search

In the previous section, we first introduced the shape histograms as geometric descriptors of RNA motifs, then we showed a useful measure (*Cos*) to compare these histograms and to extract, given a query motif, all the motif instances in the target RNA structure. We recall that each query corresponds to a single sequence of consecutive nucleotides for 1-loop motifs, or two strands of consecutive nucleotides for 2-loop motifs.

The general search procedure is composed of two main steps:

1. The query is matched against all fragments of consecutive nucleotides of the same length extracted from the target structure. The similarity is established by the cosine value of their histograms: fragments with a *Cos* value exceeding a certain threshold are selected as candidate motifs.

2. The candidates of this reduced list are superimposed to the query by minimizing the RMSD of the corresponding backbone atoms.

More details need to be added for the 2-loop case. When the query is a 2-loop motif, the search procedure described above is applied to each of the strands composing the query. This results in two lists of candidate single strands L_1 and L_2. The set of all possible pairs of fragments, with one fragment from the L_1 and the other from L_2, compose the list of candidate motifs.

At this point, for the query and for each candidate a new histogram of distances is built; it comprises the distances of the backbone atoms of the pair of fragments from their centroid. Similarity is again established as in the first step of the general procedure. The 2-loops are correctly detected by histograms comparison only, which makes the second step of the procedure, i.e., the RMSD computation, not needed for 2-loops.

We successfully applied this technique to search for known motifs on HM 23S: tetraloops, E-loops, kink-turns, S2-motifs, π-turns, and Ω-turns. All the results are shown in Table 3.1. Few considerations about the results in Table 3.1 need now to be made. We note the presence of few false positives among both the standard tetraloops and the tetraloops with deletion. The search for standard tetraloops is made using as query a GNRA standard tetraloop; indeed all the 14 GNRA tetraloops of HM 23S are correctly identified among the 15 top ranked with neither false positives nor false negatives. Furthermore only one positive match is not a tetraloop; the others are six out of the ten tetraloops with deletion and one out of the three tetraloops with insertion as reported in Hsiao et al. (2006). The results of the search for tetraloops with deletion show that all the false positives are instances of tetraloops, with the exception of the fragment starting at position 1967. Indeed TL1170 and TL1469 are standard tetraloops, TL1707 has an insertion, TL482 and TL506 have the fourth base switched with the next nucleotide in the sequence. The presence of false positives among the top ranked was not surprising since the three nucleotides in a tetraloop with deletion have high structural similarity with other tetraloops. Also the fourth base conformation often differs only slightly from the corresponding base in other tetraloops.

As shown in Table 3.1, we searched also for a single-strand version of the kink-turn. The selected single strand is the one characterized by the bulge. This was done to search for the composite kink-turns introduced in Leontis et al. (2006).

Finally, we note that the list of π-turns and Ω-turns originally introduced in Wadley and Pyle (2004) included two misclassified motifs. Despite a similar 3D conformation, the Ω-turn starting at position 245 is actually part of the characteristic bulged strand of the kink-turn KT244/259 as correctly identified by Sarver et al. (2006), as well as by our method. Similarly the π-turn π408 reported in Wadley and Pyle (2004) is part of the E-loop E380/406.

Two new motif instances (S1870 and π65) were newly discovered by our method.

3.4.2   Ribonucleic acid (RNA) junction classification

As already pointed out, the shape histogram is both simple, i.e., able to capture the structure at a low level of resolution, and powerful, i.e., able to identify characteristic structural motifs. Here we investigate the ability of shape histograms to characterize recurrent 3D conformations of more complex structures such as the k-loops ($k \geq 3$) or junctions.

These conformations are described by the shape histograms and the sequence of planar angles formed by the set of consecutive fragments of the k-loop. The histogram takes into account the distances from the centroid of

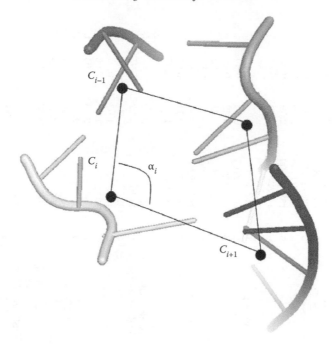

FIGURE 3.3: Triplets of consecutive centroids of the fragments composing the junction (C_{i-1}, C_i, C_{i+1}) form a planar angles (α_i). The sequence of such angles can reveal the junction eccentricity.

the backbone atoms of the whole junction. The angles are those formed by connecting the centroids of consecutive fragments, as shown in Figure 3.3.

Note that a sequential order on the fragments is naturally derived from the order of nucleotides of the RNA chain.

We consider first k-junctions with $k > 3$, in the following denoted as k^*-junction; indeed the case of the three junctions needs to be treated separately. A junction can be considered a flexible loop, then it is easy to see that the sequence of angles induced by the triplets of centroids vary depending on the loop eccentricity. Indeed the more this loop resembles a circle, the more uniform will be the angles formed by the centroids. By contrast, if the loop resembles a squeezed ellipse or band, it will generate a set of obtuse angles, along the long sides of the band, and few small angles at the bends. Thus, from the sequence of inter-fragments angles, we define two main conformations: low eccentricity E_L which indicates a circular junction and high eccentricity E_H which indicates a band. We define also an intermediate state, denoted as E_M.

At this point, we use the shape histogram to reveal the folding of the junction. Precisely, the important features of the histograms are the *range* (i.e., the variability) of the distance values, and the presence of *peaks* in the chart. Once the eccentricity is known we can predict the folding of the junction in

TABLE 3.1: Results of the search for RNA structural motifs with shape histograms.

	Rank	Fragment	RMSD	Cos		Rank	Fragment	RMSD	Cos	
Standard tetraloops	1	2412	0	1	GAAA	1	77/92	0	1	Kink-turns
	2	469	0.23	0.99	GUGA	2	1338/1311	1.89	0.97	
	3	1794	0.28	0.99	GGAA	3	936/1025	2.89	0.95	
	4	577	0.29	0.99	GCGA	4	1212/1146	1.69	0.95	
	5	2630	0.33	0.99	GUGA	5	1588/1600	2.59	0.95	
	6	1327	0.36	0.98	GAAA	6	244/259	2.76	0.95	
	7	1055	0.43	0.99	GUAA	1	212/225	0	1	E-loops
	8	691	0.56	0.99	GAAA	2	2691/2701	0.44	0.99	
	9	253	0.63	0.98	UCAC	3	174/159	0.49	0.99	
	10	2249	0.64	0.99	GGGA	4	587/568	0.71	0.99	
	11	1629	0.7	0.98	GAAA	5	1369/2053	0.42	0.99	
	12	1863	0.72	0.97	GCAA	6	357/292	0.81	0.99	
	13	2696	0.73	0.97	GAGA	7	463/475	0.74	0.99	
	14	2877	0.77	0.98	GUAA	8	380/406	1.1	0.97	
	15	805	0.79	0.98	GAAA					
	16	1469	0.82	0.99	CAAC	1	92	0	1	Kink-turns L
	17	1596	1.06	0.97	UAA-	2	1146	1.02	0.95	(single strand) L
	18	1707	1.44	0.96	GCG(A)A	3	42	1.33	0.95	C
	19	150	1.49	0.97	#	4	1600	1.61	0.94	L
	20	1198	1.56	0.96	UAAC	5	1311	1.76	0.96	L
	21	1500	1.72	0.98	UAA-	6	259	2.29	0.94	L
	22	1918*	1.74	0.96	UACA	7	2821	2.47	0.95	C
	23	2598	1.75	0.96	UAA-	8	852	3.16	0.94	
	24	1170	1.79	0.95	UAGA	9	407	3.4	0.94	
	25	1809	1.82	0.98	GCA-	10	1025	3.53	0.94	L
	26	1187	1.83	0.98	UAA-					
Tetraloops with deletion	1	1187	0	1	UAA-	1	892	0	1	S2-motif
	2	1389	1.06	0.91	GAG-	2	1983	0.48	0.95	
	3	314	1.1	0.91	GGA-	3	1163	1.13	0.93	
	4	1500	1.12	0.95	UAA-	4	1775	1.41	0.97	

Rank	ID	RMSD	Cos	Sequence		Rank	ID	RMSD	Cos
5	1809	1.15	0.9	GCA-		5	1870**	1.5	0.92
6	506	1.2	0.91	GAAA(U)↔A					
7	625	1.23	0.9	UUG-	π−turns				
8	482	1.25	0.9	GCAA↔A		1	1873	0	1
9	1596	1.26	0.9	UAA-		2	65**	1.31	0.89
10	1707	1.36	0.91	GCG(A)A		3	451	1.44	0.83
11	1170	1.45	0.95	UAGA		4	2847	1.58	0.76
12	2598	1.54	0.92	UAA-		5	1854	1.75	0.85
13	1992	1.61	0.91	UCA-	Ω−turns				
14	1469	1.76	0.93	CAAC		1	1416	0	1
15	1967	1.8	0.91	#		2	1744	0.79	0.95
16	1749	1.81	0.94	UCG-					

Tetraloops with insertion	Rank	ID	RMSD	Cos	Sequence
	1	1276	0	1	UCA(U)A
	2	1707	1.17	0.97	GCG(A)A
	3	494	1.59	0.92	GCG(A)A

Notes: The 1-loop motifs as well as the single strand 2-loops are ranked according to the RMSD value. By contrast double strand 2-loops are ranked according to the Cos value; for these motifs indeed the RMSD computation is not needed and is shown here for completeness only. The few false positives are in the table. The notation in the last column of the results for the tetraloops denotes the nucleotide sequence and the type of tetraloop as follows:

XXXX denotes standard tetraloop; (X) denotes an inserted base; - means a deletion occurs; ↔ means the two bases are switched in the 3D space; # denotes a false positive which is not a tetraloop.

The results for the kink-turns (single strand case) distinguish between *local* kink-turn (L) and *composite* kink-turns (C).

*Despite the fact this tetraloop is classified as standard in Hsiao et al. (2006), it should be considered as a tetraloop with both a deletion and an insertion: U(A)C-.

**These motif instances were newly discovered by our method.

the 3D space just looking at its shape histogram. Given an eccentric loop, if it is folded in a compact conformation it will have all the atoms at approximately the same distances from the centroid, thus a small range of distances characterized by a peak in the distribution; by contrast, the distances of the atoms of an unfolded eccentric loop from its centroid will vary uniformly over a wider range. Note that a circular junction cannot be folded, since in the 3D space the folding would obviously affect its angles.

From these consideration we can derive three main conformations: the *wrapped band* characterized by a small range and the presence of a peak, the *straight band* that presents a more uniform distribution of the histogram values over a wider range of distances, and the *flat circle*.

Interestingly we found that all the ten k^*-junctions we have analyzed have an E_H, occasionally E_M, eccentricity. Among those seven are *wrapped bands*, an example is given in Figure 3.4a, and only three can be classified as *straight bands*, an example is given in Figure 3.4b.

The histogram analysis also revealed two other interesting subcases: the *unbalanced wrapped band* characterized by a nonuniform density of the number of atoms along the band, and the *semiwrapped band* where the band is typically folded to form an angle of $\sim$90°.

The semiwrapped band histogram presents a concentrated peak with a lower yet nonnegligible tail (Figure 3.4c). In the unbalanced wrapped band, the nonuniform distribution of the atoms causes the centroid to be closer to the more dense part. The atoms can then be split in two groups such that the atoms of each group are at approximately the same distance from the centroid, but the two groups are at two different distances. This is revealed in the histograms by the relatively small range of distance values and by the presence of two peaks (Figure 3.4d).

The classification made for the k^*-junctions weakly applies to the three junctions since we could not find a clear correspondence between the sequence of angles and the eccentricity of the junction. To see this, consider the case of three similar angles. In this case, both the circular conformation and the wrapped band are possible. Indeed having only three fragments implies that the centroids of the two fragments at the opposite bends of the band are connected by one of the segments forming the three angles. This produces three similar angles as in the circle conformation. Examples of three junctions are given in Figure 3.4e through f.

3.5 Conclusions

In this chapter we have briefly reviewed the organization of RNA molecules at the different levels of representations, from the primary structure to the secondary and tertiary. Then, focusing on the tertiary structure, we have

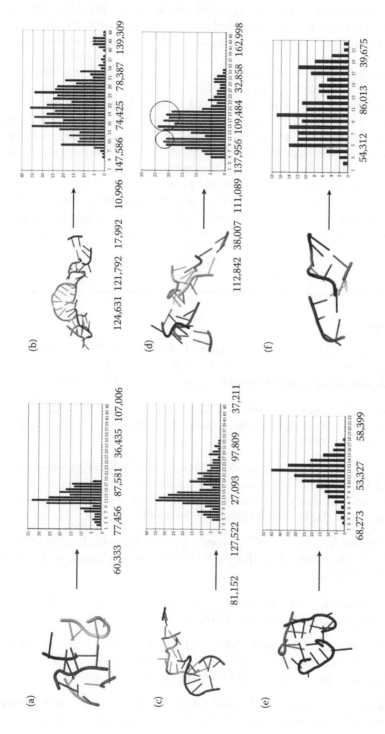

FIGURE 3.4: Examples of *k*-junctions: (a) wrapped band; (b) flat band; (c) semiwrapped band with a medium-high eccentricity; (d) unbalanced wrapped band (the two peaks are circled). Below are examples of three junctions: (e) wrapped band; (f) straight band.

considered the problem of identifying small motifs in the 3D data. After a review of the literature, we have presented our approach to motif detection and characterization based on a simple shape descriptor. We have demonstrated the effectiveness of the method by applying it to the rRNA of the *Haloarcula marismortui* large ribosomal subunit. One possibly interesting conclusion of our analysis is that the RNA tertiary structure is better described and provides more insights when observed and analyzed at a coarse level of resolution.

The amount of data currently available is limited thus we have not been able to do extensive experimentation. However the method can be extend to any new structure. The limited amount of data has played a critical role in the characterization of junctions, since the number of junctions within each molecule is much lower than other, simpler loop motifs. We believe that the current characterization may prove useful when aligning similar junctions between two different but related structures. This will be one important step in aligning two complete structures. Future work will address these challenges.

References

Apostolico, A., Ciriello, G., Guerra, C., Heitsch, C.E., Hsiao, C., and Williams, L.D. 2009. Finding 3D motifs in ribosomal RNA structures. *Nucleic Acids Research*, doi: 10.1093/nar/gkn1044.

Ban, N., Nissen, P., Hansen, J., Moore, P.B., and Steitz, T.A. 2000. The complete atomic structure of the large ribosomal subunit at 2.4 A resolution. *Science*, 289, 905–920.

Bindewald, E., Hayes, R., Yingling, Y.G., Kasprzak, W., and Shapiro, B.A. 2008. RNA Junction: a database of RNA junctions and kissing loops for three-dimensional structural analysis and nanodesign. *Nucleic Acids Research*, 36, D392–D397, doi: 10.1093/nar/gkm842.

Burks, J., Zwieb, C., Muller, F., Wower, I., and Wower, J. 2005. Comparative 3-D modeling of TmRNA. *BMC Molecular Biology*, 6, 14, doi: 10.118611471-2199-6-14.

Dror, O., Nussinov, R. and Wolfson, H. 2005. ARTS: alignment of RNA tertiary structures. *Bioinformatics*, 21(37), 47–53.

Duarte, C.M., Wadley, L.M., and Pyle, A.M. 2003. RNA structure comparison, motif search and discovery using a reduced representation of RNA conformational space. *Nucleic Acids Research*, 31, 4755–4761.

Ferre', F., Ponty, Y., Lorenz, W.A., and Clote, P. 2007. DIAL: a web server for the pairwise alignment of two RNA three-dimensional structures using nucleotide, dihedral angle and base-pairing similarities. *Nucleic Acids Research*, 35, W659–W668.

Harrison, A.M., South, D.R., Willett, P., and Artymiuk, P.J. 2003. Representation, Searching and Discovery of Patterns of Bases in Complex RNA Structures, *Journal of Computer-Aided Molecular Design*, 17(8), 537–549.

Hershkovitz, E., Tannenbaum, E., Howerton, S.B., Sheth, A., Tannenbaum, A., and Williams, L.D. 2003. Automated identification of RNA conformational motifs: theory and application to the HM LSU 23S rRNA. *Nucleic Acids. Research*, 31(21), 6249–6257.

Hsiao, C., Mohan, S., Hershkovitz, E., Tannenbaum, A., and Williams, L.D. 2006. Single nucleotide RNA choreography. *Nucleic Acids Research*, 34(5), 1481–1491.

Huang, H.-C., Nagaswamy, U., and Fox, G.E. 2005. The application of cluster analysis in the intercomparison of loop structures in RNA. *RNA*, 11, 412–423.

Klein, D.J., Schmeing, T.M., Moore, P.B., and Steitz, T.A. 2001. The kink-turn: a new RNA secondary structure motif. *The EMBO Journal*, 20(15), 4214–4221.

Leontis, N.B., Lescoute, A., and Westhof, E. 2006. The building blocks and motifs of RNA architecture. *Current Opinion in Structural Biology*, 16, 279–287.

Leontis, N.B., and Westhof, E. 1998. A common motif organizes the structure of multi-helix loops in 16 S and 23 S ribosomal RNAs. *Journal of Molecular Biology*, 283, 571–583.

Lescoute, A., and Westhof, E. 2006. Topology of three-way junctions in folded RNAs. *RNA*, 12, 83–93.

Sarver, M., Zirbel , C.L., Stombaugh, J., Mokdad, A., and Leontis, L.B. 2006. FR3D: finding local and composite recurrent structural motifs in RNA 3D structures. *Journal Math Biology*, doi: 10.1007/s00285-007-0110.

Shapiro, B.A., Yingling, Y.G., Kasprzak, W., and Bindewald, E. 2007. Bridging the gap in RNA structure prediction. *Current Opinion in Structural Biology*, 17(2), 157–165.

Tamura, M. et al. 2004. Scor: Structural classification of RNA. *Nucleic Acids Research*, 32, 182–184.

Wadley, L.M., and Pyle, A.M. 2004. The identification of novel RNA structural motifs using COMPADRES: an automated approach to structural discovery. *Nucleic Acid Research*, 32, 6650–6659.

Chapter 4

Protein Structure Classification Using Machine Learning Methods

Yazhene Krishnaraj and Chandan Reddy

Wayne State University

4.1   Introduction

Proteomics is the large-scale study of protein's structure and their functions. Protein structure prediction plays an important role in proteomics as it is useful for drug design in medicine, design of novel enzymes in biotechnology and many other fields. The Human Genome Project has created large amounts of protein sequence data, but since tertiary structure of the protein gives better understanding of the protein in terms of its function, structure prediction is one of the most important research tasks in this domain. Experimental methods and computational methods are two well known approaches used for the prediction of the tertiary structure of a protein.

4.1.1   Experimental methods

Experimental methods such as X-ray crystallography and nuclear magnetic resonance (NMR) spectroscopy are ideal ways to derive structural information from the sequence for a given protein. These methods have several disadvantages. To determine the structure of a single protein, crystallography can take several months to several years. Even though NMR is comparatively quicker than crystallography, it is still not fast enough to close the huge gap between the number of known protein sequences and the number of known tertiary structures of the protein. Both of these are very expensive as well.

4.1.2   Computational methods

Computational methods used to solve this problem can be broadly categorized into three different groups namely: homology modeling, threading and de novo approach. Both homology modeling and protein threading fall into comparative protein modeling as they both use previously solved structure as templates to predict a structure for a given protein sequence. On the other hand de novo modeling methods build the structure for the given protein sequence from scratch.

4.1.2.1   Homology modeling

In homology modeling, given a protein sequence, its structure is assigned using sequence similarity [15]. It assumes that two proteins have the same structure if their sequences have high homology. This method usually takes a template protein structure and builds the structure for the new protein if it has more than 30% of sequence similarity with the template. There are usually three steps in the homology modeling [9].

- First, sequence of a protein with an unknown three dimensional structure is compared with selected set of protein sequences with known structure. Search techniques such as FASTA and BLAST are commonly used for the template selection. In particular PSI-BLAST is the most commonly used method which can identify more distantly related homologs.

- Second, model generation is done from the chosen template and an alignment. A three-dimensional structural model of the target is generated using these information. Spatial restraint-based modeling software called MODELLER is most widely used for this purpose.

- Third, the model created is validated in the model assessment phase. Statistical potentials or physics-based energy calculations are traditionally used for this task. New methods use machine learning techniques like neural networks (NNs) and support vector machines (SVMs).

4.1.2.2 Threading

Threading aims at detecting structural similarities by comparison with a library of known structures even for low similarity protein sequences [14]. Threading also consists of three steps similar to homology modelling [9]. They are (1) structural template selection, (2) creating the alignment, and (3) model building. For the protein sequence with unknown structure, protein threading, aligns (or threads) the sequence to each template in the structural library and produces a list of scores. This scoring function is a measure of the fitness of a sequence structure alignment using both homology and structure information. The fold with the best score is chosen from the rank of scores obtained and assigned as the fold for the given new sequence. Unlike homology modeling, threading is more effective even for sequences with only distant homology found. But this method would fail if the structural library did not have a correct structure match. GenTHREADER [18], PROSPECT [26] and RAPTOR [25] are the some of the software tools that are used for threading.

4.1.2.3 De novo or ab initio approach

In de novo or ab initio protein modeling, method prediction is done from the sequence alone using some physics and chemistry laws [22]. With the use of suitable energy function, protein folding is simulated in atomic detail using methods like molecular dynamics or Monte Carlo simulations. All the energetics involved in the process of folding is modeled and then the structure with the lowest energy is found. There are two main steps involved in the ab initio prediction, (1) finding an energy (scoring) function to differentiate the structures and (2) a search method to explore the conformational space. This procedure requires very high computational resources as any fully descriptive energy function must consider interactions between all the pairs of atoms in the polypeptide chain. Clearly, as the number of amino acids in the protein sequence increases, the number of pairs of atoms to be considered grows exponentially. Since this method is computationally intensive, it can be applied only to small proteins or powerful supercomputers must be used for larger ones. Even though ab initio prediction is the most difficult among the above three methods, it is still arguably the most valued approach as it can predict the structure of the protein purely from its sequence without relying on the structure or sequence similarities with previously predicted proteins. Rosetta is a one of the leading methods for ab initio prediction.

4.1.2.4 Machine learning techniques

In recent years, machine learning techniques have shown promising results in structural classification of proteins (SCOP) from sequence alone even in cases where no significant identity exists to proteins with known structures [6]. These methods create the training data from global descriptors of a primary protein sequence in terms of the physical, chemical, and structural properties of the constituent amino acids [6]. Algorithms learn from the training data and build a classifier which will identify the three-dimensional fold of a protein using classification methods from its sequence information alone. In machine learning, fold recognition is categorized as a classification task which is one of the classical approaches of problem solving.

Even though there are many machine learning algorithms that can be used for classifying the structure of the protein, support vector machine (SVM), k-nearest neighbor (KNN), and NNs are the most popular algorithms used by the researchers because they are more accurate and compare efficiently to other algorithms in the literature. SVM performs classification by using nonlinear mapping to transform the training data to a new higher dimension, where a linear optimal separating hyperplane is created to separate the data into two classes. KNN is an instance-based method where a test object is classified based on the closest training object in the feature space. In neural network (NN), a classifier is a network with a set of connected input/output units with weighted connections between them and these weights are adjusted as the network learns during the learning phase.

The boosting meta-algorithm is relatively new, efficient, simple, and easy to manipulate additive modeling technique that can use potentially any weak learner available. In this chapter, we studied the performance of the two most popular boosting algorithms, AdaBoost and LogitBoost for this problem. Boosting algorithms combine weak learning models that are slightly better than random models. Boosting algorithms are generally viewed as functional gradient descent schemes and obtain the optimal updates based on the global minimum of the error function [8].

This chapter provides a comprehensive study of different machine learning algorithms for the problem of protein structure classification. The rest of the chapter is organized as follows: Section 4.2 gives some relevant background about protein structure, Section 4.3 gives details of protein databases and data extraction, Section 4.4 explains existing classification methods used in the literature, Section 4.5 describes the data used in the experiment, Section 4.6 summarizes the results and compares it with results obtained from other classification algorithms and Section 4.7 concludes our discussion with future research directions.

4.2   Background on Proteins

Proteins regulate functioning of a cell and also carry out many tasks that are essential to life. From the protein hemoglobin that transports oxygen to tissues, to protein collagen that strengthens skin and bone, to protein antibodies that are responsible for reacting with specific foreign substances in the body, proteins seem to do it all. Proteins are complicated molecules formed by a chain of simple building blocks called amino acids. Figure 4.1 shows the structure hierarchy of proteins [13]. There are four distinct aspects of a protein's structure:

1. *Primary structure:* 20 different amino acids bonded to one another by peptide bonds form the unique primary structure of the protein. The sequence of a protein is unique to that particular protein, and defines

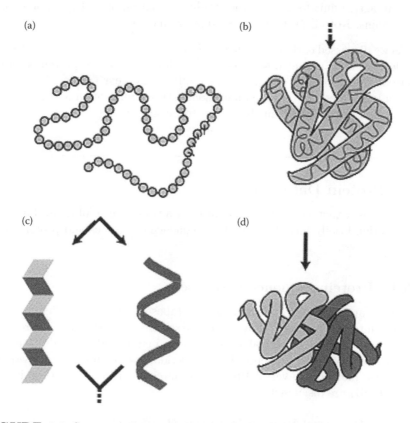

FIGURE 4.1: Structure hierarchy of proteins. (a) Primary protein structure. (b) Secondary protein structure. (c) Tertiary protein structure. (d) Quaternary protein structure. (From: National Human Genome Research Institute.)

the structure and function of the protein. The primary structure alone cannot carry out the functions of a protein but is indirectly responsible through additional levels of structure.

2. *Secondary structure*: This is an ordered structure brought about due to interactions between chemical groups of the amino acids. Though there are a few secondary structures that are possible, a few characteristic patterns occur frequently within folded proteins. They are the α helix, the β sheet (also known as β pleated sheet) and the β-turn.

3. *Tertiary structure*: This is also referred to as 'fold' of a protein, it is a three dimensional structure and is caused by the secondary structure folding back upon itself. The function of a protein is dictated by this three dimensional structure.

4. *Quaternary structure*: This final structure of a protein is formed by the interaction between several protein molecules. Quaternary structure is an active unit formed by the stable association of multiple polypeptide chains. Not all proteins have a quaternary structure.

As we discussed earlier, advances in DNA sequencing techniques in the last few decades have provided scientists with a huge amount of protein sequence information which is growing exponentially every year. Since, the function protein is dictated by its three dimensional structure, protein fold recognition is a very important problem that is yet to be solved.

4.3   Protein Databases

There is a vast amount of protein data available in biological databases. This section briefly introduces protein sequence databases and protein structure databases.

4.3.1   Protein sequence databases

There are several protein sequence databases that are publicly available over the internet. Some of them are simple sequence repositories, whereas others are databases with original sequence record enhanced with additional rich information about the sequence. In this section, two leading protein sequence databases, namely, Universal Protein Resource (UniProt) and Protein Data Bank (PDB) are discussed.

4.3.1.1   Universal Protein Resource (UniProt)

Universal Protein Resource (UniProt) is the world's most comprehensive catalog of information on proteins and it has more than two million proteins sequenced in its database. UniProt Consortium is a collaboration between the European Bioinformatics Institute (EBI), the Swiss Institute of Bioinformatics (SIB) and the Protein Information Resource (PIR). UniProt comprises four components [2]:

1. UniProt Knowledgebase (UniProtKB) contains high-quality manually annotated and nonredundant protein sequence records. It is used to access functional information on proteins.

2. UniProt Reference Clusters (UniRef) databases provide clustered sets of sequences from UniProtKB and selected UniParc records to provide complete coverage of sequence space at several resolutions.

3. UniProt Archive (UniParc) is designed to capture all publicly available protein sequence data and contains all the protein sequences from the main publicly available protein-sequence databases.

4. UniProt Metagenomic and Environmental Sequences (UniMES) is a repository specifically developed for metagenomic and environmental data.

4.3.1.2   Protein Data Bank (PDB)

Protein data bank (PDB) is the single worldwide repository of structural data of biological macromolecules [4]. PDB was established at Brookhaven National Laboratories (BNL) in 1971. It initially was an archive for biological macromolecular crystal structures and held just seven structures. But in the 1980s due to X-ray crystallography and NMR technologies the number of deposited structures increased dramatically. As of 2008, it had about 50,000 solved protein structures.

4.3.2   Protein structure databases

The proteins in the PDB are classified using different methods to provide a greater understanding of structure and function. In this section, we will discuss about SCOP, CATH and FSSP which are the most widely used comprehensive databases that use unique methods for classifying the proteins according to their structure [10]. Also, we will briefly discuss how the features can be extracted from the sequence for applying the machine learning algorithms.

4.3.2.1   Structural Classification of Proteins (SCOP)

SCOP database classifies protein of known structure from PDB based on their evolutionary, functional and structural relationship [10]. This is a completely manual process done by human experts with visual inspection and comparison of structures with the assistance of tools to classify protein structures in to different levels of hierarchy. Though many levels exist in the hierar-

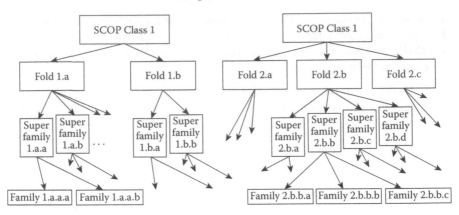

FIGURE 4.2: Hierarchical structure of SCOP database.

chy, principal levels are family, superfamily, fold and class. Family is a group of proteins clustered together if all proteins that have residue identities of 30 and greater, or proteins with lower sequence identities but whose functions and structures are very similar. Superfamily is a group of families whose proteins have low sequence identities but whose structural and functional features have a probable common evolutionary origin. Proteins from superfamilies and families are grouped into the same fold if they have the same major secondary structures in the same arrangement and with the same topological connections. Class is a group of different folds. Most of the folds are assigned to one of the five common structural classes (α, β, $\alpha + \beta$, α/β and multidomain) or few other less populated ones. Figure 4.2 shows the hierarchial structure of the SCOP database.

4.3.2.2   CATH

The CATH protein structure classification is a hierarchical classification of protein structure by semiautomatic method [10]. The name CATH is an acronym for four main levels hierarchy: class (C), architecture (A), topology (T), and homologous superfamily (H). A semiautomated procedure involving computational techniques, statistical analysis, literature reviews and expert analysis is used for setting boundaries between different levels. Automatic procedure is adopted when a given domain has sufficiently high sequence and structural similarity with a domain that has been previously classified in CATH [1]. Otherwise a manual classification is done based upon an analysis of the results derived from a range of comparison algorithms in the literature. Detailed classification in CATH differs from its rival SCOP, but they share broad features. For example, topology level and homologous superfamily in CATH is equivalent to the fold and the superfamily level, respectively.

4.3.2.3   Families of structurally similar proteins (FSSP)

FSSP is a database of structural alignments of proteins. It is a unique and fully automated method of classification based on an exhaustive 3D structure comparison of protein structures in the PDB [10]. When new structures are available in PDB, classification and structure alignments are done automatically and are updated continuously by an automatic server Dali (Distance matrix ALIgnment) that does a 3D comparison of protein structures.

4.4   Classification Methods

Researchers have used different machine learning algorithms for the problem of protein structure classification. In this section, we will describe the existing methods proposed in the literature, which include SVM, NN, KNN, and boosting.

4.4.1   Support vector machines (SVMs)

SVM is a binary classification algorithm based on the statistical learning theory originally developed by Vapnik [23] and his colleagues at Bell Laboratories, and has been improved by others. SVM constructs a decision boundary to separate the positive and negative samples. The decision boundary consists of a hyperplane that separates the classes maximizing the margin between the two hyperplanes which will give the most generalized model. For applications with linearly separable data, SVM searches for this maximal margin hyperplane. But for nonlinear cases, kernel trick is applied to create a nonlinear classifier. A kernel function is used to transform original input vectors to a new high dimensional feature space such that linear decision boundary can be obtained in the transformed space.

Ding and Dubchak [5] used polynomial kernel and Gaussian kernel instead of a linear kernel. While a polynomial kernel worked better, Gaussian kernel gave the best results [5]. To handle multiple classes they have used One-versus-Others (OvO) methods, unique One-versus-Others (uOvO) method and All-versus-All (AvA) method. If there are K classes, OvO method partitions K classes into a two-class problem by having proteins from one fold as *"true"* class and all the remaining proteins are combined to one *"others"* class [5]. K two-way classifiers are trained by repeating this procedure for each of the K classes. When a new test protein comes in, all the K classifiers are used to predict the protein fold resulting in K results. More than one classifier might predict the protein to a positive class because of the *"false positive"* problem. To overcome this, they introduced uOvO method. uOvO is an extension of

OvO where the classifiers with positive predictions are considered in the second step. A series of two-way classifiers are trained for each of the pairs and voting is done to predict the final fold. AvA method [5] constructs $K(K-1)/2$ number of two-way classifiers and voting decides the protein fold.

In the paper by Huang and his colleagues [11], SVM is used as a base classifier for their hierarchical learning architecture (HLA). HLA is a network with two levels. In the first level, protein is classified in to one of its four major classes namely: α, β, $\alpha + \beta$, and α/β. The second level of the network does the further classification of protein into 27 folds. They adopted multiclass classification SVM from another paper [17].

Multiclass classification is done using the output codes in the work by Eugene et al. [12]. In their methods, the relative weights of the OvO classifiers are learnt along with the hierarchial information of the protein structure to improve the classification. Rangwala et al. [21] gives a comprehensive study of the SVM-based multiclass classification methods. They show that the SVM with two-level learning framework (where at the first level, cross-validation methodology is used to predict the training instances using a set of k binary classifiers, and these prediction outputs are used as the input for second-level learning) are more effective than using the OvO method for multiclass classification. Results of the last two works are not discussed in this chapter as they have used different datasets derived from SCOP.

4.4.2 Neural networks (NNs)

NNs are widely used algorithms in many applications and are modeled based on biological neural systems. For classification, the model is trained in such a way that input and output nodes are connected with a set of weighted links based on input–output association of training data. More complex NNs can have one or more hidden layers with hidden nodes in between input and output nodes to model more complex input–output relationships. Depending upon the application, either a feed-forward (nodes in one layer is connected only to nodes in next layer) or recurrent (nodes can be connected to nodes in the same layer, previous layer or next layer) network is built. The relationship between input and output is defined by an activation function which along with the weighted links decide the behaviour of the NN. The activation function can be linear, sigmoid (logistic) and hyperbolic tangent function. Back-propagation algorithm is the most popular learning algorithm applied in NN to learn the model by adjusting the weights of the node by minimizing the total sum of squared errors.

Ding and Dubchak [5] used a three-layer feed-forward NN in their experiments. Conjugate gradient minimization is used for adjusting the weights. In their NN architecture, the number of input nodes were the same as the number of feature vectors (between 20 and 125) with one hidden node and two output nodes (one for *"true"* class to indicate one protein fold and *"other"* for all the other protein folds.)

Huang and his colleagues [11] used three different NN models as base classifiers for their HLA. First, a multilayer perceptron (MLP) is a feed forward NN with multiple layers and non linear activation functions that can handle the data which is not linearly separable [16]. Second, radial basis function network (RBFN) is a two-layer NN model. At the hidden layer, nodes use a Gaussian function as the activation function which is mapped from input. A linear combination of hidden layer values represent the output layer [19]. Finally, general regression neural networks (GRNN) is a four layer NN based on statistical kernel smoothing techniques.

4.4.3 K-nearest neighbors (KNNs)

KNNs is an instance-based learner, where the modeling of the classifier is deferred until the classification of the test data. The training data with n input attributes is represented in an n-dimensional space. When a test object needs to be classified, a proximity measure (like Euclidean distance) is used to find KNNs and voting is taken to assign the class. Okun et al. [20] used k-local hyperplane distance nearest neighbor algorithm (HKNN) for the protein classification. This algorithm forms a linear local hyperplane for each class in the dataset. When a test data comes in, the distance between the test point and these local hyperplanes are calculated to determine the class.

4.4.4 Boosting

Boosting is a machine learning algorithm that produces a strong classifier by iteratively learning with a set of weak learners [7]. To solve a two-class classification problem, boosting repeatedly applies a weak learner to modified versions of the data, thereby producing a sequence of weak classifiers. Each boosting iteration performs the following steps: (1) fits a weak learner to the weighted training samples and (2) computes the error and updates the weights for every datapoint. Whether it is a classification or a regression problem, the main challenges in the boosting framework are the choice of the weak learner and the complexity of the weak learner. We used decision trees as our weak learning because of their popularity. The final model obtained by boosting algorithm is a strong learner which is a linear combination of several weak learning models. There are several variations of boosting algorithm that have been developed over the years. They mainly vary in terms of weight updation of the training data at each iteration and the way they combine the weak learners to produce a strong learner.

AdaBoost is an adaptive algorithm which gives more importance for the weak classifier with lower error rate. The first iteration starts with equal weights for all the data points. As the boosting rounds progress, the data points which are misclassified by the weak learner gets higher weight, so that the weak learner in the next iteration will focus more on those hard to classify examples. Each weak learner also gets a weight proportional to its accuracy.

In the end, strong learner is formed by taking the weighted average of predictions made by each weak learner.

LogitBoost is another variant of the boosting algorithm that performs additive logistic regression [8]. In each iteration, given the weak classifier, each model maximizes the probability of the data and the models are added. In the end, for a two class problem, the strong classifier assigns the instance x to 1 if $P(1|x) < 0.5$, otherwise predicts x to 0. This basic algorithm can also be extended for a multiclass problem.

4.5   Data Description

Machine learning algorithms need relevant input features derived from the raw data (protein sequence). This section introduces the feature extraction process and describes the dataset.

4.5.1   Feature extraction

As the amino acids are the basic building blocks of a protein, a protein sequence is given as an unique sequence of amino acid residues. These amino acids have various physico-chemical and structural properties that determine the protein structure. These properties need to be extracted and transformed into the feature vectors that would be the inputs for machine learning algorithms. The feature vectors were obtained by a method which uses the combination of local and global information of amino acid sequence [6]. The feature vectors that characterized six different properties of a protein are: amino acid composition (C), hydrophobicity (H), polarity (P), predicted secondary structure (S), van der Waals volume (V), and polarizability (Z). Six parameter sets formed by these feature vectors are shown in Table 4.1.

TABLE 4.1:   Six parameter datasets and the dimension of their feature vectors.

Symbol	Parameter	Dimension
C	Amino acids composition	20
S	Predicted secondary Structure	21
H	Hydrophobicity	21
P	Normalized van der Waals volume	21
V	Polarity	21
Z	Polarizability	21

These global descriptors obtained are in terms of the physical, chemical, and structural properties of the constituent amino acids. Extraction of these parameter sets involve two important steps [6].

1. Sequences of physico-chemical and structural properties of residues are formed from the sequence of the amino acids. Twenty amino acids were divided into three groups (indexed 1, 2, 3). For each of the six different amino acid attributes, every amino acid was replaced by one of the three groups that it belongs to.

2. Three descriptors, composition (C), transition (T), and distribution (D) were calculated for a given attribute. Composition describes the global percent composition of each of the three groups in a protein. The percent frequencies with which the attribute changes its index along the entire length of the protein is described by transition. Distribution is the distribution pattern of the attribute along the sequence.

The feature vector for the amino acid composition consists of 20 dimensions where as all of the rest have 21 dimensions.

4.5.2   Dataset

We used a dataset [5] containing proteins which are classified in SCOP database and the algorithm aims to predict the fold class of these proteins. A training set was taken from the 27 most populated SCOP folds of the PDB select set, in which no two proteins share more than 35% sequence identity for aligned subsequences longer than 80 residues. This training set contained 311 proteins. They derived an independent test set from the PDB 40D set, which consists of all SCOP sequences having less than 40% sequence identity with each other. Using the same 27 SCOP folds, 383 proteins were selected, and any PDB 40D protein having more than 35% sequence identity with the proteins in the training set was excluded. Table 4.2 shows the 27 SCOP folds used in the dataset and the corresponding number of proteins in each fold for training and test dataset. First, the test was done on six parameter sets (C, S, H, P, V, Z) separately. Second, the test was done with new feature vectors ranging in size from 41 dimensions (C+S) to 125 (C+S+H+P+V+Z) by combining all the properties one after the other.

4.6   Experimental Results

Classification was done with different parameter sets using AdaBoost and LogitBoost [27]. This section explains the tools used in the experiment and the prediction accuracy which was used as the evaluation measure.

TABLE 4.2: 27 SCOP folds in the training and test dataset and the total number of proteins in each fold are given.

Fold	Index	No. training	No. test
Globin-like	1	13	6
Cytochrome c	3	7	9
DNA-binding 3-helical bundle	4	12	20
Four-helical up-and-down bundle	7	7	8
4-helical cytokines	9	9	9
Alpha; EF-hand	11	6	9
Immunoglobulin-like beta-sandwich	20	30	44
Cupredoxins	23	9	12
Viral coat and capsid proteins	26	16	13
ConA-like lectins/glucanases	30	7	6
SH3-like barrel	31	8	8
OB-fold	32	13	19
beta-Trefoil	33	8	4
Trypsin-like serine proteases	35	9	4
Lipocalins	39	9	7
(TIM)-barrel	46	29	48
FAD (also NAD)-binding motif	47	11	12
Flavodoxin-like	48	11	13
NAD(P)-binding Rossmann-fold	51	13	27
P-loop containing nucleotide	54	10	12
Thioredoxin-like	57	9	8
Ribonuclease H-like motif	59	10	12
Hydrolases	62	11	7
Periplasmic binding protein-like	69	11	4
Beta-Grasp	72	7	8
Ferredoxin-like	87	13	27
Small inhibitors, toxins, lectins	110	13	27

4.6.1 Tools

The experiments were done on a PC with a 1.73 GHz Intel Core Duo CPU and 1 GB RAM, using Windows Vista operating system. Data mining toolkit WEKA (Waikato Environment for Knowledge Analysis) version 3.4.11 is used for classification. WEKA is an open source tool kit and it consists of a collection of machine learning algorithms [24]. The AdaBoost and LogitBoost algorithm are discussed in Section 4.4. For AdaboostM1, default parameters of WEKA were changed to perform 100 iterations with re-sampling. J48 (WEKA's own version of C4.5) decision tree algorithm was used as base classifier for boosting. For LogitBoost, 100 iterations were done with resampling and decision stump was used as the weak learner. For each parameter set considered, ten fold cross-validation was performed to build the model and independent test data was evaluated with that model.

4.6.2 Prediction accuracy

The standard Q percentage accuracy [3,5] is used to measure prediction accuracy of the algorithms. If we have K number of classes with N number of test proteins such that n_1 is the number of test proteins observed in class F_1, n_2 is the number of proteins in F_2 etc., N can be expressed as $N = [n_1 + n_2, \ldots, n_k]$. If a_1 is the number of proteins that are correctly classified as class F_1, etc. (a_i belongs to diagonal entries of $K \times K$ confusion matrix) the total number of proteins that are correctly classified can be given as $A = [a_1 + a_2 + \ldots + a_k]$. The class accuracy is given by:

$$Q_i = a_i/n_i. \tag{4.1}$$

The overall accuracy is calculated by taking the weighted average of individual class accuracy. Weight for each class is $w_i = n_i/N$. Therefore, the overall accuracy for the dataset is given by:

$$Q = \sum_{i=1}^{K} w_i Q_i = A/N. \tag{4.2}$$

4.6.3 Discussion

Our experimental results clearly show that boosting outperformed all the state-of-the-art methods proposed in the literature for the protein fold recognition problem [27]. Table 4.3 shows that with independent test data, the maximum accuracy of 60.13% was achieved using LogitBoost algorithm and 58.22% was achieved using AdaBoost algorithm. AdaBoost reached its maximum accuracy with just 62 features (C+S+H) where as LogitBoost's maximum accuracy was with 104 (C+S+H+P+V) features. Table 4.4 shows comparison

TABLE 4.3: Comparison of prediction accuracy by various classifiers on combination of all six parameter datasets.

Classifier	C (%)	CS (%)	CSH (%)	CSHP (%)	CSHPV (%)	CSHPVZ (%)
OvO NN	20.50	6.80	0.60	41.10	41.20	41.80
OvO SVM	43.50	3.20	45.20	43.20	44.80	44.90
uOvO SVM	49.40	8.60	51.10	49.40	50.90	49.60
AvA SVM	44.90	52.10	56.00	56.50	55.50	53.90
HKNN	–	–	57.10	57.90	55.80	–
HLA (MLP)	32.70	48.60	47.50	43.20	43.60	44.70
HLA (RBFN)	44.90	53.80	53.30	54.30	55.30	56.40
HLA (GRNN)	–	–	–	–	–	45.20
HLA (SVM)	–	–	–	–	–	53.20
AdaBoost	51.96	57.7	58.22	57.18	57.18	57.18
LogitBoost	46.21	56.4	58.49	58.75	60.31	56.14

of individual fold accuracies with independent test data only for some of the methods discussed. Independent test accuracies for individual folds are not available for HLA (MLP), HLA (RBFN), HLA (GRNN), and HLA (SVM) [11]. For each method, most successful (highest overall accuracy) combination of parameter set is listed. The following are the combination of parameter sets that yielded highest accuracy for each of those methods.

- OvO NN : C + S + H + P + V + Z

- OvO SVM : C + S + H

- uOvO SVM : C + S + H

TABLE 4.4: Comparison of prediction accuracies of different classifiers for individual folds obtained by independent test set.

Fold index	OvO NN	OvO SVM	uOvO SVM	AvA SVM	HKNN	AdaBoost	LogiBoost
1	55.60	**87.50**	83.30	83.30	83.30	83.33	83.33
3	27.80	50.90	66.70	77.80	77.80	**88.89**	55.56
4	25.60	43.70	43.70	35.00	50.00	**60.00**	**60.00**
7	37.50	53.50	62.50	50.00	**87.50**	50.00	50.00
9	77.80	69.80	100.00	100.00	88.90	**100.00**	88.89
11	27.80	50.00	55.60	**66.70**	44.40	44.44	33.33
20	53.90	48.60	60.20	71.60	56.80	70.45	**75.00**
23	12.50	15.30	16.70	16.70	25.00	33.33	25.00
26	44.20	46.80	53.80	50.00	**84.60**	76.92	76.92
30	33.30	25.00	33.30	33.30	50.00	**50.00**	33.33
31	52.10	41.70	50.00	50.00	50.00	**75.00**	62.50
32	26.30	27.40	31.60	26.30	42.10	26.32	**47.37**
33	25.00	50.00	50.00	50.00	50.00	**50.00**	**50.00**
35	0.00	25.00	25.00	25.00	50.00	25.00	**50.00**
39	40.50	39.30	50.00	57.10	42.90	42.86	**57.14**
46	65.80	60.50	64.60	77.10	79.20	72.92	**81.25**
47	38.90	56.90	54.20	58.30	58.30	**58.33**	**58.33**
48	21.80	29.50	34.60	48.70	53.90	53.85	**61.54**
51	42.60	31.20	46.90	**61.10**	40.70	40.74	40.74
54	29.20	**47.20**	36.10	36.10	33.30	33.33	25.00
57	50.00	25.00	25.00	50.00	37.50	**50.00**	25.00
59	38.10	39.30	28.60	35.70	**71.40**	66.67	66.67
62	57.10	**78.60**	71.40	71.40	71.40	57.14	57.14
69	0.00	25.00	25.00	25.00	25.00	0.00	**25.00**
72	25.00	25.00	25.00	12.50	25.00	25.00	**37.50**
87	21.40	24.50	29.60	37.00	25.90	25.93	**44.44**
110	60.30	69.30	83.30	83.30	85.20	**100.00**	96.30
Average	41.80	45.20	51.10	56.00	57.10	58.22	**60.31**

Note: Weighted average is given at the end. For each fold, classifier with the highest accuracy rate is highlighted in bold.

- AvA SVM : C + S + H + P

- HKNN : C + S + H

- AdaBoost : C + S + H

- LogitBoost : C + S + H + P + V

Individual fold statistics in the Table 4.4 clearly shows that boosting methods give the highest accuracy in most of the folds.

4.7   Conclusion

This chapter presents an empirical study on the performance of different machine learning algorithms to solve the problem of protein structure classification which is a crucial task in proteomics. Experimental results show that the AdaBoost and LogitBoost perform better than other algorithms on given data. In addition to the improvements in the classification accuracy, the boosting approach provides two other advantages. First, boosting provides us with a better interpretable model. Only a subset of features are used in model building. These features can provide evidence for the biological relationship of those features with respect to the folds considered. These insights can provide vital feedback to SCOP or CATH databases to generate hierarchies of data primarily based on these features. Second, faster training of boosting provides very high run-time efficiency. Considering the fact that this higher accuracy is achieved using basic boosting algorithms, further work on boosting with hierarchical learning architecture or other modifications will certainly provide promising results.

References

[1] http://www.cathdb.info/

[2] The UniProt Consortium. The universal protein resource (UniProt). *Nucleic Acids Research,* 36:D190–195, 2008.

[3] P. Baldi, S. Brunak, Y. Chauvin, C. Anderson, and H. Nielsen. Assessing the accuracy of prediction algorithms for classification:an overview. *Bioinformatics,* 16(5):412–424, 2000.

[4] HM. Berman, J. Westbrook, and Z. Feng. The protein data bank. *Nucleic Acids Research,* 28(1):235–242, 2000.

[5] C.H.Q. Ding and I. Dubchak. Multi-class protein fold recognition using support vector machines and neural networks. *Bioinformatics*, 17(4):349–358, 2001.

[6] I. Dubchak, I. Muchnik, C. Mayor, I. Dralyuk, and S.H. Kim. Recognition of a protein fold in the context of the structural classification of proteins (scop) classification. *Proteins*, 35:401–407, 1999.

[7] Y. Freund, and R.E. Schapire. A decision-theoretic generalization of on-line learning and an application to boosting. *Journal of Computer and System Sciences*, 55(1):119–139, 1997.

[8] J.H. Friedman, T. Hastie, and R. Tibshirani. Additive logistic regression: A statistical view of boosting. *Annals of Statistics*, 28(2):337–407, 2000.

[9] R.A. Friesner. *Computational Methods for Protein Folding: A Special Volume of Advances in Chemical Physics*. Wiley-IEEE, Hoboken, NJ, 2002.

[10] C. Hadley, and D.T. Jones. A systematic comparison of protein structure classifications: Scop, cath and fssp. *Biological Science*, 7(9):1099–1112, 1999.

[11] C.D. Huang, C.T. Lin, and N.R. Pal. Hierarchical learning architecture with automatic feature selection for multiclass protein fold classification. *IEEE Transactions on NanoBioscience*, 2(4):221–232, 2003.

[12] E. Ie, J. Weston, W.S Noble, and C. Leslie. Multi-class protein fold recognition using adaptive codes. *ACM International Conference*, 119:329–336, 2005.

[13] National Human Genome Research Institute. http: //www. genome. gov/Pages/Hyperion/DIR/VIP/Glossary/Illustration/protein.shtml.

[14] D.T. Jones. Genthreader: an efficient and reliable protein fold recognition method for genomic sequences. *Molecular Biology*, 287(4):797–815, 1999.

[15] K. Karplus, C. Barrett, and R. Hughey. Hidden markov models for detecting remote protein homologies. *Bioinformatics*, 14(10):846–856, 1998.

[16] S. Lee and R.M. Kil. Multilayer feedforward potential function network. *International Joint Conference on Neural Networks*, 1:161–171, 1988.

[17] C. J. Lin and C.W. Hsu. A comparison of methods for multiclass support vector machines. *IEEE Transactions on Neural Networks*, 13(2):415–425, 2002.

[18] L.J. McGuffin, and D.T. Jones. Improvement of the genthreader method for genomic fold recognition. *Bioinformatics*, 19(7):874–881, 2003.

[19] J. Moody, and C.J. Darken. Fast learning in networks of locally tuned processing units. *Neural Computing*, 1(2):281–294, 1989.

[20] O. Okun. Protein fold recognition with k-local hyperplane distance nearest neighbor algorithm. *Proceedings of the 2nd European Workshop on Data Mining and Text Mining for Bioinformatics,* 47–53, 2004.

[21] H. Rangwala, and G. Karypis. Building multiclass classifiers for remote homology detection and fold recognition. *BMC Bioinformatics,* 16(7):455, 2006.

[22] B. Rost and C. Sander. Prediction of protein secondary structure at better 70% accuracy. *Journal of Molecular Biology,* 232(2):584–599, 1993.

[23] V. Vapnik. *The Nature of Statistical Learning Theory.* Springer, New York, 1995.

[24] I. Witten, and E. Frank. *Data Mining: Practical machine learning tools and techniques,* 2nd Edition. Morgan Kaufmann, Publishers Inc. San Fransisco, CA, USA, 2005.

[25] J. Xu, Y. Xu, K. Dongsup, and M. Li. Raptor: Optimal protein threading by linear programming. *Bioinformatics and Computational Biology,* 1(1):95–117, 2003.

[26] Y. Xu and D. Xu. Protein threading using prospect: design and evaluation. *Proteins: Structure, Function, and Genetics,* 40(3):343–354, 2000.

[27] Y. Krishnaraj, and C.K. Reddy. Boosting methods for protein fold recognition: An empirical comparison. *BIBM IEEE International Conference on Bioinformatics and Biomedicine,* 393–396, 2008.

Chapter 5

Protein Surface Representation and Comparison: New Approaches in Structural Proteomics

Lee Sael and Daisuke Kihara

Purdue University

5.1 Introduction

The 3D shape and physicochemical properties on protein surface carry essential information for understanding function of a protein. For example, catalytic reaction of an enzyme is realized by a set of atoms on the active site. Also residues at an interface surface region establish physical contacts in protein–protein interaction. Those local surface regions which are responsible for function tend to be better conserved than other surface regions in terms of shape and physicochemical properties, which are not detectable by conventional sequence or main-chain conformation similarity searches. Thus development of effective methods for describing and comparing protein surfaces will provide new insights into how function is realized in proteins.

Despite the importance and promise of surface-based protein characterization methods, they have not been well studied until recently partly due to its higher technical complexity. However, development of protein surface analysis methods have been highlighted recently because of its urgent need; an increasing number of structures of unknown function have been solved and accumulated by structural genomics projects in the past few years. The current protein structure database, protein data bank (PDB), contains more than 2300 structures which are categorized as "unknown function," whose function were not confidently predicted by conventional approaches. Concurrently in the computer science field, 3D object representations and searching algorithms have become a research focus in many domains, such as computer-aided design, game development, computer vision, and computational geometry. Some of developed algorithms in those domains can be readily applied to protein surface analyses. Moreover, bioinformatics resources, including databases of protein sequences and structures and classification of protein function, have been well developed. These resources enable computational analyses of the relationship of protein function and structure, facilitating development of new bioinformatics tools. Therefore now the time is ripe for extensive development of protein surface analysis methods which can provide functional annotation to proteins through surface comparison.

In the following sections, we overview evaluation criteria for methods for 3D shape analysis. Then we discuss how protein surface is defined. Next, we review 3D object analysis methods developed in the computational geometry and graphics field. What follows is a review of recent protein surface representations and comparison methods. In the last section, we introduce our recent works on surface-based fast protein structure and surface property comparison methods.

5.2　Evaluation Criteria

There are seven criteria for evaluating characteristics of general object shape and protein surface analysis methods (Tangelder and Veltkamp, 2004). In the following sections, we will refer these criteria when describing existing methods.

1. Invariance to Euclidean transformation: three Euclidean transformations, i.e., rotation, scaling, and translation, do not change the shape of

the original object. Some methods use representations that are invariant to Euclidean transformation. Although this may seem simple, it needs nontrivial mathematical derivation or often achieved by oversimplification. Euclidean transformation invariant representations are convenient in comparing objects, because their representations can be directly compared without preprocessing. When the other representations are employed, either the most similar positions of objects needed to be searched, which is time consuming, or a normalization process of object positions is needed prior to analysis.

2. Need for pose normalizing: pose normalization is to rotate, translate, and scale an 3D object to a standard position for comparison. Normalizing in terms of translation and scaling can easily be done, for example, by moving the object in a way that its center of gravity locates at the origin and then scale it into a unit sphere. However, normalization against rotation is often not robust depending on the shape of the object. Principal component analysis (PCA) is the most widely practiced method for pose normalization. PCA often fails to provide a unique robust solution when an object has symmetrical mass distribution; i.e., PCA generates many equal eigenvalues, which suggests more than one positioning of the object are possible. This is especially problematic to handling protein shapes, because they are more or less spherical.

3. Ability for partial matching: partial matching aims to find similar local regions of two objects. It is especially important in protein matching considering that a function of protein is attributed to local surface regions such as active sites and protein docking interfaces.

4. Capability of changing resolution: depending on the content of object analysis, being able to adjust the level of details of object description becomes useful. Higher resolution provides detailed information while lower resolution is more focused in the overall shape and computationally efficient.

5. Tolerance to small noises or changes: this property is also important for protein shape analysis along with the capability of changing resolution. Proteins are flexible in nature and structures are solved experimentally in different resolutions depending on the method used and experimental condition. Moreover, if a predicted structure is handled, some errors are unavoidable. Thus it is important to account for these changes by allowing resolution change or by making the method tolerant to small differences.

6. Ability to incorporate additional properties: nonshape properties such as electrostatic potential, hydrophobicity, and residue conservation are important factors in determining and analyzing function of proteins.

Thus being able to use additional properties can extend the applicability of protein surface analysis.

5.3   Surface Representation

5.3.1   General object representation

There are two types of object representations, volume- and boundary-based. Well known volume-based representations are voxels and octrees. In voxels, the volume of an object is represented by filled grid points while in octrees the object space is hierarchically subdivided. Widely used boundary-based representations include polygon mesh and point cloud. A polygon mesh is composed of nodes and edges that form triangles that are connected to completely cover the surface of an object. In point cloud, a set of (x, y, z) points on the surface are used to represent a surface.

Generally, volume-based representations are used to for experimental data, such as computed tomography scans. On the other hand, boundary-based representations are used for computer designed objects, such as ones used in computer games. Volume-based representation requires a larger space but can provide information about the interior of an object while boundary-based representation is efficient in drawing an object on the computer screen.

5.3.2   Protein surface definition

The underlying physical substance of a protein surface is the van der Waals radius of atoms of the protein. Thus an intuitive way of defining protein surface is to compute the union of boundaries of spheres of van der Waals radius of each protein atom (the van der Waals surface). Often inflated (i.e., enlarged) van der Waals radius is used for defining the surface. However, direct use of the van der Waals sphere of atoms usually leaves unoccupied spaces between atoms, making small clefts and cavities on the surface. Those small cavities, where water molecules and ions cannot enter, are negligible or often cause unnecessary noises for many applications of protein surface representation. A common way to obtain a smoother surface is to roll a probe sphere (usually of the size of a water molecule) over the van der Waals surface and to trace the center of the sphere (solvent accessible surface) or to trace the inward-facing surface of a probe sphere (solvent excluded surface or Connolly surface (Connolly, 1983)).

The other protein surface definitions include α-surface (Wang, 2001). The algorithm of α-surface connects points to construct triangle meshes, whose resolution is controlled by a parameter, α. The solvent accessible surface and the Connolly surface are also usually represented by triangle meshes. The other representations, such as point cloud and voxels are also used.

5.4   General Object Analysis Methods

This section provides a list of well known shape analysis methods in computer science and other engineering fields that are already applied or have the potential of being applied to proteins. Roughly, methods can be classified as global and local shape analysis methods.

5.4.1   Global shape analysis

Global shape descriptors represent the overall shape of objects. They can be classified into three categories, feature and feature distribution-based methods, 3D coordinate centered methods, and view-based methods.

5.4.1.1   Feature/feature distribution-based methods

Methods in this category describe an object by one or a set of features of the object, such as the volume and the area of surface. These methods are one of the earlier methods developed, and are still actively applied individually or often integrated into recent object analysis methods because of their simplicity. An advantage of these methods is that nonshape properties of objects can be easily combined with shape-based features. On the other hand, the disadvantage is that they are obviously less descriptive since an object shape is represented by a few number of features. Those features are represented as a feature vector.

Elad et al. (2000) use statistical moments, which describe the distribution of the position of vertices on a polygon mesh of an object, i.e., the center of gravity, variance, and skewness etc., as features of the object. Then the extracted moments are used as a feature vector and compared by a weighted Euclidean distance. In their method, invariance to Euclidean transformation is obtained by normalizing against the first two moments (center of gravity and variance) of the surface points. Zhang and Chen (2001) also utilize the statistical moment in addition to some other features. They propose an efficient method for computing and comparing the global features including the volume, the surface area, the volume-surface ratio, the statistical moments, and coefficients of the Fourier transform of 3D objects.

Rather than representing an object by a single value, feature distribution-based methods use a histogram of global shape features as a descriptor. An example is the shape distribution, which uses a histogram of the Euclidean distance of randomly chosen two points on the surface of an object (Osada et al., 2002). The solid angle histogram places a sphere at each representing points of an object and computes the fraction of volume of the sphere occupied by the object (Connolly, 1986).

5.4.1.2 3D coordinate centered methods

These methods directly represent 3D shapes of objects in space. There are two main approaches in this category. The first approach is to use mathematical transformation of a 3D function, which is the position of points or surface of a 3D object in the case of the shape analysis. Another approach is to compute how an object occupies the 3D space by its volume when the object is represented in voxels.

Mathematical transformations have been widely studied in 2D image processing. And the 3D version of those transformations, such as Fourier, Hough, Radon, and wavelet transformations, have been applied for 3D objects. These methods are variant to Euclidean transformations and need pose normalization prior to extraction of descriptors. More recently, spherical harmonics have been widely explored. Spherical harmonics are functions of a set of a polar angle, θ, and a colatitude angle, $\varphi : Y_l^m(\theta, \varphi)$. Since spherical harmonics form an orthonormal set of functions, a 3D function (thus, a 3D object) can be expanded as a series of spherical harmonics with a different degree l and an order m on the unit sphere.

Limitations in direct applications of spherical harmonics include its variance to Euclidean transformations and also that they can correctly capture only star like shapes, i.e., shapes that have no reentrant surfaces.

Funkhouser's group introduced a spherical harmonics-based shape descriptor, which is rotation invariant and can also be applied to nonstar like shapes (Kazhdan et al., 2003). The method first segments an object into concentric spheres and then computes spherical harmonics for each of the spheres. Since rotating a spherical function does not change its L^2 norm, combining the L^2 norm computed for each group of harmonics of the same parameters (l and m) yields a rotation invariant (thus invariant to Euclidean transformations) descriptor. Nonstar like shapes are better handled by the segmentation to concentric spheres. Application of spherical harmonics in partial matching has also been made by the same group as an extension of the spherical harmonics-based method (Funkhoser and Shilane, 2006).

The above method considers the radial information (the distance from the center of an object) by the segmentation of an object into concentric spheres. In contrast, 3D Zernike descriptors uses Zernike–Canterakis basis $Z_{nl}^m(\mathbf{x})$, which incorporates radial information into the polynomials in Cartesian coordinates $\mathbf{x} = (x, y, z)$ (Canterakis, 1999):

$$Z_{nl}^m(\mathbf{x}) = R_{nl}(r)Y_l^m(\theta, \varphi)$$

Thus, 3D Zernike descriptors are convenient to handle a 3D object described in points or voxels in Cartesian coordinates. Rotation invariance was obtained later by (Novotni and Klein, 2004) in a similar manner to what was done by (Kazhdan et al., 2003). We have applied 3D Zernike descriptors for protein surface comparison, which will be discussed in Section 5.6.

The other special functions introduced in 3D object analysis include spherical wavelets and Krawtchouk polynomials. Mathematically, spherical wavelet descriptor (Laga et al., 2006) has two advantages over spherical harmonics. First, the level of details of description can be locally controlled. Second, sampling of points are more uniform. Weighted 3D Krawtchouk descriptor (Mademlis et al., 2006) uses polynomials of discrete variables, and thus eliminates the need for spatial discretization process. Hence no numerical approximation is involved in handling a voxelized object data. A drawback of weighted 3D Krawtchouk descriptor is again the need for pose normalization.

When objects are represented by voxels, two objects can be compared by computing the difference of distribution of the occupied voxels (volumetric difference methods). Occupied voxels of an object can be represented by a tree data structure, e.g., octree. Therefore comparison is done efficiently based on the tree representation. Volumetric difference methods are still generally slower than other global methods and still need pose normalization.

An interesting idea of representing object is to compute "energies" or cost needed to morph an object to a sphere. Then comparison is done by computing the difference in the morphing energies. A method by Leifman et al. (2003) calculates sphere projection energy as $E = \int_{dist} \vec{F}.d\vec{r}$, where *dist* is the distance between the sphere and the object surface, and $\vec{F}$ is the applied force which is assumed constant. In a method proposed by Yu et al. (2003), a feature map is used to record a local energy needed to morph an object. The object is first normalized and fitted into a unit sphere. Then local energy at each point is computed, which consists of two parts: the distance from the object surface to the bounding sphere and the number of object surfaces penetrated when a ray is shot from the sphere center. This method additionally uses Fourier transform of the feature map, which is better in tolerance to noise which may have been introduced in the pose normalization process.

5.4.1.3 View-based methods

View-based methods describe a 3D object as a set of projected 2D images of the 3D object from different viewing angles. Each image contains characteristics of the object from that angle, however, relative spatial information between images from different view points is not captured.

The most well known view-based method is light field descriptors (Chen et al., 2003). In computing the light field descriptor of a 3D object, the object is first scaled and placed into a bounding sphere. Then a light field of the object is created, which consists of 20 uniformly distributed silhouettes of the object from 10 rotational positions on the bounding sphere. Subsequently, a combination of 2D Zernike moments and Fourier transforms are used as the 2D descriptor for each silhouette. To compare descriptors of two objects, basically silhouettes of the two objects are compared exhaustively to find matches.

Ohbuchi et al. (2003) proposed another view-based technique, which captures depth of an object from each angle in addition to the 2D silhouettes. Then for each image a Fourier transform-based descriptor is generated.

5.4.2 Local shape analysis

The basis of local shape analysis is to capture geometrical feature of a local region around a given point on a surface. The curve of a local surface is described using Gaussian, mean curvature, and the shape index. Among them, the shape index has been also used for protein surface analysis. The shape index (Koenderink and van Doorn, 1992) is a single-value that ranges from -1 to 1 which measures the slope of local surface using principal curvatures. The spin image is another popular method to describe a local shape (Johnson and Hebert, 1999; de Alarc et al., 2002). A spin image is a 2D histogram of distances from a central vertex to neighboring vertices. Two distances characterize spatial relationship between the two vertices; the radial distance, α, which is defined as the perpendicular distance from the central vertex to another through the surface normal, and the axial distance, β, a signed perpendicular distance to the tangent plane of the central vertex. By definition, a spin image does not change upon rotating around the norm of a central vertex. The spin image is calculated for each vertex on a surface mesh of an object.

Using surface curvature information captured at each vertex as mentioned above, a larger surface region can be described by connecting vertices as a graph. A graph captures relative spatial information of vertices and enables partial matching of two local surfaces. However, generally speaking partial graph matching has a high complexity thus often slow for comparing large graphs. Methods for global shape analysis, such as spherical harmonics-based methods, can also be used for describing a local shape around a vertex.

5.5 Protein Surface Analysis Methods

In this section, we discuss existing methods for protein surface representation and comparison. Identifying similar global and local surface shapes of proteins has application to structure-based function prediction. Protein surface representation has been also studied in the context of protein-protein docking and protein-small ligand docking, in which case complementarity of two shapes is taken into account. We first discuss three major categories of protein surface analysis methods, namely, graph-based, geometric hashing, and methods using series expansion of 3D function.

5.5.1 Graph-based methods

Graph theoretical approaches are frequently applied for protein surface comparison since some common protein surface representations, e.g., triangular mesh, can be naturally considered as a graph. In a graph representation of a protein surface, geometrical and often physicochemical features of a local region are assigned to each vertex and edges connecting vertices describe positional relationship of the vertices. A nice thing about graph representation is that partial matching of two protein surfaces can be done using existing algorithms in the graph theory.

The method proposed by Pickering et al. (2001) first generates a Connolly surface of the region of interest. Then for each vertex point, shape information, such as shape index and radius of curvatures, is calculated as well as biological features such as types of residues. The matching process involves finding the maximal common subgraph of two graphs representing the protein surfaces.

Kinoshita and his colleagues developed a database of protein surfaces of functional sites, named eF-site, and a method to search for the similar local surface sites in a query protein against the database (Kinoshita et al., 2002). Triangular meshes of the Connolly surface constitute a graph of a protein. Each vertex is assigned with the electrostatic potential and curvatures. To find similar local regions of two proteins, a clique detection algorithm on an association graph is used. An association graph of two graphs is formed first by creating a node for a pair of vertices, one from each protein, that have similar features. Then an edge connecting a pair of nodes is drawn when the spatial distance of the pair of original vertices belonging to the two nodes is similar. Next, the largest clique in the association graph, i.e., the largest fully connected subgraph, is selected. The selected clique is considered as the most similar part between the two protein surfaces.

SURFCOMP also uses a clique detection algorithm on an association graph (Hofbauer et al., 2004). In SURFCOMP, surface critical points are considered as vertices, which are either one the three classes, a convex, a concave, or a saddle point. Rather than using all the vertices in a Connolly surface, using critical points reduces vertices to be considered, making the method more efficient. The graphs are further simplified by several filters that compare surrounding shape, local arrangement of the critical points, and physicochemical properties.

Baldacci et al. (2006) further reduce the number of graph nodes by considering surface patches. A patch in a protein surface is a local circular region where residues included have homogeneous geometrical and physicochemical properties. The properties considered are geometrical curvature, the electrostatic potential, and hydrophobicity. Each patch contains at least ten amino acids and typically a protein surface is represented by less than ten patches. Patches are connected by edges, representing a protein by a spatial graph. They used the spatial graphs for classifying proteins by similarity of patterns of patches.

5.5.2 Geometric hashing

The Wolfson and Nussinov group applies the geometric hashing technique, which was originally developed for computer vision applications (Rosen et al., 1998). The method first extracts sparse critical points defined at the centers

of mesh faces abstracted as convex, concave, or saddle of the protein surface (Lin et al., 1994). The geometric hashing is composed of two stages,
a hashing stage and a recognition (matching) stage. In the hashing stage,
transformation-invariant information of protein surface shapes to be compared
against (called models) is extracted and stored in a hash table. Concretely, a
protein surface shape represented by critical points is placed relative to every
possible admissible reference frame and their position and features are stored
in the hash table. This stage can be executed off-line and the table can be
reused once created. In the recognition stage, a target protein surface is placed
relative to every possible reference frame and the hash table is accessed to find
matching model critical points. Then a vote is registered for a pair of model
and target reference frames if their critical points match. The geometric hashing allows a partial surface matching. Also a target protein can be compared
with multiple proteins at the same time once they are hashed in a table. Later
they also applied geometric hashing for protein-small ligand molecule docking,
and protein–protein docking (Fischer et al., 1995; Halperin et al., 2002).

5.5.3   Methods using series expansion of 3D function

Using mathematical transformation has become popular in protein surface
analyses as well as 3D object analysis. Here protein surface is treated as a
3D function, which is expanded in a series function. The major advantage
of these methods is the compactness in description, which allows rapid real-
time comparison against a large number of proteins. A series expansion is also
suitable in changing resolutions of surface description. Also properties on a
surface can be naturally incorporated in surface description.

An early work in this category include use of Fourier series expansion
as a shape descriptor (Gerstein, 1992). Protein surfaces are superimposed
and Fourier coefficients are extracted and compared at various resolutions.
The author used the method to compare shape of antigen-combining sites of
antibody molecules.

Thornton and her colleagues used spherical harmonics to describe the volume of ligand binding pockets of proteins (Kahraman et al., 2007). A ligand
binding pocket in a protein surface is detected using the SURFNET program, which identifies a pocket by inserting spheres of a certain size. Thus
a pocket is represented as overlapping spheres, which constitute the volume
of the pocket. Then the spherical harmonics expansion is applied to the volumetric representation of the pocket and the coefficients are taken as the
descriptor. Interesting application of their approach is direct comparison of
shape of pockets and ligand molecules, which is possible because both pockets
and ligands are represented as a closed volume. For comparison, shapes should
be pose normalized.

Spherical harmonics has been also applied for protein–protein docking
prediction (Ritchie and Kemp, 2000). By using spherical harmonics, a complete search for docking conformation over all six degrees of freedom can be

performed conveniently by rotating and translating the initial expansion co-efficients.

Recently, we employed 3D Zernike function, which is an extension of spherical harmonic expansion to compare protein global surfaces (Sael et al., 2008a,b). The most favorable features of the 3D Zernike descriptor aside from its advantages originating from spherical harmonics, are its rotation invariance and applicability to nonstar-like shapes. The two advantages are worth further attention and will be described extensively in Section 5.6.

5.5.4 Several other methods

The volumetric difference method, which was originally developed for 3D object representation, has been applied for protein surface comparison. Masek et al. (1993) defines molecular "skins," which is a thin layer of voxels composing protein surface. The method compares the shape by computing the similarity of the maximum overlap between a pair of protein skins. Another volumetric difference method utilizes a genetic algorithm to find the optimal superimposition of protein surfaces or fragments of proteins (Poirrette et al., 1997). The spin image representation is also applied to identify structurally equivalent surface regions in two proteins (Bock et al., 2007).

Shentu et al. (2008) proposed a local surface structure characterization method named context shape, which considers visible directions from critical points on a protein surface. A context shape of a critical point essentially describes the visible directions from the point to a surrounding sphere of a given radius, which is not blocked by voxels occupied by the protein volume. The context shape is represented as a binary string with 1 for blocked and 0 for visible directions. They used this method to evaluate shape complementarity of two protein surfaces in protein–protein docking prediction.

Pawlowski and Godzik (2001) proposed a method which is aimed to compare physicochemical features of protein surfaces, such as electrostatic potential and hydrophobicity. Those features are mapped on a surrounding sphere of a protein and comparison is done between spheres after they are superimposed. As obvious from its design, this method does not compare shapes of proteins but only physicochemical properties, thus, it can only analyze proteins of the same structure (e.g., protein of the same family). Nevertheless this method is interesting as it can quantify the difference of properties on the surface.

There are several methods which combines surface shape information with residue or sequence information. As sequence motifs (e.g., PROSITE database) or spatial arrangements of catalytic residues (Arakaki and Skolnick, 2004) are traditionally used in function prediction in protein bioinformatics area, these methods can take advantage of accumulated knowledge of sequence-function relationship of proteins.

The SURFACE database stores a library of functionally important residues found at pocket regions of proteins (Ferrè et al., 2005). In selecting functional

sites of proteins, pockets in protein surfaces are identified by SURFNET and
residues which reside in the pockets are referred to functional motif databases
including PROSITE. Two local sites are compared in terms of the root means
square deviation of positions of superimposed amino acids.

A method by Binkowski et al. (2003) utilizes local sequence information of
binding pockets and surface shape to predict function of proteins. The local
sequence of a pocket region is extracted by concatenating short sequences
which compose the pocket. To compare the extracted local sequence, local
sequence alignment by dynamic programming algorithm is performed.

5.6   3D Zernike Descriptors

A 3D Zernike descriptor (3DZD) is categorized as a 3D coordinate centered method using special kernel functions. Canterakis first introduced 3D
Zernike moments that combine a radial function with spherical harmonics
to describe objects in 3D Cartesian coordinate system (Canterakis, 1999).
Novotni and Klein later applied 3D Zernike moments to construct rotation
invariant descriptors of 3D objects (Novotni and Klein, 2003). 3DZD was applied to describe overall shape of small ligands by Mak et al. (2008). The first
thorough applications to describe protein shape and physicochemical property
on protein surface have been conducted by our group. This section summarizes
our works described in two recent papers (Sael et al., 2008a,b).

5.6.1   Characteristics of 3DZD

3DZD has several significant advantages regarding the comparison of protein surface. First, it represents a protein compactly allowing fast retrieval
capable for real-time database search. Second, 3DZDs are rotation invariant,
that is, protein structures need not be aligned for comparison. Related works,
such as spherical harmonics for binding pocket and ligand comparisons by
Thornton's group (Kahraman et al., 2007), need pose normalization because
the methods are not rotation invariant. Pose normalization could be problematic especially in comparison of protein shapes, which are almost globular
and the principle axes are not robustly determined. Third, the resolution of
the description of protein structures can be easily and naturally adjusted by
changing the order of 3DZDs. The rough global difference of protein structures reflects the difference of the first couple of invariants that correspond to
lower orders of the 3DZD (Sael et al., 2008b). Figure 5.1 illustrates different
resolutions of a reconstructed protein surface by changing the order of 3DZD.
Here the order is changed from 5 to 10, 15, 20, and 25. When a lower order is
used, pear-like global surface shape of this protein is highlighted, while more
description of local geometry shows up as the order becomes higher. We used

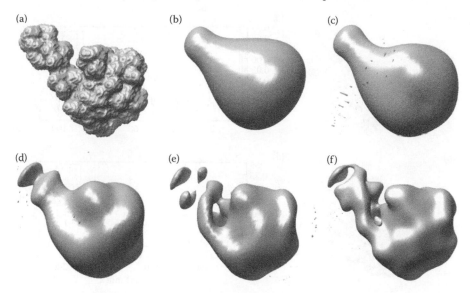

FIGURE 5.1: Resolution of 3D Zernike descriptor. (a) surface abstraction of 1ew0A, which is used as the input. (b) through (f) are reconstructed figures of 3D Zernike moments using different order from 5 up to 25 increasing with an interval of 5.

the order of 20 for our work since it yielded satisfactory results in a 3D shape retrieval benchmark by Novotni and Klein. Moreover, physicochemical properties of a protein surface, such as electrostatic potentials and hydrophobicity, can be incorporated into the description considering an appropriate 3D function (Sael et al., 2008a).

5.6.2 Steps of computing 3DZD for protein surfaces

The first steps to compute 3DZD for a protein are calculating protein surface and placing it on a cubic grid (voxelization). To represent a surface shape, each voxel is assigned 1 if it is on the surface and 0 otherwise. Real number values of other physicochemical properties can also be assigned only to the surface voxels. The resulting voxels with values are considered as the 3D function, which will be expanded in a 3DZD. Using the order of 20, a 3DZD results in 121 invariants (numbers). To convert physicochemical property which ranges from negative to positive values to 3DZD, a 3DZD for a set of voxels with a positive value assigned and those with negative value are separately computed. Then two 3DZDs are combined yielding a descriptor of 242 invariants. This is because a 3DZD recognizes the contrast of patterns of the positive and the negative value but not the value itself.

3DZDs of two proteins are compared in terms of the Euclidian distance or a correlation coefficient based distance, which is defined as 1—*correlation*

(a)

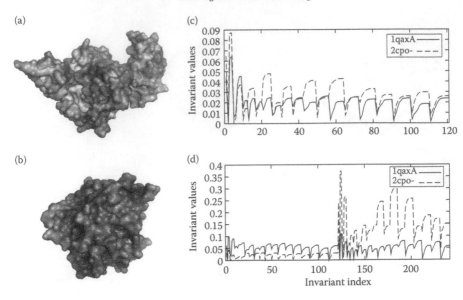

FIGURE 5.2: 3DZD of protein surface shape and electrostatic potential. Surface electrostatic potential of two proteins proteins; (a), 1qaxA; and (b), 2cpo. (c), 3DZD of surface shape; and (d), surface electrostatic potential of the two proteins is shown. The order used for both shape and electrostatic potential is 20. The number of invariants computed for shape is 121. For electrostatic potential, positive and negative regions are calculated separately forming 242 number of invariants. In (a) and (b) the gray scale ranges from −5 (black) to +5 (white) to represent the electrostatic potential.

coefficient. Figure 5.2 shows an example of 3DZDs of two proteins, 1qaxA and 2cpo. Figure 5.2c shows 3DZDs of the surface shape and Figure 5.2d shows 3DZDs of surface electrostatics of the two proteins. In Figure 5.2d, higher peaks at the 122th to the 242th invariants by 2cpo relative to 1qaxA indicate that 2cpo has a larger region with a negative electrostatics value.

5.6.3   Test results of 3DZD

We evaluated 3DZDs in two ways: (1) the ability to retrieve similar protein structures (Sael et al., 2008b) and (2) the ability to compare proteins in terms of their surface physicochemical properties (Sael et al., 2008a).

For the protein structure retrieval test, we prepared a dataset of 2432 protein structures, which are preclassified by another protein structure comparison method, the combinatorial extension (CE) method (Shindyalov and Bourne, 1998). CE compares two protein structures by their main-chain conformation as many other conventional protein structure comparison methods do. Despite the difference in structure representation, 3DZD retrieved proteins of the same conformation defined by CE in 89.6% of the cases within the top five closest structures. This level of agreement with CE is the same as between CE and another commonly used protein structure comparison method, DALI. In addition to this retrieval accuracy, the strength of 3DZD is its extremely fast computational time. Computing a 3DZD for a protein takes about 37 seconds, but once it is computed, a database search against entire PDB with over 54,000 structures takes less than a minute. In contrast, typically a pairwise structure comparison by CE takes a couple of seconds. Therefore, a search against the entire PDB would take more than a day. Thus 3DZD can dramatically make a global protein structure search efficient. Even if one still wants to find proteins in a database which have the similar main-chain conformation to a query protein, 3DZD can be used as a rapid pre-screening prior to using CE.

Moreover, we found some cases where protein surface shape is indicative to functional classes of proteins. In our paper we showed such examples of pairs of DNA binding proteins and transmembrane transporters. These protein pairs have distinct surface shape similarity but does not share detectable similarity in sequence or main-chain conformation, thus their functional relevance can not be easily identified by conventional bioinformatics methods. In the case of DNA binding proteins, they have a saddle like local surface shape which is used to mount on DNA strands.

Next, we compared protein surface physicochemical properties of several protein families using 3DZDs (Sael ct al., 2008a). We used globin proteins and three protein families with both thermophilic and mesophilic homologs as datasets. The globin family is known to have a conserved fold with a wide variety of function. A varied range of affinity to oxygen, different functions, and different environments where the globin proteins locate coincide with the relatively large distance of surface electrostatic potentials measured by 3DZDs. Thermophilic proteins have gained substantially higher thermal stabilities as compared to their mesophilic orthologs. And surface electrostatics has been identified as one of the major stabilization factors of thermophilic proteins. For the three protein families studied, we showed that 3DZDs successfully distinguish the thermophilic proteins from mesophilic proteins based on similarity of surface electrostatics. The sequence similarity and the main-chain conformation similarity cannot differentiate these two classes, because all of members of the families have more or less similar in sequence and structure. Since 3DZDs can be quantitatively compared, a tree can be drawn for a set of proteins based on similarity of their surface physicochemical property. This will be quite useful for studying protein function and evolution.

Further, we showed electrostatic potential of local regions of proteins can also be compared. Figure 5.3 illustrates a procedure for local surface analysis using 3DZDs. In this example, ligand binding sites of two TIM barrel proteins, 1fdjB and 1goc, are compared. TIM barrel is one of the most prevalent folds adopted by a variety of enzymes. Ligand binding sites of TIM barrel enzymes are usually located at the cleft with cluster of loops of the barrels. Reflecting

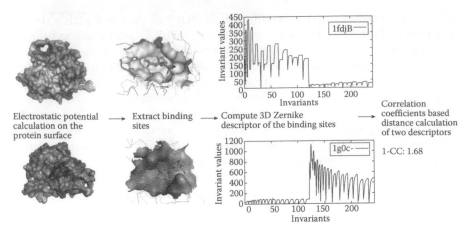

FIGURE 5.3: Local surface analysis procedure. A procedure for analyzing electrostatic potential on ligand binding region is illustrated. On the right are surface electrostatic potential of proteins 1fdjB, top, and 1g0c, bottom. Middle figure is surface electrostatic potential of extracted binding region which are the input to 3D Zernike method. The graphs are extracted 3DZD of the binding regions. The dissimilarity measures are calculated by correlation coefficients based distance of the two 3DZDs: 1.68. The gray scale ranges from −5 (black) to +5 (white) to represent electrostatic potential.

the nature of binding ligands, active sites show wide ranging behavior in terms of electrostatics, whose similarity can be quantified by using 3DZDs.

5.7 Discussion

In this chapter, we reviewed methods for 3D object shape analysis in the context of protein shape analysis. Some of the available methods are listed in Table 5.1.

Protein surface analysis is especially difficult because most of proteins have more or less sphere-like shape. Therefore a descriptor needs to differentiate relatively small differences. For the same reason, typical pose normalization methods, such as PCA, do not give a unique solution. Also nonshape features should be considered, such as physicochemical properties and residue conservation, as they are important to understand protein function. In addition, it is desired that computing similarity of two proteins is executed fast enough so that a database search is performed in a real-time.

To meet all these requirements, we applied 3DZD for protein surface comparison. Conventionally proteins have been long analyzed and classified in

TABLE 5.1: List of existing tools and computational resources.

	URL	Available materials
Light field descriptors (Chen et al., 2003)	http://3d.csie.ntu.edu.tw/	Source code/executable/dataset
Ef-site (Kinoshita et al., 2002)	http://ef-site.hgc.jp/eF-site/ http://ef-site.hgc.jp/eF-seek/	Database Web server
SURFACE (Ferre et al., 2005)	http://cbm.bio.uniroma2.it/surface/	Database
SURFCOMP (Hofbauer et al., 2004)	http://teachme.tuwien.ac.at/surfcomp/index.html	Toolkit/source code/executable
SURF'S UP (Pawlowski et al., 2001)	http://asia.genesilico.pl/surfs_up/	Web server
3d Zernike Server (Sael et al., 2008b)	http://dragon.bio.purdue.edu/3d-surfer/	Web server
Princeton Shape Retrieval and Analysis Group	http://shape.cs.princeton.edu/search.html	Web server

terms of their sequences and main-chain conformation. However, there are cases that these methods are not capable of detecting similarities and dissimilarities in a biologically meaningful way. In such cases, the 3DZD-based surface analysis can often do a better job than these methods, as it captures global and local protein surface shape, which is directly responsible for biological function, and also it is able to quantify similarity of physicochemical properties.

As more and more protein structures are experimentally solved, the need for effective and efficient methods for protein structure characterization increases. We expect the surface-based analysis will become a routine option for protein characterization besides sequence- and main-chain conformation-based methods. Biology has entered an informatics era when computational methods for retrieving useful knowledge from databases and reasoning using the knowledge become crucial. Various interdisciplinary approaches are essential and protein surface analysis will become one of such existing fields.

Acknowledgments

This work is supported by the National Institute of General Medical Sciences of the National Institutes of Health (R01GM075004). We thank Gregg Thomas for proofreading the manuscript.

References

Arakaki, A.K. and J. Skolnick. Large-scale assessment of the utility of low-resolution protein structures for biochemical function assignment. *Bioinformatics*, 20, 2004, 1087–1096.

Baldacci, L., M. Golfarelli, A. Lumini, and S. Rizzi. Clustering techniques for protein surfaces. *Pattern Recogn.*, 39, 2006, 2370–2382.

Binkowski, T. Andrew, L. Adamian, and J. Liang. Inferring functional relationships of proteins from local sequence and spatial surface patterns. *J. Mol. Biol.*, 332(2), 2003 505–526.

Bock, M. E., C. Garutti, and C. Guerra. Discovery of similar regions on protein surfaces. *J. Comput. Biol.*, 14(3), 2007, 285–299.

Canterakis, N. 3D Zernike moments and Zernike affine invariants for 3D image analysis and recognition. In *Proceedings of the 11th Scandinavian Conference on Image Analysis*, Kangerlussuaq, Greenland, 1999, 85–93.

Chen, D. Y., M. Ouhyoung, X. P. Tian, Y. T. Shen, and M. Ouhyoung. On visual similarity based 3D model retrieval. In *Proceedings of Eurographics 2003*. Granada, Spain, 2003, 223–232.

Connolly, M. L. Solvent-accessible surfaces of proteins and nucleic-acids. *Science*, 221, 1983, 709–713.

Connolly, M. L. Shape complementarity at the hemoglobin alpha I beta I subunit interface. *Biopolymers*, 25, 1986, 1229–1247.

de Alarc, P. A., A. D. Pascual-Montano, and J. M. Carazo. Spin Images and Neural Networks for Efficient Content-Based Retrieval in 3D Object Databases. In *CIVR '02: Proceedings of the International Conference on Image and Video Retrieval*. London, UK: Springer-Verlag, 2002, 225–234.

Elad, M., A. Tal, and S. Ar. Directed search in a 3d objects database using svm. Technical report, HP Laboratories, Israel, 2000.

Ferrè, Fabrizio, G. Ausiello, A. Zanzoni, and M. Helmer-Citterich. Functional annotation by identification of local surface similarities: a novel tool for structural genomics. *BMC Bioinform.*, 6, 2005, 194–208.

Fischer, D., S. L. Lin, H. L. Wolfson, and R. Nussinov. A geometry-based suite of molecular docking processes. *J. Mol. Biol.*, 248(2), 1995, 459–477.

Funkhouser, T., and P. Shilane. Partial matching of 3D shapes with priority-driven search. In *Proceedings of the Fourth Eurographics Symposium on Geometry Processing.* Carligari, Sardinia, Italy. ACM International Conference Proceeding Series. Eurographics Association, Aire-la-Ville, Swizerland, Vol. 256, June 26–28, 2006, 131–142.

Gerstein, M. A resolution-sensitive procedure for comparing protein surfaces and its application to the comparison of antigen-combining sites. *Acta Crystallogr.,* A48, 1992, 271–276.

Halperin, I., B. Ma, H. Wolfson, and R. Nussinov. Principles of docking: an overview of search algorithms and a guide to scoring functions. *Proteins,* 47, 2002, 409–443.

Hofbauer, C., H. Lohninger, and A. Aszódi. SURFCOMP: a novel graph-based approach to molecular surface comparison. *J. Chem. Inf. Comput. Sci.,* 44(3), 2004, 837–847.

Johnson, A. E., and M. Hebert. Using spin images for efficient object recognition in cluttered 3D scenes. *IEEE Trans. Pattern Anal. Mach. Intell.,* 21(5), 1999, 433–449.

Kinoshita, K., J. Furui, and H. Nakamura. Identification of protein functions from a molecular surface database, eF-site. *J. Struct. Funct. Genomics,* 2(1), 2002, 9–22.

Kahraman, A., R. J. Morris, R. A. Laskowski, and J. M. Thornton. Shape variation in protein binding pockets and their ligands. *J. Mol. Biol.,* 368(1), 2007, 283–301.

Kazhdan, M., T. Funkhouser, and S. Rusinkiewicz. Rotation invariant spherical harmonic representation of 3D shape descriptors. In *SGP '03: Proceedings of the 2003 Eurographics/ACM SIGGRAPH Symposium on Geometry Processing.* Aire-la-Ville, Switzerland. Eurographics Association, 2003, 156–164.

Koenderink, J. J., and A. J. van Doorn. Surface shape and curvature scales. *Image Vision Comput.,* 10(8), 1992, 557–564.

Laga, H., H. Takahashi, and M. Nakajima. Spherical wavelet descriptors for content-based 3D model retrieval. In *SMI '06: Proceedings of the IEEE International Conference on Shape Modeling and Applications,* Matsushima, Japan, 2006, 75–85.

Leifman, G., S. Katz, A. Tal, and R. Meir. Signatures of 3D models for retrieval. In *4th Israel Korea Bi-National Conference on Geometric Modeling and Computer Graphics,* Tel-Aviv, Israel, 2003, 159–163.

Lin, S. L., R. Nussinov, D. Fischer, and H. J. Wolfson. Molecular surface representations by sparse critical points. *Proteins,* 18(1), 1994, 94–101.

Mademlis, A., A. Axenopoulos, P. Daras, D. Tzovaras, and M. G. Strintzis. 3D content-based search based on 3D Krawtchouk moments. In *3DPVT '06: Proceedings of the Third International Symposium on 3D Data Processing, Visualization, and Transmission (3DPVT'06)*. Washington, DC: IEEE Computer Society, 2006, 743–749.

Mak, L., S. Grandison, and R. J Morris. An extension of spherical harmonics to region-based rotationally invariant descriptors for molecular shape description and comparison. *J. Mol. Graph. Model.*, 26(7), 2008, 1035–1045.

Masek, B. B., A. Merchant, and J. B. Matthew. Molecular skins: a new concept for quantitative shape matching of a protein with its small molecule mimics. *Proteins*, 17(2), 1993, 193–202.

Novotni, M., and R. Klein. 3D Zernike descriptors for content based shape retrieval. In *The 8th ACM Symposium on Solid Modeling and Applications*, Seattle, Washington, 2003.

Novotni, M., and R. Klein. Shape retrieval using 3D Zernike descriptors. *Computer-Aided Design 36*, 11, 2004, 1047–1062.

Ohbuchi, R., M. Nakazawa, and T. Takei. Retrieving 3D shapes based on their appearance. In *MIR '03: Proceedings of the 5th ACM SIGMM International Workshop on Multimedia Information Retrieval*. New York, NY: ACM Press, 2003, 39–45.

Osada, R., T. Funkhouser, B. Chazelle, and D. Dobkin. Shape distributions. *ACM Trans. Graph.*, 21(4), 2002, 807–832.

Pawlowski, K., and A. Godzik. Surface map comparison: studying function diversity of homologous proteins. *J. Mol. Biol.*, 309, 2001, 793–800.

Pickering, S. J., A. J. Bulpitt, N. Efford, N. D. Gold, and D. R. Westhead. AI-based algorithms for protein surface comparisons. *Comput. Chem.*, 26(1), 2001, 79–84.

Poirrette, A. R., P. J. Artymiuk, D. W. Rice, and P. Willett. Comparison of protein surfaces using a genetic algorithm. *J. Comput. Aided Mol. Des.*, 11(6), 1997, 557–569.

Ritchie, D. W., and G. J. L. Kemp. Protein docking using spherical polar Fourier correlations. *Proteins*, 39, 2000, 178–194.

Rosen, M., S. L. Lin, H. Wolfson, and R. Nussinov. Molecular shape comparisons in searches for active sites and functional similarity. *Protein Eng.*, 11(4), 1998, 263–277.

Sael, L., D. La, B. Li, R. Rustamov, and D. Kihara. Rapid comparison of properties on protein surface. *Proteins*, 73, 2008a, 1–10.

Sael, L., B. Li, D. La, Y. Fang, K. Ramani, R. Rustamov, and D. Kihara. Fast protein tertiary structure retrieval based on global surface shape similarity. *Proteins,* 72, 2008b, 1259–1273.

Shentu, Z., M. Al Hasan, C. Bystroff, and M.J. Zaki. Context shapes: efficient complementary shape matching for protein-protein docking. *Proteins,* 70(3), 2008, 1056–1073.

Shindyalov, I. N., and P. E. Bourne. Protein structure alignment by incremental combinatorial extension (CE) of the optimal path. *Protein Eng.,* 11(9), 1998, 739–747.

Tangelder, J. W. H., and R. C. Veltkamp. A survey of content based 3D shape retrieval methods. In *SMI '04: Proceedings of the Shape Modeling International 2004 (SMI'04).* Washington, DC: IEEE Computer Society, 2004, 145–156.

Wang, X. Alpha-surface and its application to mining protein data. In *Proceedings of the 2001 IEEE International Conference on Data Mining,* 2001, 659–662.

Yu, M., I. Atmosukarto, W. K. Leow, Z. Huang, and R. Xu. 3D model retrieval with morphing-based geometric and topological feature maps. In *Proceedings of IEEE Computer Society Conference on Computer Vision and Pattern Recognition,* Vol. 2, 2003, 656–661.

Zhang, C., and T. Chen. Efficient feature extration for 2D/3D objects in mesh representation. In *Proceedings of the 2001 International Conference on Image Processing (ICIP 2001).* Thessaloniki, Greece, 2001, October 7–10.

Chapter 6

Advanced Graph Mining Methods for Protein Analysis

Yi-Ping Phoebe Chen, Jia Rong and Gang Li

Deakin University

6.1 Introduction

As one of the primary substances in a living organism, protein defines the character of each cell by interacting with the cellular environment to promote the cell's growth and function [1]. Previous studies on proteomics indicate that the functions of different proteins could be assigned based upon protein structures [2,3]. The knowledge on protein structures gives us an overview of protein fold space and is helpful for the understanding of the evolutionary principles behind structure. By observing the architectures and topologies of the protein families, biological processes can be investigated more directly with much higher resolution and finer detail. For this reason, the analysis of protein, its structure and the interaction with the other materials is emerging as an important problem in bioinformatics. However, the determination of protein structures is experimentally expensive and time consuming. This makes scientists largely dependent on sequence rather than more general structure to infer the function of the protein at the present time. For this reason, data mining technology is introduced into this area to provide more efficient data processing and knowledge discovery approaches.

Unlike many data mining applications which lack available data, the protein structure determination problem and its interaction study, on the contrary, could utilize a vast amount of biologically relevant information on protein and its interaction, such as the protein data bank (PDB) [4], the structural classification of proteins (SCOP) databases [5], CATH databases [6], UniProt [7], and others. The difficulty of predicting protein structures, specially its 3D structures, and the interactions between proteins as shown in Figure 6.1, lies in the computational complexity of the data. Although a large number of approaches have been developed to determine the protein structures such as *ab initio* modeling [8], homology modeling [9] and threading [10], more efficient and reliable methods are still greatly needed.

In this chapter, we will introduce a state-of-the-art data mining technique, *graph mining*, which is good at defining and discovering interesting structural patterns in graphical data sets, and take advantage of its expressive power to

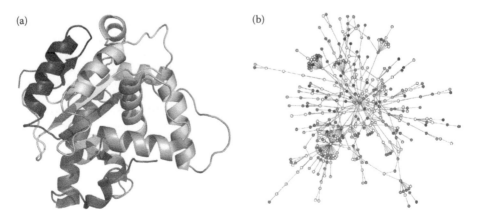

(a)   (b)

FIGURE 6.1: Graph examples for biological data. (a) Protein 3-D structure. (b) Protein–protein interaction network.

study protein structures, including protein structure prediction and comparison, and protein–protein interaction (PPI). The current graph pattern mining methods will be described, and typical algorithms will be presented, together with their applications in the protein structure analysis.

The rest of the chapter is organized as follows: Section 6.2 will give a brief introduction of the fundamental knowledge of protein, the publicly accessible protein data resources and the current research status of protein analysis; in Section 6.3, we will pay attention to one of the state-of-the-art data mining methods, graph mining; then Section 6.4 surveys several existing work for protein structure analysis using advanced graph mining methods in the recent decade; finally, in Section 6.5, a conclusion with potential further work will be summarized.

6.2 Protein Structures

6.2.1 What is protein?

The precise definition of proteins is any of a group of complex organic macromolecules that contain carbon (C), hydrogen (H), oxygen (O), nitrogen (N), and usually sulfur and are composed of one or more chains of amino acids. Over billions of years of life evolution, the intricate structure and the remarkable versatility of these macromolecules have made proteins one of the unsolved enigmas in biology. In fact, there is no other type of biological macromolecule which could possibly assume all of the functions that proteins have amassed.

Many biological tools have been developed to describe protein families, physical characteristics and cellular locations. As one of the fundamental components of all living cells, proteins include many substances, such as enzymes, hormones, and antibodies. The cell's growth and functions are promoted by the interactions between certain proteins and the cellular environment. In other words, the functions of proteins are defined by their roles in cells. However, the key to understanding a protein's functions is believed to be hidden in its structure. The distinctive structures of proteins allow for the placement of particular chemical groups in specific places in three-dimensional (3-D) space. This precision allows proteins to act as catalysts for an impressive variety of chemical reactions. Precise placement of chemical groups also allows proteins to play important structural, transport, and regulatory functions in organisms. The interactions with small molecules, as well as other proteins also expand these diversified functions of proteins.

It is still a big challenge to determine the 3-D structure of all the major protein families. However, based on the fast development of computer techniques and the great efforts of researchers who have studied proteins for decades, the protein structure can be roughly defined at four levels: primary, secondary, tertiary and quaternary structure.

6.2.1.1 The protein primary structure

The primary structure of a protein is the amino acid sequence of the peptide chains (see Figure 6.2), which is determined by the gene corresponding to the protein. Primary structure refers to the linear sequence of amino acid residues in a polypeptide chain. The amino acids are joined by peptide bonds on each side of the Cα carbon atom. Amino acids, as shown in Figure 6.3, are formed by four parts: an amino group (NH$_2$), a carboxyl groups (COOH), a hydrogen atom attached to a central alpha (α) carbon, and a side chain (or R group) attached to the α carbon [11]. A carboxyl group (CO) is a carbonyl group bonded to a hydroxyl group (OH), which can only appear at the end of a carbon chain because the carbon must make three bonds in addition to its connection to the R group. The R side chain distinguishes one amino acid from another and also confers the specific chemical properties of the amino acid. Twenty standard amino acids are incorporated into a protein based on the coded instructions and they are grouped into three classes (hydrophobic, polar, and charged) via the properties conferred by their side chains [11].

6.2.1.2 The protein secondary structure

The secondary structure, shown in Figure 6.4, is a representation from the primary amino acid sequence to secondary motifs, such as α helix, strands of β sheet, random coils and loops, which refers to the local folding pattern of the polypeptide chain [11].

The α helix is one of the most common forms in the secondary structure of proteins, which is created by a curving of the polypeptide backbone. Because the polypeptide backbone can be coiled in left or right directions, helices exhibit handedness. A helix with a rightward coil is known as a right-handed helix. Almost all helices observed in proteins are right-handed, as steric restrictions limit the ability of left-handed helices to form. Among the right-handed helices, the α helix, shown in Figure 6.4a, is by far the most prevalent. An α helix is distinguished by having a period of 3.6 residues per turn of the backbone coil. The structure of this helix is stabilized by hydrogenbonding

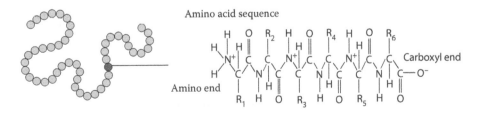

FIGURE 6.2: Protein primary structure.

interactions between the carbonyl oxygen of each residue and the amide proton of the residue, four residues ahead in the helix. Consequently, all possible backbone hydrogen bonds are satisfied within the α helix, with the exception of a few at each end of the helix, where a partner is not available. Other helices have also been observed in proteins, though much less frequently due to their less favorable geometry, such as 3_{10} helix and π helix.

Unlike helices, another important element in protein secondary structure is β sheets (see Figure 6.4b), which are formed from several individual β-strands by hydrogen bonds between adjacent polypeptide chains rather than within a single chain. β-strands represent an extended conformation of the polypeptide chain, where the Φ and Ψ angles are rotated approximately $180°$ with respect to each other. This arrangement produces a sheet that is pleated, with the residue side chains alternating positions on opposite sides of the sheet. Two configurations of β sheet are possible: parallel and antiparallel. In parallel sheets, the strands are arranged in the same direction with respect to their amino-terminal (N) and carboxy-terminal (C) ends. In antiparallel sheets, the strands alternate their amino and carboxy terminal ends, such that a given

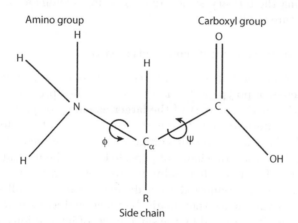

FIGURE 6.3: The structure of a prototypical amino acid.

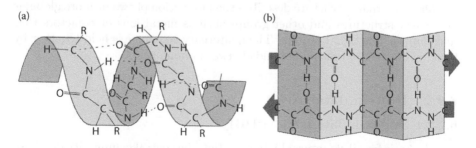

FIGURE 6.4: The protein secondary structures. (a) α helix. (b) β sheets.

strand interacts with strands in the opposite orientation. β sheets can also form in a mixed configuration, with both parallel and antiparallel sections, but this configuration is less common than the uniform types mentioned above.

6.2.1.3 The protein tertiary structure

The tertiary structure, shown in Figure 6.5a, is the 3-D structure of a single protein molecule; a spatial arrangement of the secondary structures. It is also the 3-D arrangement formed by packing secondary structure elements into globular domains. The tertiary structure describes how the secondary structure elements are arranged to form the overall 3D folding pattern. The tertiary structure is held together by hydrogen, ionic, and disulphide bonds between amino acids. It is this unique structure that gives a protein its specific function. The final 3D tertiary structure of a protein is commonly referred to as its fold. Many researchers agree that the tertiary protein structure problem can be considered as the protein folding problem [3], by which a linear polypeptide chain achieves its distinctive fold. However, the protein folding is a complex process that is not yet completely understood by biologists. Therefore, predicting the tertiary structure of proteins remains a big challenge for protein structure study.

6.2.1.4 The protein quaternary structure

The quaternary protein structure (see Figure 6.5a) is a complex of several protein molecules or polypeptide chains, usually called protein subunits in this context, which function as part of the larger assembly or protein complex. In biochemistry, quaternary structure is the arrangement of multiple folded protein molecules in a multisubunit complex. The quaternary structure is the interaction between several chains of peptide bones. The individual chains are called subunits. The individual subunits are not necessarily covalently connected, but might be connected by a disulfide bond. Not all proteins have quaternary structure, since they might be functional as monomers. The quaternary structure is stabilized by the same range of interactions as the tertiary structure. Quaternary structure involves this arrangement of several polypeptide chains.

The quaternary structure describes the interaction of two or more globular or tertiary structures and other groups such as metal ions or cofactors that make up the functional protein. The quaternary structure is held together by ionic, hydrogen, and disulfide bonds between amino acids.

6.2.2 Protein databases

6.2.2.1 Protein data bank (PDB)

The basis for all structural bioinformatics, the central community database for structural biology, is the PDB [4]. The PDB was established at Brookhaven

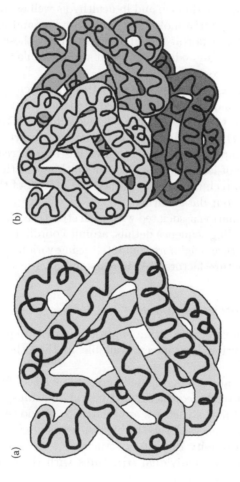

FIGURE 6.5: The protein (a) tertiary and (b) quaternary structures.

National Laboratory (BNL) (Bernstein et al., 1977) in 1971 as an archive for biological macromolecular crystal structure. PDB represents one of the earliest community-driven molecular biology data collections. Since October 1998, the PDB has been managed by the Research Collaboratory for Structural Bioinformatics (RCSB).

The PDB is the single worldwide depository of information about the 3-D structures of large biological molecules, including proteins and nucleic acids. These are the molecules of life that are found in all organisms including bacteria, yeast, plants, flies, and mice, and in healthy as well as diseased humans. Understanding the shape of a molecule helps to understand how it works.

The RCSB PDB is a portal for information about these molecules, and as such enables research and education about the molecular basis of life. The PDB is available at no cost to all users.

The PDB is growing constantly and statistics on content growth are available. In 2008, more than 50,830 determined structures were deposited from scientists all over the world (see Figure 6.6 taken from RCSB PDB website (http://www.rcsb.org/pdb/)). In addition to the huge growth in the numbers of structures, the complexity of the structures has also increased. Now there are several examples of large macromolecular machines in the database. Advances in science and technology have pushed the growth of the PDB as have changing attitudes about data sharing.

A variety of information associated with each structure is available through the RCSB PDB including sequence details, atomic coordinates, crystallization conditions, 3-D structure neighbors computed using various methods, derived geometric data, structure factors, 3-D images and a variety of links to other resources.

6.2.2.2   Other protein databases

As well as the primary information on protein structure provided by PDB, there are many additional resources available in the other databases (see Table 6.1).

The SCOP [5] database was originally created as a tool for understanding protein evolution through sequence-structure relationships and determining if new sequences and new structures are related to previously known protein structures. On a more general level, the highest levels of classification provide an overview of the diversity of protein structures. The specific lower levels are helpful for comparing individual structures with their evolutionary and structurally related counterparts.

The CATH database [6], is initially established as a domain-based database, contains sequence- and structure-based relationships between multidomain proteins and also includes families and superfamilies of multidomain proteins with links to their constituent domains. CATH is a hierarchical classification comprising four major levels: class, architecture, topology and homology. The protein class is determined by the secondary structure composition and packing using an automated approach; architecture describes the

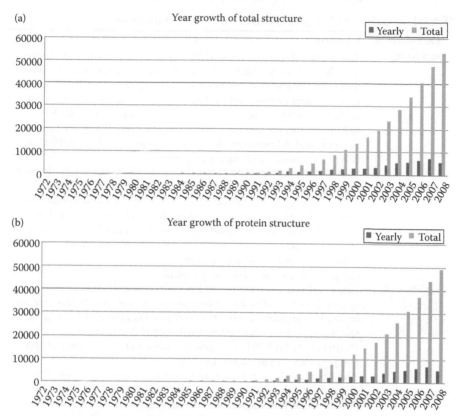

FIGURE 6.6: Yearly growth of the structures in PDB (1976–2008).

orientation of the secondary structures in 3-D space, regardless of their connectivity; while topology, both secondary structure orientation and connectivity between the secondary structures is taken into account in describing the fold of the protein; and homologous superfamilies are the groups of proteins according to whether there is sufficient evidence to support an evolutionary relationship.

Universal protein resource (UPR) [12] is the world's most comprehensive catalog of information on proteins. It is a central repository of protein sequence and function created by joining the information contained in UniProtKB/Swiss–Prot [13], UniProtKB/TrEMBL, and protein information resource (PIR). The UniProt [7] is comprised of three components, each optimized for different uses. The UniProt Knowledgebase (UniProtKB) is the central access point for extensive curated protein information, including function, classification, and cross-reference. The UniProt Reference Clusters (UniRef) databases combine closely related sequences into a single record to speed searches. The UniProt Archive (UniParc) is a comprehensive repository, reflecting the history of all protein sequences. The UniProt Metagenomic and

TABLE 6.1:   Other protein structure-based databases.

Databases	Content
SCOP [5]	The SCOP database, created by manual inspection and abetted by a battery of automated methods, aims to provide a detailed and comprehensive description of structural and evolutionary relationships between all proteins whose structure is known. As such, it provides a broad survey of all known protein folds, detailed information about the close relatives of any particular protein, and a framework for future research and classification (http://scop.mrc–lmb.cam.ac.uk/scop/).
CATH [6]	CATH database is a hierarchical classification of protein domain structures in the PDB, which clusters proteins at four major levels: class (C), architecture (A), topology (T) and homologous superfamily (H). Only crystal structures solved to resolution better than 4.0 Å are considered, together with NMR structures. Protein structures are classified using a combination of automated and manual procedures (http://www.cathdb.info/index.html).
UniProt [7]	Universal protein resource (UPR) is the world's most comprehensive catalog of information on proteins. It is a central repository of protein sequence and function created by joining the information contained in UniProtKB/Swiss–Prot, UniProtKB/TrEMBL, and protein information resource (PIR) (http://www.ebi.ac.uk/uniprot/).
Dali [14]	The Dali database is based on exhaustive, all-against-all 3-D structure comparison of protein structures in PDB. The classification and alignments are automatically maintained and regularly updated using the Dali search engine (http://ekhidna.biocenter.helsinki.fi/dali/start).
3Dee [15]	The database of protein domain definitions (3Dee) contains structural domain definitions for all protein chains in PDB, curated by the European Bioinformatics Institute (EBI) and the RCSB, that have 20 or more residues and are not theoretical models. These domains have been clustered by both sequence and structural similarity. The resulting families are stored in a hierarchy (http://www.compbio.dundee.ac.uk/3Dee/).
InterPro [16]	The InterPro database is a database of protein families, domains, repeats and sites in which identifiable features found in known proteins can be applied to new protein sequences (http://www.ebi.ac.uk/interpro/index.html).
UniProtKB /Swiss–Prot [13]	The UniProtKB/Swiss–Prot Protein Knowledgebase is an annotated protein sequence database established in 1986 by EBI, which provides a high level of annotation, a minimal level of redundancy and a high level of integration with other databases. Together with UniProtKB/TrEMBL, it consistutes the UniProt Knowledgebase, one component of the Universal Protein Resource (UniProt), a one-stop shop allowing easy access to all publicly available information about protein sequences (http://www.ebi.ac.uk/swissprot/).

Environmental Sequences (UniMES) database is a repository specifically developed for metagenomic and environmental data. As well as these most popular protein databases, there are many other open-access resources (see Table 6.1) available for the various use of research and education purposes.

6.2.3 Dimensions of protein analysis

6.2.3.1 Protein structure comparison

The 3-D structure of a protein can be determined by X-ray crystallography or nuclear magnetic resonance (NMR) spectroscopy [11]. As more and more protein structures have been determined and deposited in various protein structure databases, the prediction of protein structure by computer algorithms is becoming more feasible. When proteins of unknown structure are similar to a protein of known structure at the sequence level, the 3-D structure of the proteins can be predicted. The stronger the similarity and identity, the more similar are the 3-D folds and other structural features of the proteins. However, it should be noted that proteins with no apparent sequence similarity could also have very similar structure, and that the 3-D structure of protein is much more highly conserved than the amino acid sequence. By tracking their structural similarities, very distant evolutionary relationships between proteins may be inferred.

In bioinformatics, structural comparison of proteins is useful in several domains. For example, as protein function is intrinsically tied to a protein's structure, identifying structural similarity between a protein and other proteins whose function is known can allow the prediction of that protein's function. Over the last decade, a number of techniques for structurally comparing proteins have been developed. However, none have proved adequate across a range of applications.

6.2.3.2 Protein structure prediction

To predict the structure of a new protein it is usual to check the presence of a set of certain patterns or motifs, for example, some specific amino acid patterns or profiles with significant special structures. This type of prediction is also called comparative modeling [17], and it requires a clear sequence relationship between the target structure and the known structures. However, protein fold prediction from an amino acid sequence is still a distant goal, and most current algorithms aim at predicting only the secondary structures, such as helices, strands, and loops/coils and few are working on 3-D structures. The prediction of the secondary structure is an essential intermediate step on the way to predicting the full 3-D structure of a protein. If the secondary structure of a protein is known, it is possible to derive a comparatively small number of possible tertiary structures using knowledge about the ways that the secondary structural methods, are organized.

As we mentioned in the previous section, to understand the function of a certain protein, it is important to know its structures, especially the 3-D structure that is directly determined by the sequence of amino acids in the molecule. A major goal in bioinformatics and structural molecular biology is to understand the relationship between the amino acid sequence and the 3-D structure in protein, and to predict the fold based on the amino acid sequence alone.

6.2.3.3   Protein–protein interaction (PPI) analysis

PPI are the central part of systems biology because they are directly involved in regulation of cellular processes [18]. There are various types of protein complexes under different conditions and situations, which can be transient or permanent, obligate or nonobligate, homo- or hetero-oligomeric [19]. Although many algorithms have been developed to discover the protein interacting partners, one of the most reliable and informative methods introduced by Zhang et al. [18], is to solve the crystal structure of the protein complex. Due to the complexity of the computational process, the existing knowledge of PPI was from high throughput experimental approaches and from theoretical prediction methods.

The predictive methods for protein interaction is a combination of sequence conservation and structural analysis [18]. For sequence conservation cases, the certain residues are conserved across the family, but in the subfamily, there are different residue types which are conserved. Due to the fact that if each subfamily interacts with a different partner, then it has a somewhat different nature of contacts, this sequence conservation information can be involved in the consideration of protein interaction analysis. Besides, there is another case in which it is appropriate to use sequence conservation, to correlate interprotein mutations. Correlated mutations are particularly useful when the structures of each of the proteins exist in the apo form. On the other hand, Interactions between proteins can also be indirectly inferred from phylogenetic progiles, gene fusion, and gene neighborhoods. If two proteins are both present or both missing from the genomes of different species, they are likely to be involved in functional interactions. Species that are missing two individual genes may have them as a single fused gene, which can be found in whole genome comparison. The length of the branches and the structure of phylogenetic trees offer further help.

6.3   Graph Mining

Graph mining has a high potential to provide practical applications because the structured graphical data widely occurs in various practical fields including biology, chemistry, material science and communication networking [20]. The main objective of graph mining is to discover the potential patterns or principles in an efficient way by mining topological substructures embedded in graphical data. Although many successful applications are achieved by using graph mining technology, graph mining is still in its infancy. Some basic graph pattern mining and search problems have not been systematically examined; they are yet to be solved and improved by the great efforts of researchers in this area.

6.3.1 Problem definition of mining graphical data

Before having a close look at the graph mining algorithms, some basic knowledge of graphs and their mining process is essential to our understanding. In this section, we will start with some general definitions that are used in graphical data mining. This knowledge helps us to understand the existing graph mining problems, which need to be solved and the tasks which need to be carried out. Following that are the common measurements and the problem's complexity that we need to consider for mining evaluation.

6.3.1.1 Notation

In the common sense of a mathematical term, a graph can be represented by three elements, $G(V, E, f)$, where V is a set of vertices in the graph, E is a set of edges connecting certain vertex pairs in V, and f is a mapping function, $f : E \rightarrow V \times V$, which indicates the relationships between the vertices in V and the corresponding edges in E. For example, take the first graph in Figure 6.7a. In this graph, it contains five vertices V: $\{v_1, v_2, v_3, v_4, v_5\}$; seven edges construct the edge set E : $\{e_1, e_2, e_3, e_4, e_5, e_6, e_7\}$; and these edges can be represented using the mapping function $f(e_i) = (v_x, v_y)$, such as, $f(e_1) = (v_1, v_2), f(e_5) = (v_1, v_3)$ and $f(e_7) = (v_1, v_5)$.

The process of mining the graphical data normally starts from finding out a set of frequent subgraphs that are used as the graph patterns for further knowledge or principle discovery. Once we have an idea of what is a graph, the next problem will be what is a subgraph of this graph?

A graph g is a subgraph of a graph G if there exists a subgraph isomorphism from g to G, written $g \subseteq G$. Here it brings another concept of subgraph isomorphism that is a very important point in *graph-based theory*, and Yan [20] gave us a very good description for it:

A subgraph isomorphism is an injective function h: $V(g) \rightarrow V(G)$, such that (1) $\forall u_i \in V(g), h(u_i) \in V(G)$ and $f(u_i) = f'(h(u))$, and (2) $\forall(u, v) \in E(g), (h(u), h(v)) \in E(G)$ and $f(u, v) = f'(h(u), h(v))$, where f and f' are the label function of G_s and G, respectively. h is called an embedding (image) of g in G.

Once a set of subgraphs is generated, can we use the whole set as graph patterns to continue the mining process directly? The answer to this question

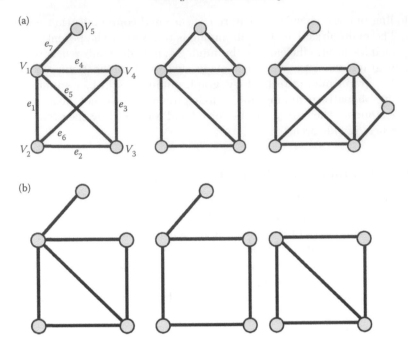

FIGURE 6.7: Graphs and frequent subgraphs. (a) Graph Examples in Graphical Data Set. (b) Frequent Graph Patterns.

is NO. There is another preprocessing phase needing be done, that is, the frequent subgraph selection processing. To do this, the frequency of each subgraph candidate needs to be calculated first. The subgraphs then are reordered by their frequencies and only the ones with higher rank will be selected. These frequent subgraphs are the important graph patterns for graph matching and searching tasks in further processing stages. In Figure 6.7, there are three examples in a graphical data set (Figure 6.7a), and three of frequent subgraphs (Figure 6.7b) depicted from these examples.

It is shown that the main work we do for graphical data mining can be summarized into two tasks: (1) generate a set of subgraphs from the given graphical data set; and (2) test and select the frequent subgraphs as the graph patterns for further processing. However, the method to test and select these frequent subgraph patterns has not been touched on. In the following section, our focus will concentrate on this point; the measurement of the frequency will be discussed first.

6.3.1.2   Mining measures and computational complexity

As a partial area of data mining, graph mining also shares similar measurement methods that are widely used in data mining, such as information entropy, information gain, gini-index, minimum description length (MDL) and others [21,22]. Among those well-developed approaches, a method to compare the support of each item from the Basket analysis [23] becomes the most popular one in the current decade. Its basic idea can be represented as follows:

Given a labeled graph data set $D = \{G_1, G_2, \ldots, G_n\}$,

$$\sup(g) = \frac{|D_g|}{|D|} = \frac{number\ of\ graphs\ including\ subgraph\ in\ D}{total\ number of\ graph\ in\ D}$$

where D_g is the number of graphs in D where g is a subgraph; $\sup(g)$ is the support data set of g; $|D_g|$ is called the (absolute) support of g; and $|Dg|$ is the total number of graphs in D. A subgraph g can be defined as a frequent subgraph if and only if it support is no less than a minimum support threshold, δ, by testing its frequency $\sup(g) = |D_g| / |D| \geq \delta$ [20].

It seems that there is no such difficulty in calculating the support of each subgraph in the graphical data set and selecting the frequent subgraph patterns. However, in fact, graph mining has a big challenge for its unknown computational complexity that has not yet found a satisfactory solution. This originates from the problem of the graph isomorphism. The problem of deciding whether two graphs have identical topological structure or whether one graph is a subgraph of another one is still NP-Complete in graph mining, and even in mathematics [24,25].

6.3.2 Current graph mining methods

In the early 1990s, Holder et al. proposed a greedy graph mining algorithm called Subdue [26], which is based on recursively finding a subgraph that provides the best compression based on the MDL principle. However, because of its recursive process, Subdue NP-Complete to discover all frequent patterns, searching the entire space of subgraphs. Unlike the greedy algorithm of Subdue, researchers tried to induce the subgraphs by mining the vertex sets only. There are two basic approaches to the frequent subgraph mining problem: (1) a priori-based approach and (2) pattern-growth approach.

6.3.2.1 A priori-based approach

A priori-based frequent subgraph mining algorithms share similar characteristics with a priori-based frequent itemset mining algorithms. The search for frequent graphs starts with graphs of small size, and proceeds in a bottom-up manner. At each iteration, the size of newly discovered frequent subgraphs is increased by one. These new subgraphs are first generated by joining two similar but slightly different frequent subgraphs that were discovered already. The frequency of the newly formed graphs is then checked. Two typical a priori-based frequent subgraph mining methods have already been discovered: AGM algorithm by Inokuchi et al. [27] and FSG by Kuramochi and Karypis [28].

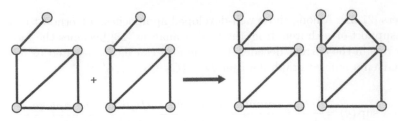

FIGURE 6.8: Frequent subgraph candidate generation in AGM.

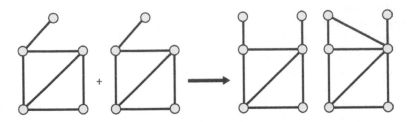

FIGURE 6.9: Frequent subgraph candidate generation in FSG.

The AGM algorithm generates the subgraph candidates by increasing the number of the vertices in each iteration. This algorithm starts from the frequent subgraph candidates with one single vertex, after each searching in bottom-up manner, an extra vertex will be added to each candidate. As shown in Figure 6.8, edges are added between the extra vertex and the vertices in the existing frequent subgraphs.

Unlike AGM, FSG is an edge-based candidate generation algorithm that each new generated subgraph candidate has one more edge than the existing ones. In each iteration, FSG generates a new candidate by merging two subgraphs with k edges if and only if these two subgraphs have the same subgraph with $k-1$ edges. Figure 6.9 shows an example of edge-based candidate generation in FSG algorithm.

6.3.2.2 Pattern-growth approach

Unlike a priori-based approach, the pattern-growth mining algorithm extends a frequent graph by adding a new edge in every possible position. A potential problem with the edge extension is that the same graph can be discovered many times. To solve this problem, Yan et al. [29] introduced a right-most extension technique, gSpan, where the only extensions take place on the right-most path. A right-most path is the straight path from the starting vertex v_o to the last vertex v_n, according to a depth-first search (DFS) on the graph.

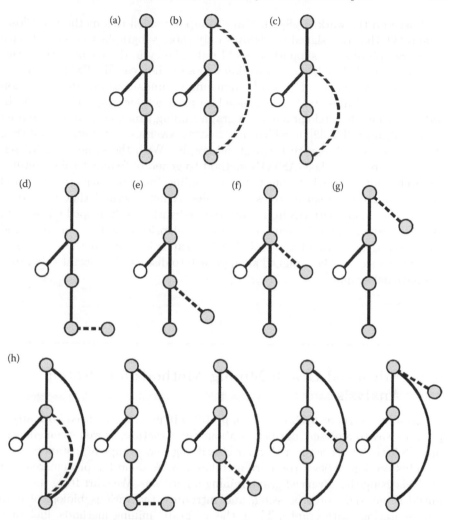

FIGURE 6.10: Right-most extension.

Figure 6.10 shows an example of right-most extension. Figure 6.10a is the existing frequent graph (parent) in which the grey vertices are linked together by the right-most path. Figure 6.10b through g are all the possible new graph candidates (children) that were generated based on Figure 6.10a by adding one edge at each position on the right-most path. Following the same method, Figure 6.10h contains all the further right-most extended graph candidates (children) for Figure 6.10b. By applying this right-most extension technique in a DFS manner, gSpan successfully derives a complete searching of the entire space of subgraphs and achieves an efficient computational performance over the earlier a priori-based approaches, such as AGM and FSG.

Based on the work of gSpan, Yan's group extended this method to Close-Graph [30] that is designed to discover only those subgraphs that do not have a supergraph of same support to avoid redundancy in the output. Moreover, another group led by Huan was working on a method called FFSM [31] to improve the performance of gSpan by reducing redundant candidate generation through a vertical search scheme based on an algebraic graph framework. In order to avoid the circle problems in graph mining, Huan et al. also presented an algorithm, SPIN [32], in which they first discover a set of subtrees and then extended these subtrees to frequent subgraphs. With the same idea, Nijssen et al. [33] proposed the GASTON method to generate frequent substructures hierarchically by starting from paths, extending frequent paths to trees, and further extending frequent trees to graphs. A new method that is used to mine approximate patterns from a massive network was developed Chen et al. [34]. Hintsanen and Toivonen introduced an algorithm that was used to find reliable subgraphs from large probabilistic graphs in 2008. At the same time, Papdopoulos et al. [35] designed an approach to discover important subgraphs in relational graphs.

6.4   Advanced Graph Mining Methods for Protein Analysis

To solve a structural classification problem in practice, we need both an expressive way to represent our beliefs about the structure, as well as an efficient probabilistic inference algorithm for classifying new groups of instances.

This section reviews the current related work done for protein relevant analysis using the advanced graph mining algorithms. We start from the early initial stage, data preprocessing, and introduce the various biological data representations with graphs. Then, the subgraph mining methods, including the algorithms from the unilateral graph mining aspect, as well as the ones with the consideration of the biological meaning, are discussed in detail. We also present the applications of graph mining techniques in the wide area of protein studies in the following context.

6.4.1   Generate the biological graph data sets

In the core of many graph-related applications, a common and critical problem at the first stage is *how to represent the raw collected data by graphs*.

There are two types of the biological data available for protein studies. One type is the biological network that represents the interaction or the known relationship between the proteins, for example the protein relevant motifs and the PPI networks (see Figure 6.1b). Huan et al. [31] used the spatial motifs that represented the sequence and proximity characteristics of a protein's amino acid residues (the protein primary structure) in the SCOP database. The network motifs have been used to represent PPI networks elsewhere [36–40].

The other type of protein data used in bioinformatics is the raw sequence or structural information available from the PDB, SCOP, CATH, or the other public online protein resources (see Table 6.1), which need further processing to convert them into graph-based formats. Most of the data sets used for protein structure comparison, prediction, clustering or classification are data of this type. Because of this, there is a need to convert these data into some graph-based structural representation, as well as to keep their original biological structural characteristics.

Graphs, in some cases even very large ones, are simple objects that are at least composed by vertices and edges. The vertices represent the parts of the system and the edges geometric and/or functional relationships between parts. The simplest way to convert protein structures into graphs is to represent the proteins based on their geometrical structures. For example, Wangikar et al. [41] represented the protein structure as a labeled and weighted undirected graph, in which the vertices are the functional atoms of the amino acid side chains, and the edges are between vertices that are within an interacting distance. Only one functional atom is considered per amino acid residue. Similar methods were used for the projects with variety research focuses [39,42,43]. Deng et al. [42] represented oxygen atoms and the pairs of these oxygen atoms apart within an O–O cutoff distance as the vertices and the edges in graphs to predict calcium-binding sites in proteins; Canutescu et al. [43] denoted vertices by the side-chains for rapid protein sidechain prediction, and any two residues that have rotamers with nonzero interaction energies were connected by an edge in their graphs. For protein active sites analysis, a set of predefined pseudocenters (vertices) were connected by the edges if their Euclidean distance is below 12.0 Å, to represent a protein binding pocket by Weskamp et al.[39].

Other work focused on 3-D structures of the proteins and defined the domains as the nodes in the graph data sets, which are connected by the domain combinations [44,45]. As described in the studies of Wuchty [44] and Ye et al. [45], a domain graph $G_D = (V_D, E_D)$) is formally defined by a vertex set V_D consisting of all domains found within proteins. Two domains are regarded as being adjacent if they occur together in one protein at least once. An undirected edge connecting these two vertices indicates this relationship. Such connections define the edges set E_D. In this graph, the degree k of a vertex is the number of other vertices to which it is linked. Two vertices are linked together if and only if both domains are present in at least one protein. Figure 6.11 shows an example of the domain graph in which each vertex represents a domain linked with the domain combinations in the proteins.

Unlike most cases that define the edges between any two vertices only by the Euclidean distances, Huan et al. [32] computed the edges in their protein graph using three different approaches. Except the physical interaction

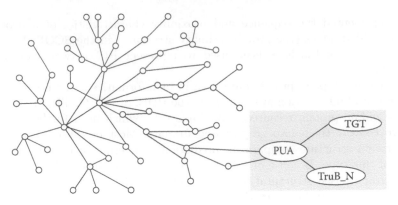

FIGURE 6.11: A domain graph.

distance used in the first method, they connect the edges using Delaunay tessellation that two vertices are linked if and only if one vertex can find an empty sphere whose boundary contains those two vertices. With the consideration of the difficulties of protein measurement, imprecision and atomic motions, an almost-Delaunay edge was denoted as the third graph representation.

6.4.2 Mine the frequent subgraphs for pattern discovery

To compare protein structures is to use graph mining theory to define the protein structure and make comparisons between pairs of graphs or subgraphs. In this section, we state the second problem of using the graph mining technique for protein structure analysis: *how to define and generate the set of the frequent subgraphs for the graph pattern discovery?*

The first common method is to find the frequent subgraphs from the given graph data set directly by using the existing graph pattern discovery methods. In this way, the process only focuses on how to find out the set of the frequent subgraphs from the graph database. Therefore, the biological problem of protein structure analysis is simplified to a theoretical graph mining problem without considering the biological meaning of the protein structure. In the current decade, many existing graph mining algorithms have been proposed to efficiently generate frequent subgraphs based on the large graph data sets.

Unlike the above unilateral graph mining methods for subgraph pattern discovery, some models combined both graph mining algorithms with the biological characteristics of the protein structures. In 2006, based on the suggestion of Wagner [46] that orthologous proteins share common interaction, Koyutürk and coworkers [38] tried to use ortholog contraction method to simplify the graph mining problem for PPI, which maintains not only the correctness by preserving the underlying frequent subgraphs in the graph database, but also the biological relevance and interpretability of the discovered patterns.

6.4.3 Protein structure and protein–protein interaction (PPI) network analysis by graph mining methods

Once the methods are well-developed, the next problem will be *how to apply the graph mining methods to the practical projects for protein structure prediction and PPI network analysis.* With the significant distribution of efficient computational cost reduction and high quality classification and prediction, graph mining techniques have appeared in many biological data mining applications [32,37,39,47], and have achieved initial success in the comparison and prediction of protein structures.

Huan et al. [32] successfully applied a frequent graph mining technique to study protein structural families. After generating the candidate subgraphs using the typical graph pattern mining method, only the best subgraphs, with low mutual information and strong correlation with their own subgraphs, were selected as the most frequent patterns in the protein structure graphs. The resulting small set of the coherent subgraphs provided high performance protein classification. The problem of characterizing functional protein families was solved by Weskamp et al. [39] by detecting conserved structural patterns (frequent subgraphs) in a given family along with discriminative patterns between different but related families. They started with identifying the frequent subgraphs in pairwise cases, then extended the approach to multiple graph comparison.

In the work of dissecting genome-wide protein–protein interactions (PPIs) in meso-scale network motifs, Chen et al. [37] developed a graph-based mining algorithm called NeMoFinder to discover unique and repeated meso-scale network motifs in a large PPI network. They generated a set of the repeated subgraphs by selecting the size k trees with higher frequency. These repeated size k trees were used to partition the network into a set of subgraphs so that the problems of finding the unique and frequent subgraphs can be efficiently reduced. Koyuturk et al. [47] proposed an innovative new graph mining method, MULE, to detect frequent subgraphs in biological networks: they observed considerably large frequent subpathways in metabolic networks and PPI networks within seconds.

6.5 Conclusion and Future Work

In this chapter, we reviewed the current progress of protein structure prediction and PPI detection. The use of the graph mining methods to study complicated protein structures and protein interaction problems, provided an intuitive picture and helped to gain useful insights into the potential development and application of advanced data mining techniques in biological research. However, the studies of proteomics is in its infancy, and some important limitations still remain.

First, the rapid increase in the number of known protein sequences hastens the development of the efficient subgraph mining and searching tools with less computational cost. And the big problem before applying any graph mining methods to the protein structure data is how to represent the original data with structured graphs. The graphs are expected to have less complex structures but more information on the biological characteristics. With different research focus, different graphs will be generated. Therefore, more work is needed to find the solution of a more efficient structure representation of the proteins.

Second, the graph mining itself is an expensive computational problem since the subgraph isomorphism is NP-complete. There is no guarantee that can be made by any researcher to give a set of frequent subgraphs that are the most effective ones. And the measurement of the effects has not been clearly defined yet. These questions are still open in the data mining research.

Another interesting topic in mining graph-based protein structures can be how to do the similarity determination by mining subgraphs among different organisms. Some proteins in two organisms may show similar structures and functions, and knowledge can help us to further explore the mystery of life.

References

[1] Jacq, B. Protein function from the perspective of molecular interaction and genetic networks. *Bioinformatics,* 2, 2001, 38–50.

[2] Hunter, L., ed. *Artificial Intelligence and Molecular Biology.* The MIT Press Classics Series and AAAI Press, Cambridge, Massachusetts, USA, 1993.

[3] Pevsner, J. *Bioinformatics and Functional Genomics.* Wiley-Liss, Hoboken, NJ, 2003.

[4] Berman, H.M., Westbrook, J., Feng, Z., Gilliland, G., Bhat, T.N., Weissig, H., Shindyalov, I.N., Bourne, P.E. The protein data bank. *Nucleic Acids Research*, 28, 2000, 235–242.

[5] Murzin, A.G., Brenner, S.E., Hubbard, T., Chothia, C. Scop:a structural classification of proteins database for the investigation of sequences and structures. *Journal of Molecular Biology,* 247, 1995, 536–540.

[6] Pearl, F., Bennett, C., Bray, J., Harrison, A., Martin, N., Shepherd, A., Sillitoe, I., Thornton, J.M., Orengo, C. The CATH: an extended protein family resource for structural and functional genomics. *Nucleic Acids Research*, 31, 2003, 452–455.

[7] Apweiler, R., Bairoch, A., Wu, C.H., Barker, W.C., Boeckmann, B., Ferro, S., Gasteiger, E., Huang, H., Lopez, R., Magrane, M., Martin, M.J., Natale, D.A., ODonovan, C., Redaschi, N., Yeh., L.S.L. UniProt: The universal protein knowledgebase. *Nucleic Acids Research*, 32, 2004, D115–D119.

[8] Simons, K.T., Strauss, C., Baker, D. Prospects for ab initio protein structural genomics. *Journal of Molecular Biology*, 306, 2001, 1191–1199.

[9] Fiser, A., Sali, A. Modeller: generation and refinement of homology models. *Methods in Enzymology*, 374, 2003, 461–491.

[10] Kelley, L.A., MacCallum, R.M., Sternberg, M.J.E. Enhanced genome annotation using structural profiles in the program 3d–pssm. *Journal of Molecular Biology*, 299, 2000, 501–522.

[11] Bourne, P.E., Weissig, H., eds. *Structural Bioinformatics*. Wiley-Liss, Hoboken, NJ, 2003.

[12] Consortium, T.U. The universal protein resource. *Nucleic Acids Research*, 36, 2008, 190–195.

[13] Bairoch, A., Apweiler, R. The SWISS–PROT protein sequence data bank and its new supplement TREMBL. *Nucleic Acids Research*, 24, 1996, 21–25.

[14] Holm, L., Sander, C. Mapping the protein universe. *Science*, 273, 1996, 595–603.

[15] Siddiqui, A.S., Dengler, U., Barton, G.J. 3Dee: a database of protein structural domains. *Bioinformatics*, 17, 2001, 200–201.

[16] Mulder, N.J., Apweiler, R., Attwood, T.K., Bairoch, A., Bateman, A., Binns, D., Bork, P., et al. New developments in the InterPro database. *Nucleic Acids Research*, 35, 2007, D224–D228.

[17] Liew, A.W.C., Yan, H., Yang, M. Data mining for bioinformatics. In: *Bioinformatic Technologies*, Y.P.P. Chen (ed.). Springer-Verlag Berlin Heidelberg Germany, 2005, 63–106.

[18] Zhang, Q., Veretnik, S., Bourne, P.E. Overview of structural bioinformatics. In: *Bioinformatic Technologies*, Y.P.P. Chen (ed.). Springer-Verlag Berlin Heidelberg Germany, 2005, 15–44.

[19] Nooren, I.M.A., Thornton, J.M. Diversity of protein–protein interaction. *EMBO Journal*, 22, 2003, 3486–3492.

[20] Yan, X. Mining, indexing and similarity search in large graph data sets. PhD Dissertation, University of Illionis, Urbana-Champaign, IL, 2006.

[21] Cook, D.J., Holder, L.B. Substructure discovery using minimum description length and back-ground knowledge. *Journal of Artificial Intelligence Research*, 1, 1994, 231–255.

[22] Yoshida, K., Motoda, H., Indurkhya, N. Graph-based induction as a unified learning frame-work. *Journal of Applied Intelligence*, 4, 1994, 297–328.

[23] Agrawal, R., Srikant, R. Fast algorithms for mining association rules. In: *Proceedings of 20th International Conference on Very Large Data Bases (VLDB'94)*, Santiago de Chile, Chile, 1994, 487–499.

[24] Cook, S. The complexity of theorem-proving procedures. In: *Proceeding of 3rd ACM Symposium on Theory of Computing (STOC'71)*, New York, NY, USA, 1971, 151–158.

[25] Washio, T., Motoda, H. State of the art of graph-based data mining. *SIGKDD Explorations Newsletter,* 5, 2003, 59–68.

[26] Holder, L.B., Cook, D.J., Djoko, S. Substructure discovery in the subdue system. In: *Proceeding of the AAAI'94 Workshop Knowledge Discovery in Databases (KDD'94)*, Seattle, WA, 1994, 169–180.

[27] Inokuchi, A., Washio, T., Motoda, H. An a priori-based algorithm for mining frequent substructures from graph data. In: *Proceedings of 2000 European Principles and Practice of Knowledge Discovery in Database (PKDD'00)*, Lyon, France, 2000, 13–23.

[28] Kuramochi, M., Karypis, G. Frequent subgraph discovery. In: *Proceedings of 2001 International Conference on Data Mining (ICDM'01)*, San Jose, CA, 2001, 313–320.

[29] Yan, X., Han, J. gSpan: Graph-based substructure pattern mining. In: *Proceedings of 2002 International Conference on Data Mining (ICDM'02)*, Maebashi, Japan, 2002, 721–724.

[30] Yan, X., Han, J. CloseGraph: mining closed frequent graph patterns. In: *Proceedings of 2003 International Conference of Knowledge Discovery and Data Mining (KDD'03)*, Washington, DC, 2003, 286–295.

[31] Huan, J., Wang, W., Bandyopadhyay, D., Snoeyink, J., Prins, J., Tropsha, A. Mining spatial motifs from protein structure graphs. In: *Proceedings of 8th Annual International Conference on Research in Computational Molecular Biology (RECOMB'04)*, San Diego, California, USA, 2004, 308–315.

[32] Huan, J., Wang, W., Bankyopadhyay, D., Snoeyink, J., Prins, J., Tropsha, A. Mining protein family specific residue packing patterns from protein

structure graphs. In: *Proceeding of 8th International Conference of Research in Computational Molecular Biology (RECOMB'04)*, San Diego, California, USA, 2004, 308–315.

[33] Nijssen, S., Kok, J.N. A quickstart in frequent structure mining can make a difference. In: *Proceedings of 10th ACM SIGKDD International Conference on Knowledge Discovery and Data Mining (KDD'04)*, Seattle, WA, USA, 2004, 647–652.

[34] Chen, C., Yan, X., Zhu, F., Han, J. gApprox: mining frequent approximate patterns from a massive network. In: *Proceeding of 7th IEEE International Conference on Data Mining (ICDM'07)*, Omaha, NE, USA, 2007, 445–450.

[35] Papadopoulos, A.N., Lyritsis, A., Manolopoulos, Y. SkyGraph: an algorithm for important subgraph discovery in relational graphs. *Data Mining and Knowledge Discovery*, 17, 2008, 57–76.

[36] Li, X.L., Tan, S.H., Foo, C.S., Ng, S.K. Interaction graph mining for protein complexes using local clique merging. *Genome Informatics*, 16, 2005, 260–269.

[37] Chen, J., Hsu, W., Lee, M.L., Ng, S.K. NeMoFinder: dissecting genome-wide protein-protein interactions with meso-scale network motifs. In: *Proceedings of the 12th ACM SIGKDD International Conference on Knowledge Discovery and Data Mining (KDD06)*, Philadelphia, PA, 2006, 106–115.

[38] Koyuturk, M., Kim, Y., Subramaniam, S., Szpankowski, W., Grama, A. Detecting conserved interaction patterns in biological networks. *Journal of Computational Biology*, 13, 2006, 1299–1322.

[39] Weskamp, N., Hullermeier, E., kuhn, K., Klebe, G. Multiple graph alignment for the structural analysis of protein active sites. *IEEE/ACM Transactions on Computational Biology and Bioinformatics*, 4, 2007, 310–320.

[40] McGarry, K., Chambers, J., Oatley, G. A multi-layered approach to protein data integration for diabetes research. *Artificial Intelligence in Medicine*, 41, 2007, 129–143.

[41] Wangikar, P.P., Tendulkar, A.V., Ramya, S., Mali, D.N., Sarawagi, S. Functional sites in protein families uncovered via an objective and automated graph theoretic approach. *Journal of Molecular Biology*, 326, 2003, 955–978.

[42] Deng, H., Chen, G., Yang, W., Yang, J.J. Predicting calcium-binding sites in proteins a graph theory and geometry approach. *Proteins: Structure, Function, and Genetics*, 64, 2006, 34–42.

[43] Canutescu, A.A., Shelenkov, A.A., Roland L. Dunbrack, J. A graph-theory algorithm for rapid protein side-chain prediction. *Protein Science*, 12, 2003, 2001–2014.

[44] Wuchty, S. Scale-free behavior in protein domain networks. *Molecular Biology and Evolution*, 18, 2001, 1694–1702.

[45] Ye, Y., Godzik, A. Comparative analysis of protein domain organization. *Genome Research*, 14, 2004, 343–353.

[46] Wagner, A. The yeast protein interaction network evolves rapidly and contains few redundant duplicate genes. *Molecular Biology and Evolution*, 18, 2001, 1283–1292.

[47] Koyuturk, M., Grama, A., Szpankowski, W. An efficient algorithm for detecting frequent subgraphs in biological networks. *Bioinformatics*, 20, 2004, i200–i207.

[48] Bernstein, F. C., Koetzle. T. F., Williams, G. J., Meyer, E. F. J., Brice, M. D., Rodgers, J. R., Kennard, O., Shimanouchi, T. and Tasumi, M. The protein data bank: a computer-based archival file for macromolecular structures. *Journal of Molecular Biology*, 112, 1977, 535–542.

Chapter 7

Predicting Local Structure and Function of Proteins

Huzefa Rangwala

George Mason University

George Karypis

University of Minnesota

7.1   Introduction

Proteins have a vast influence on the molecular machinery of life. Stunningly complex networks of proteins perform innumerable functions in every living cell. Knowing the function and structure of proteins is crucial for the development of improved drugs, better crops, and even synthetic biofuels. As such, knowledge of protein structure and function leads to crucial advances in life sciences and biology.

With rapid strides made in large scale sequencing technologies, we have seen an exponential increase in the available protein sequence information. Protein structures are primarily determined using X-ray crystallography or NMR spectroscopy, but these methods are time consuming, expensive and not feasible for all proteins. The experimental approaches to determine protein function (e.g., gene knockout, targeted mutation, and inhibitions of gene expression studies) are low throughput in nature [21]. As such, our ability to produce sequence information far out-paces the rate at which we can produce structural and functional information.

Over the past two decades several computational methods have been developed to characterize the structural and functional aspects of proteins from sequence information. Function prediction is generally approached by using inheritance through homology [21], i.e., proteins with similar sequences (common evolutionary ancestry) frequently carry out similar functions. It has also been shown that a stronger correlation exists between structure conservation and function i.e., structure implies function, and a higher correlation exists between sequence conservation and structure i.e., sequence implies structure (sequence→ structure→ function).

Also certain parts of the protein structure may be conserved and interact with other biomolecules (e.g., proteins, DNA, RNA, and small molecules) and perform a particular function due to such interactions. Thus, there is an emphasis to determine the local structural and functional characteristic of proteins i.e., determine the structure/function of every residue of a protein sequence. The task of assigning every residue with a discrete class label or continuous value is defined as a *residue annotation* problem. These structural and functional properties are local in nature i.e., they depend on the neighboring residues and correlate with the sequence information. As such, these problems are also called *local structure and function prediction problems.*

Examples of structural annotation problems include the secondary structure prediction [16], local structure prediction [7], and contact order prediction [31]. Examples of function property annotation include prediction of interacting residues [24] (e.g., DNA-binding residues, and ligand-binding residues), solvent accessible surface area estimation [26], and disorder prediction [5] (reviewed in further detail in Section 7.3). Support vector machines (SVMs) [34] along with other machine learning tools have been extensively used to successfully predict the residue-wise structural or functional properties of proteins.

This chapter aims to provide a broad overview of several residue-wise structure and function prediction problems, along with a standard machine learning based technique to solve the various defined problems. This chapter will also provide case studies on four prediction problems: (i) disorder prediction, (ii) residue-wise contact order prediction, (iii) protein–DNA interaction site prediction, and (iv) 16-state local structure alphabet prediction.

7.2 Supervised Learning Methods

Supervised learning is a common machine learning approach that attempts to generate a function for mapping inputs to desired outputs by observing labelled training data. Supervised learning plays a critical role in several bioinformatics applications [34].

Formally, given an input domain $\mathcal{X}$ and output domain $\mathcal{Y}$, supervised learning methods learn a function that maps every member of $\mathcal{X}$ to a member of domain $\mathcal{Y}$. The learning method uses the training data given by object pairs, $(X_1, Y_1) \cdots (X_n, Y_n)$ to learn a function $f : \mathcal{X} \rightarrow \mathcal{Y}$ mapping each element $X_i \in \mathcal{X}$ to its corresponding output $Y_i \in \mathcal{Y}$. The elements of domain $\mathcal{X}$ are generally provided to the supervised learning algorithm in the form of vectors. During the prediction or inference phase, given the learned function f, the algorithm attempts to predict the output $f(X_u)$ for the unknown input example X_u. The output domain $\mathcal{Y}$ of the function f can be a discrete label or a continuous value. The task of predicting a discrete class label is called as classification, and the task of predicting the continuous value is called regression.

For the supervised learning methods to produce good classification or regression models, it is assumed that there exists an underlying probability distribution $\mathcal{P}(X, Y)$ over $\mathcal{X} \times \mathcal{Y}$. This distribution is assumed to be the same for the training and test instances, but is unknown. The training and test instances assumed to be independently and identically drawn from the distribution $\mathcal{P}(X, Y)$ (i.i.d assumption).

Supervised learning methods can be categorized into types: (i) parametric learners and (ii) distribution free learners. Parametric learners explicitly model the joint distribution $\mathcal{P}(X, Y)$ or the conditional distribution $\mathcal{P}(Y|X)$ for all $\mathcal{X}$. Bayesian and Markovian learners are examples of parametric learning methods. Distribution-free learning methods do not learn the distribution $\mathcal{P}(X, Y)$, but learn a function in a specific hypothesis space for the prediction. SVMs along with other margin-based learners are examples of distribution free learners.

7.2.1 Support vector machines (SVMs)

SVMs can be used for solving both classification and regression problems. They are based on principles of maximum margin classifiers. Maximum margin classifiers simultaneously minimize the classification/regression error while maximizing the geometric margin [34].

Given a set of positive examples $\mathcal{A}^+$ and a set of negative examples $\mathcal{A}^-$, a SVM learns a classification function $f(X)$, given by

$$f(x) = \sum_{x_i \in \mathcal{A}^+} \lambda_i^+ \mathcal{K}(x, x_i) - \sum_{x_i \in \mathcal{A}^-} \lambda_i^- \mathcal{K}(x, x_i) \qquad (7.1)$$

where λ_i^+ and λ_i^- are nonnegative weights that are computed during training by maximizing a quadratic objective function, and $k(.,.)$ is the *kernel* function designed to capture the similarity between pairs of examples. Having learned the function $f(x)$, a new example x is predicted to be positive or negative depending on whether $f(x)$ is positive or negative. The value of $f(x)$ also signifies the tendency of x to be a member of the positive or negative class and can be used to obtain a meaningful ranking of a set of the test examples.

The support vector classifier is primarily a binary classifier, that differentiates between the positive and negative classes. The output domain for $f(x)$ is either $+1$ or -1, denoting the two class labels.

The task of differentiating between more than two class labels is a multiclass classification problem. Several approaches exist that can directly train a multiclass SVM-based classifier. A popular approach involves training one-versus-rest binary classifiers for each of the classes. Each of these binary SVM classifiers are trained to answer the prediction whether an unknown example belongs to a particular class or not. The final prediction i.e., class label assignment is based on ranking of the prediction outputs from the multiple binary classifiers.

SVMs are also capable of solving regression problems using different loss functions. The error insensitive support vector regression ε-SVR [34] learns a function $f(x)$ for estimation using the training instances. Given a set of training instances (x_i, y_i) where y_i is the continuous value to be estimated for example x_i, the ε-SVR aims to learns a function of the form

$$f(x) = \sum_{x_i \in \Delta^+} \alpha_i^+ \mathcal{K}(x, x_i) - \sum_{x_i \in \Delta^-} \alpha_i^- \mathcal{K}(x, x_i) \qquad (7.2)$$

where Δ^+ contains the examples for which $y_i - f(x_i) > \epsilon$, Δ^- contains the examples for which $y_i - f(xi) < -\epsilon$, and a_i^+ and α_i^- are nonnegative weights that are computed during training by maximizing a quadratic objective function. The objective of the maximization is to determine the flattest $f(x)$ in the feature space and minimize the estimation errors for instances in $\Delta^+ \cup \Delta^-$. Hence, examples that have an estimation error satisfying $|f(x_i) - y_i| < \epsilon$ are neglected. The parameter ϵ controls the width of the regression deviation or tube.

7.3 Residue-Wise Structure and Function Prediction

The prediction of structural and functional properties of protein residues using sequentially local information derived from protein sequence information is an important task in computational biology. The most popular local structure prediction problem is the secondary structure prediction problem [16,27]. Secondary structures are locally recurring substructural units that organize themselves independently from the rest of the protein. The primary secondary structure elements include the alpha-helix (a coil-like structure), and beta-sheets (parallel arrangement of strands of residues). These secondary structure elements have regular shape, and compose together to form the overall three-dimensional structure of the protein. The secondary structure prediction problem is to determine a residue of a protein to be in one of the secondary structure element states using sequence information. The successful prediction methods developed so-far [16,27] utilize sequence information from the residue in consideration, and its neighboring or local residues. Similar to the secondary structure prediction problem, there have been several other local structure and function prediction problems. Each problem has inspired its own set of methods, but the general methodology remains similar. In this section we provide few examples of interesting and important local structure and function prediction problems.

7.3.1 Disorder prediction

Some proteins contain regions which are intrinsically disordered in that their backbone shape may vary greatly over time and external conditions. A disordered region of a protein may have multiple binding partners and hence can take part in multiple biochemical processes in the cell which make them critical in performing various functions [9]. Accurate prediction of disordered regions can relieve some of the bottlenecks caused during high-throughput proteome analysis.

Several studies [33] have shown the differences in sequences for ordered and disordered regions. As such, a large number of computational approaches have been developed to predict the disordered segments using sequence information. Predicting disordered regions forms part of the biennial protein structure prediction experiment CASP.* Disorder region prediction methods mainly use physiochemical properties of the amino acids or evolutionary information. In particular IUPred [8] uses a pairwise energy function derived from amino acid composition, Poodle [11] employs a combination of different physiochemical properties as features for a SVM-based learning and prediction approach. Another disordered prediction tool, DISPro [5] utilizes a

---

* http://predictioncenter.org

combination of evolutionary and sequence-derived features within a recurrent neural network.

7.3.2 Protein–DNA interaction site prediction

When it is known that the function of a protein is to bind to DNA, it is highly desirable from an experimental point of view to know which parts of the protein are involved in the binding process. These interaction sites usually involve protein residues which come into contact with DNA and stabilize the complex due to favorable interactions with DNA. Sequence-based methods are used to identify the most likely binding residues as the full structure of the protein is rarely known. Accurate methods that do so would allow an experimentalist to alter the protein behavior by mutating only a few residues.

The usual approach for a machine learning approach is to define a cutoff distance from DNA. If parts of a protein residue are within this cutoff, it is considered an interacting residue and is otherwise considered noninteracting, a binary classification problem. DISIS [24] uses SVMs and a radial basis function kernel with PSI-BLAST [4] derived profiles (PSSMs), predicted secondary structure, and predicted solvent accessibility as input features. This is framework is directly comparable to our own along with neural network method of Ahmad and Sarai [2] which employs only PSSMs. Researchers have also utilized structure information such as the structural neighbors in DISPLAR [32] and the solvent accessibility using in the earlier work of Ahmad et al. [1].

7.3.3 Transmembrane helix prediction

Proteins which span the cell membrane have proven to bet quite difficult to crystallize in most cases and are generally too large for NMR studies. Computational methods to elucidate transmembrane protein structure are a quick means to obtain approximate topology. Many of these proteins are composed of an intercellular, extra-cellular, and membrane portions where the membrane portion contains primarily hydrophobic residues in helices. Accurately predicting these helix segments allows them to be excluded from function studies as they are usually not involved in the activity of the protein.

MEMSAT [13] in its most recent incarnation uses profile inputs to a neural network to predict whether residues in a transmembrane protein are part of a transmembrane helix, interior or exterior loop, or interior or exterior helix caps. Kernytsky and Rost have benchmarked a number of methods and maintain a server to compare the performance of new methods [18].

7.3.4 Local structure alphabets

The notion of local, recurring substructure in proteins has existed for many years primarily in the form of the secondary structure classifications. With the advent of fragment assembly methods for tertiary structure prediction, there has been increased interest in methods for predicting the backbone conformation of a fixed length section of protein. This extended local structure, usually a superset of the traditional three-state secondary structure, can be a significant first step toward full tertiary structure.

Many local structure alphabets have been generated by careful manual analysis of structures such as the alphabet of DSSP [14] while others have been derived through purely computational means. One such example are the protein blocks of de Brevern et al. [7] which were constructed through the use of self-organizing maps, a clustering technique. The method uses residue dihedral angles during clustering and attempts to account for order dependence between local structure elements which should improve predictability. Karchin et al. used neural nets to predict local structure for a variety of alphabets [15]. They found protein blocks to be the best choice according to their 'bits saved per position,' a measure of how much prediction improvement there is for the alphabet over simply predicting the most frequent character.

7.3.5   Relative solvent accessibility prediction

Solvent accessibility determines the degree to which a residue in a protein structure can interact with a solvent molecule. This is important, as it can ascertain the local shape of protein based on whether the residue is buried/exposed. The residue-wise notion of solvent accessibility is defined by DSSP [14] by determining the accessible surface area relative to the maximum possible surface area obtainable for the specific amino acid residue.

Predicting solvent accessibility can be formulated as a classification problem by defining buried or exposed classes by thresholding on the relative solvent accessibility value (normally 16% or 25%), and can also be a regression or density estimation problem of attempting to determine the percentage value using sequence information only. There are many methods available for solvent accessibility prediction that deploy a wide range of learning methods including neural networks [30], bi-recurrent neural networks [26], and SVMs [23] using the set of standard sequence derived features.

7.3.6   Residue-wise contact order prediction

Pairs of residues are considered to be in contact if their C_β atoms are within a threshold radius, generally 12 Å. Residue-wise contact order [19] is an average of the distance separation between contacting residues within a sphere of set threshold. It defines the extent to which a residue makes long-range contacts in native protein structure, and can be used to set up constraints in the overarching three-dimensional structure prediction problem. A support vector regression method [31] has used a combination of local sequence-derived information in the form of PSI-BLAST profiles [4] and predicted secondary structure information, and global information based on amino acid composition and molecular weight for good quality estimates of the residue-wise contact order value. Amongst other techniques, critical random networks [20] use PSI-BLAST profiles as a global descriptor for this estimation problem.

7.4   *Pro*SAT Overview

In this section, we present a protein residue annotation toolkit called *Pro*SAT.* This toolkit uses a SVM framework and is capable of predicting both a discrete label or a continuous value. *Pro*SAT is a tool that is designed to allow life science researchers to quickly and efficiently train SVM-based models for annotating protein residues with any desired property. The protocol for training the models, and predicting the residue-wise property is similar in nature to the methods developed for the different residue annotation problems reviewed in Section 7.3.

*Pro*SAT allows use of any type of sequence information associated with residues for the annotation. For every residue, *Pro*SAT encodes the input information in the form of fixed length feature vector. The information is extracted for the residue in consideration, as well as neighboring residue using a fixed-length subsequence. *Pro*SAT also allows for a new flexible encoding scheme that differentially weighs information extracted from neighboring residues, based on the distance to the central residue.

The *Pro*SAT implementation includes the standard kernel functions (linear and radial basis functions) along with a second-order exponential kernel function shown to be effective for secondary structure prediction and pairwise local structure prediction [16,28]. *Pro*SAT is capable of learning two-level models that use predictions from the first-level model to train a second-level model. These two level models are well suited for accounting for residue properties that depend on the property values of near-by residues (i.e., they are sequentially autocorrelated). This form of cascaded learning has been shown to perform well for secondary structure prediction [16,30].

7.4.1   Implementation

*Pro*SAT approaches the protein residue annotation problem by utilizing local sequence information (provided by the user) around each residue in a supervised machine learning framework. *Pro*SAT uses SVMs in both classification and regression formulations to address the problem of annotating residues with discrete labels and continuous values respectively. The *Pro*SAT implementation utilizes the publicly available SVM^{light} program [12].

---

* ProSAT is available at http://bio.dtc.umn.edu/prosat and a web server with trained models can be accessed at http://bio.dtc.umn.edu/monster.

*Pro*SAT consists of two separate independent programs, one for the learning phase denoted by *Pro*SAT-L, and the other for the prediction phase denoted by *Pro*SAT-P. The *Pro*SAT-L program trains either a classification or regression model for solving the residue annotation problem. For the discrete label classification problems, *Pro*SAT-L trains one-versus-rest binary classification models. When the number of unique class labels are two (e.g., disorder prediction), *Pro*SAT-L trains only one single binary classification model to differentiate between the two classes. When the number of unique class labels are greater than two (e.g., three-state secondary structure prediction), *Pro*SAT-L trains one-versus-rest models for each of the classes i.e., if there are K discrete class labels, *Pro*SAT-L trains K one-versus-rest classification models. In case of the continuous value estimation problem (e.g., solvent accessible surface area estimation problem), *Pro*SAT-L trains a single support vector regression (ϵ-SVR) model.

The *Pro*SAT-P program assigns a discrete label or estimates a continuous value for each residue of the input sequences using the trained models (generated by *Pro*SAT-L). In case of the classification problems, *Pro*SAT-P uses the K one-versus-models to predict the likelihood of a residue to be a member of each of the K classes. *Pro*SAT-P assigns the residue the label or class which has the highest likelihood value. In case of the regression problems, *Pro*SAT-P estimates a floating-point continuous value for each residue.

7.4.2 Input information

The input to *Pro*SAT consists of two types of information.

Firstly, to train the prediction models true annotations are provided to *Pro*SAT-L. For every input sequence used for training a separate file is provided. Each line of the file contains an alphanumeric class label or a continuous value i.e., true annotation for every residue of the sequence.

Secondly, *Pro*SAT can accept any general user-supplied features for prediction. For a protein *Pro*SAT accepts any information as feature matrices. Both, *Pro*SAT-L and *Pro*SAT-P accept these input feature matrices. *Pro*SAT-L uses these feature matrices in conjunction with the true annotation files for learning the predictive models, whereas *Pro*SAT-P uses the input feature matrices to make predictions for the residues.

A feature matrix F for sequence X is of dimensions $n \times d$, where n is the length of the protein sequence and d is the number of features or values associated with each position of the sequence. These values depend on the specific features. In Figure 7.1a we show the PSI-BLAST derived position specific scoring matrix of dimensions $n \times 20$. For every residue the PSI-BLAST matrix captures evolutionary conservation information by providing a score for each of the 20 amino acids. Other examples of feature matrices include the predicted secondary structure matrices.

We use F_i to indicate the ith row of matrix F, which corresponds to the features associated with the ith residue of X. *Pro*SAT can accept multiple

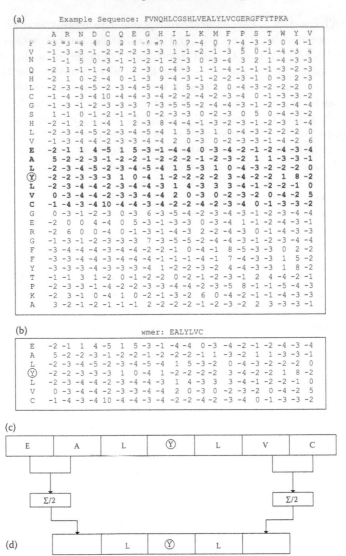

FIGURE 7.1: (a) Input example sequence along with PSI-BLAST profile matrix of dimensions $n \times 20$, with a residue circled to show the encoding steps. (b) Example $w$mer of $w = 3$ and length seven, with extracted submatrix from the original PSI-BLAST matrix. (c) Encoded vector of length 7×20 formed by linearizing the submatrix. (d) Flexible encoding showing three residues in the center using the finer representation, and two residues flanking the central residues on both sides using a coarser representation as an averaging statistic. Length of this vector equals 5×20.

types of feature matrices per sequence. When multiple types of features are considered, the lth feature matrix is specified by F^l.

7.4.3 Information encoding

In order to encode information for a residue, *ProSAT* also uses the information from neighboring residues. As seen in Figure 7.1b for the circled residue, three residues above and below are also selected and the corresponding information from the feature matrix is extracted.

Specifically, *ProSAT* uses a subsequence also called *w*mer, centered around the residue for deriving fixed length feature vectors.

Given a sequence X of length n and a user-supplied parameter w, the *w*mer at position i of $X(w < i \leq n - w)$ is defined to be the $(2w + 1)$-length subsequence of X centered at position i. That is, the *w*mer contains x_i, the w amino acids before, and the w amino acids after x_i. *ProSAT* uses a *w*mer-based encoding to capture sequence information for residue x_i to perform the residue-wise prediction. In the example shown in Figure 7.1b, for the central residue the w size is three resulting in a seven length-subsequence used to extract information from the input PSI-BLAST matrix.

ProSAT uses the $(2w + 1)$ rows of the matrix F, $F_{i-w} \ldots F_{i+w}$ to encapsulate the feature information associated with the *w*mer centered at residue x_i. This submatrix is denoted by *w*mer(F_i) and is linearized to generate a vector of length $(2w+1)d$, where d is the dimension of the matrix F. This is shown in Figure 7.1c where the linearized submatrix is shown as a vector. Now *ProSAT* can compute kernel function or similarity between residue-pairs as a function of the dot product between their vector representations. In this way *ProSAT* captures information of the residue and its neighboring residues.

7.4.4 Kernel functions

ProSAT uses a SVM-based framework, and implements several kernel functions to capture similarity between pairs of *w*mers (*w*mer representations). A kernel function computes a similarity between two objects and selection of an appropriate kernel function for a problem is key to the effectiveness of SVM learning.

7.4.4.1 Linear kernels

Given a pair of *w*mers, *w*mer(x_i) and *w*mer(y_j) a linear kernel function can be defined between their feature matrices *w*mer(F_i) and *w*mer(G_j), respectively as

$$k_w(x_i, y_j) = \sum_{k=-w}^{w} \langle F_{i+k}, G_{j+k} \rangle \tag{7.3}$$

where $\langle , \rangle$ denotes the dot-product operation between two vectors.

Some problems may require only approximate information for sequence neighbors which are far away from the central residue while nearby sequence neighbors are more important. For example, the secondary structure state of a residue is in general more dependent on the nearby sequence positions than the positions that are further away [6]. As such, *Pro*SAT allows a window encoding shown in Figure 7.1d where the positions away from the central residue are averaged to provide a coarser representation, whereas the positions closer to the central residue provide a finer representation. This two-parameter linear window-based kernel denoted by $\mathcal{W}_{m,f}$, which computes the similarity between features $wmer(F_i)$ and $wmer(G_j)$ as

$$
\mathcal{W}_{w,f}(x_i, y_j) = \sum_{k=-f}^{f} \langle F_{i+k}, G_{j+k} \rangle +
$$

$$
\left\langle \sum_{k=f+1}^{w} \langle F_{i+k}, \sum_{k=f+1}^{f} G_{j+k} \right\rangle +
$$

$$
\left\langle \sum_{k=-w}^{-f-1} \langle F_{i+k}, \sum_{k=-w}^{-f-1} G_{i+k} \right\rangle. \tag{7.4}
$$

The parameter w governs the size of the $wmer$ considered in computing the kernel. Rows within $\pm f$ contribute an individual dot product to the total similarity while rows outside this range are first summed and then their dot product is taken. In all cases f is less than w and as f approaches w, the window kernel becomes simply a sum of the dot products, the most fine-grained similarity measure considered and equivalent to a one-parameter dot product kernel that equally weighs all positions of the $wmer$ and can be defined by Equation 7.3. As such we can express kernel $\mathcal{K}_w$ as $\mathcal{W}w,w$.

The parameter f provides control over the fine-grained versus coarse-grained sections of the window. Specifying f to be lesser than w merges these distant neighbors into only a coarse contribution to the overall similarity, as it only accounts for compositional information and not the specific positions where these features occur.

7.4.4.2 Exponential kernels

*Pro*SAT implements the standard radial basis kernel function (rbf), defined for some parameter γ by

$$
\mathcal{K}^{\mathrm{rbf}}(x, y) = \exp\left(-\gamma \parallel x - y \parallel^2\right) \tag{7.5}
$$

*Pro*SAT also implements the normalized second order exponential (*soe*) kernel function shown to better capture pairwise information, and improve accuracy for secondary structure and local structure information prediction [16, 28]. Given any base kernel function $\mathcal{K}$, we define $\mathcal{K}^2$ as

$$
\mathcal{K}^2(x, y) = \mathcal{K}(x, y) + (\mathcal{K}(x, y))^2 \tag{7.6}
$$

which is a second-order kernel in that it computes pairwise interactions between the elements x and y. We then define K^{soe} as

$$k^{soe}(x, y) = \exp\left(1 + \frac{k^2 + (x, y)}{\sqrt{k^2(x, x) k^2(y, y)}}\right) \qquad (7.7)$$

which normalizes K^2 and embeds it into an exponential space.

By setting a specific γ parameter value and using normalized unit length vectors in Equation 7.5 it can be shown that the standard rbf kernel is equivalent (up to a scaling factor) to a first order exponential kernel which is obtained by replacing $K^2(x, y)$ with only the first-order term as $K(x, y)$ in Equation 7.6, and plugging this modified $K^2(x, y)$ in the normalization framework of Equation 7.7.

7.4.5   Integrating information

When multiple information in the form of different feature matrices is provided to *Pro*SAT, the kernel functions and information encoding per residue for each of the feature matrices remain the same. The final kernel fusion is accomplished using a weighted linear combination across the results of the original base kernels. The weights per feature matrices are provided by the user.

For example, we can use a fusion of second-order exponential kernels on different features of a protein sequence. Considering two sequences with features F^l and G^l for $l = 1, \ldots, k$, our fusion kernel is defined

$$k^{\text{fusion}}(x_i, y_j) = \sum_{l=1}^{k} \omega_l k^{soe}\left(F_i^l, G_j^l\right) \qquad (7.8)$$

where the weights ω_l are supplied by the user. In most cases, these weights can be set to be equal but should be altered according to domain-specific information.

For example, we can use the PSI-BLAST based profile matrix and predicted secondary structure matrix of dimensions $n \times 20$, and $n \times 3$ in conjunction within this framework, respectively.

7.4.6   Cascaded models

Several prediction algorithms like PHD [30], and YASSPP [16] developed previously for secondary structure prediction use a two-level cascaded prediction framework. This two-level framework trains two models referred as the L_1 and L_2 models, connected together in a cascaded fashion. Both the models train K one-versus-rest binary classification models for predicting a discrete label, or a single ϵ-SVR regression model for estimating a continuous value.

The predictions from the first-level L_1 model are used as an input feature matrix along with the original features for training a L_2 model. This can be easily accomplished within *Pro*SAT's framework which allows easy incorporation of any user-supplied features (also described in the earlier sections).

*Pro*SAT allows training of cascaded models in an easy and efficient manner. First, the entire training set is used to train a L_1 classification/regression model using the user provided input features. This is followed with a n-fold cross-validation step to generate predictions for the entire training set using the fold specific trained L_1 model. In each iteration, $1/n$th of the dataset is set aside for prediction whereas the remainder of the dataset is used for training. Finally, using the original input features and the predictions generated from the L_1 models, a L_2 model is trained. During the prediction stage, the fully trained L_1 model is used for generating predictions which is used as input features along with the original features for producing predictions from the L_2 model. The weights (ω) between the different features as described by Equation 7.8 can be provided by the user.

7.4.7 Predictions output

The output of *Pro*SAT's prediction program is similar to the input feature matrix. For a classification problem, *Pro*SAT outputs a space-delimited text file (also called as profile file) containing the SVM prediction scores generated for every residue by the one-versus-rest classifiers, along with a predicted label for every residue. The predicted label is the class for which the SVM classifier shows the maximum prediction value. For the regression problem the output is a text file containing the estimated value, produced by the ϵ-SVR model.

7.5 Case Studies

*Pro*SAT has already been used to develop a transmembrane-helix segment identification and orientation prediction system called TOPTMH [3], which has shown the best performance on a static independent benchmark [18]. *Pro*SAT has also been useful in predicting ligand-binding sites using sequence information [17]. Predicted ligand-binding sites have shown to be useful in homology modeling of protein regions known to interact with small molecules or ligands [17].

In this chapter we provide case studies assessing the performance of *Pro*SAT on the disorder prediction, residue-wise contact order estimation, protein-DNA site prediction, and local structure prediction problems. A comprehensive evaluation comparing the performance of the different information used, kernel functions, window-based encoding has already been done in previous studies [3,16,17,28,29].

7.5.1 Datasets

Empirical evaluations for *Pro*SAT are performed for different sequence annotation problems on previously defined datasets. Table 7.1 presents information regarding the source and key features of different datasets used in our cross validation and comparative studies. The datasets were used in previous studies, and the pairwise sequence identities for the different datasets was less than 40%.

The general protocol used for evaluating the different parameters, and features, as well as comparing to previously established studies remained fairly consistent across the different problems. In particular a n-fold cross validation methodology was used, where $1/n$th of the database in consideration was used for testing and the remaining dataset was used for training, with the experiment being repeated n times. Table 7.1 also reports the number of cross validation folds that were used for the different studies.

7.5.2 Evaluation metrics

The quality of the models were evaluated using the standard receiver operating characteristic (ROC) scores. The ROC score is the normalized area under the curve that plots the true positives against the false positives for different thresholds for classification. The ROC score reported is averaged across the different classes and folds. Other standard statistics, including precision as $TP/(TP + FP)$, and recall as $TP/(TP + FN)$ were also computed. The accuracy of K-way multiclass classification was evaluated as $Q_k = \left(\sum_{i=1}^{k} TP_i \right) / (Total\ Residues)$. Here, TP, FP, TN, FN denote the standard true positives, false positives, true negatives, and false negatives, respectively. Another classification measure, called the F_1 score was given as $2*Precision*Recall/(Precision+Recall)$.

The ROC score serves as a good quality measure in case of unbalanced classes, where measuring the accuracy or Q_k, especially in case of binary

TABLE 7.1: Problem-specific datasets.

Problem	Source	#C	#Seq	#Res	#CV	%
Disorder prediction	DisPro [5]	2	723	215612	10	30
Protein-DNA site	DISIS [24]	2	693	127240	3	20
Residue-wise contact	SVM [31]	∞	680	120421	15	40
Local structure	Profnet [25]	16	1600	286238	3	40

Note: #C, #Seq, #Res, #CV, and % denote the number of classes, sequences, residues, number of cross validation folds, and the maximum pairwise sequence identity between the sequences, respectively. ∞ represents the regression problem.

classification model may be skewed by predicting a particular class with larger number of instances. In such cases, it is essential to observe the precision and recall values, which penalize the classifiers for under-prediction as well as over-prediction. The F_1 score is a weighted average of precision and recall lying between 0 and 1, and also is a good measure for different classification problems.

The regression performance is assessed by computing the standard Pearson correlation coefficient (CC) between the predicted and observed true values for every protein in the datasets, and root mean square error (rmse) between the predicted and observed values. The results reported are averaged across the different proteins and cross validation steps. For the rmse metric, a lower score implies a better quality prediction.

For the best performing models, the *errsig* rate as the significant difference margin for Q_k and CC scores is also reported (to distinguish between two methods). *errsig* is defined as the standard deviation divided by the square root of the number of proteins ($\sigma/\sqrt{N}$), and can help in assessing how significant the differences between the best performing models and the other models, as well as serves a reference for future studies.

7.5.3   Feature matrices

The different information was provided to *Pro*SAT as feature matrices. In this section, the specifics on the different feature matrices used is explained.

7.5.3.1   Position specific scoring matrices

For a sequence of length n, PSI-BLAST [4] generates a position-specific scoring matrix $\mathcal{P}$ of dimensions $n \times 20$, where the 20 columns of the matrix correspond to the 20 amino acids.

The profiles in this study were generated using the latest version of the PSI-BLAST [4] (available in NCBI's blast release 7.2.10 using **blastpgp** −j 5 −e 0.01 −h 0.01) searched against NCBI's NR database that was downloaded in November of 2004 and contains 2,171,938 sequences.

7.5.3.2   Predicted secondary structure information

The state-of-the-art secondary structure prediction server called YASSPP [16] (default parameters) was used to generate the S feature matrix of dimensions $n \times 3$. The (i, j)th entry of this matrix represents the propensity for residue i to be in state j, where $j \in \{1, 2, 3\}$ corresponds to the three secondary structure elements: alpha helices, beta sheets, and coil regions.

Predicted secondary structure is an example of a local structure alphabet and plays a critical role in protein structure prediction. YASSPP [16] has an identical framework to *Pro*SAT and is one of the best performing secondary structure prediction methods with a reported Q_3 accuracy of 80%.

7.5.3.3   Position independent scoring matrices

A $n \times 20$ feature matrix was created where each row of the matrix is a copy of the BLOSUM62 row corresponding to the amino acid at that position in the sequence. This feature matrix is referred to as $\mathcal{B}$.

By using both PSSM- and BLOSUM62-based information, the SVM learner can construct a model that is partially based on nonposition specific information. Such a model will remain valid in cases where PSI-BLAST could not generate correct alignments due to lack of homology to sequences in the NR database.

7.5.4   Experimental results

7.5.4.1   Disorder prediction performance

*Pro*SAT was evaluated on the disorder prediction problem by training binary classification models to discriminate between residues that belong to part of disordered region or not. In particular, the PSI-BLAST profile matrix denoted by $\mathcal{P}$, BLOSUM62 derived scoring matrix denoted by $\mathcal{B}$, and predicted secondary structure matrix denoted by $\mathcal{S}$ were used both independently, and in combinations. The performance using different parameters for *Pro*SAT i.e., w and f parameters for the linear base kernel $\mathcal{W}$ alongwith kernel functions lin, rbf, and soe were studied. Table 7.2 shows the binary classification performance measured using the ROC and F_1 scores achieved on the disorder dataset after a ten-fold cross validation experiment, previously used to evaluate the DISPro prediction method.

Comparing the ROC performance of the $\mathcal{P}^{soe}$, $\mathcal{P}^{rbf}$, and $\mathcal{P}^{lin}$ models across different values of w and f used for parameterization of the base kernel ($\mathcal{W}$), it was observed that the soe kernel showed superior performance to the lin kernel and slightly better performance compared to the normalized rbf kernel. This verifies results of previous studies for predicting secondary structure [16] and predicting RMSD between subsequence pairs [28], where the soe kernel outperformed the rbf kernel.

The performance *Pro*SAT on the disorder prediction problem was shown to improve when using the $\mathcal{P}, \mathcal{B}$, and $\mathcal{S}$ feature matrices in combination rather than individually. We show results for the $\mathcal{PS}$ and $\mathcal{PSB}$ features in Table 7.2. The flexible encoding introduced by *Pro*SAT shows a slight merit for the disorder prediction problem.

The best performing fusion kernel shows comparable performance to Dis-Pro [5] that encapsulates profile, secondary structure and relative solvent accessibility information within a bi-recurrent neural network.

7.5.4.2   Contact order performance

*Pro*SAT was used to train ϵ-SVR regression models for estimating the residue-wise contact order on a previously used dataset [31]. In Table 7.3 residue-wise contact order performance results are presented by performing

TABLE 7.2: Classification performance on the disorder dataset.

	w	$f = 3$		$f = 5$		$f = 7$		$f = 9$		$f = 11$	
		ROC	F1	ROC	F1	ROC	F1	ROC	F1	ROC	F1
$\mathcal{P}^{\text{lin}}$	3	**0.800**	0.350	–	–	–	–	–	–	–	–
	7	**0.817**	0.380	0.816	0.384	0.816	0.383	–	–	–	–
	11	0.826	0.391	**0.828**	0.396	0.826	0.400	0.824	0.404	0.823	0.403
	13	0.829	0.398	0.832*	0.405	0.830	0.404	0.828	0.407	0.826	0.409
$\mathcal{P}^{\text{rbf}}$	3	0.811	0.369	–	–	–	–	–	–	–	–
	7	**0.849**	0.450	0.848	0.445	0.845	0.442	–	–	–	–
	11	0.855	0.478	0.858	0.482	**0.858**	0.480	0.855	0.470	0.853	0.468
	13	0.855	0.484	0.859	0.490	0.861*	0.492	0.860	0.487	0.857	0.478
$\mathcal{P}^{\text{soe}}$	3	**0.816**	0.379	–	–	–	–	–	–	–	–
	7	**0.852**	0.461	0.852	0.454	0.851	0.454	–	–	–	–
	11	0.856	0.482	0.860	0.491	**0.862**	0.491	0.861	0.485	0.862	0.485
	13	0.856	0.485	0.861	0.491	0.864	0.495	0.865*	0.494	0.864	0.492
$\mathcal{PS}^{\text{soe}}$	3	**0.838**	0.423	–	–	–	–	–	–	–	–
	7	**0.862**	0.476	0.860	0.473	0.859	0.468	–	–	–	–
	11	0.867	0.496	**0.868**	0.498	0.868	0.495	0.866	0.488	0.865	0.485
	13	0.867	0.503	0.870	0.503	0.871*	0.503	0.870	0.498	0.868	0.492
$\mathcal{PSB}^{\text{soe}}$	3	0.841	0.428	–	–	–	–	–	–	–	–
	7	**0.870**	0.499	0.869	0.494	0.867	0.489	–	–	–	–
	11	0.875	0.518	**0.877**	0.517	0.877	0.512	0.874	0.508	0.873	0.507
	13	0.875	0.522	0.878	0.521	0.879**	0.519	0.879	0.518	0.876	0.514

Note: DISPro [5] reports a *ROC* score of 0.878. The numbers in bold show the best models for a fixed w parameter, as measured by ROC. $\mathcal{P}$, $\mathcal{B}$, and $\mathcal{S}$ represent the PSI–BLAST profile, BLOSUM62, and YASSPP scoring matrices, respectively. soe, rbf, and lin represent the three different kernels studied using the $\mathcal{W}_{w,f}$ as the base kernel. * denotes the best classification results in the subtables, and ** denotes the best classification results achieved on this dataset. For the best model we report a Q_2 accuracy of 84.60% with an *errsig* rate of 0.33.

15-fold cross validation using different kernels and feature matrices. These results are evaluated by computing the CC and rmse values averaged across the different proteins in the dataset.

Analyzing the effect of the w and f parameters for estimating the residue-wise contact order values, we observe that a model trained with $f < w$ generally shows better CC and rmse values. The best models as measured by the CC scores are highlighted in Table 7.3. A model with equivalent CC values but having a lower f value is considered better because of the reduced dimensionality achieved by such models.

The best estimation performance achieved by our ϵ-SVR based learner uses a fusion of the $\mathcal{P}$ and $\mathcal{S}$ feature matrices and improves CC by 21%, and rmse value by 17% over the ϵ-SVR technique of Song and Barrage [31]. Their method uses the standard rbf kernel with similar local sequence-derived amino acid and predicted secondary structure features.

7.5.4.3 Protein–DNA interaction sites performance

*Pro*SAT was used to train binary classification models on a previously used dataset [24]. Similar to the DISIS evaluation [24], three-fold cross validation

TABLE 7.3: Residue-wise contact order estimation performance.

	w	f = 3		f = 5		f = 7		f = 9		f = 11	
		CC	rmse	CC	rmse	CC	rmse	CC	rmse	CC	rmse
$\mathcal{P}^{lin}$	3	**0.647**	0.747	–	–	–	–	–	–	–	–
	7	0.653	0.738	0.662	0.729	**0.662**	0.729	–	–	–	–
	11	0.653	0.737	0.663	0.728	0.664*	0.726	0.663	0.726	0.662	0.727
	15	0.652	0.738	0.663	0.728	**0.664**	0.726	0.664	0.725	0.663	0.725
$\mathcal{P}^{rbf}$	3	**0.670**	0.723	–	–	–	–	–	–	–	–
	7	0.682	0.708	**0.692**	0.701	0.690	0.703	–	–	–	–
	11	0.683	0.706	**0.694**	0.697	0.694	0.696	0.693	0.697	0.691	0.699
	15	0.682	0.706	0.695*	0.696	0.695	0.696	0.695	0.694	0.693	0.696
$\mathcal{P}^{soe}$	3	**0.672**	0.720	–	–	–	–	–	–	–	–
	7	0.688	0.702	**0.698**	0.694	0.679	0.694	–	–	–	–
	11	0.688	0.700	0.701	0.690	0.702	0.688	**0.703**	0.687	0.701	0.688
	15	0.688	0.700	0.701	0.689	0.703	0.686	0.704*	0.685	0.704	0.685
$\mathcal{PS}^{lin}$	3	**0.686**	0.718	–	–	–	–	–	–	–	–
	7	0.694	0.707	0.702	0.698	**0.703**	0.697	–	–	–	–
	11	0.695	0.703	0.704	0.694	**0.705**	0.692	0.704	0.691	0.704	0.692
	15	0.694	0.703	0.703	0.693	**0.704**	0.691	0.704	0.690	0.704	0.690
$\mathcal{PS}^{rbf}$	3	**0.707**	0.696	–	–	–	–	–	–	–	–
	7	0.716	0.680	**0.721**	0.677	0.720	0.677	–	–	–	–
	11	0.718	0.676	0.723*	0.671	0.722	0.671	0.720	0.672	0.718	0.673
	15	0.716	0.675	**0.723**	0.669	0.723	0.669	0.721	0.669	0.719	0.670
$\mathcal{PS}^{soe}$	3	**0.708**	0.692	–	–	–	–	–	–	–	–
	7	0.719	0.677	**0.723**	0.672	.722	0.672	–	–	–	–
	11	0.720	0.673	**0.725**	0.667	0.725	0.666	0.724	0.666	0.722	0.667
	15	0.719	0.672	0.726**	0.665	0.726	0.664	0.725	0.664	0.723	0.664

Note: CC and rmse denotes the average CC and rmse values. The numbers in bold show the best models as measured by CC for a fixed w parameter. $\mathcal{P}$ and $\mathcal{S}$ represent the PSI-BLAST profile and YASSPP scoring matrices, respectively. soe, rbf, and lin represent the three different kernels studied using the $\mathcal{W}_{mj}$ as the base kernel. * denotes the best regression results in the sub-tables, and ** denotes the best regression results achieved on this dataset. For the best results the *errsig* rate for the CC values is 0.003. The published results [31] uses the default rbf kernel to give CC=0.600 and rmse=0.78.

studies were performed ensuring that sequence identity between the different folds was less than 40%.

The experimental results showed that for predicting the protein–DNA interaction sites, the window kernels with $w = f$ performed best. Table 7.4 reports results by setting the w parameter equal to the f parameter. The performance of the soe and rbf kernels are very comparable, and there is an advantage in using the fusion of $\mathcal{P}$ and $\mathcal{S}$ features.

The model obtained by combining the $\mathcal{P}$ and $\mathcal{S}$ features which gives a raw Q_2 accuracy of 83%. The protein–DNA interaction site program DISIS uses a two-level approach to solve this problem [24]. The first level, which uses SVM learning with profile, predicted secondary structure, and predicted solvent accessibility as inputs, gives $Q_2 = 83\%$ to which our performance compares favorably. DISIS goes on to smooth this initial prediction using a rule-based approach which improves accuracy.

TABLE 7.4:   Classification performance on the
protein–DNA interaction site prediction.

	$w = f = 3$		$w = f = 7$		$w = f = 11$	
	ROC	F1	ROC	F1	ROC	F1
$\mathcal{P}^{\mathrm{lin}}$	**0.745**	0.453	0.749	0.462	0.744	0.453
$\mathcal{P}^{\mathrm{rbf}}$	**0.741**	0.456	**0.752**	0.465	0.761	0.473
$\mathcal{P}^{\mathrm{soe}}$	**0.741**	0.453	**0.752**	0.458	0.762	0.474
$\mathcal{PS}^{\mathrm{lin}}$	**0.756**	0.463	**0.758***	0.469	0.748	0.452
$\mathcal{PS}^{\mathrm{rbf}}$	**0.753**	0.465	**0.754**	0.462	0.759	0.466
$\mathcal{PS}^{\mathrm{soe}}$	**0.754**	0.466	**0.756**	0.468	0.763	0.468

Note: The numbers in bold show the best models for a fixed w parameter, as measured
by ROC. $\mathcal{P}$, and $\mathcal{S}$ represent the PSI-BLAST profile and YASSPP scoring matrices, respectively. soe, rbf, and lin represent the three different kernels studied using the $\mathcal{W}_{s,f}$ as
the base kernel. * denotes the best classification results in the subtables, and ** denotes
the best classification results achieved on this dataset. For the best model we report a Q_2
accuracy of 83.0% with an *errsig* rate of 0.34.

7.5.4.4   Local structure alphabet performance

*Pro*SAT was evaluated using the protein blocks [7] for local structure prediction as they were found to be the best representatives for tertiary structure
[15]. There are 16 members in this alphabet which significantly increases prediction difficulty over traditional three-letter secondary structure prediction.

*Pro*SAT was evaluated on a dataset consisting of 1600 proteins derived
from the SCOP [22] version 1.57 database, classes a to e, and where no two
protein domains have more than 75% sequence identity.

Due to the high computational requirements associated with such a large
training set, *Pro*SAT was evaluated using the *soe* kernels on a wide set of
parameters for w and f, but only on a small subset of the 1600 proteins present
in the dataset. From this experiment, it was observed that the prediction was
best when $w = f$ and used this to limit the choice of parameters for larger-
scale evaluation. Once these promising models were determined, a three-way
cross validation experiment was carried out using all 1600 proteins for each
parameter set. Table 7.5 reports the classification accuracy in terms of the Q_{16}
accuracy and average of the ROC scores for different members of the protein
blocks.

From Table 7.5 it can be seen that the *soe* kernel performs marginally
better than the rbf kernel. The addition of predicted secondary structure
information, S features does improve the Q_{16} performance marginally as
was expected for local structure prediction. The reported Q_{16} results are
very encouraging, since they are above 67%, whereas the prediction accuracy for a random predictor would be only 6.25%. Competitive methods
for local structure alphabet prediction have reported a Q_{16} accuracy of
40.7% [10].

TABLE 7.5: Classification performance on the local structure alphabet dataset.

	$w = f = 5$		$w = f = 7$		$w = f = 9$	
	ROC	Q_{16}	ROC	Q_{16}	ROC	Q_{16}
$\mathcal{P}^{\text{rbf}}$	0.82	64.9	0.81	64.7	0.81	64.2
$\mathcal{P}^{\text{soe}}$	0.83	67.3	0.82	67.7	0.82	67.7
$\mathcal{PS}^{\text{rbf}}$	0.84	66.4	0.84	66.9	0.83	67.2
$\mathcal{PS}^{\text{soe}}$	0.85	68.0	0.84	68.5	0.83	68.9**

Note: $w = f$ gave the best results on testing on few sample points, and hence due to the expensive nature of this problem, we did not test it on a wide set of parameters. ** denotes the best scoring model based on the Q_{16} scores. For this best model the *errsig* rate of 0.21.

7.6 Concluding Remarks

In this chapter, we provided a brief overview of machine learning based methods for predicting the structural and functional properties of proteins from their sequence or sequence-derived information. Specifically, we explored and defined several residue-wise structure and function prediction problems, and reviewed the commonly used methods. We then provided details about a general purpose residue annotation toolkit (*Pro*SAT), giving a flavor of the information used, encoding schemes, kernel functions, and learning protocols. We also provided case studies on four of the local structure and function problems discussed.

*Pro*SAT provides to the practitioners an efficient and easy-to-use tool for a wide variety of annotation problems. In the future, we expect use of this annotation toolkit and methodology described here toward prediction of various 1D features of a protein. These predictions could then be effectively integrated to provide valuable supplementary information for determining the 3D structure of proteins, and hence the function.

References

[1] S. Ahmad, M. Michael Gromiha, and A. Sarai. Analysis and prediction of dna-binding proteins and their binding residues based on composition, sequence and structural information. *Bioinformatics*, 20(4):477–486, 2004.

[2] S. Ahmad and A. Sarai. Pssm-based prediction of DNA binding sites in proteins. *BMC Bioinformatics*, 6:33, 2005.

[3] R. Ahmed, H. Rangwala, and G. Karypis. Toptmh: Topology predictor for transmembrane alpha-helices. In *European Conference in Machine Learning*, R. Goebel, J. Siekmann, and W. Wahlster (Eds.). Antwerp,

Belgium, 2008, Proceedings of European Conference in Machine Learning, Antwerp, Belgium, 2008, 23–28.

[4] S. F. Altschul, L. T. Madden, A. A. Schffer, J. Zhang, Z. Zhang, W. Miller, and D. J. Lipman. Gapped blast and psi-blast: a new generation of protein database search programs. *Nucleic Acids Research*, 25(17):3389–402, 1997.

[5] J. Cheng, M. J. Sweredoski, and P. Baldi. Accurate prediction of protein disordered regions by mining protein structure data. *Data Mining and Knowledge Discovery*, 11(3):213–222, 2005.

[6] G. E. Crooks, J. Wolfe, and S. E. Brenner. Measurements of protein sequence-structure correlations. *Proteins: Structure, Function, and Genetics*, 57:804–810, 2004.

[7] A. G. de Brevern, C. Etchebest, and S. Hazout. Bayesian probabilistic approach for predicting backbone structures in terms of protein blocks. *Proteins*, 41(3):271–287, 2000.

[8] Z. Dosztnyi, V. Csizmok, P. Tompa P, and I. Simon. Iupred: web server for the prediction of intrinsically unstructured regions of proteins based on estimated energy content. *Bioinformatics*, 21(16):3433–3434, 2005.

[9] A. K. Dunker, C. J. Brown, J. D. Lawson, L. M. Iakoucheva, and Z. Obradovic. Intrinsic disorder and protein function. *Biochemistry*, 41(21):6573–6582, 2002.

[10] C. Etchebest, C. Benros, S. Hazout, and A. de Brevern. A structural alphabet for local protein structures: improved prediction methods. *Proteins: Structure, Function, and Bioinformatics*, 59:810–827, 2005.

[11] S. Hirose, K. Shimizu, S. Kanai, Y. Kuroda, and T. Noguchi. Poodle-l: a two-level svm prediction system for reliably predicting long disordered regions. *Bioinformatics*, 23(16):2046–2053, 2007.

[12] T. Joachims. Advances in kernel methods: support vector learning. In *Making Large-scale SVM Learning Practical*, Joachims (ed.). MIT Press, Cambridge, 1999.

[13] D. T Jones. Improving the accuracy of transmembrane protein topology prediction using evolutionary information. *Bioinformatics*, 23(5):538–544, 2007.

[14] W. Kabsch and C. Sander. Dictionary of protein secondary structure: pattern recognition of hydrogen-bonded and geometrical features. *Biopolymers*, 22:2577–2637, 1983.

[15] R. Karchin, M. Cline, Y. Mandel-Gutfreund, and K. Karplus. Hidden markov models that use predicted local structure for fold recognition: alphabets of backbone geometry. *Proteins*, 51(4):504–514, 2003.

[16] G. Karypis. Yasspp: better kernels and coding schemes lead to improvements in protein secondary structure prediction. *Proteins*, 64(3):575–586, 2006.

[17] C. Kauffman, H. Rangwala, and G. Karypis. Improving homology models for protein-ligand binding sites. In *LSS Comput Syst Bioinformatics Conference*, number 08-012, San Francisco, CA, 2008. Proceedings of LSS Comput Syst Bioinformatics Conference, number 08–012, San Francisco, CA, 2008.

[18] A. Kernytsky and B. Rost. Static benchmarking of membrane helix predictions. *Nucleic Acids Research*, 31(13):3642–3644, 2003.

[19] A. R. Kinjo, K. Horimoto, and K. Nishikawa. Predicting absolute contact numbers of native protein structure from amino acid sequence. *Proteins: Structure, Function, and Bioinformatics*, 58(1):158–165, 2005.

[20] A. R. Kinjo and K. Nishikawa. Crnpred: highly accurate prediction of one-dimensional protein structures by large-scale critical random networks. *BMC Bioinformatics*, 7(401), 2006.

[21] D. Lee, O. Redfern, and C. Orengo. Predicting protein function from sequence and structure. *Nature Reviews Molecular Cell Biology*, 8(12):995–1005, 2007.

[22] A. G. Murzin, S. E. Brenner, T. Hubbard, and C. Chothia. Scop: a structural classification of proteins database for the investigation of sequences and structures. *Journal of Molecular Biology*, 247:536–540, 1995.

[23] M. N. Nguyen and J. C. Rajapakse. Two-stage support vector machines to protein relative solvent accessibility prediction. In *Proceedings of the 2004 IEEE Symposium on Computational Intelligence in Bioinformatics and Computational Biology*, David, W. Corne, Jagath C. Rajapakse, and L. Gwenn Volkert (Eds.). 67–72, 2004.

[24] Y. Ofran, V. Mysore, and B. Rost. Prediction of dna-binding residues from sequence. *Bioinformatics*, 23(13):347–353, 2007.

[25] T. Ohlson and A. Elofsson. Profnet, a method to derive profile-profile alignment scoring functions that improves the alignments of distantly related proteins. *BMC Bioinformatics*, 6(253), 2005.

[26] G. Pollastri, P. Baldi, P. Farselli, and R. Casadio. Prediction of coordination number and relative solvent accessibility in proteins. *Proteins: Structure, Function, and Genetics*, 47:142–153, 2002.

[27] G. Pollastri, D. Przybylski, B. Rost, and P. Baldi. Improving the prediction of protein secondary structure in three and eight classes using recurrent neural network and profiles. *Proteins: Structure, Function, and Bioinformatics*, 47:228–235, 2002.

[28] H. Rangwala and G. Karypis. frmsdpred: Predicting local rmsd between structural fragments using sequence information. *Proteins*, 72(3):1005–1018, 2008.

[29] H. Rangwala, C. Kauffman, and G. Karypis. A generalized framework for protein sequence annotation. In *Proceedings of the NIPS Workshop on Machine Learning in Computational Biology*, G. Chechik, C. Leslie, W. Noble, Gunnar Rätsch, Quaid Morris, K. Tsuda (Eds.). Vancouver, Canada, 2007.

[30] B. Rost. Phd: predicting 1d protein structure by profile based neural networks. *Methods in Enzymology*, 266:525–539, 1996.

[31] J. Song and K. Burrage. Predicting residue-wise contact orders in proteins by support vector regression. *BMC Bioinformatics*, 7(425), 2006.

[32] H. Tjong and Huan-Xiang Zhou. Displar: an accurate method for predicting dna-binding sites on protein surfaces. *Nucleic Acids Research*, 35(5): 1465–1477, 2007.

[33] V. N. Uversky, J. R. Gillespie, and A. L. Fink. Why are "natively unfolded" proteins unstructured under physiologic conditions? *Proteins: Structure, Function, and Genetics*, 41(3):415–427, 2000.

[34] Vladimir N. Vapnik. *The Nature of Statistical Learning Theory*. Springer Verlag, New Jersey, 1995.

Part II

Genomics, Transcriptomics, and Proteomics

Chapter 8

Computational Approaches for Genome Assembly Validation

Jeong-Hyeon Choi, Haixu Tang, and Sun Kim

Indiana University

Mihai Pop

University of Maryland

8.1 Introduction

Although biologists are aware that there are likely mistakes in a draft genome assembly, there is a lack of computer software to automatically detect the mistakes or to assign confidence scores in difference regions of the assembly. The problem, referred to as the *genome validation problem* herein, is particularly important owning to the wide application of the massively parallel sequencing technologies, which often generate short reads and thus increase the chance of misassembly.

A large fraction of genome misassemblies are caused by repeats, especially in the assembly of large eukaryotic genomes. The assemblers may be confused by pseudo-overlaps between repetitive reads (from different copies of nearly identical repeats) and place them together [25]. Typically, two types of misassemblies may be induced by repetitive reads: repeat collapse and large scale rearrangement [31]. Sometimes, repeat collapsing can result in base-calling errors in the consensus sequences of repeats [33]. The identification and separation of these collapsed repeats have been studied as an independent computational problem, known as the *repeat separation problem* [15]. Several probabilistic and combinatorial algorithms have been proposed to solve this problem [15,30] and the employment of paired end sequence data can improve the separation of nearly identical repeat copies. Especially, misassemblies due to segmental duplication in the draft assembly of the human has been identified by using mate-paired clones [2]. The Human Genome Structural Variation project has been launched to identify structural variation such as large insertions, deletions, and inversions of DNA. As part of this project, structural variation from eight human genomes has recently been identified by high quality fosmid end-sequence pairs and several biological experiments based on Affymetrix SNP array, BAC arrayCGH, and Illumina Human1M Genotyping BeadChip [16].

This chapter summarizes existing assembly validation techniques. Section 8.2 discusses signatures of misassemblies in the context of single nucleotide polymorphism, misplaced mate-pairs, and deviation from randomness. Section 8.3 describes an entropy-based method utilizing statistics of short patterns of fixed length in shotgun data. Section 8.4 describes a method, developed by Choi et al. [5], that combines evidences from multiple measures using machine learning techniques. The main idea of this method is that no single measure, or signature of misassembly, works best but measures are complementary to each other, thus, combining multiple measures under standard machine learning schemes could improve prediction accuracy for misassemblies.

The methods we describe in this chapter were developed in the specific context of validating the output of a genome assembler. Similar algorithmic issues arise from the increasing use of resequencing technologies as a tool to understand genomic variation. In resequencing, the assembly of a genomic sample is implicitly determined by the alignment of reads against a reference genome, and inconsistencies in this assembly highlight polymorphisms rather than errors. The techniques and software described in this chapter can be applied in a resequencing setting, by converting the mapping of reads along a reference to one of the commonly used assembly formats. Conversely, new algorithms being developed for resequencing applications will likely also find an application to detecting assembly errors.

8.2   Misassembly Signatures

Genome assembly is often likened to solving a large jigsaw puzzle, containing many pieces that are similar in shape and color (genomic repeats) and without any a priori knowledge about the picture we are trying to reconstruct. In this metaphor, assembly errors can be thought of as puzzle pieces forced together but that do not fit well in the big picture. The notion of "fit" can be more formally defined in a genomic assembly context, in both categorical and probabilistic terms. In the former, categorical case, assembly errors are identified by sequence reads that cannot correctly be placed within the genome. Such misplaced reads can represent singleton sequences (the proverbial "extra" pieces left behind after putting back together a piece of machinery), mate-pairs whose placement in the genome is inconsistent with the library constraints (see Figure 8.1), or overlapping reads whose base composition differs more than can be explained by sequencing errors (corresponding to puzzle pieces that have been forced together even though they do not fit). In the probabilistic formulation, assembly errors correspond to regions of the genome where the tiling of shotgun reads is inconsistent with the random process used to generate these sequencing. For example, sections of an assembly where more reads "pile up" than expected can indicate the collapse (coassembly) of multiple copies of a genomic repeat. Note that this probabilistic setting leads to an elegant formulation of genome assembly as the task of identifying a tiling of the shotgun reads that best matches the properties of the random process used to generate the data. This formulation, originally proposed by Myers [22], is computationally intractable in the general case, however provides a powerful tool that can both guide the assembly process and can be used to identify assembly errors.

In the following sections, we survey several signatures that can be used to identify assembly errors. We separately discuss categorical and probabilistic approaches for detecting misassemblies, as well as the ability of the different techniques to provide an insight into the correct reconstruction of a genome.

8.2.1   Categorical signatures of assembly errors

In the ideal case, the jigsaw metaphor accurately mirrors the complexity of genome assembly. Every genomic fragment provided as input to the assembler fits perfectly in a correct reconstruction of the genome. In practice, however, sequencing technologies are not perfect and a certain amount of error must be tolerated during the assembly process—corresponding to loosely fitting jigsaw pieces. Allowing imperfect overlaps between the reads during the assembly process exacerbates the problems posed by repeats—genomic regions present in identical or near-identical copies throughout the genome. Sequence reads anchored in a repeat are often "misplaced" by an assembler, usually within the incorrect copy of the repeat. In many cases, this ambiguity in the placement of individual reads can lead to an incorrect reconstruction of the genome. In this section we will describe several computational approaches for identifying inconsistencies in the placement of sequence reads within an assembly. Such inconsistencies indicate positions in the assembly that are potentially incorrect.

8.2.1.1 Polymorphism detection

Individual copies of a genomic repeat frequently differ from each other, either due to mutations acquired after a duplication event, or when the repeat represents multiple insertions of an evolving mobile element (e.g., transposon or virus). In a correct assembly of a clonal organism, every read is placed within the appropriate repeat copy leading to only a few differences between coassembled reads corresponding to sequencing errors. In a misassembled genome, however, some or all of the reads from a copy of a repeat may be incorrectly mixed with reads originating from another copy of the repeat, leading to multiple differences between coassembled reads. Furthermore, the same differences can be consistently found within multiple reads, providing a signature for this type of error.

Several approaches have been proposed for identifying correlated polymorphisms between a set of overlapping reads. The simplest relies on the base quality information provided by virtually all available sequencing technologies. This information is usually provided as a phred score—the scaled log-probability of error at a specific location within a read, i.e., $score = -10\log_{10} P(error)$. Specifically, a quality value of ten corresponds to a one in ten chance that the base call is incorrect while a score of 20 corresponds to a one in 100 chance of error. Assuming that sequencing errors are independent, and that

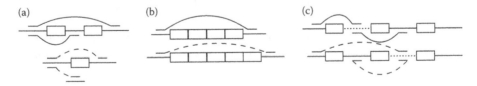

FIGURE 8.1: Mate-pair violations due to misassemblies. Top pictures represent correct genome structure while the bottom represent misassemblies. Boxes represent repeats. Mate-pairs are represented as short lines connected by thin arcs (correct arrangement) or dashed arcs (incorrect arrangement). (a) Collapsed repeat results in the creation of two contigs, a shortened mate-pair (top) and a mate-pair linking the two contigs. (b) Incorrect estimation of number of copies within a tandem repeat results in a lengthened mate-pair. (c) Rearrangement of genome around three-copy repeat (dotted and thick sections are swapped in the incorrect assembly) results in a lengthened mate-pair (top) and an incorrectly oriented pair (bottom).

the phred scores accurately reflect the quality of the underlying data, this information provides a means for evaluating whether differences identified within a column of the multiple alignment of reads can be explained by sequencing errors alone. For example, assume that we identify a heterogeneous column in the assembly, i.e., a column representing a disagreement between two or more sets of reads. We can compute the strength of the support for each of the different base-calls within this column by simply adding together the quality values (corresponding to multiplying the probabilities of error). We can then choose a quality level threshold and infer that base calls of lower quality represent sequencing errors. The remaining differences are likely caused by assembly errors. Note that this approach is sensitive even in the presence of sequencing errors—by adding the scores of agreeing reads we guarantee that the quality cut-off will be exceeded for true differences between the reads.

In practice this simple column by column approach can lead to many false positives, due to deviations from the assumptions made above. Specifically, quality scores do not perfectly reflect the probability of error, furthermore the errors are often correlated across multiple reads. To address this issue, the amosvalidate package of Phillippy et al. [24] only reports assembly errors if multiple polymorphic columns occur nearby each other. A more rigorous approach is provided by Tammi's defined nucleotide position (DNP) framework [30]. Defined nucleotide positions (DNPs) are defined as pairs of differences that cooccur within multiple reads, and which are unlikely to have occurred by chance (as estimated on the basis of quality values and Poisson statistics). Note, however, that the sensitivity of this approach is limited in the context of short-read sequencing technologies, due to the narrow range within which differences must occur in order to be detected.

Polymorphism information was also used to guide methods for correcting an erroneous assembly. The DNPTrapper program of Arner et al. [1] allows users to cluster together reads that share a common pattern of polymorphisms, thereby reconstructing individual copies of a repeat. Zhi et al. [33] describe the program Euler-AIR that rearranges misplaced reads in order to correctly reconstruct the consensus of individual repeat copies. Specifically, they use an iterative process aimed at increasing the consistency of reads aligned within a specific repeat copy.

Note that the methods described above can only be applied to validate and improve the assembly of clonal organisms due to the large number of real (vs. misassembly induced) polymorphisms present in the assemblies of out-bred eukaryotes or mixtures of organisms.

8.2.1.2 Missing reads

In all but the simplest cases, the output of a genome assembler contains a set of "singleton" reads (sometimes called shrapnel)—shotgun reads that could not be placed within the assembly. These are frequently dismissed as contaminant DNA or sequencing artifacts, however they can also indicate the

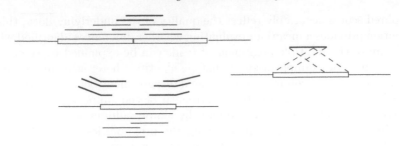

FIGURE 8.2: Alignment of reads along a genome within a tandem repeat (top). The reads spanning the boundary between the two copies (shown in bold) do not fit within an assembly where the repeat was collapsed into a single copy (bottom). Their alignment to the misassembled genome consists of two segments as shown in the middle panel.

presence of misassemblies. A typical example is the case of a collapsed tandem repeat, as shown in Figure 8.2. The reads spanning the boundary between two copies of this repeat cannot be placed within the incorrect assembly that contains only one of the copies. By mapping the singleton reads back to the assembly, this situation can be recognized from the tell-tale alignment signature shown in Figure 8.2.

This approach is implemented in the amosvalidate package of Phillippy et al., who used the program MUMmer for the alignment of singleton reads to a genomic assembly. They then identified alignment breakpoints inferred from singleton reads that had one or more partial alignments to assembly.

To account for potential sequencing artifacts (such as chimeric reads or mistrimmed sequences), only breakpoints confirmed by multiple reads were reported as potential signatures of assembly errors.

8.2.1.3 Misplaced mate-pairs

Perhaps the most useful source of information for misassembly detection is the information provided by paired-end reads. These represent pairs of sequencing reads whose relative separation within the genome is approximately known, and this information can be generated by virtually all current sequencing technologies.

The tiling of reads along a correct genome assembly should be consistent with the constraints imposed by the mate-pair information with the exception of possible errors in the estimation of mate-pair sizes, or other experimental error. Deviations from this ideal correlate with different types of assembly error (see Figure 8.3). Before briefly discussing several approaches for identifying such signatures, it is important to highlight the fact that experimental estimates of mate-pair sizes are often inaccurate, requiring the recomputation of this information from assembly data. Specifically, the mate-pair distance is computed for every set of paired reads that map within a same contig in

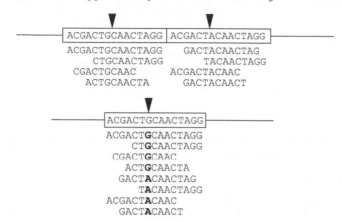

FIGURE 8.3: Two-copy tandem repeat containing a difference between the repeat copies (top). The tiling of the reads within the collapsed repeat contains a polymorphic column highlighting the collapse (bottom).

the assembly in the correct orientation. All mate-pairs that deviate too much from an original size estimate are excluded from this analysis. A new estimate is then computed of the mean and standard deviation of the library size, then the entire process is repeated in an iterative fashion until convergence. Note that in order to avoid end-contig effects only mate-pairs that are sufficiently far from the end of the contigs are considered in the calculation.

Several approaches have been proposed that use the mate-pair information for detecting misassemblies. The CE statistic of Zimin et al. [34] is used to compare, using the statistical Z test, the deviation between the distribution of mate-pair sizes within a library and the observed distribution of sizes for mates spanning a specific location in the genome. Regions that significantly deviate from the expected sizes represent potential compression (deletion of repeat copies) or expansion (addition of repeat copies) misassemblies.

The Tampa program of Dew et al. [6] relies on a computational geometric approach to identify several types of assembly errors, including rearrangements, inversions, as well as the afore-mentioned compressions and expansions. This approach converts the alignment of the end reads of a mate-pair to a two-dimensional polygon, then computes the intersection of multiple polygons (corresponding to multiple mate-pairs spanning a same error) in order to refine the boundaries of the misassembly.

The amosvalidate package also includes several mate-pair validation procedures, based on identifying clusters of mate-pairs that exhibit a same pattern when aligned to the assembly. This package also reports regions of the genome that are not spanned by any mate-pairs, corresponding to potential false joins between adjacent contigs (see Figure 8.4).

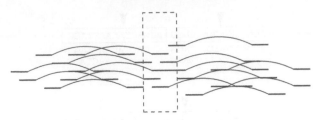

FIGURE 8.4: Weak join within an assembly. The region highlighted by the box is not spanned by any mate-pair highlighting a possible misjoin.

Finally several packages are available that allow a manual qualitative analysis of mate-pair consistency. These include consed [12], BACCardI [3], and Hawkeye [29]. Note that Hawkeye directly integrates with the amosvalidate pipeline allowing for the simultaneous examination of multiple types of misassembly signatures.

8.2.2   Identifying deviations from randomness

Shotgun sequencing process can be approximated as a random sampling process—every location of the genome is equally likely to be sampled by an individual read. Deviations from this property can be used as signatures of misassembly. For example, the collapse of two repeat copies leads to a noticeable increase in the density of the reads mapped within this repeat. Several approaches have been proposed to detect such deviations from randomness. Myers et al. [23] compute an arrival rate statistic (A-stat)—the log ratio between the probabilities that a specific genomic region (their analysis was focused on entire assembled contigs) represents a two-copy collapsed repeat versus unique sequence. These probabilities are computed using Poisson statistics, as the likelihood that k reads of length F start within a genomic region of length p in a genome of length G : $\frac{(pF/G)^k}{k!}e^{-pF/G}$.

Phillippy et al. propose a different approach that compares the repeat content of the genome as inferred from the set of reads to that inferred from the assembled genome. Specifically, they compare the frequencies of k-mers (words of length k) within the set of reads to the frequencies of the same k-mers within the genome. For a given k-mer, it's frequency within the set of reads is denoted by KR, and it's frequency within the assembly as KA. Within a correct assembly the ratio of these values, $K^* = KR/KA$, should equal the depth of coverage and be constant throughout the genome. The frequency of k-mers located within collapsed repeats is, however, lower, leading to higher values for the K^* parameter, providing a signature of the misassembly.

It is important to note that real data contain frequent deviations from randomness. Such situations are incorrectly characterized as misassemblies by the methods described above. These statistical approaches must, therefore, be combined with other validation methods in order to reduce the incidence of false-positives.

8.3   An entropy based sequence assembly validation

Kim et al. [20] developed a sequence assembly validation method by using a probabilistic function from statistics on the number of patterns per fragment, called *fragment distribution*. The probability function was designed to measure how much each aligned fragment is likely to contribute to misassembly, thus, the misassembly detection method was called as an entropy-based method.

8.3.1   Fragment distribution

The distribution of patterns of a fixed size is widely used for handling repeats in the fragment assembly problem. Since fragment selection in the shotgun data is based on the random sampling principle, patterns that occur within repeats should exhibit unusually high occurrences, compared to the average coverage of the shotgun data. Thus, patterns with unusually high occurrences are useful in detecting fragments from repeats. However, assembly is misled only when repeats are longer than fragment overlap regions. This implies that single patterns that occur unusually frequently do not necessary mislead the assembly process. The fragment distribution was one of repeat handling techniques used in a sequence assembler called AMASS [21] and it is designed to consider distributions of multiple patterns to identify sets of patterns that are likely to contribute to the misassembly of fragments. For each fragment, a fixed number of nonoverlapping patterns are selected at random positions.

Occurrences of these randomly selected patterns are searched in all fragments using a fast multiple string pattern matching algorithm [18]. The final distribution, the number of patterns in a fragment versus the number of fragments with a given number of patterns, is then generated. Note that the fragment distribution is a fragment-centric distribution that counts multiple pattern occurrences within a fragment. Figure 8.5 compares pattern and fragment distributions from four shotgun data sets: *M. gen,* G1 (a 2M bp genome), G2, and G3 (4M bp genomes). The last three genomes are yet to be published so we use G1, G2, and G3 instead of their real names. G3 had more repeats than G2, thus it was the most difficult to assemble correctly. Two patterns of 20 bp were selected from each fragment to compute the pattern and fragment distributions. *M. gen* and G1 do not have long repeats but they have many short repeats that do not present hurdles to correct fragment assembly. The length of tails in both pattern and fragment distributions indicate the complexity of assembly. The fragment distributions for *M. gen* and G1 correctly show the characteristics of the shotgun data sets with short tails in the fragment distribution while the pattern distributions show long tails. G2 and G3 were of similar size. G2 had more short repeats than G3 while G3 had longer repeats than G2; note again that only long repeats are likely to mislead the

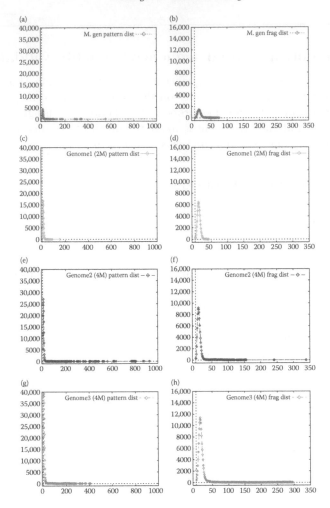

FIGURE 8.5: Comparison of the pattern and fragment distributions from four shotgun data sets. There were more repeats in the order of *M. gen,* G1, G2, and G3 (most repeat sequences). Fragment distributions were more useful in showing the complexity of assembling fragments due to repeats than pattern distributions. For example, long tails in pattern distributions of *M. gen* and G1 were due to very short repeats that did not affect assembly of fragments. In the fragment distributions of *M. gen* and G1, there were no long tails unlike in the pattern distributions. Comparison of the pattern and fragment distributions from the shotgun data sets of the similar size genomes (G2 and G3) shows the effectiveness of the fragment distribution in detecting patterns from repeats; compare the length of tails in the pattern and fragment distributions.

assembly. The fragment distributions for G2 and G3 show the characteristics of the shotgun data sets (in terms of the length of tails in the distribution) while the pattern distributions did not.

8.3.2 Probabilistic function and entropy

To measure entropy for fragments' contribution to misassembly, a probability function is needed since entropy is a sum of the negative logarithm of probability values. Since a fragment with more pattern occurrences is likely to lead to misassembly, a simple function f was used to reflect how close each fragment was to the tail region of the fragment distribution. The probability that a fragment f contributed to misassembly was computed as below:

$$\text{prob}(f_i) = 0.001 \text{ if number of patterns in } f_i < 2 \times p_v$$
$$0.01 \text{ if } 2 \times p_v \leq \text{ number of patterns in } f_i < 3 \times p_v$$
$$0.1 \text{ if } 3 \times p_v \leq \text{ number of patterns in } f_i < 4 \times p_v$$
$$0.889 \text{ if } 4 \times p_v \leq \text{ number of patterns in } f_i$$

where p_v denotes the peak value (mode) in the fragment distribution.

Once the probability function was defined, entropy at base position p in a contig was defined as below:

$$\text{entropy}(p) = \sum_{p - \delta \leq \text{pos}(f_i) \leq p + \delta} -\text{prob}(f_i) \times \log(\text{prob}(f_i))$$

where $\text{pos}(f_i)$ denotes the left end position of f_i in the contig and δ is a user input parameter (by default, it is the same as the window size used for the fragment distribution calculation).

8.3.3 Experiment with assembly of *Mycoplasma genitalium*

The authors tested their approach with the *de novo* sequence assembly of *Mycoplasma genitalium* genome [10] generated using PHRAP 2000 version [13] with default parameters. Seven contigs were assembled from shotgun data obtained from *The Institute for Genomic Research* (TIGR): C1, C3, C4, C5, C6, C7, and C8 (contigs were numbered as PHRAP generated). By aligning the contigs to the known target sequences, six misassembled regions were identified. Note that misassembly was done using a 2000 version of PHRAP; a recent test with the current PHRAP version (2007) made no mistake in the assembly.

The authors compared two approaches for detecting misassembled regions: entropy plot and fragment coverage plot. The fragment coverage approach is quite intuitive since misassembled regions corresponding to repeats exhibit coverage much higher than the average coverage of shotgun data. However, there are two issues when the coverage plot is used for detecting misassembled regions: (1) it is difficult to distinguish randomly occurring high coverage regions from repeat-induced high coverage regions, and (2) misassembly can occur in low coverage regions. Indeed, coverage plots were not effective at all in detecting misassembled regions. The entropy based approach was successful in finding all misassembled regions (100% sensitivity) and a few false positives (less than 20 peaks, representing possible misassembled regions; high specificity; an exact calculation of specificity would depend on the definition of peaks). See Kim et al. [20] for more detail.

8.4 A Machine Learning Approach to Combined Evidence Validation of Genome Assemblies

Choi et al. [5] developed a sequence assembly validation method that combines evidences from multiple measures (signatures in Section 8.2) using machine learning techniques. Several measures for predicting misassemblies were used to predict correct-assembly or misassembly. Predictions from these measures were used as attributes to train standard machine learning predictors. Trained predictors were then evaluated for misassemblies in contigs that were not used for training these predictors. We first describe how to compute individual measures for assembly validation, and explain how to combine individual measures for machine learning algorithms.

8.4.1 Individual measures for assembly validation

There are several measures for assembly validation based on coverage statistics, clone length statistics, and repeat analysis: read coverage (RC), clone coverage (CC), compression/expansion (CE), good-minus-bad (GMB) coverage, average Z-score of clone length (AZ), and average signed Z-score of clone length (ASZ). RC and CC count the number of *local* reads and clones (i.e. paired end reads) that span a specific region or segment in the genome assembly. These simple statistics are useful because a segment in a contig that is collapsed by repeats is expected to exhibit a higher coverage than the average coverage. On the other hand, a segment with a sequence inserted as a result of misassembly will exhibit a lower coverage than the average coverage. Thus, the deviation of the RC/CC from the average coverage along the entire genome would indicate a putative misassembly within this segment.

The widely used double-barrel shotgun sequencing approach utilizes paired end read information: DNA fragments sequenced at the both ends of a clone. The paired end read information is used to generate scaffold contigs to construct a longer DNA segment with gaps. After assembly and scaffolding of contigs, paired reads, simply denoted *clones* hereafter, can be classified into four groups (Figure 8.6):

- Intracontig clones: paired end reads placed within the same contig

- Intrascaffold clones: paired end reads placed within the same scaffold, yet anchored in different contigs

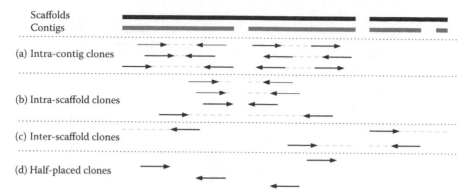

FIGURE 8.6: Clone classification. (a) intracontig clones; (b) intrascaffold clones; (c) interscaffold clones; (d) half-placed clones. The thin and thick lines represent good and bad clones, respectively.

- Interscaffold clones: paired end reads placed among different scaffolds

- Half-placed clones: one of the paired end reads placed in the assembly, but the other not.

The clone length information can be used to predict potential misassemblies as follows. If the length of a clone is smaller than the average length of all assembled clones of the same type, this would indicate a region that is potentially collapsed by repeats. On the contrary, if the length of a clone is larger than the average length of all assembled clones of the same type, this would indicate insertion of a sequence that does not belong in the region, i.e., expansion. A clone length distribution is calculated for each clone library by discarding outliers. Note that we cannot define the lengths of inter-scaffold clones and half-placed clones, thus they were not used in Choi et al. [5]. Figure 8.7 shows that a clone length distribution fits well a Gaussian model with narrow skews.

After computing a mean and a standard deviation of clone lengths for each library, the CE statistic is defined as a magnitude in standard deviations that the average length of *local* clones differs from the average length of all clones for each clone library [34]. Since clones of different types can cover a segment in a contig, the final CE statistics are computed as an average weighted by standard deviations of different clone libraries. A segment is identified as potentially misassembled if a CE statistic in the segment is smaller (say -4) or greater (say $+3$) than a preset threshold. Similar to the CE statistic, the Z-score is defined as the number of standard deviations that the length of a clone differs from the average length of all clones from the same clone library. AZ is defined as the average of the Z-scores of local clones, and ASZ are as the averages of the positive and negative Z-scores, respectively.

Kim et al. [19] have developed a measure called GMB. If the absolute Z-score of the length of a clone is smaller than a threshold, e.g., two, the intracontig or intrascaffold clone is called *good*. Otherwise the clone is called

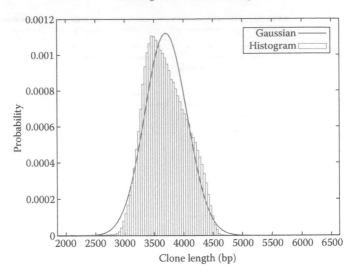

FIGURE 8.7: The distribution of clone lengths deduced from locations of paired end reads placed in the 2004 draft assembly of *Drosophila virilis*. The Gaussian distribution ($\mu = 3699$, $\sigma = 355$) fitting the library of 3704 bp is shown in the line.

bad as shown in Figure 8.6. Since the lengths of all half-placed and interscaffold clones are not defined, they are defined as *bad* clones. Clones whose paired-end reads are placed in conflicting orientations (Figure 8.6a) are also called *bad*. After classifying good and bad clones, GMB is defined as the number of bad clones minus the number of good clones that span a specific genomic segment in the assembly. Similarly, the measure RGB, at ratio of good to bad CC, is computed by a logarithmic ratio of the number of good clones to the number of bad clones.

8.4.2   Machine learning approach to combined evidences for assembly validation

In the experiments in [5], no single measure among those described in the previous section performed best in identifying misassembled regions. Thus, Choi et al. [5] developed a machine learning approach to assembly validation by combining predictions from all available measures that were described in the previous section. The machine learning approach builds a model from a given training set (contigs with assembly correctness information), and classifies a given test data (contigs that are not used for training) by predictors. Five different machine learning algorithms were used using the Weka package [32]: the decision tree (J48), the random forest (RF), the random tree (RT), the naïve Bayes classifier (NB), and the Bayesian network (BN).

The decision tree [26] constructs a tree where a node contains a rule and an edge represents a decision value of its parent rule. Final prediction is made at leaves of the tree. The RF [4] use a mode of class values predicted by many tree predictors. The generalization error for forests converges a.s. to a limit as the number of trees in the forest becomes large. The RT [7] randomizes the internal decisions of the learning algorithm when building a decision tree. It computes a certain number of best splits among those with nonnegative information gain ratio and then chooses uniformly randomly among them. The NB [8] is a simple probabilistic classifier with an assumption of independence of attributes, thus termed *naïve*. Even with the strong assumption of independence, NB often performs very well, even outperforming sophisticated classification methods. The BN [14] is a directed, probabilistic graph where a node represents a probabilistic variable and an edge represents the conditional probability between the parent and the node. On the contrary to NB, BN considers "dependence" among some attributes, those connected in a directed graph.

The performance of ML approaches often depends on which features are used for prediction. Choi et al. [5] used features in three categories: coverage statistics, length statistics, and repeat measurements. The coverage statistics include RC, CC, GMB, and good-to-bad ratio. The length statistics include AZ, average positive and negative Z-scores, and CE statistics. The repeat measurements include the number of repeats identified by RepeatMasker and by a self-comparison of the genomic segment.

To train predictors, it is necessary to have training data that shows which segments are assembled wrong. We can use as training data experimentally validated misassembly and cross species learning (Section 8.4.3). The experimentally validated data result from many data sources: cDNA sequencing, BAC fingerprinting, genetic mapping projects, genome-scale DNA tiling array [28] and optical mapping [27]. When these experimental data are available, a fraction of genome assembly can be used for training predictors and then real prediction can be made for the remaining assembly portion. When there is no such experimental data, it is possible to utilize cross species learning. A *validated* assembly of a genome that is close to the target genome can be used as training data.

8.4.3   Experimental results

8.4.3.1   Evaluation method of misassembly prediction

A published genome sequence is used to measure the performance of the predictors. A de novo assembly or a test draft assembly is generated from the shotgun data for the published genome. Then misassembled segments in contigs were identified by aligning the test draft assembly to the published genome sequence. After filtering duplicated matches of contigs to the published genome sequence, the final genome sequence contains *matched* and *unmatched*

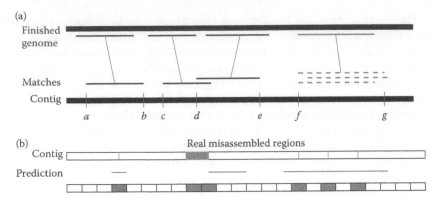

FIGURE 8.8: Determining true misassembly regions by comparing the draft assembly with a finished genome. (a) Classification of misassembly break-points. The matches between a contig in the draft assembly and the finished genome are represented by the solid and dotted lines. The first match is unique. Therefore there is no evidence of misassembly. The second and third matches along the contig have overlapping matches, which represent misassembled regions. In total, seven breakpoints (a–g) are considered true misassemblies from this evaluation. (b) Uncertainties up to 500 bp are allowed in the predicted misassembly breakpoints for a fair evaluation of the validation methods. To achieve this, a contig is split into blocks of 500 bp, and the number of blocks for true positives (TP), false positives (FP), false negatives (FN), and true negatives (TN) are counted at the block level. For this example, if the blue lines are predicted misassembled regions by a validation method and the red blocks are true misassembled blocks, we count 5 TPs, 7 FPs, 1 FNs, and 11 TNs, respectively.

regions as shown in Figure 8.8a. The *matched* regions are classified to (1) *unique* matches, e.g., the segment $[a, b]$, (2) two or more overlapping matches, e.g., the segment $[c, e]$, or (3) a match along with the other alternative matches, e.g., the segment $[f, g]$. The unmatched regions, e.g., $[b, c]$ and $[e, f]$, are defined as the breakpoints. The overlapping matches, e.g., around d with overlapping matches $[c, d]$ and $[d, e]$, are also defined as breakpoints. These breakpoints are considered *true* misassemblies. As shown in Figure 8.8b, for fair evaluation, the genome is split into fixed (say 500 bp) blocks and the number of blocks are counted instead of actual breakpoints.

8.4.3.2   Material

The draft assemblies of three *Drosophila* species, i.e. *Drosophila mojavensis*, *Drosophila erecta*, and *Drosophila virilis* were downloaded from the AAA (assembly, alignment and annotation of the now 12 sequenced *Drosophila* genomes) website at http://rana.lbl.gov/drosophila/. Table 8.1

TABLE 8.1: The statistics of draft assemblies for *Drosophila erecta,*
Drosophila mojavensis, and *Drosophila virilis.*

	2005 August	2005 August	2005 August
	D. mojavensis	*D. erecta*	*D. virilis*
N. of reads	2,717,401	2,728,578	3,320,772
L. of reads (bp)	1,838,125,872	1,855,877,180	1,939,679,395
Coverage*	9.5	12.3	9.4
N. of clones	1,074,817	1,137,308	1,085,481
N. of half-placed clones	164,347	59,658	224,235
N. of contigs	12,351	7759	19,138
Length of contigs (bp)	180,519,631	145,196,048	189,914,823
N. of N50 contigs	437	95	475
L. of N50 contigs (bp)	100,418	365,805	101,385
N. of scaffolds	5124	13,562	
L. of scaffolds (bp)	194,270,144	152,862,534	206,998,770
N. of N50 scaffolds	4	4	6
L. of N50 scaffolds (bp)	24,782,941	18,750,251	10,165,514
Total blocks	259,140	263,083	291,908
Misassembled blocks	12,959	6985	10,848

*Coverages are calculated based on the genome sizes in [11]. N and L stand for number and
length, respectively.

shows statistics of these published genome sequence assemblies. The final assemblies in comparative analysis freeze 1 (CAF1) were used as the finished genome sequences in the experiments and are reconciliations of independent assemblies performed using Arachne and the Celera Assembler [34].

8.4.3.3 Performance of nine individual measures

Figure 8.9 shows receiver operating characteristic (ROC) curves, summarizing the performance of these measures on the draft assemblies of three *Drosophila* genomes. Overall the GMB achieved the best performance (accuracy between 0.7 and 0.8). Surprisingly, the simple minimum clone coverage measure (CCN) performed in high accuracy. However, the precisions are low, typical below 10% because the predicted misassembled regions contain mostly (>90%) false positives. This approach cannot, therefore, guide finishing efforts in genome projects to correct errors in the assembly.

8.4.3.4 Performance of the machine learning approach

To measure the prediction accuracy of the machine learning method, n blocks, $n = 50\%$ of identified misassembled blocks, of the misassembled blocks and $5n$ blocks of the correctly assembled blocks were sampled to train the model. The remaining set of blocks was used for testing. Most machine learning classifiers outperformed individual measures on these data sets (see Figure 8.9) by effectively combining evidences from multiple measures. The best classifier

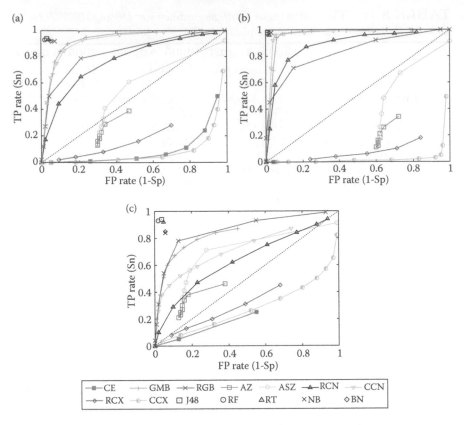

FIGURE 8.9: The ROC curves of individual measures and the combined evidence approaches based on different machine learning methods on validating the draft assemblies of (a) *Drosophila mojavensis*, (b) *Drosophila erecta* and (c) *Drosophila virilis* genomes, respectively. RCN and RCX represent minimum and maximum read coverages respectively. CCN and CCX represent minimum and maximum clone coverages, respectively.

is the RF among the five machine learning classifiers, achieving the highest precision (Table 8.2). The decision tree classifier achieved a slightly higher sensitivity than RF although its accuracy was lower than that of RF. Overall, the performance of RF is satisfactory with accuracy > 0.9 and precision ~ 0.3.

8.4.3.5 Cross species evaluation

In this experiment, one of the *Drosophila* genomes was used for training data, and another assembly was used for test misassemblies. The performance was good as shown in Table 8.3. Similar to the previous experiment, decision tree (J48) was the most accurate classifier, whereas RF achieved the highest precision.

TABLE 8.2:   The results of a cross evaluation of the machine learning approaches for validating the draft assemblies of *Drosophila* genomes.

Species	ML	TP	FN	FP	TN	PP	SN	SP
D. mojavensis	J48	6091	389	5925	201,577	0.51	0.94	0.97
	RF	6021	459	4463	203,860	0.57	0.93	0.98
	RT	6021	459	7837	201,020	0.43	0.93	0.96
	NB	5953	527	13,986	190,119	0.30	0.92	0.93
	BN	5936	544	9455	194,787	0.39	0.92	0.95
D. erecta	J48	3417	76	3386	229,717	0.50	0.98	0.99
	RF	3387	106	2600	231,082	0.57	0.97	0.99
	RT	3371	122	3861	230,365	0.47	0.96	0.98
	NB	3425	68	7054	225,139	0.33	0.98	0.97
	BN	3406	87	4490	227,872	0.43	0.98	0.98
D. virilis	J48	5118	307	9081	238,663	0.36	0.94	0.96
	RF	5030	395	6125	242,835	0.45	0.93	0.98
	RT	4979	446	11,517	237,915	0.30	0.92	0.95
	NB	4572	853	15,601	230,372	0.23	0.84	0.94
	BN	4611	814	14,485	231,404	0.24	0.85	0.94

Abbreviations: machine learning classifiers (ML), number of true positives (TP), number of false negatives (FN), number of false positives (FP), number of true negatives (TN), precision (PP), sensitivity (SN), and specificity (SP).

8.5   Summary and Conclusions

This chapter discussed characteristics of misassemblies and computational techniques for identifying misassembled regions in fragment assembly. The problem of identifying misassemblies using computational techniques is particularly relevant today mainly due to emerging massively parallel sequencing technologies such as 454 or Solexa (see Part 1 of [17] for a survey on massively parallel sequencing technologies). First, the short reads generated by the new sequencing technologies make it difficult to use several of the signatures that we discussed in this chapter, especially those relying on pattern occurrence statistics. Second, computational methods for detecting misassemblies or measuring assembly quality will be a critical prerequisite to utilizing the high throughput sequencing data since traditional, manual verification of assemblies is impractical in the context of high throughput sequencing technologies. Research on assembly validation is still in its inception and we hope this chapter provides a starting point for researchers aiming to develop new validation techniques, and provides biologists with the background necessary to incorporate measure of assembly quality in the data analysis process. Our initial work also highlights the need for good benchmark data-sets that can be used to evaluate future methods for assembly validation.

TABLE 8.3:   The results of a cross-species evaluation of the machine learning approaches for assembly validating.

Model	Test	ML	TP	FN	FP	TN	PP	SN	SP
D. mojavensis	*D. erecta*	J48	6811	174	5308	246,323	0.56	0.98	0.98
		RF	6695	290	3414	249,921	0.66	0.96	0.99
		RT	6696	289	13,737	239,968	0.33	0.96	0.95
		NB	6928	57	34,957	213,981	0.17	0.99	0.86
		BN	6907	78	13,478	235,543	0.34	0.99	0.95
D. mojavensis	*D. virilis*	J48	7622	3226	5926	272,447	0.56	0.70	0.98
		RF	6771	4077	4099	276,010	0.62	0.62	0.99
		RT	8378	2470	13,720	266,131	0.38	0.77	0.95
		NB	8902	1946	39,189	233,196	0.19	0.82	0.86
		BN	8899	1949	24,257	248,488	0.27	0.82	0.91
D. erecta	*D. mojavensis*	J48	9795	3164	5137	235,447	0.66	0.76	0.98
		RF	9427	3532	3693	239,027	0.72	0.73	0.98
		RT	9722	3237	8654	235,679	0.53	0.75	0.96
		NB	6840	6119	2988	241,949	0.70	0.53	0.99
		BN	10,215	2744	5053	233,086	0.67	0.79	0.98
D. erecta	*D. virilis*	J48	5595	5253	7185	271,088	0.44	0.52	0.97
		RF	4987	5861	5383	273,505	0.48	0.46	0.98
		RT	6087	4761	12,605	267,211	0.33	0.56	0.95
		NB	6304	4544	13,712	262,433	0.31	0.58	0.95
		BN	5836	5012	9699	267,384	0.38	0.54	0.96
D. virilis	*D. erecta*	J48	6838	147	19,555	231,898	0.26	0.98	0.92
		RF	6711	274	7748	245,201	0.46	0.96	0.97
		RT	6704	281	28,429	225,131	0.19	0.96	0.89
		NB	6913	72	27,818	221,254	0.20	0.99	0.89
		BN	6904	81	26,422	222,682	0.21	0.99	0.89
D. virilis	*D. mojavensis*	J48	12,503	456	12,890	225,864	0.49	0.96	0.95
		RF	12,137	822	8881	233,032	0.58	0.94	0.96
		RT	12,001	958	22,876	220,678	0.34	0.93	0.91
		NB	11,828	1131	9515	227,616	0.55	0.91	0.96
		BN	11,688	1271	10,924	225,591	0.52	0.90	0.95

Note: For each experiment, a pair of *Drosophila* genomes is considered, in which the first genome is used as training set, and then the second genome is tested with the model. The abbreviations and sampling follow Table 8.2.

Acknowledgments

Jeong-Hyeon Choi, Haixu Tang, and Sun Kim were supported in part by MetaCyt Microbial Systems Biology grant from the Lilly Foundations. Sun Kim was supported in part by NSF MCB 0731950 for Kim. Mihai Pop was supported in part by grant HU001-06-1-0015 from the Uniformed Services University of the Health Sciences administered by the Henry Jackson Foundation.

References

[1] E. Arner, M. Tammi, A.-N. Tran, E. Kindlund, and B. Andersson. DNPTrapper: an assembly editing tool for finishing and analysis of complex repeat regions. *BMC Bioinformatics*, 7(1):155, 2006.

[2] J. A. Bailey, Z. Gu, R. A. Clark, K. Reinert, R. V. Samonte, S. Schwartz, M. D. Adams, E. W. Myers, P. W. Li, and E. E. Eichler. Recent segmental duplications in the human genome. *Science*, 297(5583):1003–1007, 2002.

[3] D. Bartels, S. Kespohl, S. Albaum, T. Druke, A. Goesmann, J. Herold, O. Kaiser et al. BACCardI—a tool for the validation of genomic assemblies, assisting genome finishing and intergenome comparison. *Bioinformatics*, 21(7):853–859, 2005.

[4] L. Breiman. Random forests. *Machine Learning*, 45:5–32, 2001.

[5] J.-H. Choi, S. Kim, H. Tang, J. Andrews, D. G. Gilbert, and J. K. Colbourne. Machine learning approach to combined evidence validation of genome assemblies. *Bioinformatics*, 24(6):744–750, 2008.

[6] I. M. Dew, B. Walenz, and G. Sutton. A tool for analyzing mate pairs in assemblies (TAMPA). *Journal of Computational Biology*, 12(5):497–513, 2005.

[7] T. Dietterich. An experimental comparison of three methods for constructing ensembles of decision trees: bagging, boosting, and randomization. *Machine Learning*, 40(2):139–157, 2000.

[8] R. O. Duda, and P. E. Hart. Bayes decision theory. In *Pattern Classification and Scene Analysis*. John Wiley, New York, NY, 10–43, 1973.

[9] M. L. Engle, and C. Burks. GenFrag 2.1: new features for more robust fragment assembly benchmarks. *Computer Application in the Biosciences*, 10(5):567–568, 1994.

[10] C. M. Fraser, J. D. Gocayne, O. White, M. D. Adams, R. A. Clayton, R. D. Fleischmann, C. J. Bult et al. The minimal gene complement of *Mycoplasma genitalium*. *Science*, 270(5235):397–404, 1995.

[11] D. G. Gilbert. DroSpeGe: rapid access database for new *Drosophila* species genomes. *Nucleic Acids Research*, 35(suppl1):D480–485, 2007.

[12] D. Gordon, C. Abajian, and P. Green. Consed: a graphical tool for sequence finishing. *Genome Research*, 8(3):195–202, 1998.

[13] P. Green. Phrap. unpublished. http://www.phrap.org.

[14] D. Heckerman, D. Geiger, and D. Chickering. Learning bayesian networks: the combination of knowledge and statistical data. *Machine Learning*, 20:197–243, 1995.

[15] J. Kececioglu, and J. Ju. Separating repeats in dna sequence assembly. In *RECOMB '01: Proceedings of the Fifth Annual International Conference on Computational Biology*. ACM, New York, NY, 176–183, 2001.

[16] J. M. Kidd, G. M. Cooper, W. F. Donahue, H. S. Hayden, N. Sampas, T. Graves, N. Hansen et al. Mapping and sequencing of structural variation from eight human genomes. *Nature*, 453:56–64, 2008.

[17] S. Kim, H. Tang, and E. R. Mardis. *Advances in Genome Sequencing Technology and Algorithms*. Artech House, Norwood, MA, 2007.

[18] S. Kim, and Y. Kim. A fast multiple string pattern matching algorithm. In *Proceedings of 17th AoM/IAoM Conference on Computer Science*, San Diego, CA, 44–49, 1999.

[19] S. Kim, L. Liao, M. P. Perry, S. Zhang, and J.-F. Tomb. A computational approach to sequence assembly validation. unpublished, 2001. http://bio.informatics.indiana.edu/sunkim/papers/sav.ps.

[20] S. Kim, L. Liao, and J.-F. Tomb. A probabilistic approach to sequence assembly validation. In Zaki, M.J., Toivoven, H., and Wang, J.T. (Eds.) *Proceedings of the ACM SIGKDD Workshop on Data Mining in Bioinformatics (BIOKDD'01)*, San Francisco, CA, 38–43, 2001.

[21] S. Kim, and A. M. Segre. AMASS: A structured pattern matching approach to shotgun sequence assembly. *Journal of Computational Biology*, 6(2):163–186, 1999.

[22] E. W. Myers. Towards simplifying and accurately formulating fragment assembly. *Journal of Computational Biology*, 2(2):275–290, 1995.

[23] E. W. Myers, G. G. Sutton, A. L. Delcher, I. M. Dew, D. P. Fasulo, M. J. Flanigan, S. A. Kravitz, et al. A whole-genome assembly of *Drosophila*. *Science*, 287(5461):2196–2204, 2000.

[24] A. Phillippy, M. Schatz, and M. Pop. Genome assembly forensics: finding the elusive mis-assembly. *Genome Biology*, 9(3):R55, 2008.

[25] M. Pop, S. L. Salzberg, and M. Shumway. Genome sequence assembly: algorithms and issues. *Computer*, 35(7):47–54, 2002.

[26] J. R. Quinlan. *C4.5: Programs for Machine Learning*. Morgan Kaufmann, San Mateo, CA, 1993.

[27] A. Samad, E. F. Huff, W. Cai, and D. C. Schwartz. Optical mapping: a novel, single-molecule approach to genomic analysis. *Genome Research*, 5(1):1–4, 1995.

[28] M. P. Samanta, W. Tongprasit, and V. Stolc. In-depth query of large genomes using tiling arrays. *Methods Molecular Biology,* 377:163–174, 2007.

[29] M. Schatz, A. Phillippy, B. Shneiderman, and S. Salzberg. Hawkeye: an interactive visual analytics tool for genome assemblies. *Genome Biology,* 8(3):R34, 2007.

[30] M. T. Tammi, E. Arner, T. Britton, and B. Andersson. Separation of nearly identical repeats in shotgun assemblies using defined nucleotide positions, DNPs. *Bioinformatics,* 18(3):379–388, 2002.

[31] H. Tang. Genome assembly, rearrangement and repeats. *Chemistry Reviews,* 107(8):3391–3406, 2007.

[32] I. H. Witten, and F. Eibe. *Data Mining.* Hanser Fachbuch, 2001.

[33] D. Zhi, U. Keich, P. Pevzner, S. Heber, and H. Tang. Correcting base-assignment errors in repeat regions of shotgun assembly. *IEEE/ACM Transactions on Computational Biology and Bioinformatics,* 4(01):54–64, 2007.

[34] A. V. Zimin, D. R. Smith, G. Sutton, and J. A. Yorke. Assembly reconciliation. *Bioinformatics,* 24(1):42–45, 2008.

Chapter 9

Mining Patterns of Epistasis in Human Genetics

Jason H. Moore

Dartmouth Medical School

9.1 Introduction

Human genetics has a rich history of using biochemistry, biostatistics, epidemiology, molecular biology, physiology and other disciplines to determine the mapping relationship between DNA sequence information and measures of human health. This is an exciting time in human genetics because we now have access to technology that allows us to efficiently measure most of the relevant DNA sequence variations from across the human genome. We will within the next 10 years likely have access to cutting-edge technology that will deliver the entire genomic sequence for all subjects in our genetic and epidemiologic studies. Now that we have access to the basic hereditary information it is time to shift our focus toward the analysis of this data. The focus of this chapter is on the important role of computer science, and, more specifically, machine learning for mining patterns of genetic variations that are associated with susceptibility to common human diseases. Specifically, we will focus on computational methods for identifying gene–gene interactions or epistasis.

9.1.1   Epistasis and common human diseases

Human genetics has been largely successful in identifying the causative mutations in single genes that determine with virtual certainly rare diseases such as cystic fibrosis and sickle cell anemia. However, the same success has not been had for common human diseases such as cancers, cardiovascular diseases or psychiatric diseases. This is because common diseases have a much more complex etiology that requires different research strategies than were used to identify genes underlying rare diseases that follow a simpler Mendelian inheritance pattern (e.g., autosomal dominant). Complexity can arise from phenomena such as locus heterogeneity (i.e., different DNA sequence variations leading to the same phenotype), phenocopy (i.e., environmentally determined phenotypes), and the dependence of genotypic effects on environmental factors (i.e., gene–environment interactions or plastic reaction norms) and genotypes at other loci (i.e., gene–gene interactions or *epistasis*). It is this latter source of complexity, *epistasis*, that is of interest here. Epistasis has been recognized for many years as deviations from the simple inheritance patterns observed by Mendel (Bateson, 1909) or deviations from additivity in a linear statistical model (Fisher, 1918) and is likely due, in part, to canalization or mechanisms of stabilizing selection that evolve robust (i.e., redundant) gene networks (Waddington, 1942).

Epistasis has been defined in multiple different ways (e.g., Philips, 1998). We have reviewed two types of epistasis, biological and statistical (Moore and Williams, 2005). Biological epistasis when the physical interactions between biomolecules (e.g., DNA, RNA, proteins, enzymes, etc.) are influenced by genetic variation at multiple different loci. This type of epistasis occurs at the cellular level in an individual and is what Bateson (1909) had in mind when he coined the term. Statistical epistasis on the other hand occurs at the population level and is realized when there is interindividual variation in DNA sequences. The statistical phenomenon of epistasis is what Fisher (1918) had in mind. The relationship between biological and statistical epistasis is often confusing but will be important to understand if we are to make biological inferences from statistical results (Moore and Williams, 2005). Moore (2003) has argued that epistasis is likely to be a ubiquitous phenomenon in complex human diseases. The focus of the present study is the detection and characterization of statistical epistasis in human populations using machine learning and data mining methods.

To illustrate the concept difficulty, consider the following simple example of statistical epistasis in the form of a penetrance function. Penetrance is simply the probability (P) of disease (D) given a particular combination of genotypes (G) that was inherited (i.e., P[D|G]). A single genotype is determined by one allele (i.e., a specific DNA sequence state) inherited from the mother and one allele inherited from the father. For most single nucleotide polymorphisms or SNPs, only two alleles (e.g., encoded by A or a) exist in the biological population. Therefore, because the order of the alleles is unimportant, a genotype can have one of three values: AA, Aa, or aa. The model illustrated in Table 9.1

TABLE 9.1: Penetrance values for genotypes from two SNPs.

	AA (0.25)	*Aa* (0.50)	*aa* (0.25)
BB (0.25)	0	.1	0
Bb (0.50)	.1	0	.1
bb (0.25)	0	.1	0

is an extreme example of epistasis. Let's assume that genotypes *AA*, *aa*, *BB*, and *bb* have population frequencies of 0.25 while genotypes *Aa* and *BB* have frequencies of 0.5 (values in parentheses in Table 9.1). What makes this model interesting is that disease risk is dependent on the particular *combination* of genotypes inherited. Individuals have a very high risk of disease if they inherit *Aa* or *Bb* but not both (i.e., the exclusive OR function). The penetrance for each individual genotype in this model is 0.5 and is computed by summing the products of the genotype frequencies and penetrance values. Thus, in this model there is no difference in disease risk for each single genotype as specified by the single-genotype penetrance values. This epistasis example illustrates the type of concept difficulty that must be addressed by computational methods.

9.1.2 Computational challenges

Genetics and epidemiology are undergoing an information explosion and an understanding implosion. That is, our ability to generate data is far outpacing our ability to interpret it. This is especially true today where it is technically and economically feasible to measure hundreds of thousands of SNPs from across the human genome. It is anticipated that at least one SNP occurs approximately every 100 nucleotides across the $3 * 10^9$ nucleotide human genome. An important goal in human genetics is to determine which of the many thousands of SNPs are useful for predicting who is at risk for common diseases. This "genome-wide" approach is expected to revolutionize the genetic analysis of common human diseases and, for better or worse, is quickly replacing the traditional "candidate gene" approach that focuses on several genes selected by their known or suspected function.

Moore and Ritchie (2004) have outlined three significant challenges that must be overcome if we are to successfully identify genetic predictors of health and disease using a genome-wide approach. First, powerful data mining and machine learning methods will need to be developed to statistically model the relationship between combinations of DNA sequence variations and disease susceptibility. Traditional methods such as logistic regression have limited power for modeling high-order nonlinear interactions (Moore and Williams, 2002). A second challenge is the selection of genetic features or attributes that should be included for analysis. If interactions between genes explain most of the heritability of common diseases, then combinations of DNA sequence variations will need to be evaluated from a list of thousands of candidates. Filter and wrapper methods will play an important role because there are more combinations than can be exhaustively evaluated. A third challenge is the interpretation of gene–gene interaction models. Although a statistical model can be used to identify DNA sequence variations that confer risk for disease, this approach cannot be translated into specific prevention and treatment strategies without interpreting the results in the context of human biology. Making etiological inferences from computational models may be the most important and the most difficult challenge of all (Moore and Williams, 2005).

Combining the concept of interaction described above with the challenge of variable selection yields what many have called a *needle-in-a-haystack* problem. That is, there may be a particular combination of SNPs or SNPs and environmental factors that together with the right nonlinear function are a significant predictor of disease susceptibility. However, individually they may not look any different than thousands of other SNPs that are not involved in the disease process and are thus noisy. Under these models, the learning algorithm is truly looking for a genetic needle in a genomic haystack. It is now commonly assumed that at least 300,000 carefully selected SNPs may be necessary to capture all of the relevant variation across the Caucasian human genome. Assuming this is true (it is probably a lower bound), we would need to scan $4.5 * 10^{10}$ pairwise combinations of SNPs to find a genetic needle. The number of higher order combinations is astronomical. What is the optimal computational approach to this problem?

There are two general approaches to selecting attributes for predictive models. The filter approach preprocesses the data by algorithmically or statistically assessing the quality or relevance of each variable and then using that information to select a subset for analysis. The wrapper approach iteratively selects subsets of attributes for classification using either a deterministic or stochastic algorithm. The key difference between the two approaches is that the learning algorithm plays no role in selecting which attributes to consider in the filter approach. As Freitas (2002) reviews, the advantage of the filter is speed while the wrapper approach has the potential to do a better job classifying subjects as sick or healthy. We first discuss several learning algorithms that have been applied to classifying healthy and disease subjects using their DNA sequence information and then discuss filter and wrapper approaches for the specific problem of detecting epistasis or gene–gene interactions on a genome-wide scale.

9.2   Machine Learning Methods: Learning Algorithms

As discussed above, one of the early definitions of epistasis was deviation from additivity in a linear model (Fisher, 1918). The linear model plays a very important role in modern genetics and epidemiology because it has solid theoretical foundation, is easy to implement using a wide-range of different software packages, and is easy to interpret. Despite these good reasons to use linear models, they do have limitations for detecting nonlinear patterns of interaction (e.g., Moore and Williams, 2002). The first problem is that modeling interactions requires looking at combinations of attributes. Considering multiple attributes simultaneously is challenging because the available data get spread thinly across multiple combinations of genotypes, for example. Estimation of parameters in a linear model can be problematic when the data are sparse. The second problem is that linear models are often implemented such that interaction effects are only considered after independent main effects are identified. This certainly makes model fitting easier but assumes that the important predictors will have main effects. Further, it is well documented that linear models have greater power to detect main effects than interactions (e.g., Lewontin, 1974). For example, the FITF approach of Millstein et al. (2006) provides a powerful logistic regression approach to detecting interactions but conditions on main effects. Moore (2003) argues that this is an unrealistic assumption for common human diseases. The limitations of the linear model and other parametric statistical approaches has motivated the development of computational approaches such as those from machine learning and data mining that make fewer assumptions about the functional form of the model and the effects being modeled (McKinney et al., 2006). We review below decision trees that have been commonly applied in genetic studies and then discuss a novel machine learning method called multifactor dimensionality reduction (MDR) that was designed specifically for detecting epistasis in genetic studies of common diseases.

9.2.1 Decision trees and random forests (RF)

Decision trees (sometimes called classification trees) are one of the most widely used machine learning methods for modeling the relationship between one of more attributes and a discrete endpoint such as case-control status (Mitchell, 1997). A decision tree classifies subjects as case or control, for example, by sorting them through a tree from node to node where each node is an attribute with a decision rule that guides that subject through different branches of the tree to a leaf that provides its classification. The primary advantage of this approach is that it is simple and the resulting tree can be interpreted as IF-THEN rules that are easy to understand. For example, a dominant genetic model of genotype data coded $\{AA = 0, Aa = 1, aa = 2\}$ might look like IF genotype at SNP1<2 then case ELSE control. In this simple model the root node of the tree would be SNP1 with decision rule <2 and leafs equal to case and control. Additional nodes or attributes below the root node allows hierarchical dependencies (i.e., interactions) to be modeled. Random forests (RF) build on the decision tree idea and have been used to detect gene–gene and gene-environment interactions in genetic studies.

A RF is a collection of individual decision tree classifiers, where each tree in the forest has been trained using a bootstrap sample of instances (i.e., subjects) from the data, and each attribute in the tree is chosen from among a random subset of attributes (Breiman, 2001). Classification of instances is based upon aggregate voting over all trees in the forest.

Individual trees are constructed as follows from data having N samples and M attributes:

1. Choose a training set by selecting N samples, with replacement, from the data.

2. At each node in the tree, randomly select m attributes from the entire set of M attributes in the data (the magnitude of m is constant throughout the forest building).

3. Choose the best split at that node from among the m attributes.

4. Iterate the second and third steps until the tree is fully grown (no pruning).

Repetition of this algorithm yields a forest of trees, each of which have been trained on bootstrap samples of instances. Thus, for a given tree, certain instances will have been left out during training. Prediction error is estimated from these "out-of-bag" instances. The out-of-bag instances are also used to estimate the importance of particular attributes via permutation testing. If randomly permuting values of a particular attribute does not affect the predictive ability of trees on out-of-bag samples, that attribute is assigned a low importance score (Bureau et al., 2005).

The decision trees comprising a RF provide an explicit representation of attribute interaction that is readily applicable to the study of gene–gene or gene–environment interactions. These models may uncover interactions among genes and/or environmental factors that do not exhibit strong marginal effects (Cook et al., 2004). Additionally, tree methods are suited to dealing with certain types of genetic heterogeneity, since early splits in the tree define separate model subsets in the data (Lunetta et al., 2004). RF capitalize on the benefits of decision trees and have demonstrated excellent predictive performance when the forest is diverse (i.e., trees are not highly correlated with each other) and composed of individually strong classifier trees (Breiman, 2001). The RF method is a natural approach for studying gene–gene or gene–environment interactions because importance scores for particular attributes take interactions into account without demanding a pre-specified model (Lunetta et al., 2004). However, most current implementations of the importance score are calculated in the context of all other attributes in the model. Therefore, assessing the interactions between particular sets of attributes must be done through careful model interpretation, although there has been preliminary success in jointly permuting explicit sets of attributes to capture their interactive effects (Bureau et al., 2005).

In selecting functional single nucleotide polymorphism (SNP) attributes from simulated case-control data, RFs outperform traditional methods such as the Fisher Exact test when the "risk" SNPs interact, and the relative superiority of the RF method increases as more interacting SNPs are added to the model (Lunetta et al., 2004). RFs have also shown to be more robust in the presence of noise SNPs relative to methods that rely on main effects, such as Fisher's exact test (Bureau et al., 2005). Initial results of RF applications to genetic data in studies of asthma (Bureau et al., 2005), for example, are encouraging, and it is anticipated that RF will prove a useful tool for detecting gene–gene interactions. They may also be useful when multiple different data types are present (Reif et al., 2006). The primary limitation of tree-based methods is that the standard implementations condition on main effects. That is, the algorithm finds the best single variable for the root node before adding additional variables as nodes in the model. An advantage of this approach is that there are numerous open-source software packages that have implemented the algorithm for data analysis.

9.2.2 Multifactor dimensionality reduction (MDR)

MDR was developed as a nonparametric (i.e., no parameters are estimated) and genetic model-free (i.e., no genetic model is assumed) machine learning strategy for identifying combinations of discrete genetic and environmental factors that are predictive of a discrete clinical endpoint (Ritchie et al., 2001; Hahn et al., 2003; Ritchie et al., 2003; Moore, 2004; Moore et al., 2006; Moore, 2007; Velez et al., 2007; Pattin et al., 2008). Unlike most other methods, MDR was designed to detect interactions in the absence of detectable main effects and thus complements approaches such as logistic regression and RF. At the heart of the MDR approach is a feature or attribute construction algorithm that creates a new variable or attribute by pooling, for example, genotypes from multiple SNPs. The process of defining a new attribute as a function of two or more other attributes is referred to as constructive induction or attribute construction and was first developed by Michalski (1983). Constructive induction using the MDR kernel is accomplished in the following way. Given a threshold T, a multilocus genotype combination is considered high-risk if the ratio of cases (subjects with disease) to controls (healthy subjects) exceeds T, else it is considered low-risk. Genotype combinations considered to be high-risk are labeled G_1 while those considered low-risk are labeled G_0. This process constructs a new one-dimensional attribute with levels G_0 and G_1. It is this new single variable that is assessed using any classification method. The MDR method is based on the idea that changing the representation space of the data will make it easier for a classifier such as a decision tree or a naive Bayes learner to detect attribute dependencies. Cross-validation and permutation testing are used to prevent overfitting and false-positives due to multiple testing. Open-source software in Java and C are freely available from www.epistasis.org.

Numerous extensions and modifications of the MDR approach have been proposed and implemented. These include MDR for imbalanced data (Velez et al., 2007), extreme value distribution MDR for faster permutation testing (Pattin et al., 2008) and generalized MDR for continuous measures and adjustment of confounders (Lou et al., 2007), for example. In addition to the numerous methodological developments, the MDR method has been successfully applied to detecting gene–gene and gene–environment interactions for a variety of common human diseases and clinical endpoints including, for example, bladder cancer (Andrew et al., 2006, 2008). These studies demonstrate the utility of this approach.

9.3 Machine Learning Methods: Filters

As discussed above, it is computationally infeasible to combinatorially explore all interactions among the DNA sequence variations in a genome-wide association study. One approach is to filter out a subset of variations that can then be efficiently analyzed using a method such as MDR. We review below a powerful filter method based on the ReliefF algorithm and then discuss prospects for using biological knowledge to filter genetic variations.

9.3.1 ReliefF filters

There are many different statistical and computational methods for determining the quality of attributes. A standard strategy in human genetics and epidemiology is to assess the quality of each SNP using a chi-square test of independence followed by a correction of the significance level that takes into account an increased false-positive (i.e., type I error) rate due to multiple tests. This is a very efficient filtering method but it ignores the dependencies or interactions between genes. Kira and Rendell (1992) developed an algorithm called Relief that is capable of detecting attribute dependencies. Relief estimates the quality of attributes through a type of nearest neighbor algorithm that selects neighbors (instances) from the same class and from the different class based on the vector of values across attributes. Weights (W) or quality estimates for each attribute (A) are estimated based on whether the nearest neighbor (nearest hit, H) of a randomly selected instance (R) from the same class and the nearest neighbor from the other class (nearest miss, M) have the same or different values. This process of adjusting weights is repeated for m instances. The algorithm produces weights for each attribute ranging from -1 (worst) to $+1$ (best). The Relief pseudocode is outlined below:

```
set all weights W[A]=0
   for i=1 to m do begin
      randomly select an instance R_i
      find nearest hit H and nearest miss M
      for A=1 to a do
         W[A]=W[A]-diff(A, R_i, H)/m+diff(A, R_i, M)/m
   end
```

The function $\text{diff}(A, I_1, I_2)$ calculates the difference between the values of the attribute A for two instances I_1 and I_2. For nominal attributes such as SNPs it is defined as:

$\text{diff}(A, I_1, I_2) = 0$ if $\text{genotype}(A, I_1) = \text{genotype}(A, I_2)$,
   1 otherwise

The time complexity of Relief is $O(m^*n^*a)$ where m is the number of instances randomly sampled from a dataset with n total instances and a attributes. Kononenko (1994) improved upon Relief by choosing n nearest neighbors instead of just one. This new ReliefF algorithm has been shown to be more robust to noisy attributes (Kononenko 1994; Robnik-Šikonja and Kononenko, 2001, 2003) and is widely used in data mining applications.

ReliefF is able to capture attribute interactions because it selects nearest neighbors using the entire vector of values across all attributes. However, this advantage is also a disadvantage because the presence of many noisy attributes can reduce the signal the algorithm is trying to capture. Moore and White (2007a) proposed a "tuned" ReliefF algorithm (TuRF) that systematically removes attributes that have low quality estimates so that the ReliefF values if the remaining attributes can be re-estimated. The pseudocode for TuRF is outlined below:

```
let a be the number of attributes
for i=1 to n do begin
      estimate ReliefF
      sort attributes
      remove worst n/a attributes
end
return last ReliefF estimate for each attribute
```

The motivation behind this algorithm is that the ReliefF estimates of the true functional attributes will improve as the noisy attributes are removed from the dataset.

Moore and White (2007a) carried out a simulation study to evaluate the power of ReliefF, TuRF, and a naïve chi-square test of independence for selecting functional attributes in a filtered subset. Five genetic models in the form of penetrance functions (e.g., Table 9.1) were generated. Each model consisted of two SNPs that define a nonlinear relationship with disease susceptibility. The heritability of each model was 0.1 which reflects a moderate to small genetic effect size. Each of the five models was used to generate 100 replicate datasets with sample sizes of 200, 400, 800, 1600, 3200, and 6400. This range of sample sizes represents a spectrum that is consistent with small to medium size genetic studies. Each dataset consisted of an equal number of case (disease) and control (no disease) subjects. Each pair of functional SNPs was combined within a genome-wide set of 998 randomly generated SNPs for a total of 1000 attributes. A total of 600 datasets were generated and analyzed.

ReliefF, TuRF, and the univariate chi-square test of independence were applied to each of the datasets. The 1000 SNPs were sorted according to their quality using each method and the top 50, 100, 150, 200, 250, 300, 350, 400, 450, and 500 SNPs out of 1000 were selected. From each subset we counted the number of times the two functional SNPs were selected out of each set of 100 replicates. This proportion is an estimate of the power or how likely we are to find the true SNPs if they exist in the dataset. The number of times each method found the correct two SNPs was statistically compared. A difference in counts (i.e., power) was considered statistically significant at a type I error rate of 0.05. Moore and White (2007a) found that the power of ReliefF to pick (filter) the correct two functional attributes was consistently better ($P \leq .05$) than a naïve chi-square test of independence across subset sizes and models when the sample size was 800 or larger. These results suggest that ReliefF is capable of identifying interacting SNPs with a moderate genetic effect size (heritability=0.1) in moderate sample sizes. Next, Moore and White (2007a) compared the power of TuRF to the power of ReliefF. They found that the TuRF algorithm was consistently better ($P \leq .05$) than ReliefF across small SNP subset sizes (50, 100, and 150) and across all five models when the sample size was 1600 or larger. These results suggest that algorithms based on ReliefF show promise for filtering interacting attributes in this domain. The disadvantage of the filter approach is that important attributes might be discarded prior to analysis. Stochastic search or wrapper methods provide a flexible alternative. The ReliefF algorithm has been included in the open-source MDR software package available from www.epistasis.org.

9.3.2 Biological filters

ReliefF and other measures such as interaction information (Moore et al., 2006) are likely to be very useful for providing an analytical means for filtering genetic variations prior to epistasis analysis using decision trees or MDR, for example. However, there is growing recognition that we should use the wealth of accumulated knowledge about gene function to prioritize which genetic variations are analyzed for gene–gene interactions. For any given disease there are often multiple biochemical pathways, for example, that have been experimentally confirmed to play an important role in the etiology of the disease. Genes in these pathways can be selected for gene–gene interaction analysis thus significantly reducing the number of gene–gene interaction tests that need to be performed. Gene ontology (GO), chromosomal location and protein-protein interactions are all example sources of expert knowledge that can be used in a similar manner. Pattin et al. (2008) have specifically reviewed protein-protein interaction databases as a source of expert knowledge that can be used to guide genome-wide association studies of epistasis. Unfortunately, the specific methods for using expert knowledge to filter genetic variations have yet to be developed.

9.4 Machine Learning Methods: Wrappers

Stochastic search or wrapper methods may be more powerful than filter approaches because no attributes are discarded in the process. As a result, every attribute retains some probability of being selected for evaluation by the classifier. There are many different stochastic wrapper algorithms that can be applied to this problem. Moore and White (2007b) have explored the use of genetic programming (GP). GP is an automated computational discovery tool that is inspired by Darwinian evolution and natural selection (Koza 1992; Banzhaf et al., 1998). The goal of GP is evolve computer programs to solve problems. This is accomplished by first generating random computer programs that are composed of the basic building blocks needed to solve or approximate a solution to the problem. Each randomly generated program is evaluated and the good programs are selected and recombined to form new computer programs. This process of selection based on fitness and recombination to generate variability is repeated until a best program or set of programs is identified. GP and its many variations have been applied successfully in a wide range of different problem domains including bioinformatics (e.g., Fogel and Corne, 2003). It is important to note that GP differs considerably from genetic algorithms (GA) in that the solution to a problem in GP is represented by a computer program rather than a bit string as with a GA.

Moore and White (2007b) developed and evaluated a simple GP wrapper for attribute selection in the context of an MDR analysis as described below. The goal of this study was to develop a stochastic wrapper method that is able to select attributes that interact in the absence of independent main effects. At face value, there is no reason to expect that a GP or any other wrapper method would perform better than a random attribute selector because there are no 'building blocks' for this problem when accuracy is used as the fitness measure. That is, the fitness of any given classifier would look no better than any other with just one of the correct SNPs in the MDR model. Preliminary studies by White et al. (2005) support this idea. For GP or any other wrapper to work there needs to be recognizable building blocks. Moore and White (2006b) specifically evaluated whether including pre-processed attribute quality estimates using TuRF (see above) in a multiobjective fitness function improved attribute selection over a random search or just using accuracy as the fitness. Using a wide variety of simulated data, Moore and White (2007b) demonstrated that including TuRF scores in addition to accuracy in the fitness function significantly improved the power of GP to pick the correct two functional SNPs out of 1000 total attributes.

This study presents preliminary evidence suggesting that GP might be useful for the genome-wide genetic analysis of common human diseases that have a complex genetic architecture. The results raise numerous questions. How well does GP do when faced with finding three, four, or more SNPs that interact in a nonlinear manner to predict disease susceptibility? How does extending the function set to additional attribute construction functions impact performance? How does extending the attribute set impact performance? Is using GP better than filter approaches? To what extent can GP theory help formulate an optimal GP approach to this problem? Does GP outperform other evolutionary or nonevolutionary search methods? The study by Moore and White (2007b) provides a starting point to begin addressing some of these questions.

9.5   Machine Learning Methods: Interpretation

The MDR method described above is a powerful attribute construction approach for detecting epistasis or nonlinear gene–gene interactions in epidemiologic studies of common human diseases. The models that MDR produces are by nature multidimensional and thus difficult to interpret. For example, an interaction model with four SNPs, each with three genotypes, summarizes 81 different genotype (i.e., level) combinations (i.e., 3^4). How do each of these level combinations relate back to biological processes in a cell? Why are some combinations associated with high-risk for disease and some associated with low-risk for disease? Moore et al. (2006) have proposed using information theoretic approaches with graph-based models to provide both a statistical and a visual interpretation of a multidimensional MDR model. Statistical interpretation should facilitate biological interpretation because it provides a deeper understanding of the relationship between the attributes and the class variable. We describe next the concept of interaction information and how it can be used to facilitate statistical interpretation.

Jakulin and Bratko (2003) have provided a metric for determining the gain in information about a class variable (e.g., case-control status) from merging two attributes into one (i.e., attribute construction) over that provided by the attributes independently. This measure of *information gain* allows us to gauge the benefit of considering two (or more) attributes as one unit. While the concept of information gain is not new (McGill, 1954), its application to the study of attribute interactions has been the focus of a recent study (Jakulin and Bratko, 2003). Consider two attributes, A and B, and a class label C. Let H(X) be the Shannon entropy (see Pierce, 1980) of X. The information gain (IG) of A, B, and C can be written as Equation 9.1 and defined in terms of Shannon entropy (Equations 9.2 and 9.3).

$$IG(ABC) = I(A;B|C) - I(A;B) \qquad (9.1)$$

$$I(A;B|C) = H(A|C) + H(B|C) - H(A,B|C) \qquad (9.2)$$

$$I(A;B) = H(A) + H(B) - H(A,B) \qquad (9.3)$$

The first term in Equation 9.1, $I(A;B|C)$, measures the *interaction* of A and B. The second term, $I(A;B)$, measures the *dependency* or correlation between A and B. If this difference is positive, then there is evidence for an attribute interaction that cannot be linearly decomposed. If the difference is negative, then the information between A and B is redundant. If the difference is zero, then there is evidence of conditional independence or a mixture of synergy and redundancy.

These measures of interaction information can be used to construct interaction graphs (i.e., network diagrams) and an interaction dendrograms using the entropy estimates from Step 1 with the algorithms described first by Jakulin and Bratko (2003) and more recently in the context of genetic analysis by Moore et al. (2006). Interaction graphs are comprised of a node for each attribute with pairwise connections between them. The percentage of entropy removed (i.e., information gain) by each attribute is visualized for each node. The percentage of entropy removed for each pairwise MDR product of attributes is visualized for each connection. Thus, the independent main effects of each polymorphism can be quickly compared to the interaction effect. Additive and nonadditive interactions can be quickly assessed and used to interpret the MDR model which consists of distributions of cases and controls for each genotype combination. Positive entropy values indicate synergistic interaction while negative entropy values indicate redundancy.

Interaction dendrograms are also a useful way to visualize interaction (Jakulin and Bratko 2003; Moore et al., 2006). Here, hierarchical clustering is used to build a dendrogram that places strongly interacting attributes close together at the leaves of the tree. Jakulin and Bratko (2003) define the following dissimilarity measure, D (Equation 9.5), that is used by a hierarchical clustering algorithm to build a dendrogram. The value of 1000 is used as an upper bound to scale the dendrograms.

$$D(A,B) = |I(A;B;C)|^{-1} \text{ if } |I(A;B;C)|^{-1} < 1000$$

$$1000 \quad \text{otherwise} \qquad (9.4)$$

Using this measure, a dissimilarity matrix can be estimated and used with hierarchical cluster analysis to build an interaction dendrogram. This facilitates rapid identification and interpretation of pairs of interactions. The algorithms for the entropy-based measures of information gain are implemented in the open-source MDR software package available from www.epistasis.org. Output in the form of interaction dendrograms is provided.

9.6 Summary

As human genetics and epidemiology move into the genomics age with access to all the information in the genome we will become increasingly dependent on computer science for managing and making sense of these mountains of data. The specific challenge reviewed here is the detection, characterization and interpretation of epistasis or gene–gene interactions that are predictive of susceptibility to common human diseases. Epistasis is an important source of complexity in the genotype to phenotype map that requires special computational methods for analysis. We have reviewed decision trees and a powerful attribute construction method called MDR that can be used in a classification framework to detect nonlinear attribute interactions in genetic studies of common human diseases. We have also reviewed a filter method using ReliefF and a stochastic wrapper method using GP for the analysis of gene–gene interaction on a genome-wide scale with hundreds of thousands of attributes. Finally, we reviewed information theoretic methods to facilitate the statistical and subsequent biological interpretation of high-order gene–gene interaction models. These data mining and knowledge discovery methods and others will play an increasingly important role in human genetics as the field moves away from the candidate-gene approach that focuses on a few targeted genes to the genome-wide approach that measures DNA sequence variations from across the genome. We anticipate the role of expert biological knowledge will be increasingly important for successful data mining and thus learning algorithms need to adapt to exploit this information.

Acknowledgments

This work was supported by National Institutes of Health (USA) grants LM009012, AI59694, HD047447, and RR018787. We greatly appreciate the time and effort of the editors for organizing and reviewing this book.

References

Andrew, A.S., Nelson, H.H., Kelsey, K.T., Moore, J.H., Meng, A.C., Casella, D.P., Tosteson, T.D., Schned, A.R., Karagas, M.R. 2006. Concordance of multiple analytical approaches demonstrates a complex relationship between DNA repair gene SNPs, smoking, and bladder cancer susceptibility. *Carcinogenesis* 27, 1030–37.

Andrew, A.S., Karagas, M.R., Nelson, H.H., Guarrera, S., Polidoro, S., Gamberini, S., Sacerdote, C., Moore, J.H., Kelsey, K.T., Demidenko, E., Vineis, P., Matullo, G. 2008. DNA repair polymorphisms modify bladder cancer risk: a multi-factor analytic strategy. *Human Heredity* 65, 105–18.

Banzhaf, W., Nordin, P., Keller, R.E., Francone, F.D. 1998. *Genetic Programming: An Introduction : On the Automatic Evolution of Computer Programs and Its Applications.* San Francisco, CA: Morgan Kaufmann Publishers.

Bateson, W. 1909. *Mendel's Principles of Heredity.* Cambridge, UK: Cambridge University Press.

Breiman, L. 2001. Random Forests. *Machine Learning* 45, 5–32.

Bureau, A., Dupuis, J., Falls, K., Lunetta, K.L., Hayward, B., Keith, T.P., et al. 2005. Identifying SNPs predictive of phenotype using random forests. *Genetic Epidemiology* 28, 171–82.

Cook, N.R., Zee, R.Y., Ridker, P.M. 2004. Tree and spline based association analysis of gene–gene interaction models for ischemic stroke. *Statisics in Medicine* 23, 1439–53.

Fisher, R.A. 1918. The correlations between relatives on the supposition of Mendelian inheritance. *Transactions of the Royal Society of Edinburgh* 52, 399–433.

Fogel, G.B., Corne, D.W. 2003. *Evolutionary Computation in Bioinformatics.* San Francisco, CA: Morgan Kaufmann Publishers.

Freitas, A. 2001. Understanding the crucial role of attribute interactions. *Artificial Intelligence Review* 16, 177–99.

Freitas, A. 2002. Data Mining and Knowledge Discovery with Evolutionary Algorithms. New York, NY: Springer.

Hahn, L.W., Ritchie, M.D., Moore, J.H. 2003. Multifactor dimensionality reduction software for detecting gene–gene and gene–environment interactions. *Bioinformatics* 19, 376–82.

Jakulin, A., Bratko, I. 2003. Analyzing attribute interactions. *Lecture Notes in Artificial Intelligence* 2838, 229–40.

Kira, K., Rendell, L.A. 1992. A practical approach to feature selection. In: Sleeman, D.H., Edwards, P. (Eds.) *Proceedings of the Ninth International Workshop on Machine Learning* (pp. 249–256). San Francisco, CA: Morgan Kaufmann Publishers.

Kononenko, I. 1994. Estimating attributes: analysis and extension of Relief. *Proceedings of the European Conference on Machine Learning* (pp. 171–182). New York, NY: Springer.

Koza, J.R. 1992. *Genetic Programming: On the Programming of Computers by Means of Natural Selection.* Cambridge, MA: The MIT Press.

Lewontin, R.C. 1974. The analysis of variance and the analysis of causes. *American Journal of Human Genetics* 26, 400–11.

Lou, X.Y., Chen, G.B., Yan, L., Ma, J.Z., Zhu, J., Elston, R.C., Li, M.D. 2007. A generalized combinatorial approach for detecting gene-by-gene and gene-by-environment interactions with application to nicotine dependence. *American Journal of Human Genetics* 80(6), 1125–37.

Lunetta, K.L., Hayward, L.B., Segal, J., Van Eerdewegh, P. 2004. Screening large-scale association study data: exploiting interactions using random forests. *BMC Genetetics* 5, 32.

McGill, W.J. 1954. Multivariate information transmission. *Psychometrica* 19, 97–116.

McKinney, B.A., Reif, D.M., Ritchie, M.D., Moore, J.H. 2006. Machine learning for detecting gene-gene interactions: a review. *Appl Bioinformatics* 5(2), 77–88.

Michalski, R.S. 1983. A theory and methodology of inductive learning. *Artificial Intelligence* 20, 111–61.

Millstein, J., Conti, D.V., Gilliland, F.D., Gauderman, W.J. 2006. A testing framework for identifying susceptibility genes in the presence of epistasis. *American Journal of Human Genetics* 78(1), 15–27.

Mitchell, T.M. 1997. *Machine Learning.* Boston, MA: McGraw-Hill.

Moore, J.H. 2003. The ubiquitous nature of epistasis in determining susceptibility to common human diseases. *Human Heredity* 56, 73–82.

Moore, J.H. 2004. Computational analysis of gene–gene interactions in common human diseases using multifactor dimensionality reduction. *Expert Review of Molecular Diagnostics* 4, 795–803.

Moore, J.H. 2007. Genome-wide analysis of epistasis using multifactor dimensionality reduction: feature selection and construction in the domain of human genetics. In: Zhu, X., Davidson, I. (Eds.), *Knowledge Discovery and Data Mining: Challenges and Realities with Real World Data* (pp. 17–30). Hershey, PA: IGI Global.

Moore, J.H., Gilbert, J.C., Tsai, C.-T., Chiang, F.T., Holden, W., Barney, N., White, B.C. 2006. A flexible computational framework for detecting, characterizing, and interpreting statistical patterns of epistasis in genetic studies of human disease susceptibility. *Journal of Theoretical Biology* 241, 252–61.

Moore, J.H., Ritchie, M.D. 2004. The challenges of whole-genome approaches to common diseases. *Journal of the American Medical Association* 291, 1642–43.

Moore, J.H., White, B.C. 2007a. Tuning ReliefF for genome-wide genetic analysis. *Lecture Notes in Computer Science* 4447, 166–75.

Moore, J.H., White, B.C. 2007b. Genome-wide genetic analysis using genetic programming. The critical need for expert knowledge. In: Rick R., Terence S., Bill W., (Eds.). *Genetic Programming Theory and Practice IV* (pp. 11–28). New York, NY: Springer.

Moore, J.H., Williams, S.W. 2002. New strategies for identifying gene-gene interactions in hypertension. *Annals of Medicine* 34, 88–95.

Moore, J.H., Williams, S.W. 2005. Traversing the conceptual divide between biological and statistical epistasis: Systems biology and a more modern synthesis. *BioEssays* 27, 637–46.

Pattin, K.A., White, B.C., Barney, N., Gui, J., Nelson, H.H., Kelsey, K.T., Andrew, A.S., Karagas, M.R., Moore, J.H. 2008. A computationally efficient hypothesis testing method for epistasis analysis using multifactor dimensionality reduction. *Genetic Epidemiology* 33(1), 87–94.

Phillips, P.C. 1998. The language of gene interaction. *Genetics* 149, 1167–71.

Pierce, J.R. 1980. *An Introduction to Information Theory: Symbols, Signals, and Noise.* New York, NY: Dover.

Reif, D.M., Motsinger, A.A., McKinney, B.A., Crowe Jr, J., Moore, J.H. 2006. Feature selection using random forests for the integrated analysis of multiple data types. *Proceedings of the 2006 IEEE Symposium on Computational Intelligence in Bioinformatics and Computational Biology* (pp. 171–178). New York, NY: IEEE Press.

Ritchie, M.D., Hahn, L.W., Moore, J.H. 2003. Power of multifactor dimensionality reduction for detecting gene–gene interactions in the presence of genotyping error, phenocopy, and genetic heterogeneity. *Genetic Epidemiology* 24, 150–57.

Ritchie, M.D., Hahn, L.W., Roodi, N., Bailey, L.R., Dupont, W.D., Parl, F.F., Moore, J.H. 2001. Multifactor dimensionality reduction reveals high-order interactions among estrogen metabolism genes in sporadic breast cancer. *American Journal of Human Genetics* 69, 138–47.

Robnik-Šikonja, M., Kononenko, I. 2001. Comprehensible interpretation of Relief's estimates. In: Carla E. Brodley and Andrea Pohoreckyj D. (Eds.) *Proceedings of the Eighteenth International Conference on Machine Learning* (pp. 433–440). San Francisco, CA: Morgan Kaufmann Publishers.

Robnik-Šiknja, M., Kononenko, I. 2003. Theoretical and empirical analysis of ReliefF and RReliefF. *Machine Learning* 53, 23–69.

Velez, D.R., White, B.C., Motsinger, A.A., Bush, W.S., Ritchie, M.D., Williams, S.M., Moore, J.H. 2007. A balanced accuracy function for epistasis modeling in imbalanced datasets using multifactor dimensionality reduction. *Genetic Epidemiology* 31, 306–15.

Waddington, C.H. 1942. Canalization of development and the inheritance of acquired characters. *Nature* 150, 563–65.

White, B.C., Gilbert, J.C., Reif, D.M., Moore, J.H. 2005. A statistical comparison of grammatical evolution strategies in the domain of human genetics. *Proceedings of the IEEE Congress on Evolutionary Computing* (pp. 676–682). New York, NY: IEEE Press.

Chapter 10

Discovery of Regulatory Mechanisms from Gene Expression Variation by eQTL Analysis

Yang Huang*, Jie Zheng*, and Teresa M. Przytycka

National Institutes of Health

10.1   Introduction

One of the big challenges in the postgenome era is the better understating of the gene regulation process. Recently developed high-throughput genotyping and gene expression platforms have enabled a powerful new tool: expression quantitative trait loci (eQTL) analysis, where loci or markers on the genomes are associated with variations in gene expression. In such studies, gene expression is treated as a quantitative trait. DNA polymorphism at the gene, binding site or regulatory proteins, such as transcription factors

---

*These authors contributed equally to the work.

(TFs), transcription association proteins and signaling proteins, is likely to affect the expression level of the gene in an inheritable way (Monks et al., 2004; Brem and Kruglyak, 2005; Petretto et al., 2006b). Hence, a significant statistical association between a locus and the expression level of a gene suggests that the locus might regulate the gene. Since the early work of Jansen and Nap (2001), eQTL has become a widespread technique to identify such regulatory associations and has been applied to a number of species including yeast (Yvert et al., 2003; Brem and Kruglyak, 2005; Brem et al., 2005), *Arabidopsis* (DeCook et al., 2006), maize (Schadt et al., 2003), *Drosophila* (Jin et al., 2001), mouse (Klose et al., 2002; Bystrykh et al., 2005; Chesler et al., 2005), rat (Petretto et al., 2006a) and human (Monks et al., 2004; Cheung et al., 2005; Stranger et al., 2005). Many of them used the strategy of genome-wide association study (GWAS), considering loci covering the whole genome and expression profiles of all or nearly all genes identified in the organism.

eQTL mapping is a variant of the classical quantitative trait loci (QTL) mapping, which discovers associations between genotypes and organism-level phenotypes, such as heritable diseases, height, weight, etc. Compared with classical QTL analysis, eQTL analysis presents unique opportunities and challenges. While classical QTL analysis considers traits at organism-level, eQTL analysis allows us to observe the most immediate consequences of genetic polymorphism, namely, its effects on gene expression through transcription and post-transcription control. Therefore, eQTL-based approaches are often used to identify various regulation mechanisms. Moreover, combined with QTL analysis, eQTL mapping also helps to unravel genetic architecture of classical traits like complex diseases. Probably the most striking potential of eQTL studies is the number of traits being considered. Focusing on a limited number of phenotypic traits, QTL analysis often reveals a partial picture of the genetic architecture related to particular traits. Simultaneous monitoring of thousands of gene expression traits provides unique and unbiased data and opens the possibility of constructing a global view of the regulation machinery. However, eQTL analysis also faces tremendous challenges because of the huge number of genes and genomic loci. Large scale eQTL analysis must deal with large computational demand and, more seriously, loss of statistical power due to multiple-testing issue.

This review is focused on the framework to study regulatory mechanisms using eQTL analysis. We start with introduction of the statistical methods used to discover eQTL association for single locus and multiple loci. Next, we review methods to infer coregulated gene modules. Subsequently, we survey the methods for identifying causal relationship including the methods where eQTL data are combined with other biological information to identify regulator genes for differentially expressed gene(s). Finally, we discuss a computational method for discovering regulatory programs. In the concluding section, we also suggest some further readings.

10.2   Mapping of Expression Quantitative Trait Loci (eQTL)

eQTL mapping requires two types of data: sequence variation data and gene expression data for the same set of individuals. For human study, samples of families, or independent individuals are used. For the study of model organisms, two parental strains are often crossed and progeny samples (inbred, if applicable) are obtained. In fact, the methods described in this review typically focus on analyzing the second type of data referred to as crosses. The chromosomes of the segregating individuals are genotyped (i.e., the alleles at genomic markers such as microsatellites, single nucleotide polymorphisms (SNPs), etc., are determined using biological assays). Additionally, gene expression of the samples is measured and genes significantly differentially expressed in progeny strains are selected. The expression levels of these genes comprise the list of expression traits.

Given such data, putative eQTL can be obtained by associating the genotype of markers with the expression traits. The gene associated with an eQTL is often referred to as the target gene suggestive of putative regulation of the gene by the eQTL. Each trait-locus pair is assigned a score of linkage significance (e.g., a log of the odds ratio (LOD) score, or a p-value). Since there are many pairs of traits and loci in a genome, the next step is to correct for multiple-testing.

As a special type of QTL, eQTL mapping has borrowed many ideas from standard QTL mapping, and therefore, we will introduce some classic QTL mapping approaches before addressing special challenges posed by eQTL.

We start by describing a statistical model that expresses the relation between phenotype and genotype. For simplicity, let us first look at a single pair of quantitative trait and putative QTL, and later we will address more challenging problems of multiple loci.

For the ith individual, the phenotype y_i is related with genotype x_i in the linear equation

$$y_i = a + bx_i + \varepsilon,$$

where x_i is a variable indicating alleles of the locus, b measures the phenotypic effect of substituting the allele at a putative QTL, and ε is a random normal variable representing environmental contribution to the phenotype or noise, with mean 0 and variance σ^2. For the purpose of this review, we assume that the genotype at any marker takes only two values, 0 and 1. In this model, a, b and σ^2 are unknown parameters.

Within the above model, the traditional QTL mapping approach uses linear regression to test if the phenotypic effect is significantly different from 0. Since each time a single marker is analyzed, this approach is called *single-marker mapping*. This approach has a number of shortcomings. For instance, if the QTL does not lie at the marker, its phenotypic effect will be underestimated, and consequently more progenies may be required to identify such association.

To remedy the problems of the single-marker approach, Lander and Botstein (1989) proposed the approach of *interval mapping*, in which a putative QTL between two markers is located. We first formulate the linear regression in the single-marker approach as the *maximum likelihood estimates* (MLEs), and then we use it in interval mapping. The linear regression solutions of a, b and σ^2 are the values that maximize the likelihood of the full model (i.e., probability that the observed data would occur). The MLE of null model can be obtained under the restriction that $b = 0$. Let the MLEs of the full and null models be denoted L_A and L_N. The significance of linkage can be measured by the increase of likelihood assuming the presence of the linkage relative to the likelihood assuming its absence. This is naturally summarized by the LOD score defined as $\log_{10}(L_A/L_N)$. If the LOD score is higher than a threshold then the putative QTL is claimed to be significant. The method of maximum likelihood and LOD score can be adapted to interval mapping as follows. In this case the QTL genotype is unknown but the genotypes of flanking markers are known. The probability that the QTL in the ith individual takes genotype x, denoted $G_i(x)$, can be estimated conditional on the observed genotypes of the markers and the recombination fraction between the QTL and the makers using a genetic map. The likelihood for the ith individual $L_i(x)$ can be estimated as in single-marker mapping. Then, the interval version of likelihood function is

$$L = \Pi_i \left[G_i(0)L_i(0) + G_i(1)L_i(1) \right].$$

This likelihood function can be plugged into the LOD score to assess the significance of linkage. Compared with single-marker approach, the interval mapping approach is more powerful and requires fewer progenies since it takes into account of information from more markers.

Compared with traditional QTL mapping, which usually involves only a few traits and markers, eQTL has much larger number of markers or traits (in the order of thousands). The large number of traits and loci poses challenges in both computational efficiency and statistical power. One of the prominent challenges is the *multiple-testing issue* (i.e., the chance of false positive in a family of multiple hypothesis tests is higher than that of a single test). A straightforward method to correct for multiple-testing is the well-known *Bonferroni* correction, which is to inflate the significance of an individual test by the total number of tests. In fact, it controls the family wise error rate (FWER), i.e., the probability of type I error in any test of a family under simultaneous consideration. For eQTL, in which we expect only a small fraction of pairs to be true positives, Bonferroni correction may often be too stringent.

Another approach is to threshold individual significance score of linkage with a critical value in the null distribution of all significance scores (Churchill and Doerge, 1994). While the null distribution is hard to know, it is approximated by permutation tests. For each linkage between a trait and a locus, we shuffle the phenotype (i.e., randomly reassign the trait of each individual to a new individual but retain the individuals' genotypes), and we assess the significance of the association after shuffling. The results to a group of N such shuffled data provides an approximation to the null distribution for the hypothesis of no linkage. To control the overall type I error rate to no more than α, we derive the *experiment-wide* critical values as follows. For the results of each of the N shuffled data, we find the most significant test statistic over all loci; then we order these N selected values and let their $100(1 - \alpha)$ percentile be the critical value (i.e., a significant pair of trait and loci must have its test statistic more extreme than $N(1 - \alpha)$ of these values from random shuffling).

Recently, *false discovery rate* (FDR) has been frequently used as the method of choice for addressing the multiple-testing issue. By definition, FDR is the expected proportion of false positives in all the results claimed significant (Benjamini and Hochberg, 1995). Since it focuses on testing results that are claimed significant and allows the rest to be false, FDR is more powerful than Bonferroni correction. A particular approach to control for FDR is q-value (Storey and Tibshirani, 2003). Given a list of features each with a p-value to represent its significance, we calculate a q-value for each of the feature, which is equal to the FDR of the whole list when calling that feature significant. It has been shown that q-value is more powerful than the original FDR methodology.

The eQTL mapping methods described above have been successfully applied to real data analysis. To dissect the transcriptional regulation in budding yeast, Brem et al. (2002) carried out eQTL mapping in a cross between a laboratory strain and a wild strain of *Saccharomyces cerevisiae* using single-marker mapping. Schadt et al. (2003) detected microsatellite marker eQTL in maize, mouse and human, using standard interval mapping techniques and simple Bonferroni correction. Stranger et al. (2005) used HapMap data, where the expression traits are from cell-lines of HapMap human individuals and the markers are dense SNPs. They used the three methods for multiple-testing correction (i.e., Bonferroni, permutation tests, and FDR) and observed significant overlap among them.

10.3 Multiple Locus Analysis

The eQTL mapping approaches described in the previous subsection identify association between a trait and a single locus. However, it has been shown that many gene expression traits have linkage to more than one locus (Brem et al., 2002; Yvert et al., 2003; Brem and Kruglyak, 2005). Moreover, *epistatsis* (i.e., interaction among loci) is shown to be pervasive among expression traits. For instance, Brem and Kruglyak (2005) estimated that 16% of heritable yeast transcripts exhibit epistasis. Therefore, multiple locus models that explicitly consider epistasis are necessary to obtain valid estimates. For simplicity, let us assume here that we search for two loci for each trait; most approaches below can be extended to more than two loci.

A straightforward method for mapping two loci is exhaustive two-dimensional (2D) linkage scan where all pairs of loci are tested for linkage. However, this method is computationally costly and suffers from low statistical power due to the large number of tests. There are two other approaches for mapping multiple loci. One is *multiple interval mapping* which combines the previously described interval mapping with multiple regression (Zeng, 1993; Jansen and Stam, 1994). However, this method requires a priori models with a set of prechosen loci, which is difficult to formulate. Another approach is to use a *model selection* algorithm that aims to identify a subset of loci as parameters of the best model according to some optimality criterion (Zeng et al., 1999; Ball, 2001; Broman and Speed, 2002). This approach is computationally demanding since it searches over a large number of potential models. In addition, none of the above methods provides a rigorous measure of the *joint* significance of multiple loci where *all* of the identified loci are truly linked.

To remedy the pitfalls of existing approaches for multiple loci, Storey et al. (2005) proposed a sequential search approach to map two eQTLs for a gene expression trait. It includes algorithms to assess joint significance of two loci and to measure evidence for epistasis. First we define a statistical model for multiple loci as follows.

(M0) Expression=baseline level+noise,

(M1) Expression=baseline level+locus1+noise,

(M2) Expression=baseline level+locus1+locus2+locus1×locus2+noise,

where "locus1" denotes the effect of the primary locus, and "locs1×locus2" represents the epistatic interaction between the primary and secondary loci. Clearly, this model can be extended to include the third, fourth loci and so on. Using this model, we perform the following sequential search algorithm. For each gene expression trait, we first select its primary locus that gives the greatest improvement in strength of linkage by comparing model M1 with model M0. The strength of linkage is measured by the Bayesian posterior probability **Pr**(locus 1 linked | Data). Then, conditional on the primary locus, we select the secondary locus that gives the biggest improvement in **Pr**(locus 2 linked | locus 1 linked, Data) when comparing model M2 with model M1. The joint significance (i.e., probability that both loci are linked with the trait) is

Pr(locus 1 and locus 2 are linked | Data)

=**Pr**(locus 1 linked | Data)×**Pr**(locus 2 linked | locus 1 linked, Data).

The posterior probability is estimated by a nonparametric empirical Bayesian method that uses permutation tests to simulate null distributions (Churchill and Doerge, 1994; Doerge and Churchill, 1996). Then we rank the traits by the value of joint significance, and select those traits that have significance above a threshold. To obtain a reasonable threshold that controls the overall type I error and deals with multiple-testing issue, we deduce FDR from the posterior probabilities as follows. FDR is defined as the ratio of expected number of false positive traits divided by the number of traits called significant. Since the probability that a trait is a false positive is $1 - \mathbf{Pr}$(locus 1 and locus 2 are linked | Data) thus

FDR=$(\sum(1-\mathbf{Pr}$(locus 1 and locus 2 are linked | Data)))/number of significant 2-locus linkages

where the summation is over all traits called significant. The epistasis can be tested similarly by comparing the full model M2 to a purely additive model (i.e., with locus 1×locus 2 equal to 0 in model M2). Being applied to real gene expression traits in yeast (Brem et al., 2005; Storey et al., 2005), the sequential search approach is shown to be faster and more powerful than exhaustive 2D linkage scan.

There are several limitations of the sequential search approach (Storey et al., 2005). It may miss those locus pairs with primarily epistatic effects in which neither single locus has significant effect. It may not be applicable to the case when the two loci are closely linked.

10.4   Basic Properties of Expression Quantitative Trait Loci (eQTL)

Given a target gene and associated eQTL, it is convenient to distinguish two types of relationships: *cis*- and *trans*-acting eQTLs. A *cis*-acting eQTL locates in or close to the transcription region of the target gene. Such a target gene is said to be *cis*-regulated. The eQTL may exert effects on the target gene's expression in various ways. For example, DNA variation in the promoter sequence may affect TF binding. It is also possible that DNA variation in the coding sequence may affect mRNA sequence composition, splicing or secondary structure. A *trans*-acting eQTL can reside on the same chromosome with the target gene but distal to it or on a different chromosome. Such a target gene is said to be *trans*-regulated. For example, DNA variation in TF, transcription-associated proteins, such as activators, repressors and posttranscription regulation genes, can have various impacts on the expression level of the target gene. The differentiation between *cis*-and *trans*-acting eQTL is usually based on a distance threshold between an eQTL and its target gene. Note that, if only *cis*-acting eQTLs are to be identified, the multiple-testing issue discussed in Section 10.2 will not be so severe since only loci close to a gene, whose expression is used as quantitative trait, are tested. As it will be discussed in the following sections, specific computational methods or models are used to discover causal relationship in *cis*-acting and *trans*-acting eQTL (Doss et al., 2005; Kulp and Jagalur, 2006).

A *cis*-acting or *trans*-acting eQTL terminology is used to describe a single eQTL-target gene pair. A genomic region, which contains many eQTLs or to which many expression traits are mapped is referred to as an *eQTL hotspot*. eQTL hotspot is a conceptual term since there is no precise requirement about how large a hot spot should be or at least how many genes need to be mapped to a hot spot. An eQTL hotspot can regulate tens to hundreds of genes (Morley et al., 2004; Zhu et al., 2008). Since an eQTL hotspot may point to a master regulator gene in the region, identifying possible eQTL hotspots is an important element of eQTL analysis. Due to the interest in eQTL hotspots, several identification methods have been proposed. For example, Ghazalpour (2006) defined a module quantitative trait locus (mQTL) as the locus harboring a significant number of eQTLs regulating genes within a given gene module. These mQTLs can be viewed as variants of eQTL hotspots as described above. Recently, Wang et al. (2007) observed, using a simulation experiment, that real eQTL hotspots may be sometimes difficult to discern from false positives resulting from possible artifacts caused by highly correlated gene expression or linkage disequilibrium.

In addition to the properties unique to eQTL, transgressive segregation, which is a genetic phenomenon observed in classical QTL, is also observed in eQTL studies. Transgress segregation occurs, where a quantitative trait takes value more extreme relative to the values in either parental strain. Such transgressive segregation was also observed in some eQTL studies (Brem and Kruglyak, 2005; Rowe et al., 2008). Brem and Kruglyak (2005) used loci with alleles of opposite effect on expression traits to explain transgressive segregation.

10.5   Inferring Coregulated Modules

The expression level of genes in the same complex or pathway is often coregulated. Conversely, a set of coregulated genes is expected to be enriched for genes that share biological functions and/or canonical pathways. A set of coregulated, presumably related, genes is therefore usually referred as a coregulated module. The discovery of eQTL hotspots suggests that eQTL analysis can be helpful for uncovering such coregulated modules and their regulators. In particular, it has been shown that the so called *trans*-eQTL band (Wu et al., 2008), that is the set of genes linked to a common *trans*-acting eQTL, has significant enrichment for genes with common annotations from gene ontology (GO) database, KEGG database and Ingenuity Pathways Knowledge Base (Ingenuity Systems, Redwood City, CA).

The first step in inferring coregulated modules from gene expression data is, typically, identification of clusters of coexpressed genes.* This can be done by a number of ways. For example, in their pioneering study on the yeast

---

* Unless stated otherwise, the gene expression data set is referred to the data set obtained from segregating population used in eQTL study.

crosses, Yvert et al. (2003) used a simple hierarchical clustering based on similarity of expression patterns. In a more recent paper, Zhu et al. (2008) started with the so-called coexpression graph and used topological overlap distance (Ravasz et al., 2002) to identify coregulated modules. There are a number of other clustering methods that could be applied in this context. For example, Li et al. identified transcription modules in mouse by performing biclustering on two gene expression matrices (Li et al., 2006) where expression values from different parental strains were used as reference points. Namely, (i, j) element in each matrix was defined as the log ratio between the expression value of ith gene in jth progeny and the one of ith gene the corresponding parental strain.

The modules identified in these studies were often shown to be enriched for genes sharing functional annotation such as GO category and/or linkage to a common chromosomal region. For example, in the study of Yvert et al. (2003) all but one of the modules were enriched for genes linked to a common region on a chromosome, the eQTL hotspot. Such linkage is consistent with the assumption that so constructed modules are indeed coregulated.

The second step of this approach is to discover eQTL(s) that regulate the coregulated modules identified in the first step. This is usually done by using the mean expression value of the genes in a given cluster as a new quantitative phenotype to which eQTL analysis can be applied. Indeed, Yvert et al. showed that the majority of clusters retrieved by their study showed linkage to at least one locus.

Finally, the genomic region near the locus is hypothesized to contain regulatory elements that regulate target-genes in the cluster. Note that the identification of coregulated modules and eQTLs associated with such modules is not equivalent to the identification of the causal regulatory gene(s). Methods to identify such causal regulators will be discussed in the next subsection.

A different approach to find coregulated modules has been proposed by Samson and Self (2008). They associated with each gene a LOD score profile— a vector of LOD association scores of the gene with all loci. Subsequently, they used the Pearson correlation between LOD score profiles of two expression traits, ρ_L, to test if these two genes share common regulatory locus. Finally, they applied a hierarchical clustering method to identify clusters of genes associated with common loci and demonstrated that uncovered modules were enriched with genes with common GO functional annotation.

10.6 From Expression Quantitative Trait Loci (eQTL) to Causal Regulatory Relations of Genes

Mapping of eQTLs is only the first step toward discovering regulatory mechanism. Our next goal is to identify regulators, i.e., regulatory genes near the eQTL responsible for the expression traits of target genes mapped to the eQTL. This is a challenging problem due to several statistical issues. First, a putative eQTL region is usually big (a few centimorgans wide) and typically contains many genes. Thus we need to reduce the width of eQTL and the number of candidate regulators, a process called *fine mapping*. Second, neighboring markers tend to have high correlations due to linkage disequilibrium, and as a result a target gene may be linked to false positive eQTL close to the true eQTL. Third, in case that many target-genes map to an eQTL *hotspot*, it is challenging to accurately identify the complete set of coregulated target genes for the eQTL hotspot. To deal with the statistical issues and solve the problem, we mainly rely on three types of information: (1) physical locations of eQTL and genes, (2) expression traits of regulators and target genes, (3) GO information about functions of the genes.

The causal relations between genes are inferred using statistical models, and thus represent indirect connections. To fill in the gap, we further attempt to explain the causal relation with molecular interaction pathways inferred from various data, such as protein–protein interactions (PPIs), TF-DNA binding sites (TFBS), protein phosphorylation, etc. These inferred pathways can provide insight to the molecular mechanism of gene regulation. The identification of causal relations and the inference of molecular pathways are related to each other: a set of causal relations is a starting point for the inference of pathways; conversely, pathways from regulators to target genes can serve as evidence supporting the causality.

However, one has to keep in mind that correlation among gene expression values may not necessarily reflect all functional relationships among genes. A regulator can affect the expression of target genes without a change of its own expression.

10.6.1   Fine-mapping and expression-correlation-based methods

As previously mentioned, a typical eQTL is large (up to several centimorgans) and often contains many genes, among which only a few have causal effects on the expression of target genes. To narrow down the candidate causal genes, the following two steps are frequently used (1) reduce lengths of the eQTL regions by fine-mapping techniques, (2) perform expression correlation analysis between candidate regulators in an eQTL region and the target genes affected by the eQTL. Step (2) is based on the assumption that genes in the same pathway have strong correlations between their expression values. Such two-step analysis has been used, for example, in Bing and Hoeschele (2005), thus in the following we first describe their method and then mention other related work.

One way to reduce the length of eQTL regions, is to employ a bootstrap resampling method proposed by Visscher et al. (1996). Bootstrap samples are created by sampling with replacement the given set of expression values with the set of marker genotypes. For each of 1000 bootstrap data sets, Bing and

Hoeschele employ a conventional approach to identify an eQTL position with the highest test statistic. Then from the sorted physical positions of these 1000 eQTLs, they determine the 95% confidence interval of eQTL position by taking the largest and smallest values of the bottom and top 2.5%, respectively of the eQTL positions. Genes physically located inside the confidence interval are selected as candidate regulators. However, eQTL confidence intervals may still be large, especially when eQTL effect is small. A likely reason for such large eQTL confidence interval is the presence of multiple eQTL. To resolve a large eQTL interval into smaller subregions each with one eQTL, one can perform sliding three-marker regression based on theoretical properties of multiple interval mapping (Zeng, 1993). For each subregion consisting of a pair of consecutive markers, they obtain a t-statistic associated with partial regression coefficients of the two markers. More details can be found in Thaller and Hoeschele's (2000) work. If a candidate gene is found to be located in a subregion with significant t-statistic, it is identified as a strong causal candidate of regulator.

For every eQTL confidence interval containing a list of candidate regulatory genes, the list of candidate regulators can be further narrowed down using an expression correlation based approach. This correlation analysis is based on the assumption that genes belonging to the same pathway or network are more likely to have strong correlation between their expression values. First, among all genes in the confidence interval, the gene with the smallest significant p-value is identified as the primary causal gene (denoted G1). Then, for every candidate regulatory gene different than G1, the first-order partial correlation coefficients of expression profiles of that gene and the target gene, conditional on the primary regulator is computed. If there is at least one significant correlation coefficient, the most significant gene is taken as the secondary regulator (G2). The process is continued by computing second-order partial correlation and conditional on the primary and secondary regulator, etc., and then computing higher-order partial correlations until there is no more significant partial correlation coefficient. In this way we obtain a list of candidate causal genes (G1, G2, . . .). Note that this method is different from Storey's method for identifying multiple loci (Storey et al., 2005) in that it selects regulators in a set of candidate genes using expression correlation.

The existence of multiple target genes, as for example in the case of an eQTL hotspot, can be the source of additional information. It is reasonable to assume that among all possible regulators, the best candidates are those that correlate with a large number of functionally related target genes. This assumption is, for example, explored by Keurentjes et al. (2007). They start by calculating all pairwise rank-based Spearman correlations of expression between each of the functionally related target gene and candidate regulators in eQTL intervals. These correlation coefficients are then ranked so that more strongly correlated pairs of regulator-target genes are at the top of the list. Then they apply iterative group analysis (iGA) to obtain the probability of change (PC-value) for each regulator candidate (Breitling et al., 2004). Here, the PC-value of a pair (regulatory gene, target gene) is the *p*-value for testing "how likely it is to observe all target genes correlated with the particular candidate regulator at least as high on the list as the given pair, by chance." A candidate gene with a significant PC-value (adjusted for multiple hypothesis testing) is a putative regulator, and all genes contributing to the PC-value are putative target genes. In addition, we can extend the set of target genes by using expression correlation between the genes already identified and potential target genes outside the initial group of functionally related genes.

10.6.2   Likelihood-based model selection

Methods described in the previous subsection typically consist of two steps: (1) map eQTL confidence intervals, and (2) from genes physically located inside the eQTL intervals, identify candidate regulators. Kulp and Jagalur (2006) proposed an alternative approach that unified the two steps into one step. Instead of finding loci first, they directly look at the genotype in a candidate regulatory gene *and* the expression traits of the regulator, simultaneously. The genotype in the candidate regulator corresponds to putative eQTL in previous methods. Substituting loci with genes, the method is called quantitative trait gene (QTG) mapping. In QTG mapping, one can express the causal relation between a candidate regulatory gene (with genotype Q_j and expression T_j) and a target gene (with expression T_i) using a Gaussian model:

$$P(T_i|Q_j, T, \theta) = N(\beta_0 + \beta_1 T_j + \beta_2 Q_j + \beta_3 T_j Q_j, \sigma),$$

where θ is a vector that includes the parameters β and σ. Note that the term $\beta_3 T_j Q_j$ is used to model the interaction between genotype and expression value of the regulator. This conditional probability is similar to the likelihood function of the standard eQTL interval mapping. Indeed, the maximum likelihood estimation of parameters Q_j and θ can be obtained by expectation maximization (EM) algorithm. Analogous to LOD score, the strength of the causal relationship can be assessed by comparing full model with null model, as

$$\log_{10} \frac{P(T_i|Q_j, T_j, \theta)}{P(T|Q_j, T_j, \theta : \beta_1 = \beta_2 = \beta_3 = 0)}.$$

Compared to the standard two-step approach of eQTL mapping and regulator identification, the QTG method has the advantage that all evidence is integrated in a single model. Moreover, it provides us with the flexibility of choosing different null models by setting different subsets of $\{\beta_1, \beta_2, \beta_3\}$ to zero. For example, setting $\beta_1 = \beta_3 = 0$ in the null model measures the contribution of the expression values of candidate regulators to the expression traits of target genes.

After the identification of regulatory genes and their causal relation with target genes, a natural question is: how does a regulator affect the expression levels of target genes? A simple situation is that the regulator is a TF. However, it has been shown that TFs are not enriched in eQTL hotspots (Yvert et al., 2003). Nonetheless, even if the regulator is not a TF, the propagation of genetic perturbation from the regulator to target genes is likely to be mediated by TF activities. This scenario is considered by Sun et al. (2007). They estimate TF activities using the approach of Yu and Li (2005), combining data from literature and from genome-wide TF-binding study. Given an eQTL module (an eQTL hotspot together with target genes), let GC denote the expression level of one *cis*-linked gene, TA denote a TF's activity, and GT denote the expression level of any gene other than GC. Then they consider three models:

- Causal model: $GC \to TA \to GT$

- Reactive model: $GC \to GT \to TA$

- Conditional independent model: $TA \leftarrow GC \to GT$

where an arrow indicate the direction of the pairwise relation. An idea of having such three models was considered earlier by Schadt et al. to study relationships between two clinical traits (Schadt et al., 2005; Sieberts and Schadt, 2007). Each pairwise relation can be modeled by simple linear regression. From the regression parameters, conditional densities could be derived, which are further used to compute the likelihood of each model. Finally, likelihood ratio tests can be used to identify one model that is significantly better than the other two models.

Model selection methods have also been used to infer the causal relationship between expression traits and clinical traits (Schadt et al., 2005; Aten et al., 2008). Although these methods were used for a purpose different from our goal, they may be adapted to the context of finding causal relation between regulatory genes and downstream expression traits of target genes.

10.6.3 Pathway-based methods

The methods discussed in the previous subsection can only provide indirect casual relations but cannot explain such relationships on a molecular level. Realizing this problem, Tu et al. (2006) integrated eQTL mapping results with a biological network built from protein–protein interaction (PPI), protein phosphorylation and TF–DNA interaction data. In this network, they search for pathways from candidate regulatory genes in an eQTL region to target genes. The inferred pathways are subsequently used to identify most likely causal genes in the eQTL. This approach is based on two assumptions: (1) causal genes regulate target gene by affecting the activities of TFs for the target gene, (2) activities of genes on the likely pathway correlate with target gene's expression. Based on these assumptions, the computational problem is to find the pathway from a causal gene to TFs of the target gene so that the expression levels of genes on the pathway correlate with the target gene.

Tu et al. designed a stochastic backward search algorithm based on random walk starting from the target gene g_t. In the first step the algorithm picks one of the TFs regulating gene g_t from which it starts the random walk. The probability of moving from gene u, to previously unvisited gene v is proportional to the Pearson correlation coefficient of expression levels of v and the target gene. Each node can be visited only once, to make sure the path is noncyclic. The random walk is stopped when it reached a candidate regulator or when there are no more reachable unvisited nodes (hit a "dead end") or when it has reached the maximum allowed number of steps. The probability that a candidate regulator g_j is a causal gene can be approximated by $V_{t_k}(g_t)/N$, where $V_{t_k}(g_t)$ is the number of times g_j is visited and N is the number of random walks. In other words, the causal effect of g_j on g_t is modeled as the probability that g_j can be reached from g_t via random walks in the network. If there is only one TF for g_t, the candidate regulator with the largest $V_{t_k}(g_t)/N$ is taken as the best candidate. If g_t has more than one TF, a linear combination of $V_{t_k}(g_t)/N$ for different t_k is used to determine the best candidate. Then a backwards search from the regulator gene toward t_k for nodes with the largest count identifies the most probable pathway from the regulator gene to the target gene.

The random walk method, while very intuitive, is computationally intensive because a large number of random walks have to be generated. Suthram et al. (2008) proposed a more efficient approach to identify regulator genes and the pathways from regulator genes to target genes. The approach is based on the well-established analogy between random walks and electric networks. It can be proved that when one unit of current flows into an electric network, the amount of current through any intermediary node or edge is proportional to the expected number of times a random walker will pass through that node or edge. The algorithm of Suthram et al., called eQTL electrical diagrams (eQED), replaces the random walk model with current flows in electric circuits. Instead of assigning weights to nodes as in Tu et al., eQED assigns a weight to each edge (u, v), which is equal to the average of correlation coefficients of u and v with the target gene. The weights on the edges are modeled as conductance in the electric circuit. The p-value of the eQTL is used as the amount of current flowing from the locus to the target gene. Then, the problem is formulated with linear programming, where the optimal path has the maximum total sum of currents flowing. After computing current on each edge using a linear programming approach, eQED predicted the best candidate regulator gene to be the one with the highest current flowing through it.

A pathway based approach also has been used by Zhu et al. to identify casual regulator genes of a set of genes associated with an eQTL hotspot (Zhu et al., 2008). They used a gene network constructed using Bayesian network method (Zhu et al., 2004) by combining eQTL, PPI, and TFBS data (see also Section 10.7). First, genes that could be *cis*-regulated by eQTL hotspot region were identified. Next, for each gene in this set, the set of genes that could be reached by a path in the network starting from that gene were found. This set was, in turn, intersected with the set of the genes linked to the corresponding hotspot and the significance of the overlap was estimated using Fisher test corrected for multiple testing. If the overlap was significant, the corresponding gene was considered to be a regulator of the module associated with the given hotspot. Using their method the authors recovered previously identified (Yvert et al., 2003) as well as some novel hotspot regulators demonstrating added power of the integrative network reconstruction.

10.7 Using Expression Quantitative Trait Loci (eQTL) for Inferring Regulatory Networks

It may seem to be straightforward to apply the approaches discussed in the previous subsection to construct regulatory networks. As a first approximation, one can put a directed edge between a regulator gene and one of its target genes (Bing and Hoeschele, 2005; Li et al., 2005; Keurentjes et al., 2007). However approaches that infer network structure locally by adding one edge at a time may miss sophisticated regulation pattern involving a large number of genes. Therefore, methods considering the data set as a whole should be applied to construct regulatory networks.

One of such methods is the Bayesian network approach, which is a probabilistic graphical model, represented by a directed acyclic graph. In the graph, nodes represent variables and edges represent conditional probabilistic dependence between variables. There are many important applications of Bayesian network in genomics study. For example, Bayesian network can be inferred from gene expression data, where each node represents the expression level of a gene and an edge indicates that two genes' expression traits are conditionally dependent (Friedman, 2004). It is natural to interpret such a Bayesian network as a regulatory network: an edge $g_j \rightarrow g_i (g_j$ is a parent of $g_i)$ indicates the gene g_j regulates g_i. The Bayesian networks have been successfully used to infer complex causal relationship pattern among hundreds of genes and are ideal for integrating multiple data sources.

A big limitation of Bayesian network approach is its demand of large computational power. Hence it is a common practice to reduce the search space of possible networks through various biological or computational heuristics (Zhu et al., 2004; Li et al., 2005). For example, in their early work, Zhu et al. (2004) reconstruct a Bayesian network using gene expression data. In the reconstruction method, they reduce the number of possible edges in the network by assuming that one gene could have at most three regulator genes and by considering only a subset of all genes as candidate regulator genes. To select this subset they use a mutual information (MI) threshold and a measure of correlation between LOD scores similar to the one described in Section 10.5. To further reduce the computational burden, they also infer directly some causal relationship from the number of overlapping eQTLs of two genes by defining

$$\mathbf{Pr}(g_j \text{ regulates } g_i) = r(g_i, g_j)N(g_i)/(N(g_i) + N(g_j)),$$

where $r(g_i, g_j)$ measures the genetic relatedness over all chromosomes for two genes, computed by chromosome-specific LOD score (Zhu et al., 2004); $N(g_i)$ is the number of significant eQTL linked to g_i. If g_i and g_j have common eQTL, but g_j has more eQTL than g_i, a prior would be set to indicate g_j is a candidate regulator gene of g_i. Subsequently, the authors construct a consensus network from 1000 Bayesian networks. Finally, data processing inequality was applied to check if an edge $g_i \rightarrow g_k$ was over fit when there were edges $g_j \rightarrow g_i$ and $g_j \rightarrow g_k$. If MI between g_i and g_k is less than MI between g_i and g_j or g_j and g_k then the edge $g_i \rightarrow g_k$ would be removed.

More recently, Zhu et al. combined their Bayesian network with PPI data and TFBS data to construct yeast regulatory network (Zhu et al., 2008). First, using the Bayesian network method (Zhu et al., 2004), they construct a Bayesian network from the expression data. Then eQTL data was added to extend the network. Namely, genes with *cis*-eQTL are assumed to be parents of genes with coincident *trans*-eQTL. Genes derived from the same eQTL are subsequently used to infer casual/reactive or independent reaction as described in Section 10.6.2 (Schadt et al., 2005). If casual/reactive relationship could not be determined in this way, the authors break the ties using a complexity criterion that, intuitively, considers a gene with simpler and stronger eQTL signature as the causal gene (Zhu et al., 2008). Finally, PPI data are added by considering protein complexes and their regulators. Namely, if at least a half of genes in a protein complex carry a given TFBS, then all genes in the complex were included as being under the control of the corresponding transcription factor (TF). For additional details on including TFBS in the network, we refer readers to the original paper (Zhu et al., 2008). Their construction showed that integrating eQTL data with other data sources can improve the reconstruction of regulatory networks.

A different approach has been developed by Liu et al. (2008), who constructed a regulatory gene network using the SEM method (Stein et al., 2003; Kline, 2004). SEM is an extension of general linear model and is used to test and estimate causal relationships. In general, it consists of a measurement model and a structural model. The measurement model describes the relationships between latent variables, which are not directly measured, and their indicators, i.e., various measurements obtained from experiments. The structural model describes causal relationships among latent variables. In the SEM model used by Liu et al., all variables are measured, hence no measurement model is needed and the structural model describes casual relation between measured variables. Algebraically, the structural model is represented as $y_i = By_i + Fx_j + e_i$; $i \in [1..N]$ where N is the number of progenies, y_i is the expression vector, x_j is the vector of eQTL genotypes, and e_i is a vector of error terms. Matrix B models the causal relationships between gene expression traits and matrix F models the causal relationships of eQTL on gene expression traits. Here, y_i and x_j are observed variables and B, F, and e_i are model parameters. The network representing the model contains a node for every x_i and every y_i. There is an edge between nodes corresponding to kth and mth gene expression trait if $B[k, m]$ is nonzero. Similarly there is an edge between nodes corresponding to the kth eQTL and the mth gene expression trait if $F[k, m]$ is nonzero. To estimate the model, Liu et al. applied an approach combining likelihood maximization and network topology search. The quality of the model was evaluated using a criterion which considers both: a maximized likelihood function used to optimize the parameters of a given network and a penalty term for the number of free parameters to optimize over various network topologies. To reduce the search space for the network topology, the network is constrained to edges of a predefined network, which authors referred to as an encompassing directed network (EDN). To construct the EDN, several criteria were applied to include the edges that are most likely to be a part of the final model. More details about their eQTL analysis and regulator identification are beyond the scope of this review. We refer readers to the original paper. One of the advantages of the SEM method is that it allows cyclic structures, which are not allowed in Bayesian networks. However, similar to Bayesian network method, it cannot process the network of large size as it is computationally demanding.

In an interesting application related to the identification of casual relationship using eQTL data, Schadt's group (Chen et al., 2008) has recently successfully identified a gene sub-network having crucial causal relationship with complex metabolic syndromes. They first obtained several coexpression subnetworks by clustering the gene expression data set of a mouse eQTL experiment. Such subnetwork was then considered as the gene network controlling a metabolic trait if it had significant enrichment of expression traits having causal relationship with the metabolic trait.

10.8   Inferring Regulatory Programs

In the previous sections, we discussed methods to identify coregulated modules as well as regulators of such modules. Importantly, coregulated modules are often controlled by more than one regulator. In such a situation the expression of genes from the module is a function of the combination of states (expression level, phosphorylation, etc.) of the corresponding regulators. The combination of states that leads to the observed expression patterns is referred to as regulatory program. Uncovering such regulatory programs is one of the major challenges on the way to the complete understanding of regulatory networks. To address this problem, Lee et al. proposed a method that is directed toward detecting coregulated modules together with their regulatory programs

(Lee et al., 2006). The method, implemented as software called Geronemo, builds on their previous work on extracting network modules (Segal et al., 2003).

Regulatory programs targeted by Geronemo consist of TF, signaling molecules, chromatin factors and SNPs. Each such program is assumed to have hierarchical, tree-like structure, where each internal node is associated with a "regulator" and splits the gene expression of the module genes into two distinct behaviors. The two submodules correspond to two subsets of strains. Geronemo allows for two types of regulators: g-regulators and e-regulators. The g-regulators correspond to a split along a SNP and have two clearly defined split values corresponding to the two progenitor alleles. In contrast, the e-regulators define a split on the continuous set of possibilities based on their expression level. Note, however, that the e-regulators still allow for splitting the gene expression into two distinct behaviors (e.g., up-regulated and not up-regulated). In Gerenemo, the set of possible e-regulators is restricted to known TFs, signaling molecules and chromatin factors. One can think of the regulatory program as a decision tree with regulators in the decision nodes. The goal of such a decision tree is to sort the progenies into groups so that, for each group, all genes in the modules are coherently expressed. As a special case, a one-gene module with one g-regulator corresponds, roughly, to a traditional eQTL and a multigene module with one g-regulator may correspond to an eQTL hotspot. In contrast, a multigene module with one e-regulator contains genes whose expression is correlated or anticorrelated with the expression of this regulator.

The coregulated modules and their regulators are being discovered through an iterative learning approach. The learning procedure is initialized with a certain number of modules obtained by k-means clustering and then iterated over two phases: (1) assigning each gene into some regulatory module; and (2) learning the regulation program for each module. The program is learned in a recursive way by choosing, at each point, the regulator that best splits the gene expression of the module genes into two distinct behaviors. When considering a potential split, Geromeno evaluates all candidate regulators and splits values and selects the one that leads to the highest improvement in score. The procedure is iterated until convergence.

When applied to the yeast crosses, Geronemo identified a large number of modules that both have chromosomal characteristics and are regulated by chromatin modification proteins.

Compared with eQTL methods described earlier, the approach applied by Gerenomo is unique—it not only finds the coregulated modules but also identifies regulatory programs. While other approaches were focusing on finding putative regulators, Gerenomo also predicts the states of these regulators (e.g., up/down regulation of TF, SNP variant, etc). Namely, for every expression value in a gene in such a module, one can trace back the decision tree defining the regulatory program and, for an internal node on the path, read the state (e.g., up/down) of the regulator associated with this node. However, unlike the methods described in the previous section, the regulators are selected from a predefined set of genes.

10.9 Conclusions and Further Reading

The discoveries of the last decade brought a deeper appreciation of the complexity of regulatory networks. High-throughout data sets open, for the first time, the possibility of their preliminary reconstruction. The eQTL analysis provides an important step in this direction. First, it directly links genomic regions with mRNA levels of target genes *in vivo* and allows addressing the network reconstruction in a nonbiased genome-wide fashion. More importantly, genetic data can be used in a natural way to infer causal relation between genes in the region and target genes. Hence, it can provide numerous biological hypotheses about regulation to be further tested. However, even with eQTL analysis discovering causal relationships remains nontrivial. A formidable challenge is related to multiple hypothesis testing. Another difficulty is posed by the challenge related to uncovering combinatorial regulation and epistasis.

eQTL analysis involves detecting weak signals and complex relationships from large amount of data. Therefore it is not surprising that many statistical and machine learning approaches, including Bayesian networks (Zhu et al., 2004; Zhu et al., 2008), Bayesian regression (Veyrieras et al., 2008), SEM method (Liu et al., 2008), EM method (Lee et al., 2006), have been applied in eQTL analysis. However, many problems in eQTL study remain open necessitating development of new more powerful approaches.

In this review, we focused our attention on applications of eQTL analysis to uncovering regulatory mechanisms. Consequently, many other aspects of eQTL studies, as well as methods that do not apply directly on eQTL data, have not been covered. For a retrospective historical account of GWAS, we refer the readers to the article by Kruglyak (2008). Rockman and Kruglyak (2006) provide a review focusing on genetics of gene expression. Methods of combining eQTL with QTL to study complex disease traits are discussed in Schadt (2005) and Sieberts and Schadt (2007). Kendziorski's review discusses various statistical methods for eQTL in more details (Kendziorski et al., 2006).

Acknowledgments

This work was supported by the Intramural Research Program of the National Institutes of Health, National Library of Medicine. The authors thank Yoo-ah Kim (NCBI/NIH) for comments on the manuscript.

References

Aten JE, Fuller TF, Lusis AJ, Horvath S. 2008. Using genetic markers to orient the edges in quantitative trait networks: the NEO software. *BMC Syst Biol* 2: 34.

Ball RD. 2001. Bayesian methods for quantitative trait loci mapping based on model selection: approximate analysis using the Bayesian information criterion. *Genetics* 159(3): 1351–1364.

Benjamini Y, Hochberg Y. 1995. Controlling the false discovery rate: a practical and powerful approach to multiple testing. *J Roy Statist Soc Ser* B 57(1): 289–300.

Bing N, Hoeschele I. 2005. Genetical genomics analysis of a yeast segregant population for transcription network inference. *Genetics* 170(2): 533–542.

Breitling R, Amtmann A, Herzyk P. 2004. Iterative Group Analysis (iGA): a simple tool to enhance sensitivity and facilitate interpretation of microarray experiments. *BMC Bioinform* 5: 34.

Brem RB, Kruglyak L. 2005. The landscape of genetic complexity across 5,700 gene expression traits in yeast. *Proc Natl Acad Sci USA* 102(5): 1572–1577.

Brem RB, Yvert G, Clinton R, Kruglyak L. 2002. Genetic dissection of transcriptional regulation in budding yeast. *Science* 296(5568): 752–755.

Brem RB, Storey JD, Whittle J, Kruglyak L. 2005. Genetic interactions between polymorphisms that affect gene expression in yeast. *Nature* 436(7051): 701–703.

Broman KW, Speed TP. 2002. A model selection approach for the identification of quantitative trait loci in experimental crosses (with discussion). *J R Stat Soc Ser B* 64: 641–656.

Bystrykh L, Weersing E, Dontje B, Sutton S, Pletcher MT et al. 2005. Uncovering regulatory pathways that affect hematopoietic stem cell function using 'genetical genomics'. *Nat Genet* 37(3): 225–232.

Chen Y, Zhu J, Lum PY, Yang X, Pinto S et al. 2008. Variations in DNA elucidate molecular networks that cause disease. *Nature* 452(7186): 429–435.

Chesler EJ, Lu L, Shou S, Qu Y, Gu J et al. 2005. Complex trait analysis of gene expression uncovers polygenic and pleiotropic networks that modulate nervous system function. *Nat Genet* 37(3): 233–242.

Cheung VG, Spielman RS, Ewens KG, Weber TM, Morley M et al. 2005. Mapping determinants of human gene expression by regional and genome-wide association. *Nature* 437(7063): 1365–1369.

Churchill GA, Doerge RW. 1994. Empirical threshold values for quantitative trait mapping. *Genetics* 138(3): 963–971.

DeCook R, Lall S, Nettleton D, Howell SH. 2006. Genetic regulation of gene expression during shoot development in Arabidopsis. *Genetics* 172(2): 1155–1164.

Doerge RW, Churchill GA. 1996. Permutation tests for multiple loci affecting a quantitative character. *Genetics* 142(1): 285–294.

Doss S, Schadt EE, Drake TA, Lusis AJ. 2005. Cis-acting expression quantitative trait loci in mice. *Genome Res* 15(5): 681–691.

Friedman N. 2004. Inferring cellular networks using probabilistic graphical models. *Science* 303(5659): 799–805.

Ghazalpour A, Doss S, Zhang B, Wang S, Plaisier C et al. 2006. Integrating genetic and network analysis to characterize genes related to mouse weight. *PLoS Genet* 2(8): 1182–1192.

Jansen RC, Stam P. 1994. High resolution of quantitative traits into multiple loci via interval mapping. *Genetics* 136(4): 1447–1455.

Jansen RC, Nap JP. 2001. Genetical genomics: the added value from segregation. *Trends Genet* 17(7): 388–391.

Jin W, Riley RM, Wolfinger RD, White KP, Passador-Gurgel G et al. 2001. The contributions of sex, genotype and age to transcriptional variance in Drosophila melanogaster. *Nat Genet* 29(4): 389–395.

Kendziorski CM, Chen M, Yuan M, Lan H, Attie AD. 2006. Statistical methods for expression quantitative trait loci (eQTL) mapping. *Biometrics* 62(1): 19–27.

Keurentjes JJ, Fu J, Terpstra IR, Garcia JM, van den Ackerveken G et al. 2007. Regulatory network construction in Arabidopsis by using genome-wide gene expression quantitative trait loci. *Proc Natl Acad Sci USA* 104(5): 1708–1713.

Kline RB. 2004. *Principles and Practice of Structural Equation Modeling*. New York, NY: The Guilford Press.

Klose J, Nock C, Herrmann M, Stuhler K, Marcus K et al. 2002. Genetic analysis of the mouse brain proteome. *Nat Genet* 30(4): 385–393.

Kruglyak L. 2008. The road to genome-wide association studies. *Nat Rev Genet* 9(4): 314–318.

Kulp DC, Jagalur M. 2006. Causal inference of regulator-target pairs by gene mapping of expression phenotypes. *BMC Genomics* 7: 125.

Lander ES, Botstein D. 1989. Mapping mendelian factors underlying quantitative traits using RFLP linkage maps. *Genetics* 121(1): 185–199.

Lee SI, Pe'er D, Dudley AM, Church GM, Koller D. 2006. Identifying regulatory mechanisms using individual variation reveals key role for chromatin modification. *Proc Natl Acad Sci USA* 103(38): 14062–14067.

Li H, Lu L, Manly KF, Chesler EJ, Bao L et al. 2005. Inferring gene transcriptional modulatory relations: a genetical genomics approach. *Hum Mol Genet* 14(9): 1119–1125.

Li H, Chen H, Bao L, Manly KF, Chesler EJ et al. 2006. Integrative genetic analysis of transcription modules: towards filling the gap between genetic loci and inherited traits. *Hum Mol Genet* 15(3): 481–492.

Liu B, de la Fuente A, Hoeschele I. 2008. Gene network inference via structural equation modeling in genetical genomics experiments. *Genetics* 178(3): 1763–1776.

Monks SA, Leonardson A, Zhu H, Cundiff P, Pietrusiak P et al. 2004. Genetic inheritance of gene expression in human cell lines. *Am J Hum Genet* 75(6): 1094–1105.

Morley M, Molony CM, Weber TM, Devlin JL, Ewens KG et al. 2004. Genetic analysis of genome-wide variation in human gene expression. *Nature* 430(7001): 743–747.

Petretto E, Mangion J, Pravanec M, Hubner N, Aitman TJ. 2006a. Integrated gene expression profiling and linkage analysis in the rat. *Mamm Genome* 17(6): 480–489.

Petretto E, Mangion J, Dickens NJ, Cook SA, Kumaran MK et al. 2006b. Heritability and tissue specificity of expression quantitative trait loci. *PLoS Genet* 2(10): e172.

Ravasz E, Somera AL, Mongru DA, Oltvai ZN, Barabasi AL. 2002. Hierarchical organization of modularity in metabolic networks. *Science* 297(5586): 1551–1555.

Rockman MV, Kruglyak L. 2006. Genetics of global gene expression. *Nat Rev Genet* 7(11): 862–872.

Rowe HC, Hansen BG, Halkier BA, Kliebenstein DJ. 2008. Biochemical networks and epistasis shape the Arabidopsis thaliana metabolome. *Plant Cell* 20(5): 1199–1216.

Sampson JN, Self SG. 2008. Identifying trait clusters by linkage profiles: application in genetical genomics. *Bioinformatics* 24(7): 958–964.

Schadt EE. 2005. Exploiting naturally occurring DNA variation and molecular profiling data to dissect disease and drug response traits. *Curr Opin Biotechnol* 16(6): 647–654.

Schadt EE, Monks SA, Drake TA, Lusis AJ, Che N et al. 2003. Genetics of gene expression surveyed in maize, mouse and man. *Nature* 422(6929): 297–302.

Schadt EE, Lamb J, Yang X, Zhu J, Edwards S et al. 2005. An integrative genomics approach to infer causal associations between gene expression and disease. *Nat Genet* 37(7): 710–717.

Segal E, Shapira M, Regev A, Pe'er D, Botstein D et al. 2003. Module networks: identifying regulatory modules and their condition-specific regulators from gene expression data. *Nat Genet* 34(2): 166–176.

Sieberts SK, Schadt EE. 2007. Inferring causal associations between genes and disease via the mapping of expression quantitative trait loci. In: Balding DJ, Bishop M, Cannings C, editors. *Handbook of Statistical Genetics*. John Wiley & Sons, Ltd, West Sussex, England, 296–326.

Stein CM, Song Y, Elston RC, Jun G, Tiwari HK et al. 2003. Structural equation model-based genome scan for the metabolic syndrome. *BMC Genet* 4 Suppl 1: S99.

Storey JD, Tibshirani R. 2003. Statistical significance for genome-wide studies. *Proc Natl Acad Sci USA* 100: 9440–9445.

Storey JD, Akey JM, Kruglyak L. 2005. Multiple locus linkage analysis of genomewide expression in yeast. *PLoS Biol* 3(8): e267.

Stranger BE, Forrest MS, Clark AG, Minichiello MJ, Deutsch S et al. 2005. Genome-wide associations of gene expression variation in humans. *PLoS Genet* 1(6): 0695–0704.

Sun W, Yu T, Li KC. 2007. Detection of eQTL modules mediated by activity levels of transcription factors. *Bioinformatics* 23(17): 2290–2297.

Suthram S, Beyer A, Karp RM, Eldar Y, Ideker T. 2008. eQED: an efficient method for interpreting eQTL associations using protein networks. *Mol Syst Biol* 4: 162.

Thaller G, Hoeschele I. 2000. Fine-mapping of quantitative trait loci in half-sib families using current recombinations. *Genet Res* 76(1): 87–104.

Tu Z, Wang L, Arbeitman MN, Chen T, Sun F. 2006. An integrative approach for causal gene identification and gene regulatory pathway inference. *Bioinformatics* 22(14): e489–e496.

Veyrieras JB, Kudaravalli S, Kim SY, Dermitzakis ET, Gilad Y et al. 2008. High-resolution mapping of expression QTLs yields insight into human gene regulation. *PLoS Genet* 4(10): 1–15.

Visscher PM, Thompson R, Haley CS. 1996. Confidence intervals in QTL mapping by bootstrapping. *Genetics* 143(2): 1013–1020.

Wang S, Zheng T, Wang YJ. 2007. Transcription activity hot spot, is it real or an artifact? *BMC Proc, 1* Suppl 1: S94.

Wu C, Delano DL, Mitro N, Su Sv, Janes J, et al. 2008. Gene set enrichment in eQTL data identifies novel annotations and pathway regulators. *PLoS Genet* 4(5): e1000070.

Yu T, Li KC. 2005. Inference of transcriptional regulatory network by two-stage constrained space factor analysis. *Bioinformatics* 21(21): 4033–4038.

Yvert G, Brem RB, Whittle J, Akey JM, Foss E et al. 2003. Trans-acting regulatory variation in Saccharomyces cerevisiae and the role of transcription factors. *Nat Genet* 35(1): 57–64.

Zeng ZB. 1993. Theoretical basis for separation of multiple linked gene effects in mapping quantitative trait loci. *Proc Natl Acad Sci USA* 90(23): 10972–10976.

Zeng ZB, Kao CH, Basten CJ. 1999. Estimating the genetic architecture of quantitative traits. *Genet Res* 74(3): 279–289.

Zhu J, Zhang B, Smith EN, Drees B, Brem RB et al.. 2008. Integrating large-scale functional genomic data to dissect the complexity of yeast regulatory networks. *Nat Genet* 40(7): 854–861.

Zhu J, Lum PY, Lamb J, GuhaThakurta D, Edwards SW et al.. 2004. An integrative genomics approach to the reconstruction of gene networks in segregating populations. *Cytogenet Genome Res* 105(2–4): 363–374.

Chapter 11

Statistical Approaches to Gene Expression Microarray Data Preprocessing

Megan Kong, Elizabeth McClellan, and Richard H. Scheuermann

University of Texas Southwestern Medical Center

Monnie McGee

Southern Methodist University

11.1   General Overview

11.1.1   Gene expression microarrays

Over the past decade, gene expression microarray the technology has revolutionized the way we measure the level of gene transcripts in biological organisms. Traditionally, researchers used northern blotting or polymerase chain reaction (PCR) to measure the expression level of a few genes in a single experiment. Now, the expression levels of thousands of genes can be measured in one single microarray hybridization. Researchers can view the expression profiles of genes on a whole genome scale.

Gene expression microarrays consist of series of microscopic spots of DNA polynucleotides stabilized onto a slide in a defined array pattern. Each spot corresponds to one polynucleotide sequence. Each of these polynucleotide probes is designed to be complementary to a specific gene transcript, and is used to capture fluorescently or radioactively labeled polynucleotide targets derived from the specific gene transcripts under appropriate hybridization conditions. The relative abundance of target gene transcripts can then be inferred by measuring the fluorescent or radioactive signal associated with each probe.

The gene expression microarray technology has been applied to many areas of basic biological and clinical research. In basic research, using gene expression microarray, researchers can study gene function by examining gene expression pattern differences in samples from wild type and mutant organisms. Gene interaction networks, which reflect functional relationships among genes, can also be constructed using gene expression microarray data. In clinical research, gene expression microarray has been applied to biomarker discovery by comparing the expression profiles of diseased and normal tissues. It has also been used to study drug response characteristics and determinants of drug resistance. Particularly, by studying the gene expression pattern differences between different study groups, clinicians can gain insight into why some people respond to (or show resistance to) certain drug treatments, while others do not.

There are three major public gene expression microarray data repositories: GEO (http://www.ncbi.nlm.nih.gov/geo/), ArrayExpress (http://www.ebi.ac.uk/microarray-as/ae/), and the stanford microarray database (http://genome-www.5.stanford.edu). Currently, GEO has 237,664 arrays and ArrayExpress has 181,810 arrays available. These two major data sources have allowed researchers, biostatisticians and computational biologists to compare their results with results from similar studies by other groups, to develop better statistical methods for processing microarray data and to perform data mining to discover new features of the underlying biology.

Four different technology platforms have been used extensively for gene expression microarray studies, each with associated advantages and disadvantages. The initial microarrays were constructed using long cDNA copies of mRNA transcripts as the probes that are spotted on the array. The advantage to using these spotted cDNA arrays is that because the probes are relatively long the stringency of hybridization can be more easily controlled, such that the signal detected is more specific for the transcript targeted. The disadvantage is that the procedures required to synthesize and quality control the probes used is very labor intensive making it difficult to produce these arrays with acceptable reproducibility.

The second technology utilizes arrays in which oligonucleotides ($\sim$60mers) are spotted on the array support to serve as hybridization probes. While there is some loss in hybridization specificity, the advantage is that the chemical production of these synthetic oligonucleotides can be controlled more precisely than cDNA synthesis leading to improved reproducibility in probe production. The disadvantage is that spotted oligonucleotide arrays are subject to variability associated with the spotting technique used.

The third technologies utilizes arrays in which oligonucleotides are synthesized directly on the array solid support utilizing photolithography techniques, using custom masks, as in the Affymetix GeneChip™ technique, or digital micromirrors, as in the NimbleGen maskless array synthesis technique, to control light exposure during synthesis. The in situ synthesized oligonucleotide arrays offer enhanced reproducibility in array production in comparison to the spotted oligonucleotide arrays.

The fourth technology platform utilizes in situ oligonucleotide synthesis to produce the hybridization probes, but instead of using glass slides uses individual microbeads as solid support for the array. This allows the oligonucleotides to be synthesized using traditional solution-based organic chemistry approaches instead of photolithography. Microbead-based oligonucleotide arrays offer many of the same advantages as the *in situ* synthesized glass slide-based arrays, as well as high feature redundancy, and have rapidly developed as a viable alternative for gene expression studies.

Because of the unique aspects of each of these different microarray platforms, the nature of the primary data and how it should be processed to estimate gene expression levels varies between platforms. In this chapter, we focus on the preprocessing of data derived from the Affymetrix GeneChip™ platform in some detail, since this platform is the most highly represented in the data repositories described above, with some discussion of how preprocessing approach should be modified for other technology platforms addressed briefly at the end.

A major weakness of all of the current gene expression microarray technologies is high variation in data quality. Systematic uncontrollable errors unrelated to target gene expression exist due to the nature of the experiments and are reflected in the raw data, which has a significant impact on subsequent statistical interpretations. Variability contributing to these errors may occur during the array production, target preparation, hybridization, array washing, or image collection stages of the experiment. The types of challenges that must be addressed include background noise, chip-to-chip biases, and order-dependent patterns within an array. Biological variation within a subject (or between subjects) is often of interest as well and, when addressed properly by producing several replicate arrays per sample, introduces a need for additional processing. Limiting the presence and impact of such issues that can be quantitatively corrected after data collection is imperative to obtain accurate and reliable results from a microarray experiment.

Improper annotation of probes also contributes to unreliable data, such as the inaccuracy of probe set definitions for Affymetrix GeneChips due to utilization of out of date genome annotation (Dai et al., 2005). Another serious challenge with microarray experiments is systematic disturbances in chip design or manufacturing such as the order-dependent pattern discovered in Affymetrix GeneChips by Bjork and Kafadar (2007). The authors present indexed plots of typical values of interest such as expression values and variance that show a pronounced pattern consistent across several experiments.

The study concludes that transcript order may delimit expression values and variance and thus results obtained from microarray data on specific Affymetrix platforms should be interpreted with caution.

Biological variability, although independent of the microarray experimental process, should also be considered when examining challenges and sources of error. The highest level of variability in the experiment is that between the subjects from whom the sample is obtained, but is generally of interest and should be estimated rather than eliminated. However, the variation within an individual should be accounted for by assaying replicate chips per subject and adjusting the analysis to account for such differences. Challenges that occur in the laboratory phase may seriously flaw the data and must be minimized prior to any analysis in order to avoid reporting inaccurate conclusions.

11.1.2 Preprocessing of primary data

Controlling for the aforementioned challenges in a microarray experiment is imperative for production of reliable results and is typically addressed by preprocessing the probe-level data. Preprocessing occurs after feature extraction (image analysis) and alters the data in such a way that the resulting value is directly related to the expression level of a particular gene. Affymetrix chip preprocessing steps include probe annotation, background correction, normalization, and summarization. Probe annotation involves the matching of array probes with mRNA transcripts and should also include the updating of transcript definitions based on the most current genome sequences. Background correction is the process of using probe intensities to estimate the amount of background noise present on an array and adjusting for it. Normalization involves detecting and correcting for systematic nonbiological differences between microarrays. Summarization combines probe intensities within a probe set into a single expression value.

High variation in data quality must be accounted for in order to arrive at accurate and consistent conclusions. Preprocessing methods focus on variations in experimental conditions rather than natural biological variations. The examination and subsequent adjustment of these differences play a crucial role in the overall goal of realizing the true difference in gene expression between subjects. The experimenter must be able to extract meaningful data characteristics from the noisy and improperly annotated primary data for use in downstream analysis such as differential expression, clustering, and classification. For example, investigators rely on properly preprocessed data when predicting disease type. A patient may face serious consequences if given medication for the wrong disease type due to misclassification caused by an error in the background correction step of the microarray process. If the preprocessing methods are not reliable, then all subsequent analyses and conclusions may be flawed.

11.2   Current Gene Expression Microarray Preprocessing Steps and Methods—Affymetrix

11.2.1   Probe annotation

The initial design of the probes included in many of the original Affymetrix microarray chips was based on incomplete genome sequences. For example, when the Affymetrix HG-U133 chip set was designed, the human genome sequence was only about 25% complete. In addition, even for more recently designed chips, as much as 30% of the PM probes may be problematic due to potential cross-hybridization and mis-annotation in the Affymetrix HGU-133A platform (Dai, 2005; Harbig, 2005; Shi, 2006). The same annotation problem also exists with other microarray platforms (Shi, 2006).

Currently, even though the sequence of the human genome is complete, the manually curated reference sequences are being updated on a regular basis. To make sure that the probe annotation reflects the most current genome build, the Molecular and Behavioral Neuroscience Institute at the University of Michigan (BRAINARRAY, http://brainarray.mbni.med. umich.edu/Brainarray/) has established a process for the generation of periodically revised annotation files for all the Affymetrix platforms including human, mouse, rat, yeast, drosophila, and *Arabidopsis thaliana*. The new annotations remove problematic probes such as those that cross hybridize with more than one transcript and those that do not hybridize to the correct target (Dai, 2005).

We have used gene ontology (GO) term co-clustering as an evaluation approach to assess the impact of using revised annotation on Affymetrix gene expression microarray data analysis (Kong, 2007). The evaluation approach is based on the premise that genes encoding proteins involved in the same biological process or protein complex will be coordinately expressed; that is, genes that have the same GO annotations are more likely to be clustered together (Lee, 2006). We have used the BRAINARRAY revised annotation and the Affymetrix original annotation to analyze several real biological data sets. Our results demonstrate that using revised annotation, the p-values for the most significant GO terms in each gene cluster are much lower (i.e., more significant). In addition, the whole distribution of all the co-clustering p-values for all the GO terms is substantially lower when the revised annotation is used (see Figure 11.1). Thus, using revised annotation indeed produces much more significant co-clustering of related genes. Our result also supports the general approach of using GO term co-clustering as a useful evaluation metric to access preprocessing approaches with real biological data. We have applied the same methodology to assess the best microarray analysis pipeline (see detailed discussion below).

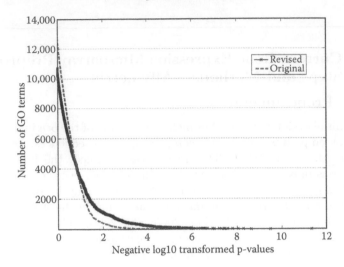

FIGURE 11.1: Comparison of GO term p-value distributions between original and revised Affymetrix probe annotation. The curves represent the distribution of all the p-values for all the GO terms in every cluster. The solid line represents the co-clustering p-values using the revised annotation. The dashed line represents the co-clustering p-values using the original annotation.

11.2.2 Background correction

It is known that the probe material that is affixed to the array carries some low level of background fluorescence. The same is true of the solid material to which the probes are attached. Therefore, there will be some fluorescence that is due to the material itself rather than the binding of the targets to the fixed probes.

It is also known that some probes bind imperfectly to their respective targets. There are two types of imperfect binding. Nonspecific hybridization (NSH) is a type of background noise that is present when RNA fragments with sequences not meant for the targeted probes bind to the probes anyway. The gene related to these probes will falsely appear expressed in the sample because the RNA bound to the unintended probes adds to the true signal. Cross-hybridization (XH) is the binding of a probe sequence to a target that is at least partially identical to the true RNA target sequence of interest. Its effect can be partially attenuated by continual revision of annotation (see the previous section) and through the establishment of appropriately stringent hybridization conditions. But even under ideal conditions both NSH and XH cause spurious signal to be attributed to a probe, therefore falsely increasing its signal intensity.

The mismatch (MM) probes included in the Affymetrix chip design were originally thought to be a measure of background noise and NSH. Since the

MM sequences differ from the perfect match (PM) sequences in the middle base, a target that bound perfectly to the PM probe would, in theory, not bind to the MM probe. Therefore, one could simply subtract the MM value from the corresponding PM value in order to remove background signal and obtain a measure of specific hybridization (or true signal) for a particular probe.

The PM–MM estimation model for specific hybridization assumes two things, neither of which are true. First, the model assumes that the hybridization conditions are perfectly balanced so that a one-base MM is sufficient to eliminate true hybridization for all probe pairs. Attaining such perfection in the conditions is impossible in practice. Second, it assumes that all MM intensities will be less than the corresponding PM intensities even if some NSH was to occur. However, it has been observed that approximately 30% of MM intensities are greater than PM intensities (Li and Wong, 2001a; Bolstad, 2004; Bolstad, 2003). The large MM intensities could be a result of cross-hybridization. Since we do not know the sequence for all genes in the human genome, it is possible that some of the MM probes are actually PMs for another gene. The MM intensities could also include effects of NSH of the MM probe to the correct target for the PM, since the structures of the MM probes are so close to that of their PM partners. If that is the case, then there should be a correlation between the intensities of PM and MM probes.

Figure 11.2 shows two plots of different probe sets from the Affymetrix HG-U133A Latin Square Spike-In experiment (see http://www.affymetrix.com/

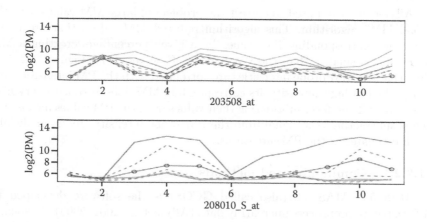

FIGURE 11.2: Log base 2 intensities for probe sets "203508_at" (top) and "208010_s_at"(bottom) from the HG-U133A latin square spike-in experiment. 203508_at is a spike-in probe, and 208010_s_at is not. The solid lines are the log base 2 PM intensities and the dashed lines are the corresponding MM intensities. Note that the MM intensities tend to track the PM intensities very closely, indicating correlation between PM and MM intensities. Furthermore, several probes in the nonspiked in probe register high intensity, which is evidence of cross-hybridization.

support/ technical/sample_data/datasets.affx for details). Four CEL files from the experiment were selected at random, and probe sets "203508_at" and "208010_s_at" were plotted for all four experiments. The top set of axes shows the intensity levels for probe set 203508_at, a spiked-in probe set. A probe set that is spike-in is one for whom the target concentrations were deliberately manipulated to achieve a known intensity for each experiment. The solid lines represent the logged PM intensities, and the dashed lines in the same color are the corresponding MM intensities. Note that for some experiments, the MM intensities are equal to or greater than the PM intensities. Furthermore, the MM intensities tend to track the PM intensities. In other words, there is evidence of correlation in the signals attributed to them most likely due to NSH.

The bottom panel of Figure 11.1 shows the PM and MM intensity values for probe 208010_s_at from the same four experiments. The probe 208010_s_at is not noted as a spike-in probe in Affymetrix's documentation for the experiment. However, probes 3–5 and 6–10 exhibit the behavior of a spiked-in probe in that they have much higher intensities than would be expected, most likely due to XH. Furthermore, the MM intensities for these probes track the PM intensities, most likely due to NSH of the cross-hybridizing target. Thus, not only does this panel depict the correlation between PM and MM intensities, but also it shows evidence of cross-hybridization for several probes within the probe set. Other probes in different probe sets exhibit similar behavior.

Affymetrix attempted to correct the problem of large MM values with the "ideal MM" algorithm. This algorithm reduced MM values that are larger than their corresponding PM values by a given percentage, creating a MM value that is smaller than the PM partner. However, from Figure 11.2, it is clear that the MM probe intensities are correlated with the PM intensities. If a PM probe has high intensity, its corresponding MM does also, with very high probability. Therefore, subtracting their values from the PM values means that one is subtracting true signal from the PM probe intensity, giving artificially small estimates of the PM intensities.

11.2.2.1 Commonly used methods

MAS 5.0: MAS 5.0 (also called GCOS) is the software developed by Affymetrix to preprocess microarray data (Affymetrix, 2002, 2003). As such, it is quite widely used, despite its poor performance compared to other methods (Li and Wong, 2001a,b; Bolstad, 2003; McGee and Chen, 2006). The background correction algorithm for MAS 5.0 is often termed local adjustment or zonal adjustment. Each array is divided into k rectangular zones (the default for k is 16). The algorithm ranks the intensities for each pixel within each zone. The mean of the smallest 2% of the intensities is chosen as the background estimate for that zone. A standard deviation of the background is also calculated. Both the background estimate and its standard deviation are smoothed, based on their distance from the center of each zone, to obtain one background estimate for the entire chip. The estimate is subtracted from each intensity measure at each point on the chip. A small fraction is added to any negative value that result from this process. The ideal MM algorithm is applied during the summarization step, to the background corrected data.

To its credit, MAS 5.0 attempts to make corrections for NSH. One might question the method, but at least the problem of NSH is acknowledged. The zonal background adjustment algorithm makes sense if one expects spatial correlation of background intensity within each chip. However, in the extreme situation of blotches and smears, the zonal adjustment algorithm will over-correct in some places and undercorrect in others. If no spatial correlation is present, the zonal adjustment algorithm runs the risk of creating it.

One can also debate the wisdom of correcting each array within an experiment separately rather than together as one group. Other algorithms (e.g., RMA and DFCM) correct all arrays at once so that intensities from the correct arrays will be on the same scale, and therefore be comparable. However, one can imagine situations, such as when arrays for the same experiment are processed on different scanners or on different days, when it would be desirable to correct each array separately.

The main complaint about MAS 5.0 is the assumption that the PM and MM probes for a given probe pair have the same error. This assumption results in poor estimation of expression values, particularly at low RNA concentrations (Irizarry, 2003a,b). The zonal adjustment algorithm also contains several arbitrary thresholds that must be determined, such as the number of zones, the smoothing parameter across the zones, and correction value for the Ideal MM. Almost everyone who uses MAS 5.0 uses the default values, but there has not been a large-scale study on the effect of changing the default values for a particular set of data.

Model based expression index (MBEI): The first challenger to MAS 5.0 was the MBEI, also known as dChip (Li and Wong, 2001a,b). It was shown to be more sensitive and specific than MAS 5.0 (Li and Wong, 2001a,b; Bolstad, 2003). Furthermore, the algorithm can detect defects in the chip and smooth them so that these aberrant intensities do not affect overall intensity estimates. The MBEI model is given by

$$y_{ij} = \text{PM}_{ij} - \text{MM}_{ij} = \theta_i \phi_j + \varepsilon_{ij}, \quad \text{where } \sum \phi_j^2 = J.$$

Here, y_{ij} is the expression value of the jth probe on the ith array, θ_i is the expression index for array i, φ_j is the probe-sensitivity index for probe j, and ε_{ij} is a random error term. The value of the probe sensitivity index determines whether or not the probe can be considered as aberrant, and therefore, its intensity can be either down-weighted or excluded from calculation of the summarized gene intensity. MBEI does not estimate background directly, but rather computes an expression index for each probe. These expression indices are later normalized and summarized (excluding the "sensitive" probes) into a single expression value for each probe set.

Unfortunately, proper estimation of the parameters θ and φ requires at least ten arrays (Li and Wong, 2001a,b). Most microarray experiments consist of much smaller numbers of arrays. The MBEI algorithm produces unstable estimates for such experiments. The normalization and summarization algorithms associated with MBEI are discussed in the next section.

Robust multichip average (RMA): the current most popular preprocessing algorithm for Affymetrix microarray data is RMA (Irizarry, 2003b). RMA, and many of its successors, are PM only algorithms, meaning that they utilize only the PM intensities from an experiment. MBEI can also be used with PM intensities only. The use of PM only was in response to the fact that the MM probes do not perform as theory predicted they would.

The background correction method used in RMA involves a convolution of an exponentially distributed signal and normally distributed noise. Specifically, it is assumed that $X = S + Y$, where X represents the observed PM intensities, S = the true signal that is exponentially distributed with rate parameter α, and Y = normally distributed noise, with parameters μ and σ. In reality, the distribution for the noise is a truncated normal. This is a normal distribution truncated at 0 so that there are no negative intensities. The point of background correction in RMA is to estimate the parameters μ, σ, and α, conditional on the knowledge of the observed intensities alone. The background correction is given by the following formula

$$E(S|X) = a + \sigma \left[\frac{\phi\left(\frac{a}{\sigma}\right) - \phi\left(\frac{x-a}{\sigma}\right)}{\Phi\left(\frac{a}{\sigma}\right) - \Phi\left(\frac{x-a}{\sigma}\right) - 1} \right]$$

where $\alpha = x - \mu - \sigma^2\alpha^2$, Φ represents the cumulative distribution function of the normal distribution, $\phi()$ is the density function of the normal distribution (Bolstad, 2004). GCRMA (Wu, 2004) is a variation of RMA in which the background estimation attempts to account for the GC content of each probe. Therefore, the background correction is not based on the exponential-normal model. The normalization and summarization algorithms for GCRMA are the same as they are for RMA. In general, GCRMA does not perform well (Chen, 2007), and it will not be discussed further.

There are two main criticisms of the exponential-normal model. First, the parameters are estimated in a very *ad hoc* manner. Second, the exponential-normal assumption is impossible to check in practice, since we cannot separate noise from signal for real data. Furthermore, even if the data follow an exponential-normal convolution model, the procedure devised to estimate the parameters results is extremely poor estimates (McGee and Chen, 2006b).

RMA-mean and RMA-75: since the estimation procedure used for RMA in the Bioconductor (Gentleman et al., 2004) package produces poor estimates, McGee and Chen (2006b) devised different methods to estimate the mean (μ) and standard deviation (σ) of the background noise while still assuming an exponential-normal convolution model. The best performing method was

TABLE 11.1: Mean squared errors for parameter estimates via the RMA ad hoc method and four other methods for estimating μ, σ, and α.

True parameter values	Ad hoc	Mean	Median	75th percentile	99.95th percentile
			Estimation method		
$\mu = 30$	233	2.62	2.62	2.62	2.68
$\sigma = 0$	92	1.35	1.35	1.51	1.39
$\alpha = 5$	45135	4.70	10.2	2.56	22.1

Smaller values for the MSE indicate better estimates. The ad hoc method (first column) gives very poor estimates. All other methods perform well in comparison. The best method is the one using the 75th percentile of all intensity values greater than the overall mode of all intensities to estimate α. The method where the mean of these values is used is a close second.

one in which a one-step iterative formula was employed using the following mathematical relationship between μ, σ, α, and the mode, x_m.

$$\phi\left(\frac{x_\mu - \mu}{\sigma} - \alpha\sigma\right) = \alpha\sigma\left[\Phi\left(\frac{x_m - \mu}{\sigma} - \alpha\sigma\right)\right].$$

The initial estimates for the iteration were given by the estimates from the *ad hoc* estimation algorithm.

RMA uses the mode of the data greater than the overall calculated made, x_m, to estimate the rate parameter of the exponential distribution (α). McGee and Chen examined the following statistics calculated from the intensities greater than x_m as replacements for the RMA estimate: the mean, the median, the 75th percentile, and the 99.95th percentile. For all estimates, the mean-squared error (MSE) between the estimated value and the known value was calculated. The MSE measures precision (variability) and accuracy (bias) simultaneously.

Table 11.1 shows the results in which an exponential-normal convolution model with parameters $\mu = 30$, $\sigma = 5$, and $\alpha = 5$ was generated. An examination of several data sets indicates these parameter values are reasonable for a variety of microarray data. The column headings in the table refer to the different methods of estimating α. Estimation for μ and σ was done via the one-step iteration method described previously. Other simulations using different values of the true parameters were done with the same qualitative results.

Since the one-step iterative estimate using either the 75th percentile or the mean of the data greater than the overall mode of all intensities to estimate α performed best, McGee and Chen devised two new background correction methods, based on the exponential normal model, but using different estimates for the parameters. The methods are called RMA-mean and RMA-75. For both methods, the mean and standard deviation of the noise distribution are estimated using the one-step iterative formula given previously. The rate parameter for the exponential distribution is estimated with the mean of all intensities greater than x_m (RMA-mean) or the 75th percentile of all intensities greater than x_m (RMA-75).

RMA-mean and RMA-75 were tested using the Affymetrix spike-in data to ascertain whether better estimates of the parameters produced better results in terms of sensitivity and specificity in the selection of differentially expressed genes. Indeed, these two methods outperformed the original RMA method (McGee and Chen, 2006b), although not by as much as expected given the clear superiority of the estimators used.

Distribution free convolution model (DFCM): Estimates that have dramatically smaller MSEs than the original ones should produce great improvements in sensitivity and specificity *if the underlying model is correct.* Using the Affymetrix latin square spike-in data sets, Chen et al. (2009) showed that the exponential distribution was too light-tailed to model the signal adequately, and that the distribution of the noise was likely non-normal. Therefore, the exponential-normal model is not a good fit for gene expression microarray data.

The idea of a convolution model for gene expression is reasonable, even if the distributional assumptions attached to it are not. A convolution suggests that noise pervades the signal at all levels, which is likely the case with gene expression data from microarrays. Therefore, Chen et al. (2009) suggested a convolution model in which the signal and noise parts of the convolution were estimated using nonparametric (or distribution free) methods. The method is called DFCM. It has the advantage that no underlying distributions of the signal and noise are assumed. It also makes some use of the MM intensities, which have been ignored in most preprocessing methods up to this point.

The DFCM procedure for background correction proceeds as follows:

1. Obtain the smallest q_1 percent of the observed PM intensities (X). ($q_1 < 30\%$ works best).

2. Obtain the smallest q_2 percent (typically 90 or 95%) of MM intensities associated with the PM intensities gathered in step 1. These MM intensities are assumed to measure background noise.

3. Use a nonparametric density estimate of the lowest q_2 percent of the MM intensities to find the mode of the noise distribution. The mode is used to estimate the mean of the noise distribution.

4. Estimate the standard deviation of the background noise by calculating the sample standard deviation of MM intensity values that are smaller than the estimated mean.

5. Correct the intensity for each probe in each probe set by subtracting the estimated mean. For very small intensities, a scaled value of the mean is subtracted.

The DFCM procedure avoids the erroneous distributional assumptions of RMA. Furthermore, it has been shown to produce good results in terms of sensitivity and specificity, in combination with various normalization and summarization methods. These results hold for real data and for spike-in data.

11.2.3 Normalization

Normalization is required in order to correct chip-specific biases influencing all probes on an array. Such biases include, but are not limited to, printing, hybridization, or scanning artifacts. In this section, we describe commonly used methods, their advantages, and their disadvantages. Schuchhardt et al. (2000) and Hartemink et al. (2001) give more complete lists of sources of obscuring variation.

Housekeeping genes: So-called "housekeeping genes" have also been used as a basis for normalizing microarray data. Affymetrix, as well as other manufacturers, include several such genes on each array. These probe sets are not supposed to catch signal from the target genes; therefore their intensities should be the same (near 0) across all arrays. However, it has been shown that housekeeping genes vary more than expected (Quackenbush, 2002). Furthermore, there are generally not enough of them to determine a reasonable function for adjusting the other probes.

Constant normalization: The earliest method of normalization was called constant or global normalization. The principle of this type of normalization is to multiply the intensity values for all arrays in an experiment by a constant so that the mean (or median) for all arrays is the same. There are many variations on this theme. For example, one can find a constant such that the resulting multiplication gives a mean of 1 for all arrays (and thus, a log base 2 mean equal to 0). This is done by dividing each value in an array by the overall mean of the array. Another method is to choose a baseline array, find its mean, and then multiply all other arrays by a constant so that they have the same mean as the baseline array. This is the normalization scheme for MAS 5.0.

Constant normalization has been shown to perform poorly compared to other normalization methods (Yang, 2002). The variation of the procedure used for MAS 5.0 has been criticized because of the necessity of choosing a baseline array, and this choice is arbitrary for most experiments (Bolstad, 2003). Furthermore, constant normalization is a linear algorithm. Therefore, it operates under the assumption that nonbiological variation adds linearly to the true intensity levels. Given the complex nature of microarray experiments, this is probably not the case.

Invariant set: To address the premise that extra-biological effects are nonlinear in nature, invariant set normalization (Li and Wong, 2001a,b) uses nonlinear regression to map values from each array in an experiment to a baseline array. The baseline array is defined as the array having "median overall brightness." Once a baseline array has been selected, normalization proceeds by finding a set of genes whose intensities do not vary in rank order from the baseline array to the comparison array (rank invariant set). Cross-validated smoothing splines are used to define a nonlinear relationship between the invariant set and the comparison array, then all probes on the comparison array are adjusted using this fit. Details of the fit and the selection of the invariant set of genes are given in Li and Wong (2001b). Invariant set normalization suffers from the same problems as do other baseline methods. In addition, it tends to be computationally expensive, as one must fit (and cross-validate) a complicated nonlinear model to each array in an experiment.

Contrast: Another nonlinear method is contrast normalization (Åstrand, 2003). It is based on the M versus A plot, where M is the difference in log expression values and A is the average of log expression values (Dudoit, 2002). The data are placed on a log scale and the basis is transformed. After transformation, a series of $n - 1$ (where n is the number of arrays) curves are fit to the M versus A curves for a pair of arrays, with the goal of adjusting the intensities for the two arrays so they are centered on the line where $M = 0$. The normalized data is obtained by back transforming to the original basis and exponentiating (Bolstad, 2003). Other than its complexity, the main disadvantage of the contrast method is that it is computationally expensive since the algorithm is performed on the set of pairwise combinations of arrays.

Local linear smoothing splines (LOESS): LOESS is more computationally expensive than the contrast method, but it seems to have enjoyed more popularity. It was introduced in the context of cDNA microarray data (Dudoit, 2002), and is also based on the M versus A plot. For LOESS, a localized linear regression function is fit to the M versus A plots for all pairwise combinations of arrays. The linear regression is used to adjust the intensities so that the data fall around a horizontal line (where $M = 0$). Usually, the normalization is performed using rank invariant sets of probes in order to decrease run time (Bolstad, 2003). Once the program has computed normalized values for each pair of arrays, these values are summarized and applied to the set of all arrays. Up to two iterations of the process are done in order to assure that further adjustments are minimal.

Variance stabilizing normalization (VSN): VSN (Huber, 2002) was devised to account for the fact that the variance of microarray data is dependent on the mean. Specifically, variability of microarray data increases as the intensity increases. For large intensities, the relationship between the mean and standard deviation is nearly linear. However, this is not the case for smaller intensities, often due to the fact that intensities below a certain arbitrary threshold are dismissed as probes that are not expressed ("absent" in the parlance of Affymetrix analysis). VSN uses a model that assumes both multiplicative and additive error. The data are transformed so that the variance of the intensities is equal regardless of the mean. The transformation is a hyperbolic sine function, which has a relatively simple mathematical relationship to the logarithm. VSN is grounded in good mathematical theory, yet it has not enjoyed the popularity of some of the more *ad hoc* methods, mainly because it has been shown to underperform on spike-in data (Bolstad, 2004). It may also be the complexity of the theory that keeps practitioners from readily employing this method.

Quantile normalization: quantile normalization is a multichip method in which all arrays are made to have the exact same distribution. The method proceeds as follows:

1. Sort the intensities for all arrays in the experiment. Find the maximum intensity in each array.

2. Substitute the median of all maximum intensities for the maximum intensity on each array.

3. Obtain the next-to-largest intensity for each array. Calculate its median. Substitute the median for each probe having that next-to-largest intensity.

4. Continue until the smallest intensity is reached.

Figure 11.3 shows a simple example of four arrays with four probes each to illustrate the method. The set of brackets to the far left represents the original data. The rows are probe intensities, and the columns represent arrays. The maximum value for each array is 9, 8, 6, 9. The median of these values is 8.5. In the second set of brackets, each of the previous maximum values have been replaced by their median: 8.5. The second-largest values are 8, 7, 6, and 8. In the third set of arrays, these values are replaced by their median: 7.5. The process continues until all of the original intensity values have been replaced.

Note that quantile normalization not only results in arrays with the same distribution, but also the arrays have the exact same values (in different places). It is possible that quantile normalization will overcorrect for nonbiological bias. There is some evidence that it greatly reduces the intensity

$$
\begin{bmatrix} 5 & 8 & 4 & 8 \\ 3 & 2 & 7 & 9 \\ 8 & 3 & 2 & 7 \\ 9 & 7 & 6 & 2 \end{bmatrix} \Rightarrow
\begin{bmatrix} 5 & 8.5 & 4 & 8 \\ 3 & 2 & 8.5 & 8.5 \\ 8 & 3 & 2 & 7 \\ 8.5 & 7 & 6 & 2 \end{bmatrix} \Rightarrow
\begin{bmatrix} 5 & 8.5 & 4 & 7.5 \\ 3 & 2 & 8.5 & 8.5 \\ 7.5 & 3 & 2 & 7 \\ 8.5 & 7.5 & 7.5 & 2 \end{bmatrix} \Rightarrow
\begin{bmatrix} 4.5 & 8.5 & 4.5 & 7.5 \\ 2.5 & 2.5 & 8.5 & 8.5 \\ 7.5 & 4.5 & 2.5 & 4.5 \\ 8.5 & 7.5 & 7.5 & 2.5 \end{bmatrix}
$$

FIGURE 11.3: A simple example of quantile normalization. The original data set with four arrays (columns) and four probes on each array (rows) is given on the far left. The algorithm begins by taking the maximum values for each array (9, 8, 7, 9, respectively). It calculates their median (8.5) and then replaces each maximum value with the median (second array). The third array shows the replacement of the second largest intensity in each array by their median (7.5). The array on the far right is the finished product.

values of probes with high intensities, and increases intensity values of probes with low ones (Calza et al., 2008), thus making differentially expressed genes more difficult to detect.

Evaluation of methods and disclaimers: Of the methods mentioned above, LOESS performs the best in terms of sensitivity and specificity, according to GO term co-clustering (Kong, 2007). Its computational complexity is not as much of a burden as it was even six years ago. However, Calza et al. (2008) note that LOESS suffers from the same problem as quantile normalization in that it tends to shrink high and low intensities toward the mean of all intensities. In order to understand the tendency of LOESS and quantile normalization to shrink estimates, it is important to understand the assumptions underlying all current normalization methods.

There are scenarios in which one might expect a large majority of the genes to vary in intensity between different arrays. However, all of the normalization algorithms described in this section assume that less than 10–20% of genes vary in intensity between arrays. In fact, all normalization algorithms must have some set of observations that are relatively stable. Otherwise, there is no standard to which to perform the normalization. The normalization methods listed above use all intensities on the arrays to obtain normalized values. This is fine as long only a modist proportion of genes change between arrays.

The second assumption is that there are equal numbers of up and down regulated genes in the experiment (symmetric differential expression) (Bolstad, 2003). If this assumption is violated, it no longer makes sense to center normalized values around a horizontal line. Therefore, in experiments where there are a large majority of up (or down) regulated genes, attempting to force their intensities to center around a horizontal line would result in grievous loss of signal, and thus a loss of power to detect differentially expressed genes.

Unfortunately, these assumptions are rarely checked in practice. In fact, it would be extremely difficult to do so. If either of these assumptions is violated, the normalization algorithms described in this section may introduce more bias than they correct. It has also been shown that choice of normalization algorithm has a great influence on the outcome of the experiment in terms of which genes are declared as differentially expressed (Hofmann, 2002). Therefore, when choosing a normalization algorithm, one must carefully consider whether the aforementioned assumptions can be expected to hold. A recent algorithm, called least variant set normalization (Calza, 2008), claims to produce more stable results when these assumptions are violated, at least for spike-in data. Since it is a new algorithm, it has not been tested on large numbers of data sets, as the other algorithms have.

11.2.4 Summarization

Summarization methods are necessary for Affymetrix arrays in order to establish a single expression value corresponding to a particular gene. Some of the more common methods are average difference (Affymetrix, 2006),

Tukey-biweight (Affymetrix, 2002; Li and Wong, 2001a,b), median polish (Bolstad, 2004) and distribution free weighted (Chen, 2007).

Average difference (AvgDiff) was the method of summarization devised by Affymetrix for MAS 4.0 (Affymetrix, 1996). It is defined as the sum of the differences between the PM probe intensities and their corresponding MM intensities for a probe set, divided by the number of "present" probes in that set. The original GeneChip software included all pairs. MAS 4.0 excluded outlier pairs of PM–MM more than three standard deviations from the mean PM–MM value. AvgDiff can be negative if MM>PM. Furthermore, it is in general unwise to subtract MM values from PM values, since the MM values tend to carry some signal (see Figure 11.1). AvgDiff is not log-transformed; therefore, the variance of the summarized values tends to increase with the mean. Affymetrix recommends the use of MAS 5.0 over MAS 4.0 (Affymetrix, 2002). In general, AvgDiff is no longer used for summarization.

MAS 5.0 employs a robust method, based on the *Tukey-biweight* function, to summarize each probe set. First, the probe intensities are background corrected, but not normalized. In MAS 5.0, normalization comes after summarization. Then an ideal MM is calculated if MM $\geq$ PM. The ideal MM is the PM intensity minus some correction value, which depends on the magnitude of the difference between the MM and PM intensities for a single probe pair. Next, the adjusted PM–MM intensities are log-transformed, and the Tukey-biweight function applied to obtain the summarized expression value. The Tukey-biweight function is a weighted calculation, meaning that values that are farther from the median log(PM–IM) intensity receive lower weight than those close to the median. Therefore, the influence of outliers is minimized, both by the use of the median as a measure of center and by the weighting scheme. However, this method has also fallen out of favor, mainly because of the use of the ideal MM.

PLIER® is the latest algorithm developed by Affymetrix (2005). The ideal model for the expected value of the observed binding for PM and MM probes is given by:

$$E(PM_{ij}) = \mu_{ij} = a_i c_j + B_{ij}$$
$$E(MM_{ij}) = B_{ij}$$

where B_{ij} is the background binding for probe pair i on array j, μ_{ij} is the binding level of probe i on array j, a_i is the binding affinity of probe i, and c_j is the concentration of RNA within sample j (Therneau and Ballman, 2008). In this formulation, the background for the PM intensities is equal to that of the MM intensities, and the MM intensities carry no signal. There are random errors associated with both PM and MM binding, ε_P and ε_M, respectively, and it is generally accepted that these errors are multiplicative (which is why log transforms are recommended). MAS 5.0 makes the assumption that $\varepsilon_P = \varepsilon_M$. The limitations of this assumption were discussed earlier.

PLIER assumes that $\varepsilon_P = 1/\varepsilon_M$, which is a biologically implausible assumption, certainly in the light of Figure 11.1, where it is clear that the PM and MM probe intensities have a positive correlation in general. Yet, PLIER outperforms MAS 5.0 for benchmark data (Cope et al., 2004). PLIER also obtains more accurate expression values for low RNA concentration levels. Therneau and Ballman (2008) examined the ability of PLIER to estimate the binding affinity for a single probe at low concentrations of RNA. The function produced by PLIER for estimating the ideal error curve (when all assumptions are true) has the correct shape, at least when considering a single probe. However, they speculate that differences in the observed error functions from the ideal error functions will be greatly increased when summarizing the intensities over an entire probe set. Therneau and Ballman (2008) go so far as to state that the improved performance of PLIER over MAS 5.0 is simply due to "good fortune." Further investigations of the performance of PLIER have shown that it is clearly outperformed by RMA, MBEI, DFCM, and many others (Chen et al., 2009).

The MBEI (dChip) algorithm computes summarized values using the following formula

$$\tilde{\theta}_i = \left(\sum_j y_{ij} \phi_j \right) / J,$$

where J =the number of probes in the probe set (excluding outliers), y_{ij} is the observed intensity level for probe j on array i, and ϕ_j is a measure of probe effect (Li and Wong, 2001b). θ_i and ϕ_j are determined using an iterative algorithm, where each value is assumed fixed while the other is estimated. The algorithm stops when changes in successive estimates of the parameters are sufficiently small. Even with outliers removed, the summarization scheme for MBEI gives the most weight to probes with the largest intensities because it is a sum of the observed intensities (i.e., it is not robust). Given the presence of NSH and XH in the data, intensity values are often overestimated. This could account for the poor performance of MBEI in comparison with more recent methods (Bolstad, 2003; McGee and Chen, 2006b).

Factor analysis for robust microarray summarization (FARMS): Hochreiter et al. (2006) developed, FARMS, another summarization method in which the observed values are modeled with a linear model. Specifically, let x =the zero-mean normalized PM intensities for a given probe set, and z =RNA concentration. The observed values are assumed to depend on concentration via the relationship

$$x = \lambda z + \varepsilon, \quad z \sim N(0,1) \quad \text{and} \quad \varepsilon \sim N(0, \Psi).$$

The error term is assumed to have a multivariate normal distribution, and Ψ is a diagonal variance-covariance matrix. The terms z and ε are assumed to be statistically independent. FARMS performs better than other competitors using benchmark data (Cope, 2004), particularly where sensitivity and specificity are concerned. It is also computationally efficient, as the EM algorithm is used to estimate the model parameters. However, Chen et al. (2007) show that FARMS does not perform as well as RMA or DFW, although this result is counter to previous claims.

Median polish: the summarization method for RMA is median polish (Mosteller and Tukey, 1977). It is a robust method for fitting the following linear model

$$\log_2(y_{ij}^{(a)}) = \mu^{(a)} + \theta_j^{(a)} + \alpha_i^{(a)} + \varepsilon_{ij}^{(a)}$$

with constraints median(θ_j) = median(α_i) = 0 and median$_i(\varepsilon_{ij})$ = median$_j(\varepsilon_{ij})$ = 0. Here, the superscript (n) represents the nth probe set on array j, y_{ij} refers to the observed intensity of the ith probe, α_i represents a probe effect, and θ_j is an array effect (Bolstad, 2004). Median polish begins by arranging the data in a matrix for each probe set n such that the probes are in rows and the arrays are in columns. A column of zeros is appended to the right of the matrix, and a row of zeroes is appended to the bottom. Next, the median of each row (ignoring the last column of zeros) is calculated, subtracted from each observation in the row and added to the final column. The procedure proceeds similarly for the columns. The algorithm continues until the changes are arbitrarily small. The end result is estimates for μ, θ, α, and ε, which are used to compute a summarized value for each probe set.

For a large number of arrays, the median polish algorithm will be computationally expensive. Median polish does not provide standard error estimates in a natural way, and it can be applied only in balanced row-column effect models (Bolstad, 2004). Furthermore, median polish gives different answers depending on whether the procedure begins with rows or columns.

Distribution free weighted (DFW) *summarization*: With the exception of MBEI, none of the summarization methods mentioned above takes into account the fact that some probes perform poorly. They either tend to cross-hybridize with the target, resulting in too much signal, or they do not catch signal at all. Evidence of both kinds of probes is seen in Figure 11.1. MBEI tends to favor probes with large intensities, which means that expression values will be overestimated. Furthermore, the MBEI model assumes independent and identically normally distributed error terms for ease of estimation. For real data, there may be multiplicative error, or the terms may be correlated.

Chen et al. (2007) devised a method that eliminates both cross-hybridizing and under-performing probes in the summarization process. The method, DFW, is a data driven method that obtains an estimate of standard deviation for each probe across all arrays, regardless of the experimental manipulation done to the array. Probes that have large standard deviations tend to be either differentially expressed between treatments, cross-hybridizing, or underperforming. The algorithm for DFW uses separate measures of variability to determine to which group a probe belongs. For cross-hybridizing or under-performing probes, a small weight is assigned during summarization so that the intensity levels assigned to these probes contribute little to the final expression value for that gene. The elimination of such probes has proved fruitful, as DFW was able to identify all differentially expressed genes in the Affymetrix latin square HG-U95Av2 spike-in experiment with no false positives (AUC=1.0), and all DEGs from the HG-U133A experiment with only two false positives (Chen, 2007). Furthermore, DFW performs best on real data, as evidenced by the GO algorithm (Kong, 2009).

11.2.5   Present/absent calls

It is not likely that all genes on a chip are represented in a given target sample and those that are not should be filtered out of the preprocessed dataset before further analysis. Reducing the abundant number of features after background correction, summarization, and normalization are carried out alleviates problems that may be encountered in large sample statistical analysis. A process that calls a gene present or absent based on its signal to noise ratio is commonly utilized to determine whether a gene is actually expressed in the sample. An absent call for a gene indicates there was not enough mRNA transcript in the sample to bind to the probes sequenced for that gene and the gene is filtered out. However, if a gene is called present then there was sufficient mRNA for the gene to be detected and the gene is kept for subsequent statistical analysis.

The current detection algorithm for Affymetrix arrays uses a discrimination score and its p-value for assigning present, marginal and absent calls. The discrimination score, $R=(PM-MM)/(PM+MM)$, where PM is perfect match intensity and MM is mismatch intensity, is calculated for a probe pair and then compared to a user-definable threshold τ (default 0.015). The one-sided Wilcoxon's signed rank test is used to obtain a p-value (Affymetrix, 2007). The user then defines cutoff values such that if a p-value is less than α_1 the gene is present, between α_1 and α_2 the gene is marginal and greater than α_2 the gene is absent, where the default values of α_1 and α_2 are 0.04 and 0.06, respectively. The experimenter may choose to use marginal genes in further analysis or consider them absent.

The Affymetrix *Statistical Algorithms Reference Guide* claims that with this method, the smaller the p-value the more likely the measured mRNA transcript is detected. However, a positive correlation often exists between the perfect and mismatch probe pairs such that a small p-value actually provides evidence of this relationship rather than evidence of detectable levels of expression as shown in Figure 11.4. Another issue with this approach is that the set of genes called marginal actually includes genes that should be called present and genes that should be called absent (see Figure 11.5). If the marginal genes are chosen for use in downstream analysis, the experimenter is unnecessarily increasing the number of false positives. If the marginal genes are filtered out, then the number of false negatives will be falsely inflated.

The discrimination score, R, uses the MM intensities as a baseline for detecting signal in PM intensities. This is problematic due to the nature of the GeneChip experiments. The MM probe is meant to be a control that measures a corresponding PM's NSH and the PM probe sequence is intended to hybridize with the mRNA from a specific gene. However, the PM probe

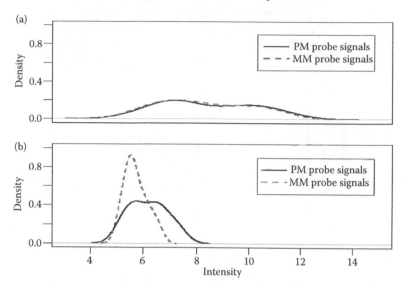

FIGURE 11.4: Intensity distributions for "present" and "absent" genes. (a) The densities of $\log_2$ PM and MM probe intensities for a single gene from a single study. Although there is signal present in this probe set, the Affymetrix detection algorithm (MAS 5) calls this gene absent because the PM and MM distributions are so similar. (b) The densities of $\log_2$ PM and MM probe intensities for a single gene from a single study. The Affymetrix detection algorithm calls this gene present because the PM and MM intensity distributions are different.

can experience nonspecific binding and the MM probe can hybridize with the mRNA meant for the PM probe. Thus, the intensity measured is not simply a signal plus background noise, rather a combination of the signal, unavoidable nonspecific binding and background noise. Simply subtracting PM–MM will not result in a properly corrected PM signal and therefore the p-value corresponding to the calculated R is not a reliable measure of detection.

A new method of calling presence or absence of a gene should utilize the fact that MM probes are not reliable estimates of background noise of PM probes because of NSH to the MM probes themselves. Warren et al. (2007) address the issue with a method called "presence-absence calls in negative probe sets" that utilizes Affymetrix-reported probes that should not bind to intended sample RNA sequences. Although the authors claim the method performs better than the current Affymetrix detection algorithm, it was developed for only two Affymetrix platforms. A method that may be applied to any platform, including those without the MM feature of Affymetrix chips, was suggested by Wu and Irizarry (2005) but uses only PM probe information. This *half-price* method makes detection calls without the use of MM probes and therefore may be less reliable because not all available information is used.

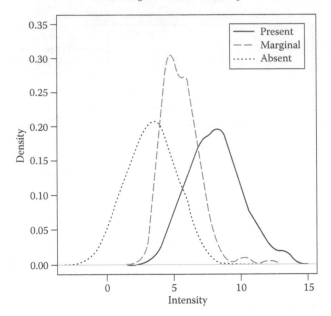

FIGURE 11.5: Distributions of log$_2$ intensities for genes called present, absent and marginal genes from a single experiment. The two peaks in the marginal genes indicate there should further separation of present and absent genes. Note there are negative values after taking the logarithm of a measured signal that is less than 1.

Additional approaches to detecting gene expression in microarrays should be formulated and tested to provide microarray users with more accurate methods of filtering genes with low signal to noise ratio.

11.3   Preprocessing Steps and Methods for Other Platforms

Unique aspects of each of the platforms require unique approaches to preprocessing. For example, the spotted arrays utilize several pins to spot transcripts onto slides, whereas the photolithography methods construct the oligonucleotides one nucleotide at a time directly on the array. A pin (or print-tip) bias may be present in spotted arrays, and must be accounted for, but is absent in in situ arrays. Also, background estimation requires image analysis software due to the nature of the spotting. The spots are not uniform within or between spotted arrays, introducing issues with defining the foreground and background. See Gentleman et al. (2005) for more details on preprocessing methods for spotted cDNA and long oligonucleotide

arrays. Microbead-based microarrays also require platform specific preprocessing steps which are discussed thoroughly by Dunning et al. (2008).

11.4 Preprocessing Methods Evaluation

Currently, there are many microarray analysis methods available, which
makes it challenging for scientists to decide which method to use for their data
analysis. The quality and the credibility of these methods when applied need
to be assessed fairly. To assess the performance of these algorithms, a series
of data sets were produced in which a group of known transcripts were mixed
at known quantities and their levels in the mixture measured using standard
microarray methodologies (Irizarry, 2003). These so-called "spike-in" data sets
provide the ability to assess sensitivity and specificity by comparing the receiver operating characteristic (ROC) curves produced (McGee and Chen,
2006b). While this approach is useful, there is some concern that analytical
approaches that work well with spike-in data may not work as well with data
derived from real, complex biological samples. ROC analysis is not feasible
for real biological data since the true expression values for target mRNA's
are rarely known, and so methods of comparison other than ROC curves are
needed.

The Gene Ontology (GO), one of the most successful biomedical ontologies, includes biological process, molecular function and cellular component
terms linked together in a directed acyclic graph with "is_a" and "part_of"
relationships. The GO has been used extensively to annotate prokaryotic and
eukaryotic gene products based on information described in the scientific literature. Based on the premise that an improvement in any step of microarray
data analysis should be reflected in improved co-clustering of related genes,
we have successfully applied GO term co-clustering as a comparative tool and
assessed the impact of using revised annotation on Affymetrix gene expression
microarray data analysis using real biological data (Kong, 2007).

We have applied the same methodology to evaluate various background
correction, normalization and summarization methods. We have chosen to assess several popular or newly developed methods including five background
correction methods, five normalization methods and four summarization
methods using real biological data. Interestingly, there exist some interdependencies among these methods. We have assembled about 300 pipelines which
represent all possible combinations of background correction, normalization,
and summarization methods. Through evaluating their GO term co-clustering
characteristic, we conclude that RMA, RMA-mean, RMA-75, and DFCM are
better background correction methods compared to MAS. For normalization,
loess and invariant set outperform constant, contrasts, and quantiles. Median
polish and DFW are the best summarization method compared to FARMS,

and MAS5 (Kong, 2009). For the dataset that we tested, the best analysis pipeline is RMA-75 as background correction, invariant set as normalization and DFW as summarization method. Our future goal is to use a data-driven approach to provide the best analysis pipeline for different distributions of microarray data, which will meet the need of the scientific community.

***

References

Affymetrix. 2007. *Statistical Algorithms Reference Guide, Data Analysis Fundamentals Technical Manual.* Affymetrix, Inc., Santa Clara, CA.

Affymetrix. 2005. *Guide to Probe Logarithmic Intensity Error (PLIER) Estimation.* Affymetrix, Inc., Santa Clara, CA.

Affymetrix. 2003. Technical note: design and performance of the GeneChip Human Genome U133 plus 2.0 and Human Genome U133A plus 2.0 arrays. Affymetrix, Inc., Santa Clara, CA.

Affymetrix. 2002. *GeneChip Expression Analysis: Data Analysis Fundamentals.* Affymetrix, Inc., Santa Clara, CA.

Affymetrix. 1996. *Microarray Analysis Suite Version 4 User Guide,* Affymetrix, Inc., Santa Clara, CA.

Åstrand M. 2003. Contrast normalization of oligonucleotide arrays. *Journal of Computational Biology* 10(1):95–102.

Bjork K, Kafadar K. 2007. Order dependence in expression values, variance, detection calls and differential expression in Affymetrix GeneChips. *Bioinformatics* 23(21):2873–2880.

Bolstad BM. 2004. Low level analysis of high-density oligonucleotide array data: Background, normalization and summarization [dissertation]. University of California at Berkeley.

Bolstad BM, Irizarry RA, Astrand M, Speed TP. 2003. A comparison of normalization methods for high density oligonucleotide array data based on variance and bias. *Bioinformatics* 19:185–193.

Calza S, Valentini D, Pawitan Y. 2008. Normalization of oligonucleotide arrays based on the least-variant set of genes. *BMC Bioinformatics* 9:140: doi:10.1186/1471-2105-9-140.

Chen Z, McGee M, Liu Q, Scheuermann RH. 2007. A distribution free summarization method for Affymetrix GeneChip arrays. *Bioinformatics* 23(3):321–327.

Chen Z, McGee M, Liu Q, Kong M, Deng Y, Scheuermann RH. 2009. A distribution free convolution method for background correction of oligonucleotide microarray data. *BMC Genomics* (in press).

Cope LM, Irizarry RA, Jaffee H, Wu Z, Speed TP. 2004. A benchmark for Affymetrix GeneChip expression measures. *Bioinformatics* 1(1):1–13.

Dai M, Wang P, Boyd AD, Kostov G, Athey B, Jones EG, Bunney WE, et al. 2005. Evolving gene/transcript definitions significantly alter the interpretation of GeneChip data. *Nucleic Acids Research* 33(20):e175.

Dunning MJ, Barbosa-Morais NL, Lynch AG, Tavaré S, Ritchie ME. 2008. *BMC Bioinformatics* 9:85: doi:10.1186/1471-2105-9-85.

Dudoit S, Yang YH, Callow MJ, and Speed TP. 2002. Statistical methods for identifying genes with differential expression in replicated cDNA microarray experiments. *Statistica Sinica* 12(1):111–139.

Gentleman RC, Carey VJ, Bates DM, Bolstad BM, Dettling M, Dudoit S, Ellis B, Gautier L, Ge Y, Gentry J, Hornik K, Hothorn T, Huber W, Lacus S, Irizarry R, Leisch F, Li C, Maechler M, Rossini AJ, Sawitzki G, Smith C, Smyth G, Tierney L, Yang JYH, Zhang J. 2004. Bioconductor: open software development for computational biology and bioinformatics. *Genome Biology* 5:R80, doi:10.1186/gb-2004-5-10-r80.

Gentleman RC, Carey VJ, Huber W, Irizarry RA, Dudoit S (eds). 2005. *Bioinformatics and Computational Biology Solutions Using R and Bioconductor.* New York, NY: Springer.

Harbig J, Sprinkle R, Enkemann SA. 2005. A sequence-based identification of the genes detected by probesets on the Affymetrix U133 plus 2.0 array. *Nucleic Acids Research* 18:33(3):e31.

Hartemink AJ, Gifford DK, Jaakkola TS, Young RA. 2001. Maximum-likelihood estimation of optimal scaling factors for expression array normalization. *Proc. SPIE 4266*, 132–141.

Hochreiter S, Clevert DA, Obermayer K. 2006. A new summarization method for affymetrix probe level data. *Bioinformatics* 22(8):943–949.

Hoffmann R, Seidl T, Dugas M. 2002. Profound effect of normalization on detection of differentially expressed genes in oligonucleotide microarray data analysis. *Genome Biology* 3(7): doi:10.1186/gb-2002-3-7-research0033.

Huber W, Von Heydebreck A, Sültmann H, Poustka A, Vingron M. 2002. Variance stabilization applied to microarray data calibration and to the quantification of differential expression. *Bioinformatics* 18(Suppl. 1):S96–S104.

Irizarray RA, Bolstad BM, Collin F, Cope LM, Hobbs B, Speed TP. 2003a. Summaries of Affymetrix GeneChip probe level data. *Nucleic Acids Research* 31:e15.

Irizarry RA, Hobbs B, Collin F, Beazer-Barclay YD, Antonellis KJ, Scherf U, Speed TP. 2003b. Exploration, normalization, and summaries of high density oligonucleotide array probe level data. *Biostatistics* 4(2):249–264.

Kong YM, Chen Z, Cai J, Scheuermann R. 2007. Use of gene ontology as a tool for assessment of analytical algorithms with real data sets: Impact of revised Affymetrix CDF annotation. 7th International Workshop on Data Mining. *Bioinformatics*, 64–72.

Kong YM, Chen Z, Qian Y, McClellan E, McGee M, Scheuermann RH. 2009. Objective selection of the optimal microarray analysis pipeline. In preparation.

Lee JA, Sinkovits RS, Mock D, Rab EL, Cai J, Yang P, Saunders B, Hsueh RC, Choi S, Subramaniam S, Scheuermann RH. In collaboration with the Alliance for Cellular Signaling. 2006. Components of the antigen processing and presentation pathway revealed by gene expression microarray analysis following B cell antigen receptor (BCR) stimulation. *BMC Bioinformatics* 7:237.

Li C, Wong HW. 2001a. Model-based analysis of oligonucleotide arrays: expression index computation and outlier detection. *Proceedings of the National Academy of Sciences.* 98:31–36.

Li C, Wong HW. 2001b. Model-based analysis of oligonucleotide arrays: model validation, design issues and standard error application. *Genome Biology* 2: doi:10.1186/gb-2001-2-8-research0032.

McGee M, Chen Z. 2006a. New Spiked-In Probe Sets for the Affymetrix HGU-133A Latin Square Experiment. *COBRA* Preprint Series. Article 5. http://biostats.bepress.com/cobra/ps/art5.

McGee M, Chen Z. 2006b. Parameter estimation for the exponential-normal convolution model for background correction of Affymetrix GeneChip data. *Statistical Applications in Genetics and Molecular Biology* 5(1): Article 24. DOI:10.2202/1544-6115.1237.

Mosteller F, Tukey J. 1977. *Data Analysis and Regression.* Reading, MA: Addison-Wesley.

Quackenbush J. 2002. Microarray data normalization and transformation. *Nature Genetics* 32:496–501.

Schuchhardt J, Beule D, Malik A, Wolski E, Eickhoff H, Lehrach H, Herzel, H. 2000. Normalization strategies for cDNA microarrays. *Nucleic Acids Research* 28(10):e47.

Shi L, Reid LH, Jones WD, Shippy R, Warrington JA, Baker SC, Collins PJ, et al. 2006. The MicroArray Quality Control (MAQC) project shows inter- and intraplatform reproducibility of gene expression measurements. *Nature Biotechnology* 24(9):1151–1161.

Therneau TM, Ballman KV. 2008. What does PLIER really do? Cancer Informatics: 6, 423–431.

Warren P, Taylor D, Martini PGV, Jackson J, Bienkouska J. 2007. PANP—a new method of gene detection on oligonucleotide expression arrays. *Proceedings of the 7th IEEE International Conference on Bioinformatics and Bioengineering*, 108–115.

Wu Z, Irizarry RA, Gentleman R, Martinez-Murillo F, Spencer F. 2004. A model-based background adjustment for oligonucleotide expression arrays. *Journal of the American Statistical Association* 99:909–917.

Wu Z, Irizarry RA. 2005. A statistical framework for the analysis of microarray probe-level data. Johns Hopkins University, Department of Biostatistics Working Papers. Working Paper 73 http://www.bepress.com/jhubiostat/paper73.

Yang YH, Dudoit S, Luu P, Lin DM, Peng V, Ngai J, Speed TP. 2002. Normalization for cDNA microarray data: a robust composite method addressing single and multiple slide systematic variation. *Nucleic Acids Research* 30(4):e15; doi:10.1093/nar/30.4.e15.

Chapter 12

Application of Feature Selection and Classification to Computational Molecular Biology

Paola Bertolazzi and Giovanni Felici

Consiglio Nazionale delle Ricerche

Giuseppe Lancia

University of Udine

12.1   Introduction

Modern technology, where sophisticated instruments are coupled with the massive use of computers, has made molecular biology a science where the size of the data to be gathered and analyzed poses serious computational problems. Very large data sets are ubiquitous in computational molecular biology: The European Molecular Biology Laboratory (EMBL) nucleotide sequence database has almost doubled its size every year in the past 10 years, and, currently, the archive comprises over 1.7 billion records covering almost 1.7 trillion base pairs of sequences. Similarly, the protein data bank (PDB) has seen an exponential growth, with the number of protein structures deposited (each of which is a large data set by itself) currently at over 50,000. An assembly task can require to reconstruct a large genomic sequence starting from hundreds of thousands of short (100–1000 bp) DNA fragments. Microarray experiments produce information about the expression of hundreds of thousands of genes in hundreds of individuals at once (data set in the order of Gigabytes), and the list goes on and on.

This abundance of large bodies of biological data calls for effective methods and tools for their analysis. The data, both structured or semistructured, have in many cases the form of two-dimensional arrays, where the rows correspond to *individuals* and the columns are associated with some *features*. While in other fields (see for instance medical data) a data set contains a large number of individuals and a small set of features in the field of molecular biology this situation is reversed, and the number of individuals is small while the number of features is very large. This is mainly due to the cost of the experiments (for instance, the DNA sequencing procedure or the phasing of genotypes require a lot of time and very complex computational procedures) and to the complexity of the molecules.

These large data sets must be analyzed and interpreted to extract all relevant information they can provide, thus separating it from extra information of little practical use. *Feature selection* (FS) and *classification* techniques are the main tools to pursue this task. FS techniques are meant at identifying a small subset of important data within a large data set. Classification techniques are designed to identify, within the analyzed data, synthetic models that are able to explain some of the relevant characteristics contained therein. The two techniques are indeed strongly related: the selection of few relevant features among the many ones available can be considered, per se, already a simple model of the data, and thus an application of learning; on the other hand, FS is always used on large bodies of data to identify those features on which to apply a classification method to identify meaningful models.

In this chapter we consider three different problems arising in computational molecular biology: *classification, representation,* and *reconstruction.*

12.1.1 Classification

When the experiments are divided into two or more classes associated with some characteristics of the individuals, it is of interest to identify those rules that put the values of the features of an individual in relation with its class; such rules are then used to shed light on the studied characteristic (e.g., if the individuals are divided into *healthy* and *ill* classes, one may want to know what genes are over-expressed in the ill individuals while being *under-expressed* in the healthy ones). Besides, when the rules exhibit a sufficient precision, they can be used to classify new individuals, e.g., to *predict* the class of an individual of unknown class. This is the case of *classification* or *supervised learning* problems. Here the relevance of the features to be selected is related with their ability to discriminate between individuals that belong to different classes; in addition, a classification method may be needed to identify with more precision the classification rules on the selected features.

12.1.2 Representation

When a set of individuals has to be analyzed according to the values of its features to discover relevant phenomena, a crucial step is to restrict the analysis to a small and treatable number of relevant features. In this case the final scope of the analysis is not known in advance; the relevance of the features to be selected is then related to the way they *represent* the overall information gathered in the experiments; such type of problems are usually referred to as *unsupervised learning.*

12.1.3 Reconstruction

This class of problems arise when the objective is to find a possibly small subset of the available features from which it is possible to derive the values of all the remaining features with a good degree of precision. Here the FS method has to be combined with a *reconstruction* algorithm, i.e., some rules that link the values of a non-selected feature with the values of one or more selected features.

The chapter is organized as follows. We give a general overview on FS methods both for unsupervised and supervised learning problems in Section 12.2. Then we review the main classification methods that are used in biological applications (Section 12.3). We then describe in greater detail two biological data analysis problems, where FS and classification methods have been used with success: in Section 12.4 we consider *microarray analysis*, where the profiles (expression levels) of mRNA (transcriptomics), proteins (proteomics), or metabolites (metabolomics) in a given cell population are studied to identify those genes (or proteins or metabolites) that appear to be correlated with the studied phenomena; Section 12.5 is dedicated to *species discrimination through DNA barcode*, a method that aims at using genotypes and haplotypes information in place of phenotype information to classify species. Finally we consider, in Section 12.6 a typical problem of representation and reconstruction: *TAG-single nucleotide polymorphisms (SNPs) selection*, that concerns the problem of finding a subset of the loci of a chromosome whose knowledge allows to derive all the others. Given the breadth and the complexity of the material considered, we provide the reader with a minimal level of technical details on the methods described, and give references to the appropriate literature.

12.2   Feature Selection Methods

As already pointed out, the identification of a subset of relevant features among a large set (FS) is a typical problem in the analysis of biological data. The selected subset has to be small in size and must retain the information that is most useful for the specific application. The role of FS is particularly important when the data collection process is difficult or costly, as it is the case in most molecular biology contexts.

In this section we look at FS as an independent task that preprocesses the data before classification, reconstruction, or representation algorithms are employed. However, the way the target features are chosen strongly depends on the scope of the analysis. When the analysis is of *unsupervised type*—as it is the case for representation or reconstruction—the ideal subset of features should be composed by features that are pairwise uncorrelated and able to differentiate each element from the others; or, put it differently, to have a high degree of variability over the individuals. In the absence of other requirements, the best that can happen is that the information (i.e., the variability over the columns of the original data matrix) is retained in the matrix projected over the columns that represent the selected features. On the other hand, in the case of *supervised* analysis (e.g., classification), good features are those that have different values in individuals that belong to different classes, while having equal (or similar) values for individuals in the same class. Obviously, also in this case it is appropriate to require the selected features to be pairwise uncorrelated: the discriminating power of a pair of strongly correlated feature would be in fact equivalent to that of only one of the feature in that pair.

In both cases, the main difficulty of FS lies in the fact that its goal is to select a subset of a larger set that has some desirable properties, where such properties strongly depend on the whole subset and it is thus not always appropriate to measure them by means of simple or low order functions in the elements. Moreover, the number of candidate subsets is exponential in the size of the original set. Many successful methods thus propose heuristic approaches, typically greedy, where the final subset is not guaranteed to be the best possible one, but is verified, by some proper method, to function "well." On the other hand, optimal approaches, that guarantee the minimization of some quality function, need some approximation in the evaluation function to become tractable.

In the remainder of this section we introduce a general paradigm for FS, and then we illustrate some of the most established search methods for FS in supervised or unsupervised settings (part of the material of this section is derived from [25]). In order to give a general overview of the methods available, we refer to the work of [49], according to whom a FS method is based on four main steps, as follows:

1. Generation procedure

2. Evaluation function

3. Stopping criterion

4. Validation procedure

The *generation procedure* is in charge of generating the subsets of features to be evaluated. From the computational standpoint, the number of possible subsets from a set of N features is 2^N. It is therefore very crucial to generate good subsets by trying to avoid the exploration of the entire search space, using heuristic strategies, that at each step select a new feature amongst the available ones to be added to the existing set, or random strategies, where a given number of subsets is generated at random, and the one with the best evaluation value is chosen. The generation starts with the empty set, and then adds a new feature at each iteration (*forward strategy*). Alternatively, it may start from the complete set of features removing one feature at each step (*backward strategy*). Finally, some methods propose to start from a randomly generated subset to which forward or backward strategy is applied.

The *evaluation function* is used to measure the quality of a subset. Such value is then compared with the best available value obtained, and the latter is updated if appropriate. More specifically, the evaluation function measures the relevance of a single feature—or of a subset of features—with respect to the final task of the analysis. An interesting classification for FS methods in the case of supervised learning is given by [23], who propose four classes based on the type of evaluation functions:

- *Distance* measures: given two classes C_1 and C_2, feature X is preferred to Y if $P(C_1|X) - P(C_2|X) > P(C_1|Y) - P(C_2|Y)$, that is, if X induces a larger increase in the class conditional probabilities with respect to Y

- *Information* measures, that tend to indicate the quantity of information retained by a given feature. For example, feature X is preferred to feature Y if the improvement in the entropy function obtained by adding X is larger that the one obtained by adding Y

- *Dependence* or *correlation* measures: they indicate the capability of a subset of features to predict the value of other features. In this setting, X is preferred to Y if its correlation with the class to be predicted is larger. These measures may also indicate redundancies in the features, based on the cross-correlation between the features themselves

- *Consistency* measures: their purpose is to evaluate the capacity of the selected feature to separate the objects in different classes. For example, a particular feature may be considered uninteresting if two elements of the data set have the same value for that feature but belong to different classes.

The *stopping criterion* is needed to avoid time consuming exhaustive search of the solution space without a significant improvement in the evaluation function. The search may be stopped when a given number of features have been selected, or when the improvement obtained by the new subset is not relevant.

Finally, the *evaluation procedure* measures the quality of the selected subset. This is typically accomplished by running classification or reconstruction methods algorithms by using only the selected features on additional data. According to the type of evaluation function adopted, FS methods are divided into two main groups: *filter methods* and *wrapper methods*. In the former, the evaluation function is independent from the classification or reconstruction algorithm that is to be applied. In the latter, the algorithms are, to a certain extent, the essence of the evaluation function: each candidate subset is tested by using the algorithm and then evaluated on the basis of its performance.

In Figure 12.1, the general design of a filter and a wrapper method is depicted, where the algorithm is represented by a generic classifier. The opinion that wrapper methods can provide better results in term of final accuracy is widely shared by the scientific community. However, these methods are extremely expensive from the computational standpoint.

The computational complexity of filter methods turns out to be very important in the analysis of biological large datasets, where the dimensions involved forbid the application of certain sophisticated classification techniques which use the complete set of features. In such cases, wrapper methods could not be applied in the first place.

For both FS approaches different methods to evaluate the quality of a partial or complete solution are proposed. Such methods are part of the FS algorithm itself, as they direct the search for a solution or the choice of a best solution among the ones available. Nonetheless, their relative importance in wrapper and filter methods is different: in the former, the evaluation function is not used to select the final solution among many, but only to obtain each of them; for this approach, in fact, the effective classification performances are those that drive the choice of the solution. In the latter (filter methods) the choice of the feature set cannot benefit from the result of the classification

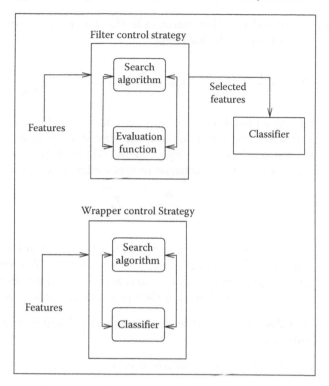

FIGURE 12.1: Wrappers and filters.

algorithm, and such task is entirely accomplished by the evaluation function. For this reason it is appropriate to stress the importance of a good choice of the evaluation function (and thus of the FS method). Below, we analyze a few of the more relevant methods—the list far from being exhaustive.

12.2.1 Methods based on consistency

Such methods have been conceived for classification problem (supervised learning). The main idea behind this class of methods is that of searching for the smallest subset of the available features that is as consistent as possible with the class variable. The work [3] proposes FOCUS, a method conceived for Boolean domains. The method searches the solution space until the feature subset is such that each combination of feature values belongs to one and only one class. The main drawback of this approach is the explosion of the dimension of the search space when the number of original features increases. They also propose some variants of the original algorithm to speed-up the search procedure. One of them is sketched below. Given a subset of features S, the available data is divided into a number of groups, each of them having the same values for the features in S. Assuming that p_i and n_i represent the number of positive and negative examples in the ith group, respectively, N is the number of individuals, the formula:

$$E(S) = -\sum_{i=0}^{2^{|S|}-1} \frac{p_i + n_i}{N} \left[\frac{p_i}{p_i + n_i} \log_2 \frac{p_i}{p_i + n_i} + \frac{n_i}{p_i + n_i} \log_2 \frac{n_i}{p_i + n_i} \right] \quad (12.1)$$

is used to evaluate all candidate features that may be added to the current subset and then select the one that shows the lowest value of $E(S)$.

A similar approach has been exploited by [51] with the las vegas filter (LVF) algorithm, where they measure inconsistency through the following formula:

$$I(S) = \sum_{g=1}^{G_s} \frac{n_g - f_g}{n}, \quad (12.2)$$

where G_s is the number of different groups of objects defined by the features in S, n_g is the number of objects in group g, f_g is the number of objects in group g that belong to the most frequent class, and n is the total number of objects in the training set. The LVF algorithm then proceeds with the following steps:

- The best subset B is filled with all the original features, and $I(B)$ is then computed

- A random subset S of the features is chosen

- If the cardinality of S is less than or equal to the cardinality of B, then $I(S)$ is computed

- If $I(S) \leq I(B)$, then $B \leftarrow S$, and iterate.

This method can have good behavior in the presence of noisy data, and may be efficient in practice. It may although be misled by features that take on a large number of different values in the training set; in these cases such a feature would provide a high contribution to the consistency measure, but would not be particularly effective for prediction purposes. Similar techniques have been investigated also by [65] and [56].

12.2.2   Methods based on information theory

Also in this case, we present methods designed to operate in the presence of a supervised learning task, where a class is value-associated to each individual and the final task is to learn how to predict such class from the selected features. A measure of the information conveyed by a subset is used to direct the search of the final features. Good examples of such methods are the minimum description length method (MDLM) [67] and the probabilistic approach by [46], that we briefly describe below.

The main idea in [46] is that a good subset of the features should present a class probability distribution as close as possible to the distribution obtained with the original set of features. More formally, let C be the set of classes, V the set of the features, X is a subset of V, $v = (v_1, \ldots, v_n)$ the values taken on by the features V, and v_x the projection of v on X. Then, FS should aim at finding a subset S such that $Pr(C|X = v_x)$ is as close as possible to $Pr(C|V = v)$.

The proposed algorithm starts with all the features and applies backward elimination. At each step, it removes the feature that minimizes the distance between the original and the new class probability distribution. Such distance is measured by means of *cross-entropy*, defined as follows:

$$D(Pr(C|V_i = v_i, V_j = v_j), Pr(C|V_j = v_j))$$
$$= \sum_{c \in C} p(c|V_i = v_i, V_j = v_j) \log_2 \frac{p(c|V_i = v_i, V_j = v_j)}{p(c|V_j = v_j)}. \quad (12.3)$$

Features are then removed iteratively until the desired number of features is reached. Given the nature of the formulas involved, the method must operate on binary features, and thus may require additional transformations of the data.

12.2.3 Methods based on correlation

The FS process for classification problems is strongly related to the correlation among the features and to the correlation of the features with the class attribute, as in [31]. Thus, a feature is useful if it is highly correlated with the class attribute. In this case, it will have a good chance of correctly predicting its value. Conversely, a feature will be redundant if its value can be predicted from the values of other features, that is, if it is highly correlated with other features. Such considerations lead to the claim that a good subset of features is composed of those features that are strongly correlated with the class attribute and very poorly correlated amongst themselves. One example of such methods is the correlation-based feature selector method (CFS), proposed in [36], where features are selected on the basis of the correlation amongst nominal attributes. A similar method is presented in [25], where the final set of features is found by searching for a particular maximum-weight, k—*subgraph* in a graph whose weighted nodes are associated with the features and whose arcs represent correlations between the features.

12.2.4 Combinatorial approaches to feature selection (FS)

In [19] the following combinatorial problem is analyzed and discussed: given a set S, select a subset K such that a number of properties $\prod_I$, $i=1, \ldots, n$ held by S are maintained in K. According to the nature of the problem, the dimension of K is to be maximized or minimized. They consider such problem a fundamental model for FS, and state two main variants:

- Subspace selection: S does not satisfy some $\prod$; identify the largest subset $K \in S$ such that $S_{|K}$ (S projected onto K) satisfies all $\prod$

- Dimension reduction: S satisfies all $\prod$; identify the smallest subset K such that $S_{|K}$ satisfies all $\prod$.

Such setting appears to be very interesting from the formal point of view. Among the different approaches, the idea of formulating the FS problem as a mathematical optimization problem where the number of selected features is to be minimized under some constraints has received some attention in the literature, and proven to be effective in many situations. In [18] the authors adopt such an approach for the selection of TAG SNPs; the mathematical model adopted turns out to be a linear problem with binary variables whose structure is well known in the combinatorial optimization literature as the *set covering problem*. Several similar models where also treated in [30], where large set covering models where proposed (a.k.a. the *test cover* problems). An interesting characteristic of this class of model is that it can be used profitably for FS both in supervised and unsupervised learning tasks, simply by changing the set of individual pairs on which the constraints of the model are defined: in the unsupervised case a constraint is generated for each pair of individuals, while in the supervised case only pairs of individuals that belong to different classes are associated with a constraint. The main drawback of this approach, and of the many variants that have been then proposed, lies in the fact that it uses one constraint of the integer programming problem for each pair of elements of the data set that belong to different classes. This implies a rapid growth of the problem dimension, and thus of its intractability, that then requires the use of nonoptimal solution algorithms. In [7] the computational problems related with the large dimension of the integer programming formulations used for FS are tackled by the use of an efficient GRASP heuristic that provides solutions of good quality in reasonable time. In the same work, an alternative and simplified model based is proposed for the supervised learning case. Such method has been used in some of the applications later described in this chapter, and we give below a short description of it assuming that the available individuals are described by aligned DNA fragments.

For simplicity, we assume the individuals to belong to only two classes, class A and class B. Given a feature f_j, we define $P_A(j, k)$ and $P_B(j, k)$ be the proportion of individuals where feature f_j has value k (for $k \in (A, C, G, T)$) in sets A and B, respectively. If $P_A(j, k) > P_B(j, k)$ (respectively $P_B(j, k) > P_A(j, k)$), then the presence of f_j with value k is likely to characterize items that belong to class A (respectively B). To better qualify the strict inequality between $P_B(j, k)$ and $P_A(j, k)$ we introduce an additional parameter $\lambda > 1$, and then define, for each feature j and for each individual i in class A the vector d_{ij} as follows.

$$
d_{ij} = \begin{cases}
1, & \text{if } f_{ij} = k \text{ and } P_A(j,k) \geq \lambda\, P_B(j,k); \\
0, & \text{if } f_{ij} = k \text{ and } \lambda P_A(j,k) \leq P_B(j,k); \\
1, & \text{if } f_{ij} \neq k \text{ and } \lambda P_A(j,k) \leq P_B(j,k); \\
0, & \text{if } f_{ij} \neq k \text{ and } P_A(j,k) >\geq \lambda\, P_B(j,k);
\end{cases}
$$

While, for individuals i in class B, the value of d_{ij} will be:

$$d_{ij} = \begin{cases} 1, & \text{if } f_{ij} = k \text{ and } \lambda P_A\,(j,k) \leq P_B\,(j,k)\,; \\ 0, & \text{if } f_{ij} = k \text{ and } P_A\,(j,k) \geq \lambda P_B\,(j,k)\,; \\ 1, & \text{if } f_{ij} \neq k \text{ and } P_A\,(j,k) \geq \lambda P_B\,(j,k)\,; \\ 0, & \text{if } f_{ij} \neq k \text{ and } \lambda P_A\,(j,k) \leq P_B\,(j,k)\,; \end{cases}$$

In the practical application, the parameter λ directly influences the density of the matrix composed of d_{ij} and can be adjusted to obtain a reasonable value for the density itself (say 20%). According to this definition, we assume that the number of ones in vector d_j is positively correlated with the capability of feature f_j to discriminate between classes A and B. We would then like to select a subset of the features that exhibits, as a set, a good discriminating power for all the items considered, so that we may use more features combined together to build rules that perform a complete separation between A and B. The purpose of the FS model is then to select a given and small number of features that guarantee a good discriminating power for all the elements of the data sets. This can be formally stated asking to select a given number of features (say, β) that maximize the minimum of the discriminating power over all the items. As is the case in the combinatorial FS models, we define the binary decision variable $x_j = \{0, 1\}$ with the interpretation that $x_j = 1$ (respectively $x_j = 0$) means that feature J is selected, (respectively, is not selected). The binary integer optimization problem can then be defined as follows:

$$\begin{aligned} \max \quad & \alpha \\ s.t. \quad & \sum_{\substack{i=1 \\ j=1}}^{m} d_{ij}x_j \geq \alpha \quad i = 1, \ldots, n \\ & \sum_{m} x_j \leq \beta \\ & x_j \in \{0, 1\} \quad j = 1, \ldots, m, \end{aligned} \qquad (12.4)$$

The optimal solution of the above problem would then select the β features that guarantee the largest discriminating power over all the elements in the data (we note that β is a parameter of the problem, and not a variable).

The number of variables of the problem is given by the number of features (m), and the number of rows by the number of individuals (n), keeping the size of the problem in a linear relation with the size of the data.

The use of the integer optimization models proposed above makes it possible to orient the type of the solution identified by the model. As pointed out before, FS aims at identifying a small number of features with high discrimination power. The fact that these two elements are in a natural position of tradeoff is well represented by the integer programming (IP) models above, where the dimension of the feature set and the discriminating power (approximated by the right hand side values α in Equation 12.4) may be, in turn, an objective or a constraint.

In real applications one has no clear clue of what would be the right threshold value to assign to one parameter while optimizing the other. For example, in solving Equation 12.4, one may obtain a value of $\alpha=3$ having fixed the parameter value of $\beta=10$ Clearly, a subsequent run of the optimization algorithm with a larger value of β would provide a solution with $\alpha\geq3$, while a smaller value of β should result in $\alpha\geq3$.

On the other hand, if the objective is to guarantee the separability of the sets with the smallest number of features, the classical *minimal test collection* model (as described, among others, in [30]) is the natural choice. In this case the discriminating power is kept to a minimum ($\alpha=1$) and such a fact may result in harder classification problems or more complicated separating models. In other cases the limitation on the number of features that can be selected are made available already by the application, e.g., one may desire to find models that use no more than β^* features. Here the proper route would be to maximize the discriminating power under the constraints that no more than β^* features are selected.

It is important to point out in this context that the ideal solution for a given problem—in terms of number of features and discriminating power—may need few iterations of search on the value of the parameter (be it α or β), and may strongly depend on the nature of the problem and on the type of classification method used after the FS has taken place.

For example, in the applications discussed at the end of this chapter, the winning choice has been to select a value of β sufficiently small to produce simple logic models provided that the associated α exhibits some degrees of redundancy in the discriminating power. The wisdom of this choice is eventually confirmed by the good predictive performances of the models extracted.

12.3   Classification Methods

Let us assume that the objects to classify are vectors of R^n (this is equivalent to saying that, for each object, we know the status, or the level, of n features). Assume furthermore that there are m classes, and that each point x belongs to exactly one class. Given an object $x \in R^n$, let us denote its "true" class by $c(x) \in \{1, \ldots, m\}$.

Typically, a classification problem is formulated as follows. Given t points $x_1,\ldots,x^t$, whose class is known, build a *classifier* and use it to classify new points, whose class is unknown. A classifier is a function $\tilde{c} : R^n \mapsto \{1, \ldots, m\}$. The goal is to have $\tilde{c}(x) = c(x)$, for all x, i.e., the hypothesized class should be the true class. Clearly, it is easy to guarantee that $\tilde{c}(x^i) = c(x^i)$ for all $i = 1, \ldots, t$, but it may be hard to guess the true c by the observation of only t points.

A classification function is then *learned from*, or *fitted to*, a training data set. In the above example, for instance, we could take any subset of t' points and construct a function $\tilde{c}'$ from it, and test the validity of the function on the remaining $t - t'$ points. Given that the true state of some points is known and that the function is built by exploiting this knowledge, classification methods of this type are so called *supervised*.

In other situations, the class label for the testing data set is unknown, or voluntarily ignored. In this case, the classification algorithm intuitively tries to label "similar" points (e.g., points whose distance is small under some metric) with the same label. The resulting classification methods of this type are called *unsupervised*, and the classification problem is also called *clustering*. Basically, the classification function defines a set of clusters. New points are then assigned to the cluster to which they have, e.g., the smallest average distance.

We now give a very brief, schematic description, of some of the most widely used classification and clustering procedures. In particular, we pay attention to those procedures that have been widely used in bioinformatics applications (see, e.g., Section 12.4 on microarrays in this chapter).

12.3.1 Support vector machines (SVMs)

Support vector machines (SVMs) [70,20] are particularly suited for binary classification tasks. In this case, the input data are two sets of n-dimensional vectors, and a support vector machines (SVMs) tries to construct a separating hyperplane, i.e., one which maximizes the margin (defined as the minimum distance between the hyperplane and the closest point in each class) between the two data sets.

More, formally, given a training data T consisting of t points, where $T = \{(x^i, c(x^i)) \mid x^i \in R^n, c(x^i) \in \{-1, 1\}\}_{i=1}^t$ (in this case, the class for each point is labeled –1 or 1, rather than 1 or 2) we want to find a hyperplane $ax = b$ which separates the two classes. Ideally, we want to impose the constraints $ax^i - b \geq 1$ for x^i in the first class and $ax^i - b \leq -1$ for x^i in the second class. These constraints can be relaxed by the introduction of non-negative slack variables s_i and then rewritten as

$$c(x^i)(ax^i - b) \geq 1 - s_i \qquad (12.5)$$

for all x^i.

A training datum with $s_i = 0$ is closest to the separating hyperplane $\{x: ax = b\}$ and its distance from the hyperplane is called margin. The hyperplane with the maximum margin is the optimal separating hyperplane. It turns out that the optimal separating hyperplane can be found through a *quadratic optimization* problem, i.e., minimize $\frac{1}{2} \|a\| 2 + C \sum_{i=1}^t s_i$ over all (a, b, s) which satisfy Equation 12.5. Here, C is a margin parameter, used to set a tradeoff between the maximization of the margin and minimization of classification error. The points that satisfy equality in Equation 12.5 are called support vectors.

A variant of SVMs, called *discrete support vector machines* (DSVMs) is introduced in [76] and is motivated by an alternative classification error function which counts the number of misclassified points. The proposed approach has been later extended for several classification tasks (see [77]).

SVMs can also be used, albeit less effectively, for multiclass classification. Basically, in the multiclass case, the standard approach is to reduce the classification to a series of binary decisions, decided by standard SVM. Two such approaches are *one-vs-the-rest* and *pairwise comparison* (other, similar approaches exist in the literature). Assume there are m classes. In the one-vs-the-rest approach, we build a binary classifier that separates each single class from the union of the remaining classes. Then, to classify a new point x, we run each of the m classifiers and look at the output. For each class k, assuming x is in k, the answers may have some inconsistencies. Let $\varepsilon(x,k)$ be the number of errors that the m classifiers made under the hypothesis that x does in fact belong to k. Then, x is eventually assigned to the class $\widehat{k}$ for which $\varepsilon(x,\widehat{k})$ is minimum (i.e., the class consistent with most of the answers given by the classifiers). This classification approach has several problems and therefore, although widely used, has also been widely criticized [2].

In the pairwise comparison method, one trains a classifier for each pair of classes, so that there are $m(m-1)$ independent binary classifiers. To classify a new data point x, each of these classifiers is run, and its output is viewed as a vote to put x in a certain class. The point is eventually assigned to the class receiving most votes (ties are broken randomly). As for the previous one, also for this classification method several problems have been pointed out.

12.3.2   Error correcting output codes (ECOC)

The ideas underlying error correcting output codes (ECOC) classification stem from research on the problem of data transmission over noisy channels. The method is based on d binary classifiers, each of which defines a binary partition of the union of the m classes. For each input object, these classifiers produce as output a d-ary vector over $C = \{-1, 1\}^d$, which can be seen as the "codeword," or "signature," of the object. Each class is assigned a unique codeword in C. The class codewords can be arranged in an $m \times d$ matrix, which has the property that (i) no column is made of only 1s or -1s (that classifier would not distinguish any class from the others); (ii) no two columns are identical or complementary (they would be the same classifier). Under these conditions, the maximum number of classifiers (columns) possible is $2^{m-1} - 1$. If the minimum Hamming distance between any two matrix rows is h, then even if a row undergoes up to $\lceil (h-1)/2 \rceil$ bit-flips, its original correct value can still be recovered.

The classification is performed as follows. Given a new input object x, its codeword $w(x)$ is computed via the d classifiers. Then, the object is assigned to the class whose codeword has smallest Hamming distance to $w(x)$. The method performs its best with codewords of maximum length (exhaustive coding). In this case, codewords for each class can be built so that the Hamming distance between each two codewords is $(2^{m-1} - 1)/2$. The method was proposed in

[24]. One of its limitations is that the number of classifiers is exponential in the number of classes.

12.3.3 Decision trees

A decision tree (sometimes also called classification tree) is a binary tree whose leaves are labeled by the classes, and whose internal nodes are associated with binary predicates related to the features defining the objects. At each node, one of the outgoing branches corresponds to objects whose features satisfy the predicate, while the other corresponds to objects which do not satisfy the predicate. The path from the root to each node corresponds therefore to a set of conditions which must be met by all objects associated to the node.

Decision trees are built (learned) recursively from the training data via some standard procedures. One such rule (entropy rule) is to find at each tree node a predicate which optimizes an entropy function, or information gain, of the partition it induces (intuitively, a predicate which splits a group in roughly two halves has a good information content). Popular tree decision software packages, widely used in bioinformatics applications, such as C4.5 [62], are based on entropy rules.

To predict the class label of a new input object, the predicates are applied to the input starting at the tree root, one at a time. The responses to the predicates define a path from the root to one leaf. The object is then classified with the class labeling the leaf.

12.3.4 K-nearest neighbor (KNN)

The K-Nearest neighbor (KNN) algorithm is a classifier based on the closest training data in the feature space [21]. This is a very simple classifier which has been applied to a large number of classification problems, from document retrieval, to computer vision, bioinformatics, computational physics, etc.

Given a training set of objects whose class label is known, a new input object is classified by looking at its k-closest neighbors (typically in the Euclidean distance) of the training set. Each of these neighbors can be imagined as casting a vote for its class. The input is eventually assigned to the class receiving the most votes (if $k = 1$, then the object is simply assigned to the class of its nearest neighbor).

k is a positive integer, usually small, and its best value depends upon the data. While larger k reduce the influence of noise in classification, they tend to degrade clear-cut separation between classes. Usually the best k for a particular problem is found by some heuristic ad hoc approach.

One of the main drawbacks of KNN is that it may be biased, i.e., classes which are over-represented in the training data set tend to dominate the prediction of new vectors. To counter this phenomenon, there are correcting strategies that take this bias effect into account for better classification.

12.3.5   Boosting

Boosting is the process by which a powerful classifier is built by the in-cremental use of weak classifiers. The underlying idea is that, even if none of the weak classifiers performs particularly well by itself, when they are taken together (perhaps weighted with different weights), they yield quite accurate classifiers. Each time a new weak classifier is added, the training data is reweighted. In particular, objects that are misclassified are given more importance (higher weight) and objects correctly classified see their weight decreased. This way, each next weak classifier must focus more on the objects that the preceding weak classifiers found difficult.

A popular boosting algorithm, which has found also many applications in bioinformatics, is AdaBoost ([28]).

12.3.6   Extraction of classifying logic formulas

The learning of propositional formulas able to execute a correct classification task is performed by several methods presented in the literature. For instance, the already discussed decision trees may be viewed as propositional formulas whose clauses are organized in a hierarchy, and indeed one of their first implementations was designed to deal with data expressed in binary or qualitative form. A more recent alternative is the more sophisticate logical analysis of data (LAD) system, originally proposed in [14], and the greedy approach originally proposed in [79]. *Lsquare* also belongs in this category, described in [26]. The basic idea of this method is that the rules are determined using a particular problem formulation that turns out to be a well know and hard combinatorial optimization problem, the *minimum cost satisfiability problem*, or MINSAT. The disjunctive normal form (DNF) formulas identified have the property of being created by conjunctive clauses that are searched for following the order with which they cover the training set. Therefore, they are formed by few clauses with large coverage (the interpretation of the trends present in the data) and several clauses with smaller coverage (the interpretation of the outliers). *Lsquare* has been applied with success to different bioengineering problems and further references to these applications will be given in the next sections of this chapter.

12.4   Microarray Analysis

Microarrays (also called DNA arrays) are semiconductor devices used to measure the expression level of thousands of genes within a single, massively parallel, experiment. A microarray consists of a grid with several rows and columns. Each grid cell contains some ad hoc probe DNA sequence, chosen so as to hybridize, by Watson–Crick complementarity, to the DNA (in fact, the mRNA) of a target gene. In the experiment, the mRNA sequences carrying the gene message, i.e., the instructions on which amino acids compose a particular protein, are amplified, then fluorescently tagged, and poured on the array. After hybridization, the array is scanned by a laser to quantify the amount of fluorescent light in each grid cell. This quantity measures the expression level of the particular gene selected by the cell probe.

Each microarray experiment yields a large amount of information as its output. Typically, a microarray is organized in a rectangular grid, where each row is associated with a sample, e.g., tissue cells, and each column is associated to a target gene. Usually, the number of rows is in the order of hundreds, and is much smaller than the number of columns that can be as large as a hundred thousands. It follows that the experiment output can be an array of as many as a billion numbers (or bits, if some binarization has been applied). This type of data size naturally calls for FS and classification algorithms, if one wants to make any meaningful use of the experiment's output.

One of the main uses of microarray experiments is for *tissue classification* with respect to some tumors. Different tissue cells from both healthy individuals and diseased patients are examined to see their expression profiles with respect to several genes, some of which may be related to the disease under study, while the others may be totally unrelated. The analysis of the expression profiles should hint as to which genes correlate the most with the disease (maybe at different stages). The tissue classification should also be able to allow classification of a new sample as healthy or diseased (and in the latter case, at which stage). All of the works that we survey in this section have chosen cases of tissue classification for the experimental results proving the effectiveness of the proposed methods.

Microarrays are also used in the pipeline to drug discovery and development [32,54]. They can be used to identify drug targets (the proteins with which drugs actually interact) and can also help to select individuals with similar biological patterns. This way, drug companies can choose the most appropriate candidates for participating in clinical trials of new drugs. Finally, microarrays could help in finding protein or metabolic patterns through the analysis of coregulated genes.

12.4.1   Data sets

The recent years have seen an explosion of interest in classification approaches and applications to microarray experiments. The number of published papers on this topic now probably exceeds a hundred. These papers are, generally speaking, very similar to each other, both in the techniques employed and in the data sets which are utilized.

One of the first such data sets concerns the expression of 2000 genes measured in 62 epithelial colon samples, both tumoral and healthy [4]. Another popular data sets is the Lymphoma data set [1], which concerns tumor and healthy samples in 47 patients, measured across 4026 genes. Table 12.1 reports the characteristics of the most common data sets used by the majority of papers on classification and FS problems in gene expression analysis. For all these data sets, the samples are classified in just two classes, tumoral and normal.

12.4.2 Literature review

As we just mentioned, the number of papers on classification approaches for microarray data is very large. Almost invariably, all these papers employ some of the classification procedures that we described in Section 12.3 (in particular, the use of SVMs is very popular), and then proceed to propose some variants of these procedures or some new ad hoc methods.

In [44], Hu et al. consider the standard classification methods and run a comparative study of their effectiveness over gene expression data. They consider classification approaches such as SVMs, decision trees (C4.5), and boosting (AdaBoost), and compare their performances over the data sets reported in Table 12.1 (with the exception of data set #7). Generally speaking, boosting methods perform better than the other procedures on these data sets. By using the raw original data, the rate of classification success for C4.5 (decision trees) is 75.3%, for AdaBoost it is 76.7% and for SVMs it is 67.1%. The paper also shows how FS preprocessing, i.e., extracting a certain number of genes and using only them for classification purposes, can greatly improve the performance of the classifiers. The approach used for FS here is *information gain ratio*, an entropy-based measure of the discrimination power of each gene. In preprocessing, the authors select 50 genes with the highest information gain, and then classification is performed over the data restricted to these genes. The results show the effectiveness of FS. In particular, the rate of classification success improves to 89.6% for C4.5, to 94.1% for AdaBoost, and to 88.3% for SVMs.

The work by Ben-Dor et al. [6] study some computational problems related to tissue classification by gene expression profiles. The authors investigate some scoring procedures and clustering algorithms used to classify tissues with respect to several tumors, using data sets 2, 3, and 7, of Table 12.1. The

TABLE 12.1: Microarray data sets.

Data set	Citation	# Samples	# Genes
1. Breast cancer	Veer et al. [71]	97	24,481
2. Colon	Alon et al. [4]	62	2,000
3. Leukemia	Golub et al. [32]	72	7,129
4. Lung cancer	Gordon et al. [33]	181	12,533
5. Lymphoma	Alizadeh et al. [1]	47	4,026
6. Ovarian	Petricoin et al. [61]	253	15,154
7. Ovarian	Schummer et al. [66]	32	$\simeq$100,000
8. Prostate	Singh et al. [68]	21	12,600

data sets are preprocessed via a simple FS scheme, whose goal is to identify a subset of genes relevant for the tumors under study. The authors use a measure of "relevance" to score each gene. The intuition is that an informative gene should have quite different values in the two classes (normal and tumor), and the relevance quantifies the gene's discriminating power. Classification is done via SVMs, boosting, KNN, and via a clustering-based classifier proposed by the authors. The main results of the paper is that FS is extremely important in the classification of tumor samples, as some of the original features "can have a negative impact on predictive performance." The paper demonstrates success rate of at least 90% in tumor vs normal classification, using sets of selected genes.

SVMs are used for classification of gene expression data in [16,35,29], DSVMs are used in [78], while a variant of KNN is used in [72] for cancer classification using microarray data. In this work, the authors propose to use a specific distance function learned from the data, instead of the traditional Euclidean distance for KNN classification. The experiments show that the performance of the proposed KNN scheme is competitive to those of more sophisticated classifiers such as SVMs and the uncorrelated linear discriminate analysis (ULDA) in classifying gene expression data.

In [50], Li et al. perform yet another comparative study of FS and classification methods for gene expression data. Among the classification methods that are employed, there are SVMs, KNN, Decision trees, and ECOC. The results are inconclusive, in that no method clearly emerges as a winner from this comparison. The paper deals mainly with multiclass classification problems, which are shown to be much more difficult than binary classification for gene expression datasets. The paper also employs several different FS methods. One of the data sets used is the acute lymphoblastic leukemia (ALL) and acute myeloblastic leukemia (AML) leukemia data set (#3 in Table 12.1) which encompasses gene expression profiles of two acute cases of leukemia: ALL and AML. As the ALL part of the data comes from two types of cells (B-cell and T-cell), and the AML part is split into bone marrow samples and peripheral blood samples, the data can be seen as a four-class data set. Among the conclusion drawn in the paper, there are (i) it is difficult to ascertain a clear winner for the FS method; (ii) the interaction of FS and classification methods seems quite complex; (iii) the final accuracy of classification seems to depend more on the classification method employed than on the FS method; (iv) although the results are good for datasets with few classes, they degrade quickly as the number of classes increase.

Another work [73] compares the performance of several FS and classification methods for gene expression data. The paper also proposes a somewhat innovative approach for FS based on a variant of linear discriminate analysis. This variant is particularly suited for "undersampled" problems, i.e., problems in which the number of data points is much smaller than the data dimension (the number of features). Notice that microarray data are naturally undersampled.

The work of Bertolazzi et al. [7] uses logical classification and a reduction to the set covering problem in order to extract the most relevant features from the leukemia data set (#3 in Table 12.1). The FS problem is solved by using an ad hoc metaheuristic of the GRASP type. The data is preprocessed in order to become discrete/binary, so that it can be used by the logical classifier *Lsquare*. This classifier produces a set of Boolean formulas (rules) according to which an object can be classified as belonging to one of two classes. Overall, the method achieves a performance of $\simeq 92\%$ success rate in classification on this data set. One important result of the paper is that it shows that the knowledge of as few as two or three features (levels of expression of two or three genes) is sufficient to define extremely accurate rules.

We finally mention the work of Umpai and Aitken [45], who utilize evolutionary algorithms for the selection of the most relevant genes in microarray data analysis. The authors remark on the importance of FS for classification, and propose a genetic algorithm for FS whose performance is evaluated using a low variance estimation technique. The computational methods developed perform robustly and accurately (with 95% classification success rate on the leukemia data set), and yield results in accord with clinical knowledge. The study also confirms that significantly different sets of genes are found to be most discriminatory as the sample classes are refined.

12.5   Species Discrimination Through DNA Barcode

12.5.1   Problem definition and setting

Species identification based on morphological keys shows its limitations for several reasons. Among these reasons, we find the lack of a sufficient number of taxonomists—and consequently of the expertise required to recognize the morphological keys—and the fact that the chosen observable keys could be absent in a given individual, because they are effective only for a particular life-stage or gender. However the main reason is that the small size of the organisms precludes easy visual identification, as much of their important morphology may be at scales beyond the resolution of light microscopy. According to [10] species identification based on morphological keys have described 1.5 million unique taxa to the species level, but the total number of species is likely to be in the region of 10 million (as of May 1988). Most of the relative large organisms (body sizes greater than 10 mm) have been described. However, the vast majority of organisms on the Earth have body sizes less than 1 mm [10] hence for these groups the taxonomic deficit is much higher.

The use of a specified DNA sequence to provide taxonomic identification (fingerprinting) for a specimen is a technique that should be applicable to all cellular (and much viral) life (see [11–13] for a systematic setting of the subject).

These DNA sequences were also called *molecular operational taxonomic units* (M-OTU), where an M-OTU is a terminal node (an organism) in coalescent trees obtained by sequencing an informative sequence of DNA. To be informative, the segment of DNA must be known to be orthologous between species (as paralogues will define gene—rather than organism—trees), and the segment must encompass sufficient variability to allow discrimination between M-OTU. The comparison between the between-taxon difference rate and the within-taxon variation and error rates will define the accuracy and specificity of the M-OTU measurement.

Two problems are generally addressed, i.e., identification of specimens (classification), given a training set of species, and identification of new species. In both cases a set S of M-OTU for which the species is known and a new M-OTU x are given. In the classification setting one would know the most likely species of x. In the species identification case the problem is to find the most likely species of x or determine that x is likely to belong to a new species.

A crucial parameter in this approach is the length of the DNA sequence. In [43] it is shown that while long sequences are needed to obtain correct phylogenetic trees and to identify new species, smaller sequences are sufficient to classify specimens.

Identification methods based on small DNA subsequences are first proposed for least morphologically distinguished species like bacteria, protists and viruses [55,57], and then extended to higher organisms [15,17].

More recently, in his first paper on this topic [40] Hebert proposes a new technique, *DNA barcoding*, that uses a specific short DNA sequence from a *standardized and agreed-upon* position in the genome as a molecular diagnostic for species-level identification. He identifies a small portion of the mitochondrial DNA (mt-DNA), the gene cytochrome c oxidase I (COI), to be used as a taxon "barcode," that differs by several percent, even among closely related species, and collects enough information to identify the species of an individual. This molecule, previously identified by [63] as a good target for analysis, is easy to isolate and analyze and it has been shown [53] that resumes many properties of the entire mt-DNA sequence. Since 2003, COI has been used by Hebert to study fish, birds, and other species [41]; one of the most significant results concerns the identification of cryptic species among insect parasitoids [69]. For sake of completeness we remind that another mt-DNA subsequence (gene), cytochrome b, was proposed as a common species-level marker, while COI is specific for animal species [39].

On the basis of these results the Consortium of Barcode of Life (CBOL)* was established in 2003. CBOL is an international initiative devoted to developing DNA barcoding as a global standard for the identification of biological

---

* http://www.barcoding.si.edu/

species, and has identified data analysis issue as one of the central objectives of the initiative. In particular:

(a) Optimize sample sizes and geographic sampling schemes, as barcodes are not easy to measure, and large samples are very expensive

(b) Consider various statistical techniques for assigning unidentified specimens to known species, and for discovering new species

(c) Stating similarity among species using character-based barcodes and identify what are the character-based patterns of nucleotide variation within the sequenced region

(d) Identify small portions of the barcode that are relevant for species classification, as sequencing long molecules is expensive (shrinking the barcode).

The last three topics deal mainly with data mining: problems (b) and (c) with classification and (d) with FS.

A last observation must be made; the word "barcode" has been used not only to indicate a specific DNA sequence, as in Hebert and CBOL approach, but it could identify other "strings" related to the DNA sequence. In some approaches arrays of gene expression derived through DNA microarrays experiments are used as fingerprinting, in other approaches the amino acid chains derived from given DNA sequences are considered, while in other approaches the barcode is associated to a set of particular substrings of the DNA chain.

12.5.2   Methods

A few approaches for barcode analysis have been proposed until now. They are either based on the concept of *distance* between M-OTUs or *character* based. Distance-based methods solve the classification and identification problems but do not fully belong to the category described in Section 12.3; they do not have a training phase and are mainly based on the construction of taxonomies. The character-based methods always comprise a FS phase and follow the schema presented in Sections 12.2 and 12.3.

12.5.2.1   Distance-based methods

In distance-based methods, the assumption is made that all the DNA sequences are aligned. For these methods different type of distance measures are used. The *Hamming distance* between two aligned barcodes is defined as the number of positions where the two sequences have different nucleotides. In these cases the new sequence is assigned to the species of the closest sequence or to the species with minimum average distance. The *convex-score similarity* between two aligned barcode sequences is determined from the positions where the two sequences have matching nucleotides by summing the contributions of consecutive runs of matches, where the contribution of a run is convexly increasing with its length. A new sequence is assigned to the species containing the highest scoring sequence. Another type of distance is the one defined on a coalescent tree. In this case, M-OTU are analyzed by first creating M-OTU profiles (i.e., identifying those loci variable enough that two unrelated individuals are unlikely to have the same alleles) and then using the neighbor joining (*NJ*) method [64] to obtain a phylogenetic tree (the NJ tree), so that each species is identified as represented by a distinct, non overlapping cluster of sequences in this tree: no FS is done. The principle of the NJ tree is to find pairs of M-OTUs that minimize the total branch length at each stage of clustering of OTUs starting with a star-like tree.

12.5.2.2 Character-based methods

A first set of character-based methods are those which solve the so called *string barcoding problem*, that allow, given a data base of "known" strings (where known in our case means belonging to a class) to extract which of the strings in the database the unknown one is most similar to. These methods require the alignment of the input DNA sequences. In this case one could simply use similarity searching programs such as BLAST [2] to identify the unknown string, but it is not this simple both for computational complexity and experimental costs. Instead, we can only test for the presence of some particular substring in the unknown string (*substring test*). What we need is a set of substring tests such that on every string in the known set, the set of answers (yes or no) that we receive is unique with respect to any other string in the known set. Then we perform the entire set of tests on the unknown string and compare the answers we receive with the answers for every known string. Since in biology each test is quite expensive, the number of tests must be minimized. So the string barcoding problem is solved with an approach that starts with a set of known strings and builds a minimum cardinality set of substrings that allow us to identify an unknown string by using substring tests.

This problem is introduced in [34] where they use suffix trees to identify the critical substrings, integer-linear programming (ILP) to express the minimization problem, and a simple idea that reduces the size of the ILP, allowing it to be solved efficiently by the commercial ILP solver CPLEX for sequences of about 100,000 base pairs. A more efficient highly scalable algorithm, based on the same approach, is due to [22], that deals with DNA sequence much larger than those of [34]. It uses a greedy setcovering algorithm for a problem instance with On^2 elements corresponding to the pairs of sequences. In [8] the substrings have unit length and the problem is solved through FS methods based on integer programming and the logic classification tool *Lsquare* (both already introduced in the initial sections of this chapter). The limited number of features of this application (the loci of the COI barcode sequence are less that 700) makes it possible to solve the FS model at optimality with a commercial integer programming code, and then extract the logic rules from the small set (20–30 loci) of selected features. A one-against-all strategy is used to solve multiclass problems, and the results obtained show a high degree of precision with combinations of very compact logic formulas.

Other character-based methods methods have been proposed by the data analysis working group of CBOL, in particular in [47] string kernel methods for sequence analysis are applied to the the problem of species-level identification based on short DNA barcodes. This method does not require DNA sequences to be aligned; sorting-based algorithms for exact string k-mer kernels and a divide-and-conquer technique for kernels with mismatches are proposed. Similarity kernel representation opens the possibility for building highly accurate predictors (they propose SVM classifier), clustering of unknown sequences, alternative tree representations with NO alignment and alternative visualization of data. This method is applied on three datasets (*Astraptes fulgerator*, *Hesperiidae*, and fish larvae, and shows a very high performance w.r.t. traditional suffix tree-based approaches).

12.5.2.3 Computational performance

Not many computational results and performance analysis are reported in the literature. In [58] an approach based on combining several distance- and character-based classifiers is proposed. The comparison among the performance of this approach and the behavior of all the selected methods run alone is presented. The proposed approach maintains a good classification accuracy (98%) even if the percentage of barcodes removed from each species and used for testing is equal to 50% (the accuracy ranges between 97.4 and 92.4% in the other methods). In order to test the accuracy of new species detection and classification a regular leave-one-out procedure is devised. More precisely a whole species is deleted and from each remaining species 0–50% of the barcodes are randomly deleted, to obtain a test set with deleted sequences and a training set with the remaining sequences. Also in this case the combined approach achieves an accuracy of 97% even if the percentage of barcodes removed from each species and used for testing is equal to 50%, while the other methods accuracy ranges between 63 and 89.7%.

The data analysis group of CBOL is building a web portal that will allow users to widely analyze the performance of several classification methods.

12.6 TAG Single Nucleotide Polymorphism (SNP) Selection and Genotype Reconstruction

12.6.1 Problem definition and setting

It is well known that genetic variation among different individuals is limited to a small percentage of positions in DNA sequences (99% of two DNA molecules being identical). These positions are called single nucleotide polymorphisms (SNPs) and are characterized by the fact that two possible values (alleles) of the four bases (T, A, C, G) are observed across a population at such sites, and that the minor allele frequency is at least 5%. The knowledge of such polymorphisms is considered crucial in disease association studies over a population of individuals, and is the target of the HapMap project, that has already released a public data base of one million SNPs *genotyped* from four populations of three geographical areas (Africa, East Asia, and Europe). Genome-wide association studies aim at identifying common genetic factors that influence health and disease. A genome-wide association study is defined as any study of genetic variation across the entire human genome that is designed to identify genetic associations with observable traits (such as blood pressure or weight), or the presence or absence of a disease or condition. Even if the number of SNPs identified in human genome is low (about seven millions of sites) w.r.t. the complete sequence, the costs for extracting this knowledge is prohibitive and one of the major research challenges arisen in the last years has been to find a selected number of SNPs (*TAG SNPs*) that are representative of all the other SNPs. This approach is supported by the observation that DNA molecules have a *block structure* [60,74]. Blocks are subsequences of DNA that have been transmitted during the evolution without splits in the sequence. A result of block transmission is *linkage disequilibrium* (LD), a parameter related to some combinations of alleles or genetic markers that in a population can occur more or less frequently than would be expected from a random formation of haplotypes from alleles based on their frequencies. In case of two SNPs a measure of LD is given by $\delta = p_1 p_2 - h_{12}$, where $p_1 p_2$ denote the marginal allele frequencies at the two loci and h_{12} denotes the haplotype frequency in the joint distribution of both alleles. A block is a region of the DNA where for each pair of SNPs $\delta \neq 0$; this means that the information contained in a block is redundant and suggests that it is possible to find a small set of SNPs (one single nucleotide polymorphism (SNP) for each block for instance) able to predict all the others. Such SNPs are commonly called TAG SNPs and the problem is called *TAG SNP selection* (TSS).

In the following we illustrate the two main variants of the problem: the first one, TSS, aims to select a minimum set of TAG SNPs able *to represent* all the other SNPs or, in other words, to maintain enough statistical power to identify phenotype-genotype association. The second one, *SNP reconstruction*, developed since 2006 mainly focuses on the reconstruction problem, i.e., the problem of computing the allelic values of all the other SNPs from the set of TAG SNPs.

Before introducing the two problems we need some additional notation.

In this context, it is usual to view DNA molecules as structured into subsequences called chromosomes. A *genotype* is a combination of alleles located in homologous chromosomes of a DNA molecule of a given individual; in case of diploid organisms, characterized by two chromosomes (the maternal one and the paternal one), the genotype is a sequence of pairs of alleles located in certain loci, corresponding to SNPs, of the DNA sequence. If the two alleles

h_1	C	C	T	A	T	G	C
h_2	A	C	T	A	G	G	A
g_{12}	C/A	C	T	A	T/G	G	C/A

FIGURE 12.2: Examples of genotypes.

are identical, the locus is called *homozygous*, if the two alleles are different, it is called *heterozygous*. In case of heterozygous loci, *the phase* (i.e., the value of the locus associated to maternal and paternal chromosomes) may or may not be known. When the phase is known, then the genotype is split in two sequences, called *haplotypes* (see Figure 12.2). A genotype of length m is usually represented by a $\{0, 1, 2,\}$ sequence where 0 and 1 stand for homozygous types $\{0, 0\}$, $\{1, 1\}$ (0 is usually associated to the most frequent allele) and 2 stands for a heterozygous type.

Let $G = \{g_1, \dots, g_n\}$ be the set of input genotypes, where each of the n elements is an m-dimensional vector (m is the number of SNPs). We use $g_{i,j}$ to denote the jth component (0, 1, or 2) of the genotype g_i.

A *phasing* of a genotype g_i is a pair of haplotypes $h_i^1, h_i^2 \in \{0,1\}^m$ such that $h_{i,k}^1 \neq h_{i,k}^2$, if $g_{i,k} = 2$ and $h_{i,k}^1 = h_{i,k}^2 = g_{i,k}$ if $g_{i,k} \in \{0,1\}$.

Given the genotype matrix G, the aim of TSS can be declared as follows: define a partition of the SNPs set in two sets, *TAG* and non-*TAG*, so that a SNP in non-*TAG* can be computed from one or more SNPs in *TAG*.

12.6.2　TAG single nucleotide polymorphism (SNP) selection (TSS)

A good review of TAG selection methods can be found in [37]. The authors, after a discussion of basic assumptions, describe selection algorithms utilizing a three-step unifying framework: (1) determining predictive neighborhoods; (2) defining a quality measure describing how well TAG SNPs captures the variance of the full set; (3) minimizing the number of TAG SNPs. The three steps are quite independent of each other, and in fact they could be combined in different ways.

12.6.2.1　Determining predictive neighborhoods

One of the crucial parameters in determining predictive neighborhoods is the dimension of the genomic region. If the region is large, an interval must be defined where tagging SNPs can be selected to predict a given SNP. In fact the problem of TSS is NP-hard in general case but becomes easy when a finite number of neighborhoods is used to predict a given SNP. Moreover, when the region is large, long range correlations may be seen by random chance, and hence the region must be partitioned into intervals, to avoid taking into account these correlations. If the genomic region is small, i.e., a single gene, it is not necessary to select an interval.

In [37] different approaches to the problem are presented. One effective method is based on partitioning a chromosome into blocks of SNPs exhibiting low haplotype diversity (high LD) and select within each block a set of SNPs that represent all the other SNPs within that block. However in [37] the authors observe that it is not easy to define block boundaries, there are many methods that produce different blocks and there is no metric to measure the quality of different decompositions. Another method uses a sliding window of a fixed number of positions to define neighborhoods of a given SNP. Finally one of the best methods uses the metric LD maps of Maniatis et al. [52] to evaluate distances between two SNPs: only those SNPs whose distance is below a certain threshold in this metric are considered correlated.

12.6.2.2 Defining a quality measure

Defining a quality measure aims at evaluating how well a set of TAG SNPs captures the variation in the complete set. In [37] a simple metric to evaluate the correlation between two SNPs is proposed whose description demands the introduction of additional notation. Let $H = \{h_1, \ldots, h_n\}$ be the matrix of all haplotypes and let $S = \{s_1, \ldots, s_m\}$ be the set of all SNPS, denote by T the set of all TAG SNPs, by t a TAG SNP, by h_i^T the reduced haplotype h_i corresponding to the SNPs in T and by H^T the set of all these reduced haplotypes. Finally, define $\phi_h(\phi_g)$ as the frequency of haplotype h in H (genotype g in G), respectively.

The proposed pairwise metric correlates one single SNP with any other SNP and is based on LD between two SNPs.

$$r_h^2(s,t) = 1 - \frac{\frac{phi_{s_1,t_1}}{\phi_{s_1}}\left(1 - \frac{\phi_{s_1,t_1}}{\phi_{s_1}}\right)}{2\phi_{t_1}(1 - \phi_{t_1})} \quad \text{if} \quad t \neq s$$

$$r_h^2(s,t) = 1 \quad \text{if} \quad t = s$$

where s is the predicted SNP, t is the TAG SNP, ϕ_{s1} and ϕ_{t1} are the frequencies of allele 1 in s and t, and $\phi_{s1,t1}$ is the frequency of haplotypes having 1 in both s and t.

If $r^2 = 1$ then one of the two SNPs can fully predict the other (no loss of information). If $r^2 \neq 1$ then a threshold is defined; if this threshold is low, less TAG SNPs are required with a significant loss of predicting power.

To maintain the predicting power we have to augment the sample dimension. Additional details on this topic can be found also in [5]. The same authors propose also a more sophisticated multivariate metric, based on the correlation of a single SNP with a set of SNPs (for details see [37]).

12.6.2.3 Minimizing the number of TAG single nucleotide polymorphisms (SNPs)

Greedy methods are the most used for selecting a reduced number of TAG SNPs. These methods add to a set of TAG SNPs a new SNP for which the quality function has the largest increase. Obtaining the optimal value of the quality function is a NP-hard problem in most cases. Branch and bound techniques are often used in case of pairwise metrics and can find optimal solution even with hundreds of SNPs. The same is true when the problem is restricted to blocks. They become very inefficient for large set of SNPs.

Clustering is another possibility for pairwise metrics, even if it does not guarantee to find the minimum set of TAG SNPs. Dynamic programming is used to partition haplotypes into blocks, thus reducing the search space. The same technique is also used to predict TAG SNPs without partitioning haplotypes into blocks.

12.6.3   Reconstruction problem

A crucial aspect in association studies is that of finding a set of TAG SNPs and to design a *reconstruction method* such that all the other SNPs can be predicted from the TAG SNPs through the proposed method. This topic has been recently addressed by a number of studies (see [59] for a complete bibliography). In the following we present the approaches proposed in [9,38,42,59]

First we introduce some more definitions and notation, as proposed in [38]. In a genotype setting, a *prediction function* is a function $f : \{0,1,2\}^T \to \{0,1,2\}^m$. Denote by f_j the jth component of a predicted genotype. For a given vector $q \in \{0,1,2\}^T$ of TAG SNPs values let $f_j(g)$ denote the jth component of that vector. Observe that $f_j = f_j(q)$, $\forall j \in T$.

To better qualify the objective of reconstruction, the notion of *prediction error* associated with a pair of TAG and non-TAG sets of a given set of individuals is introduced. The *prediction error* is the proportion of the number of alleles that are wrongly reconstructed on the total number of alleles in SNPs of the non-TAG set. Once defined the prediction error the number of tags could be minimized subject to upper bounds on prediction error measured in leave-one-out cross-validation experiments [37,38]. Some other approaches search for a partition of the SNPs set and a reconstruction function for which the prediction η error is minimized. Since the frequencies of the genotypes are not known, a learning problem is formulated, where the available data is split into a training set and a testing set. The training set is used to learn the distribution of haplotypes. At this point the problem becomes that one of finding a set of TAG SNPs T of size t such that η is minimized when the haplotype is randomly chosen from the training set. In this way the training set is used to search the partition of the SNPs set and the reconstruction function, while the test set is used to compute the *prediction error*. Most of the methods presented follow this scheme. A fundamental role is played by the reconstruction function, that characterizes the different approaches.

In [38], a dynamic programming algorithm, STAMPA, is described that takes as training set a given set of DNA sequences and basest he prediction function on the biological hypothesis that each SNP is strongly related only with the two neighboring TAG SNPs (hypothesis strictly related to the

LD structure of the DNA molecules). This results in an algorithm whose computational complexity is sufficiently small. The prediction function uses a majority vote (described in the following) in order to determine which value is more likely to appear in the unknown position. The dynamic procedure is based on the following. In the paper, a random algorithm to find the TAG SNPs is also proposed.

In [42] the prediction method is based on rounding of multivariate linear regression (MLR) analysis in sigma-restricted coding and a dependence is shown between the reconstruction method and the TAG SNP selection. We illustrate the method in the haplotype setting, using the notation introduced in the above sections. Let S be the sample population and let h be the haplotype restricted to the TAG SNP set. We want to predict a non-TAG SNP s from the value of the TAG SNPs of individuals in S. S and x are represented as a matrix M with $n+1$ rows (the sample individuals and the unknown individual) and $t + 1 columns$ corresponding to the t TAG SNPs and a single non-TAG SNP. All the values in M are known, except the value of s in x. The MLR SNP prediction method consider all possible resolution of s together with the set of TAG SNPs T and choose the closest to T.

In [59] an algorithm based on linear algebra is proposed, that identifies a set of TAG SNPs and a set of linear formulas on these TAG SNPs able to predict all the other SNPs, for genotype data sets. The reconstruction function is a linear function of the set of TAG SNPs. SNPs genotype data are converted to numerical data. The algorithm is tested on a large set of data from public data bases and from association studies. In this approach, since each predicted SNP is a function of all the TAGSNPs, it is possible to keep into account possible LD between distant SNPs.

In [9] the selection of TAG SNPs is accomplished via a direct FS approach: a combinatorial model like the one described in Section 12.2 (see also [7]) is used in its unsupervised version, where the role of the features is played by the SNPs and the subset selected by the integer programming model are exactly the TAG SNPs. The assumption behind this approach is that the TAG SNPs must retain the most information contained in the data in order to predict the values of the other SNPs. This assumption is then tested using two different reconstruction techniques. The first one is the *majority vote*, proposed by Shamir in [38]. The general principle of that method is that genotypes that are similar on the TAGSNPs are also likely to be similar on the remaining SNPs. Therefore, the non-TAG SNPs of an new individual are assigned their value according to the most frequent values that are present in the individuals of the training set that mostly agree with that individual over the TAG SNPs. The second method adopted uses a classification approach: each non-TAG SNPs plays in turn the role of the class variables, and a classification formula is learned from the training individuals; such formula is constructed to predict with the highest possible precision the value of the class (e.g., the non-TAG SNP) from the values of the TAG SNPs. This way, each non-TAG gets its own prediction model that is used for its reconstruction in new individuals. The method used to perform the classification task is *Lsquare*(see [26]), already described in Section 12.3.

The second reconstruction method appears to be more precise but requires a higher computational cost, as a classification problem has to be solved for each non-TAG SNPs.

12.7   Conclusions

This chapter is devoted to the analysis of FS and classification algorithms for computational molecular biology. Its main purpose is to bring to evidence the difficult challenges of data analysis in this application context, mainly due to the large dimension of the data sets that need to be analyzed and to the lack of established methods in a field where technology, knowledge and problems change at a very fast pace.

We have first defined three relevant problems that arise in computational biology: classification, representation, and reconstruction. In the rest of the chapter, some methods and algorithms from the literature are put in the perspective of these three problems. This part is divided into two main blocks: the first devoted to FS and classification (the methods of analysis), the second to the description of three different applications where the methods are applied (microarray analysis, barcode analysis, TAG SNPs selection).

From the material we presented, three main conclusions may be drawn:

(i) It appears that the current state of the technology in molecular biology strongly demands the deployment of sophisticated analysis tools that rely heavily on computational power and on good-quality algorithms.

(ii) Different methods with different characteristics are available and no simple rule can orient the scientist in the selection of the proper method for a given problem; rather, such choice requires a good knowledge and comprehension of both the method and the problem.

(iii) Finally, when the proper knowledge is available (as is the case for the applications considered and discussed in the second part of this chapter), then a significant contribution to the advance of science in molecular biology can be provided.

References

[1] Alizadeh A.A., Eisen M.B., Davis R.E., Ma C., Lossos I.S., Rosenwald A., Boldrick J.C. et al. Distinct types of diffuse large B-cell lymphoma identified by gene expression profiling. *Nature*, 403(6769):503–511, 2000.

[2] Allwein E., Schapire R., and Singer Y. Reducing multiclass to binary. *Journal of Machine Learning Research*, 1:113–141, 2000.

[3] Almuallim H., and Dietterich T.G. Learning with many irrelevant features. In *Proceedings of the 9th National Conference on Artificial Intelligence*. MIT Press, Cambridge, MA, 1991.

[4] Alon U., Barkai N., Notterman D.A., Gish K., Ybarra S., Mack D., and Levine A.J. Broad patterns of gene expression revealed by clustering analysis of tumor and normal colon tissues probed by oligonucleotide arrays. *Proceedings of the National Academy of Sciences*, 96(12):6745–6750, 1999.

[5] Bafna V., Halldórsson B.V., Schwartz R., Clark A.G., and Istrail S. Haplotypes and informative SNP selection algorithms: Don't block out information. In *Proceedings of the 7th Annual International Conference on Research in Computational Molecular Biology. RECOMB'03*, Berlin, Germany, 19–27, 2003.

[6] Ben-Dor A., Bruhn L., Friedman N., Nachman I., Schummer M., and Yakhini Z. Tissue classification with gene expression profiles. *Journal of Computational Biology*, 7(3–4):559–83, 2000.

[7] Bertolazzi P., Felici G., Festa P., and Lancia G. Logic classification and feature selection for biomedical data *Computer and Mathematics with Applications*, 55(5):889–899,2008.

[8] Bertolazzi P., and Felici, G. Learning to classify species with barcodes. IASI Technical Report, 665, 2007.

[9] Bertolazzi P., Felici G., and Festa P. Logic based methods for SNPs tagging and reconstruction. *Computer and Operation Research*, revision in process, 2007.

[10] Blaxter M., Mann J., Chapman T., Thomas F., Whitton C., Floyd R., and Abebe E. Defining operational taxonomic units using DNA barcode data. *Philosophical Transactions of the Royal Society B*, 360(1462):1935–1943, 2005.

[11] Blaxter M. Molecular systematics: counting angels with DNA. *Nature* 421:122–124, 2003.

[12] Blaxter M. The promise of a molecular taxonomy. *Philosophical Transactions of the Royal Society B*, 359:669–679, 2004.

[13] Blaxter M., and Floyd R. Molecular taxonomics forbiodiversity surveys: already a reality, *Trends Ecology Evolution*, 18:268–269, 2003.

[14] Boros E., Ibaraki T., and Makino K. Logical analysis of binary data with missing bits. *Artificial Intelligence*, 107:219–263, 1999.

[15] Brown B., Emberson R.M., and Paterson A.M. Mitochondrial COI and II provide useful markers for Weiseana (Lepidoptera, Hepialidae) species identification. *Bulletin of Entomological Research*, 89:04, 287–293, 1999.

[16] Brown M., Grundy W.N., Lin D., Christianini N., Sugnet C.W., Furey T.S., Ares M. (Jr.), and Haussler D. Knowledge-based analysis of microarray gene expression data by using support vector machines. *Proceedings of the National Academy of Sciences*, 97(1):262–267, 2000.

[17] Bucklin A., Guarnieri M., Hill R.S.,Bentley A.M., and Kaartvedt S. Taxonomic and systematic assessment of planktonic copepods using mitochondrial COI sequence variation and competitive, species-specific PCR. *Hydrobiology*, 401:239–254, 1999.

[18] Chang C-J., Huang Y-T., and Chao K-M. A greedier approach for finding tag SNPs. *Bioinformatics*, 22(6):685–691, 2006.

[19] Charikar M., Guruswami V., Kumar R., Rajagopalan S., and Sahai A. Combinatorial feature selection problems. In *Proceedings of the 41st Annual Symbosium on FOCS* 2000, IEEE Computer Society, Washington, DC, USA, 631.

[20] Cristianini N., and Shawe-Taylor J. *An Introduction to Support Vector Machines and Other Kernel-based Learning Methods*. Cambridge University Press, Cambridge, UK, 2000.

[21] Dasarathy B.V. (ed). *Nearest Neighbor (NN) Norms: NN Pattern Classification Techniques*. IEEE Computer Society, Los Alamitos, CA, 1991.

[22] DasGupta B., Konwar K.M., Mandoiu I.I., and Shvartsman A.A. Highly scalable algorithms for robust string barcoding. *International Conference on Computational Science* 2:1020–1028, 2005.

[23] Dash M., and Liu H. Feature selection for classification. *Intelligent Data Analysis*, I(3):131–156, 1997.

[24] Dietterich T.G., and Bakiri G. Solving multiclass learning problems via error-correcting output codes. *Journal of Artificial Intelligence Research*, 2:263–286, 1995.

[25] Felici G., de Angelis V., and Mancinelli G. Feature selection for data mining. In *Data Mining and Knowledge Discovery Approaches Based on Rule Induction Techniques*, G. Felici and E. Triantaphyllou (eds). Springer, New York, USA, 227–252, 2006.

[26] Felici G., and Truemper K. A minsat approach for learning in logic domains. *INFORMS Journal on Computing*, 13(3):1–17, 2001.

[27] Felici G., and Truemper K. The Lsquare system for mining logic data. In *Encyclopedia of Data Warehousing and Mining*, J. Wang (ed.), vol. 2. Idea Group Inc., Hershey PA, USA, 693–697, 2006.

[28] Freund Y., and Schapire R.E. A decision-theoretic generalization of on-line learning and an application to boosting. *Journal of Computer and System Sciences*, 55(1):119–139, 1997.

[29] Furey T.S., Christianini N., Duffy N., Bednarski D.W., Schummer M., and Haussler D. Support vector machine classification and validation of cancer tissue samples using microarray expression data. *Bioinformatics*, 16(10):906–914, 2000.

[30] Garey M.R., and Johnson D.S. *Computer and Intractability: A Guide to the Theory of NP-Completeness.* Freeman, San Francisco, CA, 1979.

[31] Gennari J.H., Langley P., and Fisher D. Models of incremental concept formation. *Artificial Intelligence*, 40:11–61, 1989.

[32] Golub T.R., Slonim D.K., Tamayo P., Huard C., Gaasenbeek M., Mesirov J.P., Coller H. et al. Molecular classification of cancer: class discovery and class prediction by gene expression monitoring. *Science*, 286(5439):531–537, 1999.

[33] Gordon G.J., Jensen R.V., Hsiao Li-Li, Gullans S.R., Blumenstock J.E., Ramaswamy S., Richards W.G., Sugarbaker D.J., and Bueno R. Translation of microarray data into clinically relevant cancer diagnostic tests using gene expression ratios in lung cancer and mesothelioma. *Cancer Research*, 62:4963–4967, 2002.

[34] Rash, S., and Gusfield, D. String barcoding: Uncovering optimal virus signatures. In *Proceedings 6th Annual International Conference on Computational Biology*, Washington, DC, USA, 254–261, 2002.

[35] Guyon I., Weston J., Barnhill S., and Vapnik V. Gene selection for cancer classification using support vector machines. *Machine Learning*, 46(1–3):389–422, 2002.

[36] Hall M.A. Correlation-based feature selection for machine learning. In *Proceedings of the 17th International Conference on Machine Learning.* Morgan Kaufmann, CA, 2000.

[37] Halldrsson B.V., Istrail S., and De La Vega F.M. Optimal selection of SNP markers for disease association studies. *Human Heredity*, 58:190–202, 2004.

[38] Halperin E., Kimme G., and Shamir R. Tag SNP selection in genotype data for maximizing SNP prediction accuracy. *Bioinformatics*, 21:195–203, 2005.

[39] Hajibabaei M., Singer G.A.C., Clare E.L., Paul D.N., and Hebert P.D.N. Design and applicability of DNA arrays and DNA barcodes in biodiversity monitoring. *BMC Biology*, 5(24):1–7, 2007.

[40] Hebert P.D.N., Cywinska A., Ball S.L., and deWaard J.R. Biological identifications through DNA barcodes. *Proceedings of the Royal Society of London B*, 270:313–321, 2003.

[41] Hebert P.D.N., Penton E.H, Burns J.M, Janzen D.H., and Hallwachs W. Ten species in one: DNA barcoding reveals cryptic species in the Neotropical skipper butterfly Astraptes fulgerator. *Proceedings of the National Academy Sciences USA*, 101:14812–14817, 2004.

[42] He J., and Zelikovsky A. Tag SNP selection based on multivariate linear regression, International Conference on Computational Science (2). *Lecture Notes in Computer Science* 3992:750–757, 2006.

[43] Jia Min X. and Hickey D.A. Assessing the effect of varying sequence length on DNA barcoding of fungi. *Molecular Ecology Notes* 1, 7(3):365–373, 2007.

[44] Hu H., Li J., Plank A., Wang H., and Daggard G. A comparative study of classification methods for microarray data analysis. In *Proceedings of the 5th Australasian Conference on Data Mining and Analytics*, Sydney, Australia 61:33–37, 2006.

[45] Jirapech-Umpai T., and Aitken S. Feature selection and classification for microarray data analysis: Evolutionary methods for identifying predictive genes. *BMC Bioinformatics*, 6:148, 2005.

[46] Koller D., and Sahami M. Hierachically classifying documents using very few words. In *Machine Learning: Proceedings of the 14th International Conference on Machine Learning*, Morgan Kaufmann Publishers Inc., San Francisco, CA, USA, 170–178, 1997.

[47] Kuksa P., and Pavlovic V. Kernel methods for DNA barcoding. In *Snowbird Learning Workshop*, San Juan, Puerto Rico, March 19–22, 2007.

[48] Langley P. Selection of relevant features in machine learning, *Artificial Intelligence*, 97, 245–271, 1997.

[49] Chengliang Z., Li T., and Mitsunori O. A comparative study of feature selection and multiclass classification methods for tissue classification based on gene expression. *Bioinformatics*, 20(15):2429–2437, 2004.

[50] Liu H., and Setiono R. A probabilistic approach to feature selection: A filter solution. In *Proceedings of the 13th International Conference on Machine Learning*, Bari Italy, July 27, 1996. Morgan Kaufmann Publishers Inc., San Francisco, CA, USA, 319–327, 1996.

[51] Maniatis N., Collins A., Xu C.F., Mcfhy L.C., Hewett D.R., Tapper W., Ennis S., Ke X., Morton N.E. The first linkage disequilibrium (LD) maps: delineation of hot and cold blocks by diplotype analysis. *Proceedings of the National Academy of Sciences USA*, 99:2228–2233, 2002.

[52] Min X.J., and Hickey D.A. DNA barcodes provide a quick preview of mitochondrial genome composition. *PLoS ONE*, 2(3):e325, 2007.

[53] Montgomery D., and Undem B.L. Drug discovery. CombiMatrix' customizable DNA microarrays. *Genetic Engineering News*, 22(7):32–33, 2002.

[54] Nanney, D.L. Genes and phenes in Tetrahymena. *Bioscience*, 32:783–740, 1982.

[55] Oliveira A.L., and Vincetelli A.S., Constructive induction using a non-greedy strategy for feature selection. In *Proceedings of the 9th International Conference on Machine Learning*. Morgan Kaufmann, Aberdeen, Scotland, 355–360, 1992.

[56] Pace N.R. A molecular view of microbial diversity and the biosphere. *Science*, 276:734–740, 1997.

[57] Pasaniuc B., Kentros S., and Mandoiu I.I. Boosting assignment accuracy by combining distance- and character-based classifiers. In *The DNA Barcode Data Analysis Initiative: Developing Tools for a New Generation of Biodiversity Data*. Paris, France, July 6–8, 2006, unpublished presentation, http://dna.engr.uconn.edu/?page_id=21&year=2006.

[58] Paschou P., Mahoney M.W., Javed A., Kidd J.R., Pakstis A.J., Gu S., Kidd K.K. and Drineas P. Intra- and interpopulation genotype reconstruction from tagging SNPs. *Genome Research*, 17:96–107, 2007.

[59] Patil N. Blocks of limited haplotype diversity revealed by high-resolution scanning of human chromosome 21. *Science*, 294:1719–1723, 2001.

[60] Petricoin E.F., Ardekani A.M., Hitt B.A., Levine P.J., Fusaro V.A., Steinberg S.M., Mills G.B., Simone C., Fishman D.A., Kohn E.C., and Liotta L.A. Use of proteomic patterns in serum to identify ovarian cancer. *Lancet*, 359(9306):572–577, 2002.

[61] Quinlan, J.R. *C4.5: Programs for Machine Learning*. Morgan Kaufmann Publishers, San Francisco, CA, 1993.

[62] Saccone C., DeCarla G., Gissi C., Pesole G., and Reynes A. Evolutionary genomics in the Metazoa: the mitochondrial DNA as a model system. *Gene*, 238:195–210, 1999.

[63] Saitou N., and Nei M. The Neighbour-joining method: a new method for reconstructing phylogenetic trees. *Molecular Biology Evolution*, 4(4):406–425, 1987.

[64] Schlimmer J.C. Efficiently inducing determinations: a complete and systematic search algorithm that uses optimal pruning. In *Proceedings of the 10th International Conference on Machine Learning*. Morgan Kaufmann, Amherst, MA, 284–290, 1993.

[65] Schummer M., Ng W.V., Bumgarner R.E., Nelson P.S., Schummer B., Bednarski D.W., Hassell L., Baldwin R.L., Karlan B.Y., and Hood L. Comparative hybridization of an array of 21,500 ovarian cDNAs for the discovery of genes overexpressed in ovarian carcinomas. *Gene*, 238(2):375–385, 1999.

[66] Sheinvald J., Dom B., and Niblack W. Unsupervised image segmentation using the minimum description length principle. In *Proceedings of the 11th IAPR International Conference on Pattern Recognition*, 2:709–712, 1992.

[67] Singh D., Febbo P.G., Ross K., Jackson D.G., Manola J., Ladd C., Tamayo P. et al. Gene expression correlates of clinical prostate cancer behavior. *Cancer Cell*, 1(2):203–209, 2002.

[68] Smith M. A., Woodley N. E., Janzen D. H., Hallwachs W., and Hebert P.D.N. DNA barcodes reveal cryptic host-specificity within the presumed polyphagous members of a genus of parasitoid flies (Diptera: Tachinidae). *Proceedings of the National Academy of Sciences*, 103:3657–3662, 2006.

[69] Vapnik V.N. *Statistical Learning Theory*. Wiley, New York, NY, 1998.

[70] Veer L.J., Dai H., van de Vijver M.J., He Y.D., Hart A.A., Mao M., Peterse H.L. et al. Gene expression profiling predicts clinical outcome of breast cancer. *Nature*, 415(6871):530–536, 2002.

[71] Xiong H., and Chen X. Kernel-based distance metric learning for microarray data classification. *BMC Bioinformatics*, 7:299, 2006.

[72] Ye J., Li T., Xiong T., and Janardan R. Using uncorrelated discriminant analysis for tissue classification with gene expression data. *IEEE/ACM Transactions on Computational Biology and Bioinformatics*, 1(4):181–190, 2004.

[73] Zhang K., Qin Z.S., Liu J.S., Chen T., Waterman M.S., and Sun F. Haplotype block partitioning and Tag SNP selection using genotype data and their applications to association studies. *Genome Research*, 14:908–916, 2004.

[74] Zhang K., Qin Z.S., Liu J.S., Chen T., Waterman M.S., and Sun F. HapBlock: haplotype block partitioning and Tag SNP selection software using a set of dynamic programming algorithms. *Bioinformatics*, 21:131–134, 2005.

[75] Orsenigo C., and Vercellis C. Discrete support vector decision trees via tabu-search. *Journal of Computational Statistics and Data Analysis*, 47:311–322, 2004.

[76] Orsenigo C., and Vercellis C. Accurately learning from few examples with a polyhedral classifier. *Computational Optimization and Applications*, 38:235–247, 2007.

[77] Orsenigo C. Gene selection and cancer microarray data classification via mixed-integer optimization. In: Evolutionary computation, machine learning, data mining in bioinformatics 6th European Conference, EvoBIO 2008, Naples, Italy, 2008. *Lecture Notes in Computer Science*, 4973:141–152, 2008.

[78] Triantaphyllou E., The OCAT approach for data mining and knowledge discovery. Working Paper, IMSE Department, Louisiana State University, Baton Rouge, LA, 70803-6409, 2001.

Chapter 13

Statistical Indices for Computational and Data Driven Class Discovery in Microarray Data

Raffaele Giancarlo, Davide Scaturro, and Filippo Utro

University of Palermo

13.1　Introduction

The problem of discovering new taxonomies (classifications of objects according to some natural relationships) from data has received considerable attention in the statistics and machine learning community. In this chapter, we are concerned with a particular type of taxonomy discovery, namely, cluster analysis, the discovery of distinct and nonoverlapping subpopulations within a larger population, the member items of each subpopulation sharing some common features or properties deemed relevant in the problem domain of study. This type of unsupervised analysis is of particular significance in the emerging field of functional genomics and microarray data analysis.

The most fundamental issue to be addressed when clustering data consists of the determination of the number of clusters. Related issues are how to assign confidence levels to the selected number of clusters, as well as to the induced cluster assignments. Those issues are particularly important and difficult in microarray data analysis, where the problem of a relatively small sample size is compounded by the very high dimensionality of the data available, making the clustering results especially sensitive to noise and susceptible to over-fitting. Although cluster determination and assessment are well-known in the classification and statistics literature and the corresponding methodologies rest on solid mathematical ground, they are rarely applied to microarray data analysis tasks. Of relevance here are the so called validation indices, external, internal, and relative, that should assess the biological relevance of the clustering solutions found.

Despite the vast amount of knowledge available in this area in the general data mining literature, e.g., [3,7,10,13,16–18], gene expression data provide unique challenges, in particular with respect to internal validation indices. Indeed, they must predict how many clusters are really present in a dataset, an already difficult task, made even worse by the fact that the estimation must be sensible enough to capture the inherent biological structure of functionally related genes. Despite their potentially important role, both the use of classic internal validation indices and the design of new ones, specific for microarray data, do not seem to have great prominence in bioinformatics, where attention is mostly given to clustering algorithms. The excellent survey by Handl et al. [11] is a big step forward in making the study of those techniques a central part of both research and practice in bioinformatics, since it provides both a technical presentation as well as valuable general guidelines about their use for postgenomic data analysis. Although much remains to be done, it is, nevertheless, an initial step.

For instance, in the general data mining literature, there are several studies, e.g., [22], aimed at establishing the intrinsic, as well as the relative, merit of an index. To this end, the two relevant questions are:

(A) What is the precision of an index, i.e., its ability to predict the correct number of clusters in a dataset? That is usually established by comparing the number of clusters predicted by the index against the number of clusters in the true solution of several datasets, the true solution being a partition of the dataset in classes that can be trusted to be correct, i.e., distinct groups of functionally related genes.

(B) Among a collection of indices, which is more accurate, less algorithm-dependent, etc.? Precision versus the use of computational resources, primarily execution time, would be an important discriminating factor.

Having well in mind the two questions above, and in particular for microarray data, this chapter offers a presentation of some indices that have been recently proposed and tested on microarray data and that have gained prominence in the literature. Obviously, there is no attempt to be exhaustive and some of the choices of which indices to include here may be judged as being subjective, although we account for the three most widely known families. Namely, indices based on (a) hypothesis testing in statistics; (b) bootstrapping and resampling techniques and (c) jackknife techniques.

The chapter is organized as follows: Section 13.2 gives a mathematical formulation for the classification problems of relevance here. Section 13.3 introduces the notion of null model and describes its importance in hypothesis testing in statistics. External indices are defined in Section 13.4, while internal and relative indices are the subject of the next section. Section 13.6 presents the set-up we have used for our experiments. In particular, a brief description is given of ValWorkBench, the software system we have engineered to perform this study. A detailed description of ValWorkBench is given at the ValWork-Bench home page [2], where user manual and installation instructions can be found. The remaining sections present the experiments and, based on the analysis of the results, some conclusions are drawn. Those two sections have also the role to offer a hands-on illustration of the methodologies presented in the sections preceding them.

13.2   Basic Mathematical Problem Formulations

Consider a set of n items $G = \{g_1, \ldots, g_n\}$, where g_i is specified by m numeric values, referred to as features or conditions, $1 \leq i \leq n$. That is, each g_i is an element in m-dimensional space. Let D be the corresponding $n \times m$ data matrix. Based on D, i.e., the conditions characterizing the items, we are interested in partitioning G into groups, so that the items in each group are "similar" according to the value of a distance or similarity function on their conditions.

A typical scenario for microarray data is to have a set of genes (the items), subject to expression level measurements in m different experimental conditions or time periods (the features or conditions). A typical problem of interest is to divide the set of genes in groups homogenous for expression levels across experiments. That would highlight groups of genes that are, for instance, functionally correlated or that have the same response to medical treatment. An example is reported in Figure 13.1, where the data matrix of the reduced yeast cell cycle (RYCC) dataset is shown. It contains 384 genes and 17 experimental conditions and the genes can be divided in five biological functional classes. The boundary of each class is marked by a white line in Figure 13.1. Additional features of this dataset, as well as its uses in clustering studies, is given in [5].

The specification and formalization of similarity and distance between items in G, via mathematical functions, depends heavily on the application domain. In fact, the choice of a good similarity function is one of the key steps in clustering and, for microarray data, the object of considerable debate. A compendium of those functions, particularly useful for microarray data, is presented in [14]. The state of the art, as well as some relevant progress in the identification of good distance functions for microarrays, is well presented in [25].

Usually, the partition of the items in G into groups is accomplished by means of a clustering algorithm A. In this chapter, we limit ourselves to the class of clustering algorithms that take as input D and an integer k and return a partition P_k of G into k groups. A survey of classic as well as more innovative clustering algorithms, specifically designed for microarray data, is given in [28].

Qualitatively, the fundamental question one faces is to establish how many clusters are present in G. A more sensible biological question would be, for instance, to find out how many functional groups of genes are present in a dataset (see Figure 13.1 again). Since the presence of "statistically significant patterns" in the data is usually indication of their biological relevance [20], it makes sense to ask whether a division of the items into groups is statistically significant. This qualitative question can be put on a formal ground in two different ways [16]. In order to state them, we need some notation and definitions.

Let C_j be a reference classification for G consisting of j classes. That is, C_j may either be a partition of G into j groups, usually referenced to as the *gold standard*, or a division of the universe generating G into j categories, usually referenced to as *class labels*. An *external index E* is a function that takes as input a reference classification C_j for G and a partition P_k of G and returns a value assessing how close the partition is to the reference classification. It is external because the quality assessment of the partition is established via criteria external to the data, i.e., the reference classification. Notice that it is not required that $j = k$. An *internal index I* is a function defined on the set of all possible partitions of G and with values in $\Re$. It should measure the quality of a partition according to some suitable criterion. It is internal because the quality of the partition is measured according to information contained

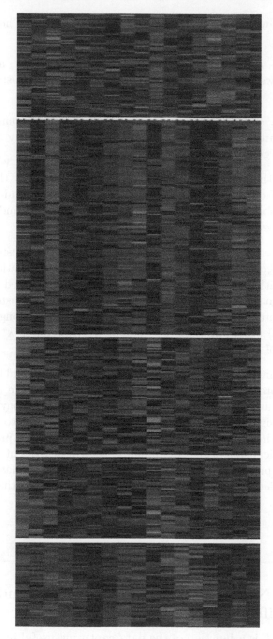

FIGURE 13.1: The RYCC dataset divided in five biologically meaningful classes.

in the dataset without resorting to external knowledge. The two questions now are:

(Q.1) Given C_j, P_k, and E, measure how far is P_k from C_j, according to E. That is, we are asking whether k is the number of clusters one expects in D.

(Q.2) Given P_k and I, establish whether the value of I computed on P_k is unusual and therefore surprising. That is, significantly small or significantly large.

We notice that the two questions above try to assess the quality of a clustering solution P_k consisting of k groups, but give no indication on what the "right number" of clusters is. In order to get such an indication, we are interested in the following:

(Q.3) Assume we are given: (Q.3.a) A sequence of clustering solutions $P_1, \ldots, P_s$, obtained for instance via repeated application of a clustering algorithm A; (Q.3.b) a function R, usually referred to as a *relative index* that estimates the relative merits of a set of clustering solutions. We are interested in identifying the partition P_{k*} among the ones given in (Q.3.a) providing the best value of R. We refer to k^* as the optimal number of clusters according to R.

The clustering literature is extremely rich in mathematical functions suited for the three problems outlined above [11]. The crux of the matter is to establish quantitatively the threshold values allowing us to say that the value of an index is significant enough. That naturally brings us to briefly mention hypothesis testing in statistics, from which one can derive procedures to assess the statistical significance of an index. As will be evident in the following sections, those procedures are rarely followed in microarray data analysis practice, being preferred to less resource-demanding heuristics that are validated experimentally.

13.3   The Null Hypothesis, Null Models, and Cluster Significance for a Given Statistic

A *statistic* T is a function of the data capturing useful information about it, i.e., it can be one of the indices mentioned earlier. In mathematical terms, it is a random variable and its distribution describes the relative frequency with which values of T occur, according to some assumptions. In turn, since T is a random variable, we implicitly assume the existence of a background or reference probability distribution for its values. That, in turn, implies the existence of a sample space. A *hypothesis* is a statement about the frequency of events in the sample space. It is tested by observing a value of T and by deciding how unusual it is, based on the probability distribution we are assuming for the sample space. In what follows, we assume that the higher the value of T, the more unusual it is, the symmetric case being dealt with similarly.

The most common hypothesis tested for clustering is the *null hypothesis* H_0: there is no structure in the data. Testing for H_0 with a statistic T in a dataset D means to compute T on D and then, based on that value, decide whether to reject or not to reject H_0. In order to decide, we need to establish how significant is the value found with respect to a background probability distribution of the statistic T under H_0. That means we have to formalize the concept of "no structure" or "randomness" in our data. Among the many possible ways, generally referred to as *null models*, we introduce the two that find common use in microarray data analysis [30], in addition to another closely related and classic one:

(M.1) *The Poisson model* (Ps). The items can be represented by points that are randomly drawn from a region R in m-dimensional space. In order to use this model, one needs to specify the region within which the points are to be uniformly distributed. The simplest regions that have been considered are the m-dimensional hypercube and hypersphere enclosing the points specified by the matrix D. Other possibilities, in order to make the model more data-dependent, is to choose the convex hull enclosing the points specified by D.

(M.2) *The Poisson model aligned with principal components of the data* (Pc). This is basically as (M.1), except that the region R is a hypercube aligned with the principal components of the data matrix D. In detail, assume that the columns of D have mean 0 and let $D = UXV^T$ be its singular value decomposition. Transform via $D' = DV$. Now, use D' as in (M.1) to obtain a data set Z'. Back transform via $Z = Z'V^T$ to obtain a new dataset.

(M.3) *The permutational model* (Pr). Given the data matrix D, one obtains a new data matrix D' by randomly permuting the elements within the rows and/or the columns of D. In order to properly implement this model, care must be taken in specifying a proper permutation for the data since some similarity and distance functions may be insensitive to permutations of coordinates within a point. That is, although D' is a random permutation of D, it may happen that the distance or similarity among the points in D' is the same as in D, resulting in indistinguishable datasets for clustering algorithms. This latter model may not be suitable for microarray data with very small sample sizes, i.e., conditions, since one will not obtain enough observations, i.e., data points, to estimate the null model even if one generates all possible permutations.

Once that a null model has been agreed upon, one would like to obtain closed form expressions of the statistic T under the null model, i.e., easy to compute formulas giving the value of T under the null model and for a specific set of parameters. Unfortunately, not too many such formulae are available. In fact, in most cases, one needs to resort to a Monte Carlo simulation, a procedure that we present next, applied to the context of assessing the significance of a partition of the data in k clusters. In technical terms, it is a p-value test assessing whether in the dataset there exist k clusters, based on T and the null model for H_0. We nickname it MECCA, an abbreviation for Monte Carlo Confidence Analysis. The procedure is also a somewhat more general version of a significance test proposed and studied by Gordon [9,10] for the same problem. It takes as input an integer ℓ (the number of iterations in the Monte Carlo simulation), a clustering algorithm A, a dataset D, the function T, a partition P_k of D obtained via algorithm A and a parameter α in [0,1] indicating a level of "significance" for the rejection of H_0. It returns a value p in [0,1]. If $p < \alpha$, the null hypothesis of no cluster structure in the data is to be rejected at significance level α. Else, it cannot be rejected at that significance level.

Procedure MECCA (ℓ, A, D, T, P_k, α):

1. For $1 \leq i \leq \ell$, compute a new data matrix D_i, using the chosen null model, and partition it into a set of k clusters $P_{i,k}$ using algorithm A

2. For $1 \leq i \leq \ell$, compute T on $P_{i,k}$ and let SL be the nondecreasing sorted array of those values

3. Let V denote the value of T computed on P_k. Let p be the proportion of the values in SL with value larger than V. RETURN p

A few remarks are in order. As pointed out by Gordon, significance tests aiming at assessing how reliable is a clustering solution are usually not carried out in data analysis. Microarrays are no exception, although sophisticated statistical techniques specific for those data have been designed, e.g., [7]. One of the reasons is certainly the high computational demand of those tests. Another, more subtle, reason is that researchers expect that "some structure" is present in their data. Nevertheless, a general procedure like MECCA is quite useful as a paradigm illustrating how one tries to assess cluster quality via a null model and a statistic T. Indeed, one computes the observed value of T (on the real data). Then, one computes, via a Monte Carlo simulation, enough values of T, as expected from our formalization of H_0 via the null model. Finally, one checks how "unusual" is the value of the observed statistic with respect to its expected value, as estimated by a Monte Carlo simulation. As we will see, two methods that we describe in this chapter resort to the same principles and guidelines of MECCA, although they are very specific on the type of statistic that is important in order to identify the number of clusters in a dataset.

13.4 External Indices

In this section we present external indices, namely, formulae that establish the level of agreement between two partitions. Usually, for a given dataset, one of the partitions is a reference classification of the data while the other is provided as output by a clustering algorithm.

External indices are usually defined via a contingency table that we introduce next. Let $C = \{c_1, \ldots, c_r\}$ be a partition of the items in G into r classes and $P = \{p_1, \ldots, p_t\}$ be another partition of G into t clusters. With the notation of Section 13.2, C is an external partition of the items, derived from the reference classification, while P is a partition obtained by some clustering method. Let $n_{i,j}$ be the number of items in both c_i and p_j, $1 \leq i \leq r$ and $1 \leq j \leq t$. Moreover, let $|c_i| = n_{i.}$ and $|p_j| = n_{.j}$. Those values can be conveniently arranged in a *contingency table* (see Table 13.1).

13.4.1 Adjusted Rand index

Let a denote the number of pairs of items that are placed in the same class in C and in the same cluster in P; let b be the number of pairs of items placed in the same class in C but not in the same class in P; let c be the number of pairs of items in the same cluster in P but not in the same cluster in C; let d be the number of pairs of items in different classes and different clusters in both partitions. The information needed to compute a, b, c, and d can be derived from Table 13.1.

$$a = \sum_{i,j} \binom{n_{i,j}}{2}, \tag{13.1}$$

$$b = \sum_{i} \binom{n_{i.}}{2} - a, \tag{13.2}$$

$$c = \sum_{j} \binom{n_{.j}}{2} - a. \tag{13.3}$$

TABLE 13.1: Contingency table for comparing two partitions.

Class/cluster	p_1	p_2	$\cdots$	p_t	Sums
c_1	$n_{1,1}$	$n_{1,2}$	$\cdots$	$n_{1,t}$	$n_{1.}$
c_2	$n_{2,1}$	$n_{2,2}$	$\cdots$	$n_{2,t}$	$n_{2.}$
$\vdots$		$\vdots$			$\vdots$
c_r	$n_{r,1}$	$n_{r,2}$	$\cdots$	$n_{r,t}$	$n_{r.}$
Sums	$n_{.1}$	$n_{.2}$	$\cdots$	$n_{.t}$	$n_{..} = n$

Moreover, since $a + b + c + d = \binom{n}{2}$, we have:

$$d = \binom{n}{2} - (a + b + c). \tag{13.4}$$

Based on those quantities, the Rand index R is defined as [26]:

$$R = \frac{a + d}{a + b + c + d} \tag{13.5}$$

Notice that, since $a + d$ is the number of pairs of items in which there is agreement between the two partitions, R measures the agreement of the two partitions in percentage points. Therefore, it has a value in $[0,1]$ and the closer the index is to one the better the agreement between the two partitions. The main problem with R is that its value on two partitions picked at random does not take a constant value, say 0. So, it is difficult to establish, given two partitions, how significant (distant from randomness) is the concordance between the two partitions as measured by the value of R. In general, given an index, it would be appropriate to take an adjusted version of it according to the following general scheme, ensuring that its expected value is 0 when the partitions are selected at random and 1 when they are identical:

$$\frac{index - expected\ index}{maximum\ index - expected\ index} \tag{13.6}$$

where *index* is the formula for the index, *maximum index* is its maximum value and *expected index* is its expected value derived under a suitably chosen model of random agreement between two partitions, i.e., the null hypothesis. The adjusted Rand index R_A is derived from Equations 13.5 and 13.6 using the generalized hypergeometric distribution as the null hypothesis. That is, it is assumed that the row and column sums in Table 13.1 are fixed, but the two partitions are picked at random. We have [15]:

$$R_A = \frac{\sum_{ij} \binom{n_{i,j}}{2} - \dfrac{\left[\sum_i \binom{n_{i.}}{2} \sum_j \binom{n_{.j}}{2} \right]}{\binom{n}{2}}}{\frac{1}{2} \left[\sum_i \binom{n_{i.}}{2} + \sum_j \binom{n_{.j}}{2} \right] - \dfrac{\left[\sum_i \binom{n_{i.}}{2} \sum_j \binom{n_{.j}}{2} \right]}{\binom{n}{2}}}.$$

R_A has a maximum value of 1, indicating a perfect agreement between the two partitions, while its expected value of 0 indicated a level of agreement due to chance. Moreover, R_A can take on a larger range of values with respect to R and, in particular, may be negative [33]. So, for two partitions to be in significant agreement, R_A must assume a nonnegative value substantially away from 0. We also notice that R_A is a statistic on level of agreement between two partitions of a dataset (see Section 13.3) while R is a simple indication of percentage agreement, following the intuition that higher that percentage is, closer the two partitions are to each other. To illustrate this point, consider two partitions of a set of 29 items giving rise to Table 13.2. Then $R = 0.677$, indicating a good percentage agreement while $R_A = -0.014$ and, being close to its expected value under the null model, it indicates a level of significance in the agreement close to random. In fact, the entries in the table have been picked at random. The adjusted Rand index R_A is a statistic recommended in the classification literature [23] to compare the level of agreement of two partitions.

13.4.2 Fowlkes and Mallows (FM) index

As with the adjusted Rand index, the Fowlkes and Mallows (FM)-index [8] is also derived from the contingency Table 13.1, as follows:

$$FM_k = \frac{T_k}{\sqrt{U_k \cdot V_k}} \tag{13.7}$$

where:

$$T_k = \sum_{i=1}^{k} \sum_{j=1}^{k} n_{ij}^2 - n \tag{13.8}$$

$$U_k = \sum_{i=1}^{k} n_{i.}^2 - n \tag{13.9}$$

$$V_k = \sum_{j=1}^{k} n_{.j}^2 - n \tag{13.10}$$

The index has values in the range [0,1], with an interpretation of the values in that interval analogous to that provided for the values of R. An example can be obtained as in Section 13.4.1. Indeed, for Table 13.2, we have $FM_5 = 0.186$ indicating a low level of agreement between the two partitions.

TABLE 13.2: Contingency table example.

Class/cluster	p_1	p_2	p_3	p_4	v_5	Sums
c_1	1	4	2	1	2	10
c_2	0	1	1	0	1	3
c_3	1	2	0	2	0	5
c_4	2	1	0	1	2	6
c_5	1	0	1	0	3	5
Sums	5	8	4	4	8	$n = 29$

13.4.3   The Fowlkes (F)-index

The Fowlkes (F)-index [27] combines notions from information retrieval, such as precision and recall, in order to evaluate the agreement of a clustering solution P with respect to a reference partition C. Again, its definition can be derived from the contingency Table 13.1.

Given c_i and p_j, their relative precision is defined as the ratio of the number of elements of class c_i within cluster p_j, divided by the size of cluster p_j. That is:

$$Prec(c_i, p_j) = \frac{n_{i,j}}{n_{.j}} \qquad (13.11)$$

Moreover, their relative recall is defined as the ratio of the number of elements of class c_i within cluster p_j, divided by the size of class c_i. That is:

$$Rec(c_i, p_j) = \frac{n_{i,j}}{n_{i.}} \qquad (13.12)$$

The F-index is then defined as an harmonic mean that uses the precision and recall values, with weight b:

$$F(c_i, p_j) = \frac{(b^2 + 1) \cdot Prec(c_i, p_j) \cdot Rec(c_i, p_j)}{b^2 \cdot Prec(c_i, p_j) + Rec(c_i, p_j)} \qquad (13.13)$$

Equal weighting for precision and recall is obtained by setting $b = 1$. Finally, the overall F-index is:

$$F = \sum_{c_i \in C} \frac{n_{i.}}{n} \max_{p_k \in P} F(c_i, p_k) \qquad (13.14)$$

F is an index with value in the range [0,1], with an interpretation of the values in that interval analogous to that provided for the values of R. An example can be obtained as in Section 13.4.1. Indeed, for Table 13.2, we have $F = 0.414$, indicating some level of agreement between the two partitions.

13.5   Internal and Relative Indices

Internal indices should assess the merits of a partition, without any use of external information. Then, a Monte Carlo simulation can establish if the value of the index on the given partition is unusual enough for the user to gain confidence that the partition is good. Unfortunately, this methodology is rarely used in data analysis for microarrays [11]. Internal indices are also a fundamental building block in order to obtain relative indices that help select which of a given number of partitions is the "best" one. In this section, we present several relative indices. We start with the ones based on within-cluster sum of errors, then present methods that are based on resampling techniques and finally we conclude with a method based on the jackknife approach.

13.5.1 Within cluster sum of squares (WCSS)

One internal index that gives an assessment of the level of compactness of each cluster in a clustering solution is the within cluster sum of squares (WCSS). Let $C = \{c_1, \ldots, c_k\}$ be a clustering solution, with k clusters, for G. Formally, let

$$D_r = \sum_{j \in c_r} \|g_j - \bar{g}_r\|^2 \tag{13.15}$$

where $\bar{g}_r$ is the centroid of cluster c_r. Then, we have:

$$\text{WCSS}(k) = \sum_{r=1}^{k} D_r \tag{13.16}$$

Assume also that we want to use WCSS to estimate, based on the given solutions in $[1, k_{\max}]$, the real number k^* of clusters in the data set. Such a number is usually unknown to us. Intuitively, as we get closer and closer to the real number of clusters in the data, the compactness of each cluster should substantially increase, causing a substantial decrease in WCSS. On the other hand, for values of $k^* > k$, the compactness of the clusters will not increase as much, causing the value of WCSS not to decrease as much. In other words, one should observe a decreasing marginal improvement in terms of cluster compactness after k^*. The following heuristic approach comes out [13]: plot the values of WCSS, computed on the given clustering solutions, in the range $[1, k_{\max}]$; choose as k^* the abscissa closest to the "knee" in the WCSS curve. Figure 13.2 provides an example. Indeed, the dataset in Figure 13.2a has two natural clusters and the plot of the WCSS curve in Figure 13.2b indicates $k^* = 2$. As it will be clear in Section 13.8.1.1, the prediction of k^* with WCSS is not so easy on real datasets.

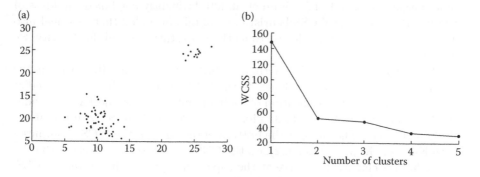

FIGURE 13.2: (a) Dataset. (b) Plot of the values of WCSS.

13.5.2 KL: the Krzanowski and Lai index

Elaborating on an earlier proposal by Marriot [21], Krzanowski, and Lai [19] proposed an internal index, which we refer to as KL. It is based on WCSS, but it is automatic, i.e., a numeric value for k^* is returned. Let

$$DIFF(k) = (k-1)^{2/m}\text{WCSS}(k-1) - k^{2/m}\text{WCSS}(k) \qquad (13.17)$$

Recall from Section 13.5.1 the behavior of WCSS, as a function of k and with respect to k^*. Based of those considerations, we expect the following behavior for *DIFF(k)*:

(i) For $k^* < k$, both $DIFF(k)$ and $DIFF(k+1)$ should be large positive values

(ii) For $k^* > k$, both $DIFF(k)$ and $DIFF(k+1)$ should be small values, and one or both might be negative

(iii) For $k = k^*$, $DIFF(k)$ should be large positive, but $DIFF(k+1)$ should be relatively small (might be negative)

Based on these considerations, Krzanowski and Lai proposed to choose the estimate on the number of clusters as the k maximizing:

$$KL(k) = \left| \frac{DIFF(k)}{DIFF(k+1)} \right| \qquad (13.18)$$

That is,

$$k^* = argmax_{2 \le k \le k_{\max}} KL(k) \qquad (13.19)$$

Notice that *KL(k)* is not defined for the important special case of $k = 1$, i.e., no cluster structure in the data.

13.5.3 The Gap statistics

The indices presented so far are either useless or not defined for the important special case $k = 1$. Tibshirani et al. [30] brilliantly combined the ideas of Section 13.3 with the WCSS heuristic, to obtain an index that can deal also with the case $k = 1$. It is referred to as the Gap statistics and, for brevity, we denote it as Gap.

The intuition behind the method is brilliantly elegant. Recall from the previous subsection that the "knee" in the WCSS curve can be used to predict k^*. Unfortunately, the localization of such a point may be subjective. Consider the WCSS curves in Figure 13.3. That is, the plot is obtained with use of the WCSS(k) values. The curve at the bottom of the figure is the WCSS computed with K-means-R (see Section 6.2), on the CNS rat dataset (see Section 13.6.1), which has six classes. The curve at the top of the figure is the *average* WCSS, computed on 10 datasets generated from the original data via the Ps null model. As it is evident from the figure, the curve on the top has a nearly

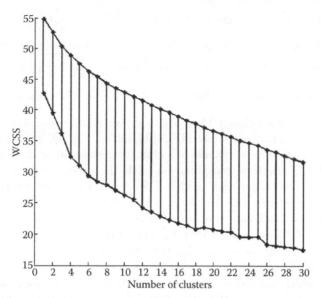

FIGURE 13.3: A geometric interpretation of the Gap statistics.

constant slope: an expected behavior on datasets with no cluster structure in them. The vertical lines indicate the gap between the null model curves and the curve computed by K-means-R, which supposedly captures "cluster structure" in the dataset. Since WCSS is expected to decrease sharply up to k^*, on the real dataset, while it has a nearly constant slope on the null model datasets, the length of the vertical segments is expected to increase up to k^* and then to decrease. In fact, in the figure, if we take as the prediction for k^* the first local maximum of the gap values (data not shown), we get $k^* = 7$, a value very close to the number of classes in the dataset. Normalizing the WCSS curves via logs and accounting also for the simulation error, such an intuition can be given under the form of a procedure, which is strikingly similar to MECCA, as discussed shortly. The first three parameters are as in that procedure, while the last one states that the search for k^* must be done in the interval $[1, k_{\max}]$

Procedure GP ($\ell, A, D, k_{\max}$):

(1) For $1 \leq i \leq \ell$, compute a new data matrix D_i, using the chosen null model. Let D_0 denote the original data matrix.

(1.a) For $0 \leq i \leq \ell$ and $1 \leq k \leq k_{\max}$, compute a clustering solution $P_{i,k}$ on D_i using algorithm A.

(2) For $0 \leq i \leq \ell$ and $1 \leq k \leq k_{\max}$, compute $\log(\text{WCSS}(k))$ on $P_{i,k}$ and store the result in matrix $SL[i, k]$.

(2.a) For $1 \leq k \leq k_{\max}$, compute $\mathrm{Gap}(k) = \frac{1}{\ell} \sum_{i=1}^{\ell} SL[i,k] - SL[0,k]$.

(2.b) For $1 \leq k \leq k_{\max}$, compute the standard deviation $sd(k)$ of the set of numbers $\{SL[1,k] \ldots SL[\ell, k]\}$ and let $s(k) = \left(\sqrt{1 + \frac{1}{\ell}} \right) sd(k)$.

(3) Return as k^* the first value of k such that $\mathrm{Gap}(k) \geq \mathrm{Gap}(k+1) - s(k+1)$.

Gap follows quite closely the statistical paradigm of assessing significance of a statistic via a procedure like MECCA. Indeed, $\log(\mathrm{WCSS}(k))$ is the statistic T used to assess how reliable is a clustering solution with k clusters. The value of that statistic is computed on both the observed data and on data generated by the chosen null model. Then, rather than returning a p-value, the procedure returns the first k for which "the gap" between the observed and the expected statistic is at a local maximum. With reference to step (3) of procedure GP, we remark that the adjustment due to the $s(k + 1)$ term is a heuristic meant to account for the Monte Carlo simulation error in the estimation of the expected value of $\log(\mathrm{WCSS}(k))$ [30].

As discussed in Section 13.1, the prediction of k^* is based on running a certain number of times the procedure GP taking then the most frequent outcome as the prediction. We also point out that further improvements and generalizations of Gap have been proposed in [32].

13.5.4   Clest

Clest is an internal index, proposed by Dudoit and Fridlyand [6]. It estimates k^* by iterating the following: randomly partition the original dataset in a *learning* set and *training* set. The learning set is used to build a classifier C for the data, then to be used to derive "gold standard" partitions of the training set. That is, the classifier is assumed to be a reliable model for the data. It is then used to assess the quality of the partitions of the training set obtained by a given clustering algorithm. Clest generalizes in many respects an approach proposed by Breckenridge [3] and can be regarded as a clever combination of the MECCA hypothesis testing paradigm and resampling techniques. We detail it in the following procedure. With reference to GP, the four new parameters it takes as input are: (a) an external index E (see Section 13.4); (b) $p_{\max}$, a "significance level" threshold; (c) $d_{\min}$, a minimum allowed difference between "computed and expected" values; (d) H, the number of resampling steps; (e) a classifier C used to obtain the "gold standard" partitions of the training set.

Procedure Clest (ℓ, H, A, E, D, C, $k_{\max}$, $p_{\max}$, $d_{\max}$):

(1) For $2 \leq k \leq k_{\max}$, perform steps **(1.a)–(1.d)**.

(1.a) For $1 \leq h \leq H$, split the input dataset in L and T, the learning and training sets, respectively. Cluster the elements of L in k clusters using algorithm A and build a classifier C. Apply C to T in order to obtain a "gold solution" GS_k. Cluster the elements of T in k groups GA_k using algorithm A. Let $m_{k,h} = E(GS_k, GA_k)$.

(1.b) Compute the observed similarity statistic $t_k = median(m_{k,1}, \ldots, m_{k,H})$.

(1.c) For $1 \leq b \leq \ell$, generate (via a null model), a data matrix $D^{(b)}$, and repeat steps **(1.a)** and **(1.b)** on $D^{(b)}$.

(1.d) Compute the average of these H statistics, and denote it with t_k^0. Finally, compute the p-value p_k of t_k and let $d_k = t_k - t_k^0$ be the difference between the statistic observed and its estimate expected value.

(2) Define a set $K = \{2 \leq k \leq k_{\max} : p_k \leq p_{\max} \text{ and } d_k \geq d_{\min}\}$.

(3) Based on K return a prediction for k^* as: if K is empty then $k^* = 1$, else $k^* = argmax_{k \in k} \, d_k$.

Again, Clest is quite close in form to MECCA. Here the statistic of interest is the similarity between the partition obtained by the algorithm and that obtained by the classifier. Moreover, in order to estimate k^*, a full-fledged p-value test is performed. Finally, we point out that Clest can deal with the case $k^* = 1$ of no structure in the data.

13.5.5 Consensus clustering

The Consensus index, proposed by Monti et al. [24], is based on resampling techniques. In analogy with WCSS, one needs to analyze a suitably defined curve in order to find k^*. It is best presented as a procedure. With reference to GP, the two new parameters it takes as input are a resampling scheme *Res*, i.e., a way to sample from D to build a new data matrix, and the number H of resampling iterations.

Procedure Consensus (*Res, H, A, D, $k_{\max}$*):

(1) For $2 \leq k \leq k_{\max}$, initialize to empty the set M of connectivity matrices and perform steps **(1.a)** and **(1.b)**

(1.a) For $1 \leq h \leq H$, compute a perturbed data matrix $D^{(h)}$ using resampling scheme *Res*; cluster the elements in k clusters using algorithm A and $D^{(h)}$. Compute a connectivity matrix $M^{(h)}$ and insert it into M

(1.b) Based on the connectivity matrices in M, compute a consensus matrix $M^{(k)}$

(2) Based on the $k_{\max} - 1$ consensus matrices, return a prediction for k^*

The resampling scheme in this case extracts, uniformly and at random, a given percentage of the rows of D. As for the connectivity matrix $M^{(h)}$, one has $M^{(h)}(i,j) = 1$ if items i and j are in the same cluster and 0 otherwise. Moreover, we also need to define an indicator matrix $I^{(h)}$ such that $I^{(h)}(i,j) = 1$ if items i and j are both in $D^{(h)}$ and it is 0 otherwise. Then, the consensus matrix $M^{(k)}$ is defined as a properly normalized sum of all connectivity matrices in all perturbed datasets:

$$M^{(k)} = \frac{\sum_h M^{(h)}}{\sum_h I^{(h)}} \qquad (13.20)$$

Based on $M^{(k)}$, Monti et al. define a value $A(k)$ measuring the level of stability in cluster assignments, as reflected by the consensus matrix. Formally,

$$A(k) = \sum_{i=2}^{n} [x_i - x_{i-1}] CDF(x_i)$$

where CDF is the empirical cumulative distribution defined over the range $[0,1]$ as follows:

$$CDF(c) = \frac{\sum_{i<j} l\{M(i,j) \le c\}}{N(N-1)/2}$$

with l equal to 1, if the condition is true and 0 otherwise. Finally, based on $A(k)$, one can define:

$$\Delta(k) = \begin{cases} A(k) & k = 2 \\ \frac{A(k+1) - A(k)}{A(k)} & k > 2 \end{cases}$$

Moreover, Monti et al. suggested the use of the function Δ' for nonhierarchical algorithms. It is defined as Δ although one uses $A'(k) = \max_{k' \in [2,k]} A(k')$. Based on the Δ or Δ' curve, the value of k^* is obtained using the following intuitive idea, based also on experimental observations.

(i) For each $k \le k^*$, the area $A(k)$ markedly increases. This results in an analogous pronounced decrease of the Δ and Δ' curves

(ii) For $k > k^*$, the area $A(k)$ has no meaningful increases. This results in a stable plot of the Δ and Δ' curves

An example of this behavior is given in Figure 13.4, which is derived from a dataset that has six clusters (the CNS rat dataset), with use of the K-means-R clustering algorithm. The plots of the CDF curves are shown in Figure 13.4a, yielding a monotonically increasing value of A as a function of k. The plot of the Δ' curve are shown in Figure 13.4b, where the flattening effect referred to in (ii) above is evident for $k \ge 6$, which is the number of clusters in the dataset. The prediction coming from the curves is $k^* = 6$. In conclusion, the "rule of thumb" is: take as k^* the abscissa corresponding to the smallest nonnegative value where the curve starts to stabilize; that is, no big variation in the curve takes place from that point on.

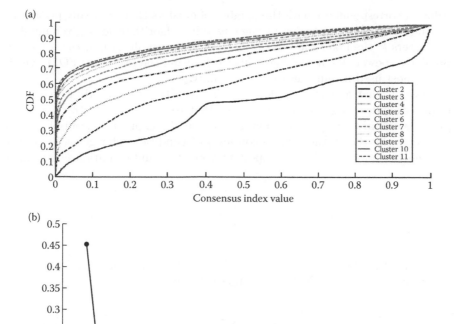

FIGURE 13.4: (a) *CDF* curves on the CNS rat dataset, computed with K-means-R for $2 \leq k \leq 11$. From bottom to top, the rank of each curve corresponds to the number of clusters in the solution used to compute it, starting with two clusters. (b) Plot of the values of the Δ' function.

13.5.6 Figure of Merit (FOM)

Figure of Merit (FOM) is a family of internal validation indices introduced by Ka Yee Yeung et al. [34], specifically for microarray data. Such a family is based on the jackknife approach and it has been designed for use as a relative index assessing the predictive power of a clustering algorithm, i.e., its ability to predict the correct number of clusters in a dataset. It has also been extended in several directions by Datta and Datta [4]. Experiments by Ka Yee Yeung et al. show that the FOM family of indices satisfies the following properties, with a good degree of accuracy. For a given clustering algorithm, it has a low value in correspondence with the number of clusters that are really present in the data. Moreover, when comparing clustering algorithms for a given number of clusters k, the lower the value of FOM for a given algorithm, the better its predictive power. We now review this work, using the two-norm FOM, which is the most used instance in the FOM family.

Assume that a clustering algorithm is given the data matrix D with column e excluded. Assume also that, with that reduced dataset, the algorithm produces k clusters $c_1, \ldots, c_k$. Let $D(g, e)$ be the expression level of gene g and $m_i(e)$ be the average expression level of condition e for genes in cluster c_i. The two-norm FOM with respect to k clusters and condition e is defined as:

$$\text{FOM}(e, k) = \sqrt{\frac{1}{n} \sum_{i=1}^{k} \sum_{x \in c_i} (D(x, e) - m_i(e))^2} \qquad (13.21)$$

Notice that $\text{FOM}(e, k)$ is essentially a root mean square deviation. The aggregate two-norm FOM for k clusters is then:

$$\text{FOM}(k) = \sum_{e=1}^{m} \text{FOM}(e, k). \qquad (13.22)$$

Both Equations 13.21 and 13.22 can be used to measure the predictive power of an algorithm. The first gives us more flexibility, since we can pick any condition, while the second gives us a total estimate over all conditions. So far, Equation 13.22 is the formula mostly used in the literature. Moreover, since the experimental studies conducted by Ka Yee Yeung et al. show that $\text{FOM}(k)$ behaves as a decreasing function of k, an adjustment factor has been introduced to properly compare clustering solutions with different numbers of clusters. A theoretical analysis by Ka Yee Yeung et al. provides the following adjustment factor:

$$\sqrt{\frac{n - k}{n}}. \qquad (13.23)$$

When Equation 13.23 divides Equation 13.21, we refer to Equations 13.21 and 13.22 as *adjusted* FOMs. We use the adjusted aggregate FOM for our experiments and, for brevity, we refer to it simply as FOM.

The use of FOM in order to establish how many clusters are present in the data follows the same heuristic methodology outlined for WCSS, i.e., one tries to identify the "knee" in the FOM plot as a function of the number of clusters. An example is provided in Figure 13.5, where the FOM curve is computed on the Leukemia dataset with K-means-R. In this case, it is easy to see that the predicted value is $k^* = 3$, corresponding to the correct number of clusters in the dataset.

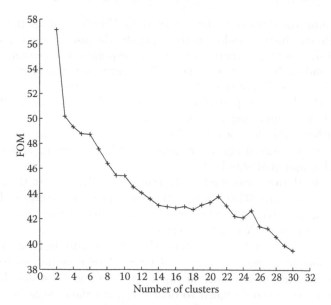

FIGURE 13.5: FOM curve on the Leukemia dataset, computed with K-means-R.

13.6 Experimental Set-Up

This section describes the experimental set-up we have used, with particular attention to the software system that has been built to perform the experiments and that can be used as a tool to apply the methodologies presented in this chapter, both in the lab and in the classroom.

13.6.1 Datasets

We use four publicly available datasets that have been used as benchmarks for experimentally validating either clustering algorithms or internal indices or both. Each dataset is a matrix, in which each row corresponds to an element to be clustered and each column to an experimental condition. The four datasets, together with the acronyms used in this chapter, are as follows and are available for download at the supplementary material web site [1].

CNS rat: the dataset gives the expression levels of 112 genes during rat central nervous system development. It is a 112×17 data matrix studied in [31] to obtain a division of the genes into six classes, four of which are composed of biologically, functionally related genes. This division is assumed to be the reference classification and gold standard. The dataset has been used for the validation of FOM and Genclust [5].

Leukemia: the dataset is the one used by Handl et al. in their survey of computational cluster validation to illustrate the use of some indices. It is a 38×100 data matrix, where each row corresponds to a patient with acute leukemia and each column to a gene. The reference classification and gold standard is a partition of the patients into three groups.

Yeast: the dataset is part of that studied by Spellman et al. [29] and has been used by Shamir and Sharan for the benchmarking of several clustering algorithms [28]. It is a 698 × 72 data matrix that can be divided into five functionally related classes of genes, which is taken to be the reference classification and gold standard.

PBM: the dataset was used by Hartuv et al. [12] to test their clustering algorithm. It contains 2329 cDNAs with a fingerprint of 139 oligos. This gives a 2329×139 data matrix. The reference classification and gold standard consists of a division of the cDNAs into 18 classes.

All of the mentioned datasets have cluster structure in them. However, it is also important to test the ability of internal/relative indices to deal with the case of absence of structure in the data. Gap and Clest are the only two indices that have been designed to deal also with the case of "no structure" in the data. Therefore, in order to assess their performance in that case, we use three datasets with no cluster structure in them: Poisson1, Poisson2 and Poisson3, which have been generated via the (M.1) null model from CNS rat, Leukemia, and Yeast, respectively.

13.6.2 Algorithms

Among the many clustering algorithms available in the literature, we have experimented with the simplest and most well known of them: hierarchical and K-means [16]. Among the hierarchical methods [16]: Hier-A (average link), Hier-C (complete link), and Hier-S (single link). Moreover, we use K-means, both in the version that starts the clustering from a random partition of the data and in the version where it takes as part of the input an initial partition produced by one of the chosen hierarchical methods. The acronyms of the versions for which we report experimental results are K-means-R and K-means-A, respectively, where R stands for random initialization and A for average link initialization.

13.6.3 Similarity/Distance functions

All of our algorithms use Euclidean distance in order to assess similarity of single elements to be clustered. Such a choice is natural and conservative, as we now explain. It places all algorithms in the same position without introducing biases due to distance function performance, rather than to the algorithm. Moreover, time course data have been properly standardized (mean equal to 0 and variance equal to 1), so that Euclidean distance would not be penalized on those data. This is standard procedure, e.g., [34], for those data. The results we obtain are conservative since, assuming that one has a provably much better similarity/distance function, one can only hope to get better estimates than ours (else the used distance function is not better than Euclidean distance after all). As it is clear from the upcoming discussion and conclusions, such better estimates will cause no dramatic change in the general picture of our findings. The choice is also natural, in view of the debate regarding the identification of a proper similarity/distance function for clustering gene expression (see Section 13.2).

13.6.4 The ValWorkBench software system

We now present a toolkit, nicknamed ValWorkBench, that allows experimentation with all the indices and algorithms presented in this chapter. Its goal is to provide a system for class discovery on datasets coming from different applications, although microarray data have been the ones on which it has been used the most. The overview of the toolkit given here is very limited, due to space constraints. Full details can be found in the user manual, available for download at [2]. At the same web address, the software system is available for download and it can be used under the GNU GPL license.

ValWorkBench has a GUI that can be conveniently divided in two parts: computation of validation indices and visualization of results. The first part can be used to choose indices, algorithms and their parameters and then to actually start the computation. An example screenshot is given in Figures 13.6 and 13.7. Its most salient features can be summarized as follows:

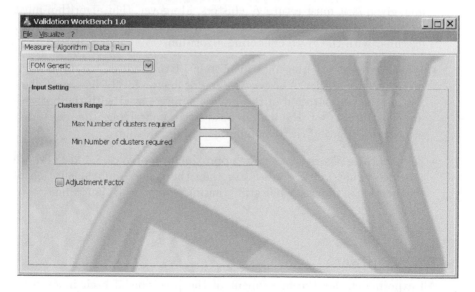

FIGURE 13.6: The ValWorkBench GUI. The measure tab with the FOM index selected.

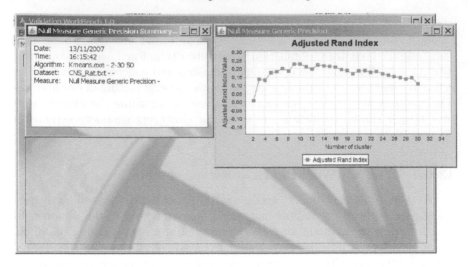

FIGURE 13.7: Example of the visualization of the adjusted Rand index.

- It is supported by an open source Java Library that can be used as a building block to implement, in a relatively short amount of time, additional validation indices to be included in ValWorkBench.

- Each index is *generic* in the sense that it takes a clustering algorithm as one of its input parameters. The clustering algorithm must be partitional and, given k, it must return K clusters. Moreover, that algorithm must satisfy the input/output formats of ValWorkBench. The index is then executed with the given clustering algorithm.

- Whenever possible, the implementation of the indices has been engineered to be efficient. In particular, for hierarchical clustering algorithms, the computation of the indices is interleaved with the bottom-up construction of the hierarchical tree underlying the clustering algorithms, to avoid costly computational duplications.

- Fast heuristic algorithms for the approximation of WCSS and FOM are provided. That is, they are faster than the generic implementation of the mentioned indices, while still granting a very close approximation to them.

13.6.5 Hardware

All experiments for the assessment of the precision of each index were performed in part on several state-of-the-art PCs and in part on a 64-bit

AMD Athlon 2.2 GHz bi-processor with 1 GB of main memory running Windows Server 2003. All the timing experiments reported were performed on the bi-processor, using one processor per run. The usage of different machines for the experimentation was deemed necessary in order to complete the full set of experiments in a reasonable amount of time. Indeed, as detailed later, some indices require weeks to complete execution on PBM, the largest of the datasets we have used. We also point out that all the operating systems supervising the computations have a 32 bits precision. We also mention that we report, for each experiment, the time it took in terms of milliseconds, i.e., an experiment that took 2.7 seconds is reported as taking 2.7×10^3 milliseconds. As will be evident later, such a "time scale" highlights very well the difference in timing among the various indices, which spans several orders of magnitude.

13.7   Experiments: Use of External Indices to Validate Algorithms and Internal/Relative Indices

External indices can be very useful in evaluating the performance of algorithms and internal/relative indices, with the use of datasets that have a gold standard solution. We briefly illustrate the methodology for the external validation of a clustering algorithm, via an external index that needs to be maximized. The same methodology applies to internal/relative indices, as discussed in [34]. For a given dataset, one plots the values of the index computed by the algorithm as a function of K, the number of clusters. Then, one expects the curve to grow to reach its maximum close or at the number of classes in the reference classification of the dataset. After that number, the curve should fall. We now present the results of our experiments, with the use of the indices discussed in Section 13.4. For each dataset and each clustering algorithm, we compute each index for a number of cluster values in the range [2,30].

Adjusted Rand index R_A: for our experiments, the relevant plots are in Figure 13.8. Based on them, we see that Hier-S is a poor performer. The remaining algorithms do well on the Leukemia and Yeast datasets. On the CNS rat dataset, the performance is somewhat mixed, while they are not as precise on the fourth and largest dataset.

The FM-index: the relevant plots are in Figure 13.9. Hier-S is still the worst among the algorithms, however now we have no consistent indication about the other algorithms.

The F-index: the relevant plots are in Figure 13.10. Hier-S is still the worst among the algorithms and the indications we get in this case about the other algorithms are essentially the same as in the case of R_A.

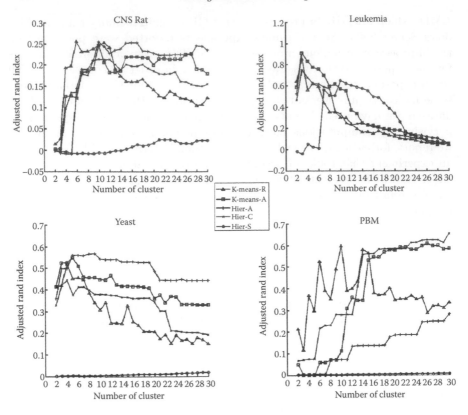

FIGURE 13.8: The adjusted Rand index curves, for each of the datasets. In each figure, the plot of the index, as a function of the number of clusters, is plotted differently for each algorithm.

In concluding this section, it should be mentioned that although all of the three indices have solid statistical justifications, R_A seems to be the best performer while the FM-index is somewhat disappointing.

13.8 Experiments: Use of Internal/Relative Indices to Identify the Number of Clusters in a Dataset

In this section, we present experiments with the aim to shed some light on Questions (A) and (B) stated in Section 13.1, regarding the internal/relative indices presented in Section 13.5, on benchmark microarray datasets.

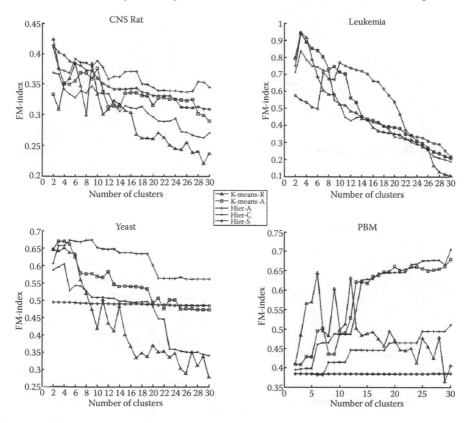

FIGURE 13.9: The FM-index curves, for each of the datasets. In each figure, the plot of the index, as a function of the number of clusters, is plotted differently for each algorithm.

13.8.1 Question (A): intrinsic ability of an index to predict the correct number of clusters

13.8.1.1 Within cluster sum of squares (WCSS)

For each algorithm and each dataset, we have computed WCSS for a number of cluster values in the range [2,30]. The relevant plots are shown in Figure S1 in the supplementary material web site [1]. Recall from Section 13.1 that, given the relevant WCSS curve, k^* is predicted as the abscissa closest to the "knee" in that curve. The values resulting from the application of this methodology to the relevant plots are reported in Table 13.3 together with timing results, those latter limited to the first two datasets. Throughout this chapter, for each cell in a table displaying precision results, a number in a circle with a black background indicates a prediction in agreement with the number of classes in the dataset; a number in a circle with a white background

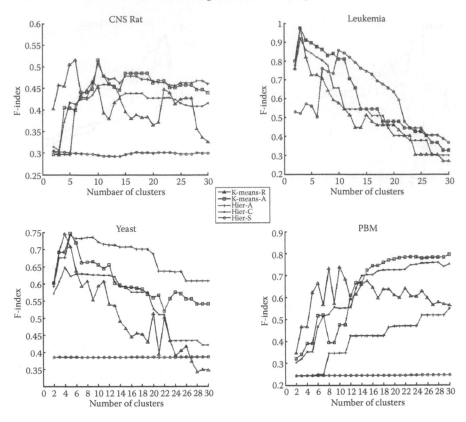

FIGURE 13.10: The F-index curves, for each of the datasets. In each figure, the plot of the index, as a function of the number of clusters, is plotted differently for each algorithm.

indicates a prediction that differs, in absolute value, by 1 from the number of classes in the dataset; a number in a square indicates a prediction that differs, in absolute value, by 2 from the number of classes in the dataset-when the prediction is one cluster, i.e. Gap statistics, this rule is not applied because the prediction means no cluster structure in the data; a number in black indicates the remaining predictions. As detailed in each table, cells with "S" indicate that the experiment was stopped because of its high computational demand.

We have that WCSS performs well with K-means-R and K-means-A, on the first three datasets, while it gives no reasonably correct indication on PBM. It is consistently poor when used in conjunction with Hier-S. Those facts give strong indication that WCSS is algorithm-dependent. Finally, the failure of WCSS, with all algorithms, to give a good prediction for PBM indicates that WCSS may not be of any use on large datasets having a large number of clusters. The results regarding the best performing algorithms are reported in Table 13.11 for comparison with the performance of the other indices.

TABLE 13.3: A summary of the results for WCSS on all algorithms and on all datasets.

	Precision				Timing	
	CNS rat	Leukemia	Yeast	PBM	CNS rat	Leukemia
K-means-R	④	❸	④	8	2.4×10^3	2.0×10^3
K-means-A	④	❸	❺	6	2.3×10^3	1.3×10^3
Hier-A	10	❸	❺	8	1.1×10^3	4.0×10^2
Hier-C	10	❸	❾	9	7.0×10^2	4.0×10^2
Hier-S	⑧	10	24	30	2.6×10^3	6.0×10^2
Gold solution	6	3	5	18	−	−

The columns under the label precision indicate the number of clusters predicted by WCSS, while the remaining two indicate the timing in milliseconds for the execution of the corresponding experiment.

13.8.1.2 The Krzanowski and Lai (KL) index

Following the same experimental set-up of WCSS, we have computed the KL index, for each dataset and each algorithm. The results are summarized in Table 13.4. The best performing algorithm seems to be K-means-A, whose results are reported in Table 13.11 for comparison with the performance of the other indices. The same conclusions drawn for WCSS seem also to apply to KL.

13.8.1.3 The gap statistics

For each dataset and each clustering algorithm, we compute three versions of Gap, namely Gap-Ps, Gap-Pc and Gap-Pr, for a number of cluster values in the range [1,30]. Gap-Ps uses the Poisson null model (M.1), Gap-Pc the Poisson null model (M.2) aligned with the principal components of the data while Gap-Pr uses the permutational null model (M.3). For each of them, we perform a Monte Carlo simulation, 20 steps, in which the index returns an estimated number of clusters for each step. Each simulation step is based on the generation of ten data matrices from the null model used by the index. At the end of each Monte Carlo simulation, the number with the majority of estimates is taken as the predicted number of clusters. Occasionally, there are ties and we report both numbers. The relevant histograms are displayed at the supplementary material web site [1]: Figures S2 through S5 for Gap-Ps, Figures S6 through S8 for Gap-Pc and Figures S9 through S12 for Gap-Pr. The results are summarized in Table 13.5. We remark that Gap-Pc on PBM resulted to be very slow, for each clustering algorithm, to the point of having made no substantial progress in a week. Therefore, the corresponding experiments were terminated. Accordingly, no value is reported in Table 13.5.

A few comments are in order, the first one regarding the null models. Tibshirani et al. find experimentally that, on simulated data, Gap-Pc is the clear winner over Gap-Ps (they did not consider Gap-Pr). Our results show that, as the dataset size increases, Gap-Pc becomes computationally unfeasible, due

TABLE 13.4: A summary of the results for KL on all algorithms and on all datasets.

	Precision				Timing	
	CNS rat	Leukemia	Yeast	PBM	CNS rat	Leukemia
K-means-R	24	③	10	5	2.7×10^3	2.3×10^3
K-means-A	4	③	3	6	2.5×10^3	1.4×10^3
Hier-A	⑦	③	3	8	1.2×10^3	5.0×10^3
Hier-C	10	③	16	30	9.0×10^2	3.0×10^2
Hier-S	8	7	24	⑱	2.8×10^3	5.0×10^2
Gold solution	6	3	5	18	–	–

The columns under the label precision indicate the number of clusters predicted by KL, while the remaining two indicate the timing in milliseconds for the execution of the corresponding experiment.

to the repeated data transformation step. Moreover, on the smaller datasets, no null model seems to have the edge and, both Gap-Ps and Gap-Pr, give no useful indication on PBM. Some of the results are also somewhat puzzling. In particular, although the datasets have cluster structure, many algorithms return an estimate of $k^* = 1$, i.e., no cluster structure in the data. An analogous situation was reported by Monti et al. In their study, Gap-Ps returned $k^* = 1$ on two artificial datasets. Fortunately, an analysis of the corresponding Gap curve showed that indeed the first maximum was at $k^* = 1$ but a local maximum was also present at the correct number of classes, in each dataset. We have performed an analogous analysis of the relevant Gap curves to find that, in analogy with Monti et al., also in our case most curves show a local maximum at or very close to the number of classes in each dataset, following the maximum at $k^* = 1$. An example curve is given in Figure 13.11. From the above, one can conclude that inspection of the Gap curves and *domain knowledge* can greatly help in disambiguating the case $k^* = 1$. We also report that experiments conducted by Dudoit and Fridlyand and, independently by Yan and K. Ye, show that Gap tends to overestimate the correct number of clusters, although this does not seem to be the case for our datasets and algorithms.

Moreover, in order to test the ability of Gap to deal with the case of "no structure" in the data, we compute it on the Poisson1, Poisson2 and Poisson3 datasets, with the same experimental setup used for real data. The results are summarized in Table 13.6, where we see a perfect performance of Gap-Pc, with Gap-Ps closely following.

As for real datasets, K-means-A seems to be the best performer, but with the Poisson null model. Such a fact indicates that Gap is algorithm-dependent *and* null-model-dependent. Moreover, Gap-Pc has severe time demand limitations on large datasets. The results for Gap-Ps-K-means-A are reported in Table 13.11 for comparison with the performance of the other indices.

TABLE 13.5: A summary of the results for Gap on all algorithms and all datasets, with use of three null models.

	Precision				Timing	
	CNS rat	Leukemia	Yeast	PBM	CNS rat	Leukemia
Gap-Ps-K-means-R	**(6)** or (7)	(4) or [5]	9	7	8.4×10^5	5.0×10^5
Gap-Ps-K-means-A	(7)	**(3)**	[7]	9	6.1×10^5	4.7×10^5
Gap-Ps-Hier-A	1	(4)	[3]	1	2.7×10^5	1.4×10^5
Gap-Ps-Hier-C	1 or 2	(4)	[7]	15	2.3×10^5	1.1×10^4
Gap-Ps-Hier-S	1	1	1	1	6.1×10^5	1.9×10^5
Gap-Pc-K-means-R	2	1	(4)	S	4.9×10^5	8.0×10^5
Gap-Pc-K-means-A	2	(4)	[3]	S	3.8×10^5	7.0×10^5
Gap-Pc-Hier-A	1	**(3)** or (4)	1 or 2 or [3]	S	3.2×10^5	3.7×10^5
Gap-Pc-Hier-C	1	(4)	[3]	S	1.9×10^5	1.9×10^5
Gap-Pc-Hier-S	1	1	1	S	7.7×10^5	3.1×10^5
Gap-Pr-K-means-R	**(6)**	(4)	8	8	8.6×10^5	5.4×10^5
Gap-Pr-K-means-A	[8]	(4)	13	4	7.4×10^5	5.0×10^5
Gap-Pr-Hier-A	3	(4)	[3]	1	2.5×10^5	1.6×10^5
Gap-Pr-Hier-C	(7)	(4)	16	1	1.3×10^5	1.4×10^5
Gap-Pr-Hier-S	1 or **(6)**	1	1	2	6.8×10^5	1.9×10^5
Gold solution	6	3	5	18	–	–

The columns under the label precision indicate the number of clusters predicted by Gap, while the remaining two indicate the timing in milliseconds for the execution of the corresponding experiment.

13.8.1.4 Clest

For CNS rat and Yeast and each clustering algorithm, we compute Clest for a number of cluster values in the range [2,30] while, for Leukemia, the range [2,10] is used due to the small size of the dataset. Moreover, although experiments have been started with PBM, no substantial progress was made

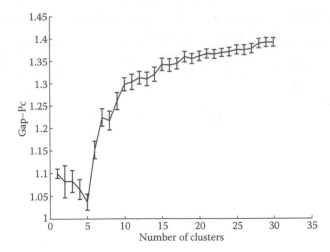

FIGURE 13.11: The Gap-Pc curve on the CNS rat dataset, with use of the Hier-A algorithm. At each point, error bars indicated the variation of the curve across simulations. The curve shows a first maximum at $k = 1$, yielding a prediction of $k^* = 1$. However, there is a local maximum at $k = 7$, which is close to the number of classes $k = 6$.

TABLE 13.6: A summary of the results for Gap on all algorithms and all simulated datasets, with use of three null models.

	Poisson1	Poisson2	Poisson3
Gap-Ps-K-means-R	1	1	1
Gap-Ps-K-means-A	1	2	1
Gap-Ps-Hier-A	1	1 or 2	1
Gap-Ps-Hier-C	1	1	1
Gap-Ps-Hier-S	1	1	1
Gap-Ps-K-means-R	1	1	1
Gap-Ps-K-means-A	1	1	1
Gap-Ps-Hier-A	1	1	1
Gap-Ps-Hier-C	1	1	1
Gap-Ps-Hier-S	1	1	1
Gap-Ps-K-means-R	3	2	4
Gap-Ps-K-means-A	1	2	2
Gap-Ps-Hier-A	1	2	2
Gap-Ps-Hier-C	3	2	1
Gap-Ps-Hier-S	1	1	1
Gold solution	1	1	1

The columns indicate the number of clusters predicted by Gap for each dataset.

after a week of execution and, for each clustering algorithm, the corresponding experiment was terminated. Following the same experimental set-up of Dudoit and Fridlyand, for each cluster value k, we perform 20 resampling steps and 20 iterations. In each step, 66% of the rows of the data matrix are extracted, uniformly and at random, to create a learning set, to be given to the clustering algorithm to be clustered in k groups. As external index E, we use each of the indices described in Section 13.4. Finally, the DLDA with diagonal covariance matrix is used as classifier. The results are summarized in Table 13.7, where the timing results for the Leukemia experiments were included only for completeness. Indeed, those experiments were performed on a smaller interval of cluster values with respect to CNS rat. Therefore, the timing results obtained on the two datasets are not comparable.

In order to test the ability of Clest to deal with the case of "no structure" in the data, we compute it on the Poisson1, Poisson2 and Poisson3 datasets, with the same experimental setup used for real data. The results are summarized in Table 13.8, where we see that Clest with use of the F-index is the best performer, with the FM-index closely following.

On real datasets, the results show that Clest has severe time demand limitations on large datasets. Moreover, it is clearly algorithm-dependent, with K-means-R being the best performer with both the FM and the F-index. Those results are reported in Table 13.11 for comparison with the performance of the other indices.

13.8.1.5 Consensus

For each of the first three datasets and each clustering algorithm, we compute Consensus for a number of cluster values in the range [2,30]. Following the same experimental set-up of Monti et al., for each cluster value k, we perform 500 resampling steps. In each step, 80% of the rows of the matrix are extracted uniformly and at random to create a new dataset, to be given to the clustering algorithm to be clustered in k groups. The prediction of k^* is based on the plot of two curves, $\Delta(k)$ and $\Delta'(k)$, as a function of the number k of clusters. As suggested by Monti et al., the first curve is suitable for hierarchical algorithms while the second suits nonhierarchical ones. We did not experiment for PBM, since Consensus was very slow (we stopped execution on each algorithm after a week). Contrary to Monti et al. indication, we have computed the $\Delta(k)$ curve for all algorithms on the first three datasets, for reasons that will be evident shortly. The corresponding plots are available at the supplementary material web site [1] as Figures S13 through S15. We have also computed the $\Delta'(k)$ curve for the K-means algorithms, on the first three datasets. Since those curves are identical to the $\Delta(k)$ ones, they are omitted. In order to predict the number of clusters in the datasets, we have used, for all curves, the rule suggested by Monti et al.: take as k^* the abscissa corresponding to the smallest nonnegative value where the curve starts to stabilize; that is, no big variation in the curve takes place from that point on. We per-

TABLE 13.7: A summary of the results for Clest on all algorithms and all datasets, with use of three external indices.

	Precision				Timing	
	CNS rat	Leukemia	Yeast	PBM	CNS rat	Leukemia
Clest-FM-K-means-R	[8]	(4)	(4)	S	1.2×10^6	3.5×10^5
Clest-FM-K-means-A	18	7	13	S	1.4×10^6	4.9×10^5
Clest-FM-K-means-A	10	6	24	S	1.1×10^6	3.8×10^5
Clest-FM-Hier-C	10	(4)	8	S	1.1×10^6	3.8×10^5
Clest-FM-Hier-S	20	10	1	S	1.1×10^6	4.0×10^5
Clest-Adj-K-means-R	(5)	(4)	2	S	1.1×10^6	3.6×10^5
Clest-Adj-K-means-A	12	●3	●5	S	1.1×10^6	3.8×10^6
Clest-Adj-Hier-A	13	●3	11	S	1.1×10^6	3.8×10^5
Clest-Adj-Hier-C	9	(4)	(4)	S	1.1×10^6	3.8×10^5
Clest-Adj-Hier-S	[4]	7	26	S	1.1×10^6	3.8×10^5
Clest-F-K-means-R	●6	●3	(4)	S	1.2×10^6	3.7×10^5
Clest-F-K-means-A	[8]	6	11	S	1.4×10^6	4.8×10^5
Clest-F-Hier-A	(7)	7	27	S	1.1×10^6	3.7×10^5
Clest-F-Hier-C	9	●3	●5	AS	1.1×10^6	3.9×10^5
Clest-F-Hier-S	28	10	1	S	1.1×10^6	3.8×10^5
Gold solution	6	3	5	18	–	–

The columns under the label precision indicate the number of clusters predicted by Clest, while the remaining one indicates the timing in milliseconds for the execution of the corresponding experiment. For Leukemia, the timing results are reported for completeness only. Indeed, experiments were performed on a smaller interval of cluster values with respect to CNS rat. Therefore, the timing results obtained on the two datasets are not comparable.

formed such an analysis on the $\Delta(k)$ curves and the results are summarized in Table 13.9, together with the corresponding timing results.

As for the precision of Consensus, all algorithms perform well, except for Hier-S. The results also show that the Δ curve may be enough for all algorithms. This contradicts the recommendation by Monti et al. and a brief explanation of the reason follows. With reference to the notation in Section 13.5.5, $A(k)$ is a value that is expected to behaves like a nondecreasing func-

TABLE 13.8: A summary of the results for Clest on all algorithms and all simulated datasets, with use of three external indices.

	Poisson1	Poisson2	Poisson3
Clest-FM-K-means-R	2	1	1
Clest-FM-K-means-A	3	10	1
Clest-FM-Hier-A	4	10	1
Clest-FM-Hier-C	1	1	1
Clest-FM-Hier-S	1	1	1
Clest-Adj-K-means-R	1	3	1
Clest-Adj-K-means-A	2	1	1
Clest-Adj-Hier-A	3	1	1
Clest-Adj-Hier-C	3	1	1
Clest-Adj-Hier-S	4	1	1
Clest-F-K-means-R	1	1	1
Clest-F-K-means-A	4	1	1
Clest-F-Hier-A	5	1	1
Clest-F-Hier-C	1	2	1
Clest-F-Hier-S	1	1	1
Gold solution	1	1	1

The columns indicate the number of clusters predicted by Clest for each dataset.

TABLE 13.9: A summary of the results for Consensus on all algorithms and the first three datasets.

	Precision				Timing	
	CNS rat	Leukemia	Yeast	PBM	CNS rat	Leukemia
K-means-R	⑥	④	⑥	S	1.0×10^6	1.1×10^6
K-means-A	⑥	④	⑥	S	1.3×10^6	1.2×10^6
Hier-A	⑥	④	⑥	S	9.6×10^5	8.0×10^5
Hier-C	⑥	④	⑥	S	1.1×10^6	9.5×10^5
Hier-S	2	8	10	S	9.9×10^5	9.6×10^5
Gold solution	6	3	5	18	–	–

The columns under the label precision indicate the number of clusters predicted by Consensus, while the remaining two indicate the timing in milliseconds for the execution of the corresponding experiment.

tion of k, for hierarchical algorithms. Therefore $\Delta(k)$ would be expected to be positive or, when negative, not too far from 0. Such a monotonicity of $A(k)$ is not expected for nonhierarchical algorithms. Therefore, another definition of Δ is needed to ensure a behavior of this function analogous to the hierarchical algorithms. We find that, for the partitional algorithms used, $A(k)$ displays exactly the same monotonicity properties of the hierarchical algorithms. The end result is that the same definition of Δ can be used for both types of algorithms. To the best of our knowledge, Monti et al. defined the function Δ',

TABLE 13.10:   A summary of the results for FOM on all algorithms
and on all datasets.

	Precision				Timing	
	CNS rat	Leukemia	Yeast	PBM	CNS rat	Leukemia
K-means-R	⑦	❸	④	14	2.9×10^4	1.9×10^5
K-means-A	⑦	❸	④	12	2.2×10^4	9.3×10^4
Hier-A	⑦	❸	⑥	14	1.6×10^3	7.5×10^3
Hier-C	10	④	❺	22	1.6×10^3	7.7×10^3
Hier-S	3	7	22	20	1.6×10^3	7.4×10^3
Gold solution	6	3	5	18	–	–

The columns under the label precision indicate the number of clusters predicted by FOM,
while the remaining two indicate the timing in milliseconds for the execution of the corresponding experiment.

but their experimentation was limited to hierarchical algorithms. The results
of the best performing algorithms are reported in Table 13.11 for comparison
with the other indices.

13.8.1.6   Figure of Merit (FOM)

The same methodology outlined for WCSS for the prediction of k^* applies
here also (see Section 13.8.1.1). The relevant plots are in Figure S16 at the
supplementary material web site [1]. The values resulting from the application
of this methodology to the relevant plots are reported in Table 13.10 together
with timing results, those latter are limited to the first two datasets. The same
conclusions drawn for WCSS also apply here. The results regarding the best
performing algorithms are reported in Table 13.11 for comparison with the
performance of the other indices.

13.8.2   Question (B): relative merits of each index

The discussion here refers to Table 13.11. It is evident that the K-means
algorithms have superior performance with respect to the hierarchical ones.
Moreover, both Consensus and FOM seem to be the most stable across algorithms. All measures have limitations on large datasets with a large number
of clusters, as PBM shows. That is either due to high computational demand
or to lack of precision. It is also obvious that, when one takes computer time
into account, there is a hierarchy of indices, with WCSS being the fastest and
Consensus the slowest.

Among the indices presented here, only Gap and Clest can deal with the
important special case of "no cluster structure" in the data. In that setting,
Gap seems to be a better performer than Clest, although it has a natural
tendency to predict no structure in data, even when the structure is present.
Therefore, a conservative choice would be to use both indices.

TABLE 13.11: A summary of the best performances obtained by each index.

	Precision				Timing	
	CNS rat	Leukemia	Yeast	PBM	CNS rat	Leukemia
WCSS-K-means-R	[4]	**❸**	④	8	2.4×10^3	2.0×10^3
WCSS-K-means-A	[4]	**❸**	**❺**	6	2.3×10^3	1.3×10^3
KL-K-means-A	[4]	**❸**	[3]	6	1.1×10^3	5.0×10^2
KL-Hier-A	⑦	**❸**	[3]	8	1.1×10^3	5.0×10^2
FOM-K-means-R	⑦	**❸**	④	14	2.9×10^4	1.9×10^5
FOM-K-means-A	⑦	**❸**	④	12	2.2×10^4	9.3×10^4
FOM-Hier-A	④	**❸**	⑥	14	1.6×10^3	7.5×10^3
Gap-Ps-K-means-A	⑦	**❸**	[7]	9	6.1×10^5	4.7×10^5
Clest-FM-K-means-R	[8]	④	④	S	1.2×10^6	3.5×10^5
Clest-F-K-means-R	**❻**	**❸**	④	S	1.2×10^6	3.7×10^5
Consensus-K-means-R	**❻**	④	⑥	S	1.0×10^6	1.1×10^6
Consensus-K-means-A	⑦	**❸**	⑥	S	1.3×10^6	1.2×10^6
Consensus-Hier-A	⑦	**❸**	**❺**	S	9.6×10^5	8.0×10^5
Consensus-Hier-C	**❻**	④	⑥	S	1.1×10^6	9.5×10^5
Gold solution	6	3	5	18	–	–

The timing results for Clest on the Leukemia dataset were performed on a smaller interval of cluster values with respect to CNS rat and so they are not comparable.

Putting all those facts together and accounting also for time performance, it seems advisable to use Gap and Clest when it is not clear that there is cluster structure in the data. For the remaining cases, it seems reasonable to use either WCSS or KL to get a quick and fast limitation of the range of values in which to search. Then, among the slowest indices, Consensus can be used on the range of cluster values so obtained. In fact, its stability across algorithms ensures to have robust indications on the proper number of clusters in a dataset.

One final consideration is in regard to the performance of the various indices, across datasets. As it is to be expected, their performance depends quite a bit on the dataset. Our experiments show that all indices perform well on the Leukemia dataset, an important fact for future experimentations. Indeed, novel classifiers/indices that do poorly on that dataset can be considered as poor performers.

13.9   Conclusions and Open Problems

We have provided a survey of techniques with grounds in the statistics literature and that are of use in estimating the number of clusters in a dataset. We have also demonstrated some of their uses for the analysis of microarray data. The evaluation resulting from the extensive experiments give additional insights, with respect to the state of the art, about the use of the mentioned indices. In particular, the comparative analysis shows that Consensus is the method of choice, due to its stability across algorithms. However, such a performance comes at a very steep computational price. Another conclusion that can be drawn from our experiments is that all indices that have been discussed show great limitations on large datasets. In view of this finding, data reduction techniques such as filtering and dimensionality reduction become even more important for class discovery in microarray data. Another promising avenue of research is to design fast approximate algorithms for the computation of the slowest indices, in particular Consensus. Finally, we remark that Gap, Clest and Consensus have various parameters that a user needs to specify. Those choices may affect both time performance and precision. Yet, no experimental study, addressing the issue of parameter selection for those methods, seems to be available in the literature.

Acknowledgments

Raffaele Giancarlo was partially supported by Italian MIUR grants PRIN "Metodi Combinatori ed Algoritmici per la Scoperta di Patterns in Biosequenze" and FIRB "Bioinformatica per la Genomica e La Proteomica" and Italy-Israel FIRB Project "Pattern Discovery Algorithms in Discrete Structures, with Applications to Bioinformatics." Davide Scaturro was supported in full and Filippo Utro was partially supported by Italy-Israel FIRB Project "Pattern Discovery Algorithms in Discrete Structures, with Applications to Bioinformatics."

References

[1] Supplementary material web site. http://www.math.unipa.it/~raffaele/suppMaterial/chapterDM/.

[2] Validation Work Bench: Valworkbench web page. http://www.math. unipa.it/ ∼raffaele/valworkbench/.

[3] J. N. Breckenridge. Replicating cluster analysis: Method, consistency, and validity. *Multivariate Behavioral Research*, 24(2):147–161, 1989.

[4] S. Datta and S. Datta. Comparisons and validation of statistical clustering techniques for microarray gene expression data. *Bioinformatics*, 19:459–466, 2003.

[5] V. Di Gesú, R. Giancarlo, G. Lo Bosco, A. Raimondi, and D. Scaturro. Genclust: A genetic algorithm for clustering gene expression data. *BMC Bioinformatics*, 6:289, 2005.

[6] S. Dudoit and J. Fridlyand. A prediction-based resampling method for estimating the number of clusters in a dataset. *Genome Biology*, 3, 2002.

[7] S. Dudoit and J. Fridlyand. Bagging to improve the accuracy of a clustering solution. *Bioinformatics*, 19:1090–1099, 2003.

[8] E. B. Fowlkes and C.L. Mallows. A method for comparing two hierarchical clusterings. *Journal of the American Statistical Association*, 78:553–584, 1983.

[9] A. D. Gordon. Null models in cluster validation. In W. Gaul and D. Pfeifer (Eds.), *From Data to Knowledge: Theoretical and Practical Aspects of Classification*. Springer Verlag, Berlin, 32–44, 1996.

[10] A. D. Gordon. Clustering algorithms and cluster validation. In P. Dirschedl and R. Ostermann, editors. *Computational Statistics*. Physica-Verlag, Heidelberg, Germany, 497–512, 1994.

[11] J. Handl, J. Knowles, and D. B. Kell. Computational cluster validation in post-genomic data analysis. *Bioinformatics*, 21(15):3201–3212, 2005.

[12] E. Hartuv, A. Schmitt, J. Lange, S. Meier-Ewert, H. Lehrach, and R. Shamir. An algorithm for clustering of cDNAs for gene expression analysis using short oligonucleotide fingerprints. *Genomics*, 66:249–256, 2000.

[13] T. Hastie, R. Tibshirani, and J. Friedman. *The Elements of Statistical Learning*. Springer, Berlin, 2003.

[14] M. J. L. De Hoon, S. Imoto, and S. Miyano. *The C Clustering Library for cDNA Microarray Data*. Laboratory of DNA Information Analysis Human Genome Center, Institute of Medical Science, University of Tokyo, 2007.

[15] L. Hubert and P. Arabie. Comparing partitions. *Journal of Classification*, 2:193–218, 1985.

[16] A. K. Jain and R. C. Dubes. *Algorithms for Clustering Data.* Prentice-Hall, Englewood Cliffs, NJ, 1988.

[17] L. Kaufman and P. J. Rousseeuw. *Finding Groups in Data: An Introduction to Cluster Analysis.* Wiley, New York, NY, 1990.

[18] M. K. Kerr and G. A. Churchill. Bootstrapping cluster analysis: assessing the reliability of conclusions from microarray experiments. *Proceedings of the National Academy of Sciences USA*, 98:8961–8965, 2001.

[19] W. Krzanowski and Y. Lai. A criterion for determining the number of groups in a dataset using sum of squares clustering. *Biometrics*, 44:23–34, 1985.

[20] M-Y. Leung, G. M. Marsch, and T. P. Speed. Over and underrepresentation of short DNA words in Herphesvirus genomes. *Journal of Computational Biology*, 3:345–360, 1996.

[21] F. H. C. Marriot. Practical problems in a method of cluster analysis. *Biometrics*, 27:501–514, 1971.

[22] G. W. Milligan and M. C. Cooper. An examination of procedures for determining the number of clusters in a data set. *Psychometrika*, 50:159–179, 1985.

[23] G. W. Milligan and M. C. Cooper. A study of the comparability of external criteria for hierarchical cluster analysis. *Multivariate Behavioral Research*, 21:441–458, 1986.

[24] S. Monti, P. Tamayo, J. Mesirov, and T. Golub. Consensus clustering: A resampling-based method for class discovery and visualization of gene expression microarray data. *Machine Learning*, 52:91–118, 2003.

[25] I. Priness, O. Maimon, and I. Ben-Gal. Evaluation of gene-expression clustering via mutual information distance measure. *BMC Bioinformatics*, 8:111, 2007.

[26] W. M. Rand. Objective criteria for the evaluation of clustering methods. *Journal of the American Statistical Association*, 66:846–850, 1971.

[27] C. Van Rijsbergen. *Information Retrieval*, second edition. Butterworths, London, UK, 1979.

[28] R. Shamir and R. Sharan. Algorithmic approaches to clustering gene expression data. In T. Jiang, T. Smith, Y. Xu, and M. Q. Zhang, editors. *Current Topics in Computational Biology.* MIT Press, Cambridge, MA, 120–161, 2003.

[29] P. T. Spellman, G. Sherlock, M. Q. Zhang, V. R. Iyer, K. Anders, M. B. Eisen, P. O. Brown, D. Botstein, and B. Futcher. Comprehensive identification of cell cycle regulated genes of the yeast Saccharomyces cerevisiae by microarray hybridization. *Molecular Biology of the Cell*, 9:3273–3297, 1998.

[30] R. Tibshirani, G. Walther, and T. Hastie. Estimating the number of clusters in a dataset via the gap statistics. *Journal Royal Statistical Society B*, 2:411–423, 2001.

[31] X. Wen, S. Fuhrman, G. S. Michaels, G. S. Carr, D. B. Smith, J. L. Barker, and R. Somogyi. Large scale temporal gene expression mapping of central nervous system development. *Proceedings of the National Academy of Science USA*, 95:334–339, 1998.

[32] M. Yan and K. Ye. Determining the number of clusters with the weighted gap statistics. *Biometrics*, 63:1031–1037, 2007.

[33] K. Y. Yeung. Cluster analysis of gene expression data. PhD Thesis, University of Washington, WA, 2001.

[34] K. Y. Yeung, D. R. Haynor, and W. L. Ruzzo. Validating clustering for gene expression data. *Bioinformatics*, 17:309–318, 2001.

Chapter 14

Computational Approaches to Peptide Retention Time Prediction for Proteomics

Xiang Zhang

University of Louisville

Cheolhwan Oh and Catherine P. Riley

Purdue University

Hyeyoung Cho

Purdue University and KAIST

Charles Buck

Purdue University

14.1 Introduction

Proteomics is a powerful, still-emerging research technology for qualitative and quantitative comparison of protein composition under different conditions to interrogate biological processes. Although genomics gave rise to the concept of a proteome, the proteome differs conceptually from the genome in several ways beyond the fact that proteins are translated from genes. The proteome has a unique complexity compared with the genome. About 25,000 protein coding-genes have been identified in human but more than 67,000 proteins are derived from these genes because of alternative splicing and protein degradation (IPI human protein database: http://www.ebi.ac.uk/IPI/IPIhuman .html). In addition, more than 100 types of posttranslational modifications

(PTMs) can present in the proteome and these are not encoded in the genome. In contrast to the genes, proteins in human proteome exist in a very large concentration range. It is estimated that protein concentration differences range to 10^{7-8}-fold in human cells, and to at least 10^{12}-fold in human plasma. Highly abundant proteins dramatically complicate proteome analyses, the 12 most abundant proteins constitute approximately 95% of the total protein mass of human plasma or serum (Anderson and Anderson, 2002). This complicates analysis of the remaining, less abundant proteins. Finally, the human proteome is highly dynamic and subject to large biological variations. A proteome differs from cell to cell and is in a state of constant flux from biochemical interactions with the genome and the environment. One organism has radically different protein expression patterns in different parts of its body, at different stages of its life cycle and in different environmental conditions.

This complexity makes analyses and understanding of a proteome tremendously challenging. However, since proteins play such a central role in biology, proteomics will continue to inform our understanding of biological systems; providing for example, discovery of biomarkers indicative of a particular disease. Significant experimental efforts have developed robust and sensitive analysis platforms for proteomics. In recent years, several mass spectrometry (MS) based technologies have emerged that enable identification of large numbers of proteins in a biological sample, mapping protein interactions in a cellular context, and analysis of biological activities. Typically, proteins are digested into short peptides by breaking at specific sites along the backbone of the protein with a protease such as trypsin. The peptide mixture is then chromatographically fractionated prior to MS analysis to circumvent the dynamic range limitation introduced by ion suppression in electrospray ionization and at the MS detector. To maximize MS analyses of component peptides, a common approach is to fractionate peptide mixtures with multiple chromatography columns coupled in tandem (Qiu et al., 2007). Fractionated individual peptides are isolated and further fragmented by collision-induced dissociation (CID) to produce information about the mass-to-charge ratios (m/z) of resulting fragment ions (MS/MS of parent peptide). Since peptides usually break at a peptide-bond with CID, the resulting tandem mass spectrum (MS/MS) contains information about the amino acid composition of peptide, and therefore, the MS/MS spectra can be used for peptide identification.

Although numerous chromatographic approaches have been introduced to fractionate and separate peptide mixtures prior to MS analysis in proteomics, reversed phase chromatography (RPC) is nearly always employed as the last separation step prior to MS. Peptide separation by RPC is similar to the extraction of different compounds from water into an organic solvent, where more hydrophobic (nonpolar) compounds preferentially extract into the nonpolar phase. The RPC column (typically, a silica support modified with a C18 bonded phase termed as stationary phase) is less polar than the water-organic mobile phase. Sample molecules partition between the polar mobile phase and nonpolar C18 stationary phase, and more hydrophobic (nonpolar) compounds are retained more strongly on the RPC column. For a given mobile-phase composition, the result is a differential retention of samples according to their hydrophobicity, with a resulting chromatogram. Hydrophobic (polar) compounds are less strongly held and elute from the column first; more hydrophobic (nonpolar) compounds elute later. It is well known that the PRC retention of a compound is determined by its polarity and experimental conditions such as mobile phase, column, and temperature. However, the detailed nature of reversed-phase retention is not understood completely even though it appears that retention can be approximated by a partition process.

Regardless of the fundamental basis of retention, the consequences of changes in experimental conditions have been well studied and can lead to a systematic approach to RPC method development. Therefore, RPC coupled with MS is now a routine analytical platform in proteomics. There is little variation in the stationary phase and elution conditions of RPC compared to other chromatographic separation methods that may be employed prior to RPC–MS. The consistency of RPC makes possible interlaboratory comparison of RPC–MS experiments in larger scale studies and stimulates the development of mathematical approaches to predict peptide retention time in this matrix. A major application of the predicted peptide retention time in modern omics research is to assist in peptide/protein identification. In the following sections, we will review current developments in peptide retention time prediction and application of this approach for protein identification.

14.2 Methods of Peptide Retention Time Prediction

14.2.1 Retention coefficient approach

Peptide retention prediction in RPC was a subject of research even before the term *proteome* was coined. Early efforts to predict peptide retention time are retention coefficient–based (Meek et al., 1980; Guo et al., 1986). In this approach, empirically determined amino acid residue retention coefficients are subjected to summation. The assumption is that the chromatographic behavior of peptides is mainly or solely dependent on amino acid composition, i.e., the contributions of each amino acid residue to peptide retention time are additive.

Chabanet and Yvon (1992) proposed that peptide retention time of charged, polar, and non-polar amino acid residues is differentially affected by peptide length. It was assumed that the contribution of each residue to peptide retention time (A_j) is a decreasing function of peptide length (l_i). The decreasing function was selected with a slope equal to zero when $l_i = 0$, and an inflection point and a lower asymptote described as:

$$t_i = \sum n_{i,j} A_j (l_i) + b_0 + \varepsilon_i$$

The term $n_{i,j}$ is the number of amino acid residues j in the peptide I, b_0 is a retention time coefficient for α-NH$_2$ and α-COOH terminal functions, ε_i is independent error, and $A_j(l)$ is defined as follows:

For nonpolar residues Gly, Ala, Val, Met, Ile, Leu, Phe, and Trp

$$A_j(l) = (a_j - I_j)\, e^{-b_1 l^2} + I_j,\, I_j = \frac{a_j}{k_1}$$

For polar residues Asp, Asn, Thr, Ser, Glu, Gln, Pro, Tyr, and His

$$A_j(l) = (a_j - I_j)\, e^{-b_2 l^2} + I_j,\, I_j = \frac{a_j}{k_2}$$

And for the charged residues Lys and Arg

$$A_j(l) = (a_j - I_j)\, e^{-b_3 l^2} + I_j,\, I_j = \frac{a_j}{k_3}$$

This model was developed using 104 peptides with one to 64 amino acids. The model was tested with the retention times of 47 peptides varying in length from two to 58 residues. Relatively good agreement was found with a linear correlation coefficient (R^2) of 0.97, and the maximum error was 7.9% ACN (the concentration of ACN can be converted into RPC retention time based on the slope of gradient).

Recently, Krokhin et al. (2004) have employed separate retention coefficients for amino acids at the N-terminus of the peptide and peptide length for modeling peptide retention times. Peptide hydrophobicity was calculated as:

$$H = K_L * \left(\sum R_c + 0.42 R_{cNt}^1 + 0.22 R_{cNt}^2 + 0.05 R_{cNt}^3 \right) \quad \text{if} \quad H < 38$$

and

$$H = K_L * \left(\sum R_c + 0.42 R_{cNt}^1 + 0.22 R_{cNt}^2 + 0.05 R_{cNt}^3 \right)$$
$$- 0.3 \left(K_L * \left(\sum R_c + 0.42 R_{cNt}^1 + 0.22 R_{cNt}^2 + 0.05 R_{cNt}^3 \right) - 38 \right) \quad \text{if}$$
$$H \geq 38$$

where R_c is the retention coefficient of each amino acid residue in the peptide, R_{cNt}^x is the retention coefficient compensating for hydrophobicity at the N-terminus, and K_L is a correction coefficient for peptide length. Using this model, the correlation between predicted peptide hydrophobicity and experimentally measured peptide retention time is $R^2 = 0.939$ for a set of 346 peptides chosen for optimization from tryptic digestion of a 17 protein mixture. Applying this model to different chromatographic conditions and for analysis of real samples provides relatively robust R^2 values between 0.90 and 0.95.

The same group further improved the retention coefficient model by including overall hydrophobicity, isolelectric point, nearest neighbor effect of charged side chains (Lys, Arg, His), and the propensity to form helical structure (Krokhin, 2006a). With the improved model, a correlation with $R^2 \sim 0.98$ was obtained for a 2,000 peptide optimization set. With tests on a number of real samples, improved correlations in the range of 0.95–0.97 R^2 value have been observed.

14.2.2 Machine learning approach

Retention coefficient models work fairly well for small peptides (up to about 20 amino acid residues), but are inadequate for proteomic applications where tryptic peptides can exceed 40 amino acid residues in length. Furthermore, with the retention coefficient approach, isomeric peptides are predicted to elute at the same time, which is not the case. To overcome the limitations of the retention coefficient approach, several machine learning models have been proposed for prediction of peptide chromatography retention times.

Petritis et al. (2003) used artificial neural networks (ANNs) to predict the retention time of peptides in RPC. A genetic algorithm (GA) was employed to normalized peptide retention times into a specific range (between 0 and 1) to improve the quality of data. Each peptide was subsequently represented as a vector with 20 elements corresponding to the 20 amino acid residues and an artificial neural network (ANN) constructed to predict peptide retention time. This work was used to improve the confidence of peptide identification results in Strittmatter et al. (2004). The model was used to generate the theoretical peptide elution time using the peptide sequence information. The retention time parameter was then incorporated into a discriminant function for the analysis of tandem MS data analyzed by the commonly used protein database search algorithm SEQUEST. The structure of this ANN was later modified to enable utilization not only of amino acid composition information but also the sequence of amino acid residues in a peptide to improve performance (Petritis et al., 2006). Multiple sequence/structural peptide descriptors were tested such as peptide length, sequence, predicted secondary peptide structure (α-helix, β-sheet, or coil), and hydrophobic moment. They found that the predicted secondary peptide structural states did not improve the performance of the ANNs. The final artificial neural networks model included 1052 input nodes, 24 hidden nodes, and one output node. Among the 1052 input nodes, two were assigned to the normalized values of peptide length and hydrophobic moment. The other 1050 input nodes were assigned to represent maxima of 50 amino acids in the peptide. More specifically, each peptide was encoded into a 21-dimensional binary vector: one with a value of one and the others with values of zero, depending on what amino acid is encoded. It should be noted that this model was trained to handle the peptides with alkylated cysteines as well as 20 proteinogenic amino acids. When the length of the peptide is less than

50, the 21-dimensional binary vectors are placed in the two edges of the 1050 input nodes and the middle input nodes occupied with zeros. A similar work can be found in Shinoda et al. (2006).

Prediction of peptide separation in liquid chromatography step gradient was recently explored in our group (Oh et al., 2007). We modeled peptide separation in strong anion exchange (SAX) chromatography wherein each peptide is assigned to either the "flow-through" or "elution" fraction with SAX separation. An ANN based pattern classification technique was implemented for the prediction. Six peptide features were evaluated for their contributions as significant features; sequence index, charge, molecular weight, length of amino acid residue sequence, N-pK$_a$ value, C-pK$_a$ value. Interestingly, the sequence index, which was devised to represent the order of amino acid residues in each peptide, turned out to be the most significant feature for modeling peptide retention in SAX.

The detailed structure of the ANN is shown in Figure 14.1. The ANN is composed of the input, hidden, and output layers. Each layer contains neurons where nonlinear transformation is performed. Each neuron in each layer is connected to every neuron in the adjacent layer(s). The training or testing vectors are presented to the input layer and processed by the hidden and output layers. The output of the network is

$$z_k = g \left(\sum_{j=1}^{m} w_{kj}^o y_j - t_k^o \right),$$

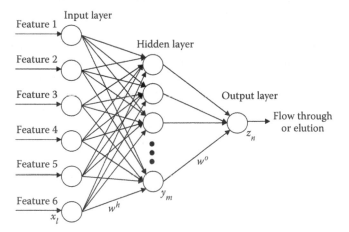

FIGURE 14.1: Artificial neural network used as a pattern classifier.(x_l: input, y_m: hidden layer, z_n: output, w^h: weight matrix between input layer and hidden layer, w^o: weight matrix between hidden layer and output layer).

where

$$y_j = g \left(\sum_{i=1}^{l} w_{ji}^h x_i - t_j^h \right),$$

and g is a nonlinear activation function of the form,

$$g(v) = \frac{1 - e^{-v}}{1 + e^{-v}}.$$

The values of the network parameters, w_{ji}^h, w_{kj}^o, t_j^h, and t_k^o are determined by the training procedure. A GA was employed for a training method, rather than the conventional back-propagation algorithm. The back-propagation method produces local optimizers because it uses a gradient of values in the optimization process; whereas the GA does not use the gradient to find the optimizer and therefore provides better opportunity to obtain global optimizers for the neural network parameters.

Another machine learning approach that is gaining popularity is support vector machine (SVM) (Vapnik, 1998, 2000). SVM was devised to handle pattern classification problems, but its function has been extended to solve regression problems. Song et al. (2002) developed a quantitative structure-retention relationship (QSRR) model to predict protein retention time in an anion exchange chromatography system using an algorithm based on support vector regression (SVR). In this model, SVR approaches select only the features relevant to anion-exchange protein retention. Selected features are used to produce an ensemble of nonlinear SVM regression models that are combined using bootstrap aggregation techniques. QSRR modeling with SVR has also been employed for peptide chromatography modeling by Ladiwala et al. (2005).

Machine learning approaches employing different methodologies for modeling peptide chromatography retention have also been reported. Baczek et al. (2005) used a QSRR regression model that sums retention times of the amino acids composing the peptide, van der Waals volumes of the peptide, and the theoretically calculated n-octanol-water partition coefficient to predict peptide retention times. A variable selection method called uninformative variable elimination by partial least squares (UVE-PLS) has also been implemented with QSRR to improve performance of this model (Put et al., 2006).

14.3   Application of Predicted Peptide Retention Time for Protein Identification

Database searching and de novo sequencing are the two methods currently used for protein identification from peptide proteomic information.

Typically ~20% of tandem spectra from LC-MS/MS analyses can be confidently identified; the remaining spectra are generally ignored or discarded. Several factors contribute to this inefficiency: imperfect scoring schema, criteria that constrain the protein database search, incorrect mass and charge assignment, and low-quality MS/MS spectra. Many efforts are underway to overcome this inefficiency and these have met with varying degrees of success. One approach is to incorporate predictions of peptide retention time to assist in peptide identification.

Predicted peptide retention time has been used to improve protein identification by peptide mass fingerprinting (Palmblad et al., 2002). In this case, peptide retention time t_{cal} is predicted based only upon the peptide amino acid composition using the following method:

$$t_{cal} = \sum_{i=1}^{20} n_i c_i + t_0$$

where c_i are the retention coefficients for the 20 amino acids, n_i are the number of each amino acid, and t_0 compensates for void volumes and a delay between sample injection and acquisition of mass spectra. These parameters were fitted by the least-squares method to experimental data from ~70 bovine serum albumin (BSA) peptides or ~100 human serum albumin (HSA) and transferrin peptides putatively identified with accurate mass measurement and high relative intensities in the mass spectra. It should be noted that both BSA and HSA are present in many biological samples used for proteomic analysis. The use of an "internal standard," e.g., BSA and HSA, can be expected to make predictions less dependent on chromatographic conditions such as pressure, flow rate, mobile-phase composition, or pH. However, the accuracy of this predictor is poor, about 8–10% even though the "internal standards" were used to train the prediction model. The poor prediction accuracy of this model may result from ignorance of peptide sequence information and also from the relatively simple mathematical model employed.

To apply predicted peptide retention for protein identification, Palmblad et al. selected candidate peptides that are those tryptic peptides within 5 ppm of mass variation between the experimentally measured mass and the theoretical mass of a in silico tryptic peptide digested from a protein sequence in a database containing 150 human body fluid proteins and BSA. The candidates are ranked by combining mass measurement error and the deviation from predicted retention by calculating a total χ^2 value, where

$$\chi^2 = \frac{(m_{cal} - m_{exp})^2}{\sigma_m^2} + \frac{(t_{cal} - t_{exp})^2}{\sigma_t^2}$$

and m and t are the mass and retention time, respectively. Standard deviations (σ) are calculated from experimental data. The likelihood of all matching peptide masses and retentions are summed to a total likelihood score, which is then used to discriminate between true and random matches. Within a 5 ppm mass variation, it has been shown that 11 proteins can be identified from one cerebrospinal fluid (CSF) run with >95% significance when information from predicted retention is applied, whereas only six proteins can be identified from the same experimental data with the same confidence level when the predicted peptide retention information is not used.

Gilar and coworkers developed a regression model to predict peptide retention in RPC based on peptide amino acid composition and length (Gilar et al., 2007). The model was utilized to identify and exclude false positive (FP) peptide identifications obtained from database searching. The predicted peptide retention time, t_i, for the i-th peptide takes the form:

$$t_i = \left(1 - c\ln\left(\sum_{j=1}^{20} n_{i,j}\right)\right)\left(b_0 + \sum_{j=1}^{20} n_{i,j}b_j\right)$$

where c is the parameter optimized iteratively during the regression to maximize the squared correlation coefficient, $n_{i,j}$ is the number of j-th amino acid in the i-th peptide, b_j is the retention coefficient of j-th amino acid, and b_0 denotes an intercept in the model.

This model was tested using human serum samples analyzed on LC/MS in triplicate. The MASCOT database search results from 87 fractions were combined and analyzed. It has been observed that 8% of peptides with MASCOT score larger than 40 are rejected using ±20% of predicted retention time window for peptide acceptance (Palmblad et al., 2002). The majority of the rejected peptides are confined to two chromatogram regions: peptides with short retention times, and long peptides exhibiting the greatest retention. The first case is explained as failure of the model to predict weakly retained peptides eluting from the column under isocratic conditions, before the actual gradient starts. The second case is likely to be related to general difficulties in predicting behavior of long peptides. Retention of these peptides is presumably more affected by their secondary structure.

To study the contribution of predicted peptide retention time to peptide identification, FP was defined in two ways: (i) FP peptides elute outside of ±20% of predicted retention time window, and (ii) MS/MS data were searched against both forward and decoy protein databases. The identified random peptides from the decoy protein database are considered FP identifications. The results indicate that the majority of random peptides (FP) elute outside of the ±20% retention time acceptance window, at least for data with low MASCOT score. For example, the FP rate reduces from 1.1% to 0.7% at MASCOT 30 and from 4.2 to 2.1% at MASCOT score 25. Thus incorporation of predicted peptide retention time can decrease the FP rate for peptide identification (Palmblad et al., 2002).

Predicted peptide retention time has also been employed for protein identification from off-line RPC-MALDI MS/MS (Krokhin et al., 2006b). This protein identification by MS/MS and peak exclusion based on mass-retention time (MART) method executes as follows: the top five abundance peaks were chosen and their MS/MS spectra submitted for database search. The proteins identified were digested in silico, allowing one missed cleavage. Peptide hydrophobicities were calculated using a sequence-specific retention calculator (Spicer et al., 2007), and further converted to predicted retention times using correlation $t = 8.8609 + (0.9752 \times hydrophobicity)$, derived from a library set of peptides and tryptic peptides recovered from a K562 cell affinity-purified protein mixture. Since these proteins are already identified, the peptides from these proteins are excluded from further consideration on the basis of two criteria: ± 15 ppm mass tolerance and ± 2 min deviation from the predicted retention time. The next five most abundant peaks from the treated peak list are selected and their MS/MS spectra combined with the first five to perform the second database search. If new proteins are identified, their potential tryptic peptides were excluded from the combined peak list. This procedure is repeated until all MS/MS spectra are either analyzed by database search or excluded from the peak list.

An RPC-MALDI data set containing MS spectra of 70 1-min fractions, a combined peak list, and 661 MS/MS spectra of the most abundant parent ions in the 900–4000 Da mass range was used to test the MART protocol. The procedure described above was repeated 57 times with 285 of the 661 spectra evaluated. The resulting list of identified proteins included the same 70 proteins found in the original MS/MS analysis without the exclusion procedure. The number of MS/MS spectra needed to identify the same set of proteins, therefore, was decreased by 57%. This should enable MS/MS analysis of lower abundance peptides and an increase in the number of identified proteins. This approach is particularly effective for the identification of large proteins with a limited number of post-translational modifications and nonspecific cleavage sites. It is also suitable for the RPC–MALDI combination because two steps of the analysis, i.e., RPC and MALDI–MS, are coupled offline. In addition, there is potential for application of this method to online ESI–MS/MS, which requires greater speed and performance of MS/MS identification algorithms.

Although numerous mathematical models have been developed to predict peptide chromatography retention time, this technique is still not widely employed in proteomics. Adoption of any of these models is effected by different chromatography systems because peptide retention time can be affected by so many variables. Further, none of existing models are able to handle the multitude of post translational modifications that may occur on a peptide. This is a major shortcoming that hinders the application of predicted peptide retention in proteomics. Finally, peptide retention time prediction is not incorporated into existing database search engines. Many proteomics laboratories may not have the bioinformatics capability to incorporate predicted peptide retention time methods into their protein identification platform. However, as proteomics continues to advance, more proteomics scientists will realized the value of applying predicted peptide retention times for protein identification, especially in cases where mass value alone can not provide confident identification.

14.4 Conclusions

LC/MS is a widely adopted analytical platform in proteomics and RPC is always employed as the last separation step prior to MS. Multiple mathematical models have been developed to predict peptide retention in RPC and this information can assist with protein identification by reducing the FP identification rate and by increasing identification efficiency. The resolving power and accuracy of chromatographic separations are much lower than in MS. However, peptide information from chromatography is complementary to, and can inform mass spec data. This information can be incorporated with negligible computational cost and at no extra experimental cost.

References

Anderson, N. L. and Anderson, N. G. 2002. The human plasma proteome: history, character, and diagnostics prospects. *Mol. Cell. Proteomics*, 1.11, 845–867.

Baczek, T., Wiczling, P., Marszatt, M., Heyden, Y. V., and Kaliszan, R. 2005. Prediction of peptide at different HPLC conditions from multiple linear regression models. *J. Proteome Res.*, 4, 555–563.

Chabanet, C., and Yvon, M. 1992. Prediction of peptide retention time in reversed-hpase high-performance liquid chromatography. *J. Chormatogr. A* 599, 211–225.

Gilar, M., Jaworski, A., Olivova, P., and Gebler, J. C. 2007. Peptide retention time prediction applied to proteomic data analysis. *Rapid Commun. Mass Spectrom.*, 21, 2813–2821.

Guo, D., Mant, C. T., Taneja, A. K., Parker, J. M. R., Hodges, R. S. 1986. Prediction of peptide retention times in reversed-phase high-performance liquid chromatography I. Determination of retention coefficients of amino acid residues of model synthetic peptides. *J. Chromatogr.*, A359, 499–518.

Krokhin, O. V., Craig, R., Spicer, V., Ens, W., Standing, K. G., Beavis, R. C., and Wilkins, J. A. 2004. An improved model for prediction of retention times of tryptic peptides in ion pair reversed-phse HPLC. *Mol. Cell. Proteomics*, 3, 908–919.

Krokhin O. V. 2006a. Sequence-sepcific retention calculator. Agorithm for peptide retention prediction in ion-pair RP-HPLC: application to 300- and 100-Å pore size C18 sorbents. *Anal. Chem.* 78, 7785–7795.

Krokhin, O. V., Ying, S., Cortens, J. P., Ghosh, D., Spicer, V., Ens, W., Standing, K. G., Beavis, R. C., and Wilkins, J. A. 2006b. Use of peptide retention time prediction for protein identification by off-line reversed-phase HPLC-MALDI MS/MS. *Anal. Chem.*, 78, 6265–6269.

Ladiwala, A., Xia, F., Luo, Q., Breneman, C. M., and Cramer, S. M. 2006. Investigation of protein retention and selectivity in HIC systems using quantitative structure retention relationship models. *Biotechnol. Bioeng.*, 93, 836–850.

Meek, J. L. 1980. Prediction retention time in high-pressure liquid chromatography on the basis of amino acid composition. *Proc. Natl. Acad. Sci. USA.* 77, 1632–1636.

Spicer, V., Yamchuk, A., Cortens, J., Sousa, S., Ens, W., Standing, K. G., Wilkins, J. A., and Krokhin, O. V. 2007. Sequence-specific retention calculator. A family of peptide retention time prediction algorithms in reversed-phase HPLC: applicability to various chromatographic conditions and columns. *Anal. Chem.* 79, 8762–8768.

Oh, C., Żak, S. H., Mirzaei, H., Buck, C., Regnier, F. E., and Zhang, X. 2007. Neural network prediction of peptide separation in strong anion exchange chromatography. *Bioinformatics*, 23, 114–118.

Palmblad, M., Ramstrom, M., Markides, K. E., Hakansson, P., and Bergquist, J. 2002. Prediction of chromatographic retention and protein identification in liquid chromatography/mass spectrometry. *Anal. Chem.*, 74, 5826–5830.

Petritis, K., Kangas, L. J., Ferguson, P. L., Anderson, G. A., Paša-Toli, L., Lipton, M. S., Auberry, K. J., Strittmatter, E. F., Shen, Y., Zhao, R., and Smith, R. D. 2003. Use of artificial neural networks for the accurate prediction of peptide liquid chromatography elution times in proteome analysis. *Anal. Chem.*, 75, 1039–1048.

Petritis, K., Kangas, L. J., Yan, B., Monroe, M. E., Strittmatter, E. F., Qian, W.-J., Adkins, J. N., Moore, R. J., Xu, Y., Lipton, M. S., Camp II, D. G., and Smith, R. D. 2006. Improved peptide elution time prediction for reversed-phase liquid chromatography-MS by incorporating peptide sequence information. *Anal. Chem.* 78, 5026–5039.

Put, R., Daszykowski, M., Baczek, T., and Heyden, Y. V. 2006. Retention prediction of peptides based on uninformative variable elimination by partial least squares. *J. Proteome Res.*, 5, 1618–1625.

Qiu, R., Zhang, X., Regnier, F. E. 2007. A method for the identification of glycoproteins from human serum by a combination of lectin affinity chromatography along with anion exchange and Cu-IMAC selection of tryptic peptides. *J. Chromatogr. B.*, 845, 143–150.

Shinoda, K., Sugimoto, M., Yachie, N., Sugiyama, N., Masuda, T., Robert, M., Soga, T., and Tomita, M. 2006. Prediction of liquid chromatographic retention times of peptides generated by protease digestion of the *Escherichia coli* proteome using artificial neural networks. *J. Proteome Res.*, 5, 3312–3317.

Song, M., Breneman, C. M., Bi, J., Sukumar, N., Bennett, K. P., Cramer, S., and Tugcu, N. 2002. Prediction of protein retention times in anion-exchange chromatography systems using support vector regression. *J. Chem. Inf. Comput. Sci.*, 42, 1347–1357.

Strittmatter, E. F., Kangas, L. J., Petritis, K., Mottas, H. M., Anderson, G. A., Shen, Y., Jacobs, J. M., Camp II, D. G., and Smith, R. D. 2004. Application of peptide LC retention time information in a discriminant function for peptide identification by tandem mass spectrometry. *J. Proteome Res.*, 3, 760–769.

Vapnik, V. N. 1998. *Statistical learning theory.* Wiley, New York.

Vapnik, V. N. 2000. *The nature of statistical learning theory,* 2nd ed. Springer-Verlag, New York.

Part III

Functional and Molecular Interaction Networks

Chapter 15

Inferring Protein Functional Linkage Based on Sequence Information and Beyond

Li Liao

University of Delaware

15.1 Overview

It has become increasingly evident that genes and proteins are not isolated entities in the cell; rather they are part of the pathways and networks of biochemical reactions and processes occurring inside the cell, and their functions depend upon the cellular context in addition to their individual properties. Correspondingly, the focus of annotation efforts in bioinformatics has recently shifted from assigning functions to individual proteins to identifying proteins that are functionally linked.

> Definition: two proteins are functionally linked if both their participation is needed for fulfilling a cellular process. Functional linkage, loosely defined as such, exists in a broad range of activities, from proteins interacting with each other (protein–protein interaction, PPI) in a signaling transduction pathway, to proteins participating as enzymes in the same metabolic pathway, to proteins that are coregulated or one protein regulating another.

From a theoretical point view, this chapter touches on some fundamental issues related to learning from data, and is focused on statistical and machine learning methods that are effective, expressive and interpretable, via incorporating domain specific knowledge of biological systems. There is an intrinsic tension in machine learning approaches: in the context of PPI prediction, more sensitive prediction tends to require extensive information, e.g., phylogenetic information, and more specific prediction tends to require more detailed information, e.g., the structural information. It is, therefore, important to be able to extract information hidden in the data and to transfer knowledge across different sources and at different levels while maintaining a balance between sensitivity and specificity. This reflects in the strategy and theme that the title of this chapter suggests: go beyond the primary sequences to garner more information, and then come back to build a method that may just require the primary sequences to make a prediction. To this end, the limits for the existing learning paradigms are tested, and new techniques and models are explored such as applying transductive learning to least squares support vector machines and formulating a new definition of Fischer scores in sufficient statistics of interaction profile hidden Markov models (ipHMMs) to facilitate feature selection.

Specifically, we focus on inference and prediction of protein functional linkage (PFL) from the following three perspectives. Evolutionary perspective: methods and models that are developed to extract and incorporate evolutionary information, such as distance matrices and phylogenetic profiles. Relation between the information content and the phylogenetic tree topology is studied and exploited for the use of regularization. Mechanisms are devised to detect nonstandard evolutionary events such as lateral gene transfer and its implication is accounted for in PFL prediction. Structural perspective: structural information such as binding residues is incorporated into models that can be used for predicting PFL, particularly PPI, for proteins that do not have structural information. Feature selection is explored and exploited for identifying structural and functional signatures and for suggesting effective knockout mutations. Network perspective: methods and models are developed for prediction of interactions involved with multiple proteins, and graph-theoretic approaches that incorporate network contexts.

Most current computational methods in bioinformatics and computational biology fall into one of the following two types.

Ab initio: use the first principles of molecular dynamics to explain and predict bio-molecular processes. A classic example is the *ab initio* approaches developed for protein folding problems.

Phenomenological: examine "phenotypic" features exhibited by the functionally linked domains (or partners) and build models to differentiate them from the ones that are not functionally linked.

Although ideally the problems of PFL should be addressed ultimately by *ab initio* methods, there is often lacking of sufficient detailed information needed by rigorous molecular dynamics solutions for connecting the dots, and even when the information is available the computation is still beyond the capacity of the current computing hardware. By and large, the methods developed for PFL, like many bioinformatics methods, belong to the phenomenological category, in the sense that we do not go down to a level where we need to calculate the electrostatic potential for putting two molecules together. Rather, we stay at a relatively high and symbolic level, i.e., abstracting proteins as a sequence of 20 alphabets corresponding to 20 amino acids, while trying to leverage available information from different sources, which can be up to the genomic scale, or down to the residue positions in a protein's X-ray crystallography structure. For example, if a group of proteins are known to have the same function, we can align their primary sequences, residue by residue, in order to detect common features. Because of the evolutionary processes such as mutations sustained by these proteins, their sequences are not identical. On the other hand, these structural and functional domains should be relatively stable (otherwise will lose the function) and are expected to show high similarity in the sequence alignment. Probabilistic models can be built and trained on these proteins with the known domain, and then are used for identifying the domain in unknown proteins. This simplified scenario illustrates the typical logic essentially shared by many informatics approaches, which turn out to be surprisingly useful and effective in extracting knowledge out of the vast volume of data generated by various high throughput technologies in genomics and proteomics projects.

15.2 Biological Background and Motivations

Recent advance in high throughput technologies has enabled studying molecular entities in the cell in a large scale and in the context of biological processes. For example, the DNA microarray experiments can measure expression levels for tens of thousands of genes simultaneously; examining the gene expression under different conditions and at different time points can provide useful insights into the molecular mechanisms of essential processes, such as gene regulation. And the method combing 2D gel and mass spectrometer is a promising technology in proteomics for directly measuring protein expression levels. However, no reliable high throughput methods are available yet to directly detect PFL, such as if two proteins interact, yeast two-hybrid (Y2B) system, the current method in this category, is known to be noisy and error-prone. Because of these difficulties and the high cost associated with experimental approaches, it has become pressing to develop effective and efficient computational methods to infer functional linkage from the tons of data accumulated either directly and derived from these high throughput technologies in large scale experiments, including DNA and protein sequences

(SwissProt and Genbank), structural info (PDB, SCOP, and CATH), evolutionary (phylogenetic profiles), expression profiles, pathway and network info (KEGG), protein localization info, and literature.

Predicting PFL is a highly nontrivial and difficult task, and has many aspects in it. To appreciate the complexity of the problem, take PPI as an example. Even if two proteins can physically interact, they may or may not actually interact, depending on whether the two proteins will be localized in the same compartment of the cell and at the same time. The core of the problem, namely, whether two proteins can interact, is a process that ultimately has to be determined by a biophysical and biochemical study: whether the binding of two protein molecules is energetically favorable and functionally complementary. When two proteins interact, the residues (typically on the surface of the molecules) that are involved in the interaction form a region. An example is shown in Figure 15.1, where the interface region is highlighted. Such regions may possess unique characteristics as a structural and/or functional unit to appear in all proteins that have the same structure and/or function. In other words, these regions become a definitive signature for the corresponding structure and/or function, and therefore the term "domain" is dubbed to indicate their relevance. In the study of protein interactions, there are two major tasks: one is identifying the residues or domains implicated in the interaction, and the other is predicting the interaction partners. While we use physical interaction between two proteins as an example, the concept of domains existing in proteins as signatures indicative of certain functions applies to functional linkage as well, that is to say, functional linked proteins may be identified via their related domains, though understandably with less direct and more subtle nexuses.

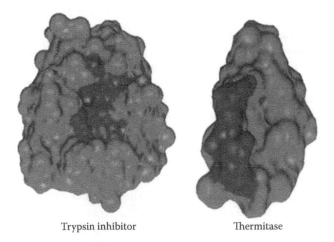

Trypsin inhibitor   Thermitase

FIGURE 15.1: Structural compatibility between interacting proteins. Dark highlighted areas indicate the interacting surfaces.

One subtle connection between functionally linked proteins is (pellergini et al., 1999) via the so-called phylogenetic profiles. Much like why the functional and structural domains need to retain their sequence compositions, functional linkage of proteins also exert selection pressure on evolution, and as a result, these functionally linked partners tend to coevolve: one protein cannot function properly without its partners. An example explaining how protein interactions can exert coevolution pressure is given in Figure 15.2. At a higher level, the evolutionary history of a protein is typically represented as

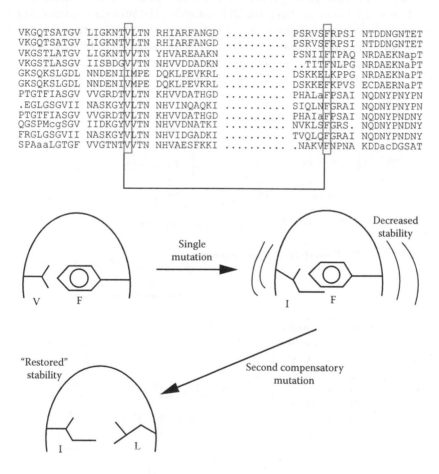

FIGURE 15.2: Coevolution imposed by protein interaction. Interaction involves forming a hydrogen bond between valine and phenylalanine. This binding is damaged by a single mutation changing valine to isoleucine in one protein, and can be retained by a mutation from phenylalanine to leucine in the other protein since isoleucine and leucine can form the needed hydrogen bond. (From Valencia, A. and F. Pazos, *Structural Bioinformatics*, edited by P.E. Bourne and H. Weissig, Wiley-Liss, Inc., 2003.)

a phylogenetic profile, a vector where each component corresponds to a specific genome and takes a value of either 1 or 0, with one (zero) indicating the presence (absence) of a significant homology of that protein in the corresponding genome. If two proteins coevolve, their phylogenetic profiles should reflect this fact by being similar or even identical, as shown in the example in Figure 15.3. Various methods have been developed to cope with the ramifications to this principle of coevolution. For example, having identical phylogenetic profiles may be a too strong requirement, since the assertion of a protein's presence or absence in a genome may not be 100% reliable, which is often based on a significant hit from BLAST searching the protein sequence against the genomic sequence. Therefore, a method based on such stringent criteria may incorrectly predict two interacting partners as not-interacting, resulting in a false negative (FN). Proteins of multiple domains also complicate the task of asserting their presence and absence, though in a different way.

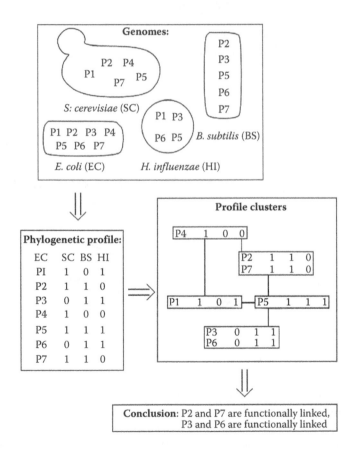

FIGURE 15.3: Phylogenetic profiles for inferring functional linkage. (From Pellegrini, M., E.M. Marcotte, M.J. Thompson, D. Eisenberg, and T.O. Yeates, *Proc Natl Acad Sci USA* 96, 4285–4288, 1999.)

15.3 Computational Methods

To infer PFL and predict PPI has become a central task in reverse engineering the biological networks, and it has practical applications in drug design and delivery. High throughput experimental approaches such as DNA microarray and Y2B are still very costly and often generating indirect and/or unreliable noisy data, and therefore should be complemented with computational methods. This situation has motivated vigorous development of computational methods for inferring PFL and predicting PPI with high accuracy and interpretability.

15.3.1 Predicting protein functional linkage (PFL) based on evolutionary information

Many efforts have been made to tap into the coevolutionary information exhibited in functionally related proteins since the concept was first proposed for functional annotations in the late 1990s. One direction is with respect to extracting and representing coevolutionary information, whereas the other direction is to develop more powerful computational tools for analyzing the information. The original phylogenetic profiles are binary valued (Pellegrini et al., 1999), using 0 and 1 to represent the absence and presence of homologous proteins in reference genomes. Real values are then used in order to reflect probabilistic nature in making a call for homology and to avoid the information loss. To better handle multidomain proteins, methods have been developed to construct phylogenetic profiles from alignments not of entire sequences but of domains or even at residue level. To more accurately capture the cause for PFL, the phylogenetic profiles are further augmented by including orthologous proteins (mirrortree) and the phylogeny for those reference genomes (tree kernel).

This ever richer information about proteins coevolution calls for (Valencia and Pazos, 2003) more sophisticated methods for mining the relationships and making accurate prediction. Efforts in this direction of research are centered at measuring the "similarity." Hamming distance was adopted as early attempt to measure similarity between the binary profiles and was successful in detecting some PFL. Yet, the shortcomings of Hamming distance are obvious; besides being too coarse grained for measuring subtle differences it cannot effectively handle real valued profiles. Mutual information can measure codependence between two proteins if their phylogenetic profiles are interpreted as distribution of probabilities that the proteins occur in the reference genomes. Calculation of the probability distributions often requires binning the values for the profile components, which can be problematic when the number of reference genomes is not sufficiently large. Pearson correlation proves to be a good alternative, though it lacks of the probabilistic interpretation. For profiles augmented with phylogenetic tree information, advanced methods have been developed, including tree kernel, mirror tree, TreeSec, least squares SVM, and transductive learning.

15.3.2 Various phylogenetic profiles

The original phylogenetic profiles shown in Figure 15.3 can be constructed by searching the protein sequence against the reference genomes for homology using sequence alignment tools such as BLAST. If the best alignment returned by BLAST has an E-value better than a preset threshold, a putative homolog is considered existing and a value 1 is assigned to the profile component corresponding to the genome; a value of 0 is assigned otherwise. However, despite its initial success, this black and white view of proteins evolutionary history based on some empirical threshold on the E-value is necessarily crude, error-prone, and suffering from information loss. A quick and convenient remedy is to use the E-values or some conversion of them instead of zeros and ones. One example conversion of E-value is $-1/\log(E)$, as used in Equation 15.1 below.

Most proteins function via domains, and some proteins may contain multiple domains of different functions. These multidomain proteins pose a difficulty to the original phylogenetic profiles where the presence/absence is determined based on the whole sequence of a protein; the profile for a multidomain protein thus will likely to get many ones, more than that of a protein which contains just one of these domains. The apparent differences between phylogenetic profiles of the multidomain protein and its single-domain partner can be challenging for the machine learning algorithms to detect their true relationship. This prompts the development of methods for constructing profiles at the domain level, or even at the residue level which can eliminate the frequent errors associated with the former caused by using a sliding window in detecting domains.

The residue level coevolutionary information is collected from BLAST search against reference genomes. Specifically, given a protein P_i, a BLAST search is made against a reference genome G_a and let $A(P_i, G_a)$ be the set of significant local alignments returned from the BLAST search with an E-value less than or equal to 0.5. Any residue in the protein may appear in more than one alignment in $A(P_i, G_a)$, and when that is the case, the best (i.e., lowest) E-value $E(A)$ is selected to represent how well the residue is conserved locally in the reference genome. For residues that do not appear in any alignment, a default value 1.0 is assigned to the corresponding residues. Then, for each amino acid residue r in protein P_i, a phylogenetic profile is defined over m reference genomes as follows.

$$\psi^r(a) = \min_{A \in A(Pi,Ga) \text{ and r is in } A}(-1/\log E(A)), \quad 1 \le a \le m. \quad (15.1)$$

The similarity between two phylogenetic profiles $\psi^r(a)$ and $\psi^s(a)$ can be measured by mutual information $I(\psi^r, \psi^r)$ defined as follows.

$$I(\psi^r, \psi^r) = H(\psi^r) + H(\psi^s) - H(\psi^r, \psi^r) \quad (15.2)$$

where

$$H(\psi^r) = \Sigma_{\alpha=1 to B} f_\alpha \log(f_\alpha) \tag{15.3}$$

is the Shannon entropy, and f_α is the frequency that, for $1 \leq a \leq m$, $\psi^r(a)$ has a value in the bin α. That is, the range of $\psi^r(a)$ is divided evenly into B subranges and each subrange is a bin. And the joint entropy $H(\psi^r, \psi^r)$ is similarly defined. While mutual information has been widely used in measuring "similarity" between two distributions, its implementation requires binning the real values into discrete set, which introduce an extra parameter—the number of bins —into the method, and optimizing on this parameter, if possible at all, can be time consuming and data dependent. Alternatively, Pearson correlation coefficient can be instead used as a measure of "similarity" between two phylogenetic profiles $\psi^r(a)$ and $\psi^s(a)$:

$$\rho^{rs} = \Sigma_{a=1 \text{ to } m}(\psi^r(a) - \psi^r)(\psi^s(a) - \psi^s)/\sqrt{\Sigma_a(\psi^r(a) - \psi^r)^2 \Sigma_a(\psi^s(a) - \psi^s)^2}, \tag{15.4}$$

where $\psi^r = (1/m)\Sigma_{a=1 \text{ to } m}\psi^r(a)$ is the mean value for the phylogenetic profile ψ^r and $\psi^s = (1/m)\Sigma_{a=1 \text{ to } m}\psi^s(a)$ is the mean value for the phylogenetic profile ψ^s. It is noted that no binning is needed and all values in the two phylogenetic profiles are accounted for. Improved accuracy from using this similarity measurement has been reported by Kim et al. (2006) and Craig et al. (2008).

Recently, methods have been developed to tap into extra information, including correlated mutations (Pazos et al., 1997; Pazos and Valencia, 2001) and phylogenetic tree information (Craig and Liao, 2007b). The similarity of phylogenetic trees of functionally linked proteins was first qualitatively observed in the context of PPI for a couple of individual cases and then statistically quantified in for large sets of interacting proteins. The hypothesis is that functionally linked proteins would be subject to a process of coevolution that would be translated into a stronger than expected similarity between their phylogenetic trees. As shown in Figure 15.4, for two proteins R and S, their phylogenetic trees can be constructed from the multiple sequence alignments (MSAs) of the corresponding homologous proteins in a group of reference genomes. If the two proteins are related, their phylogenetic trees will be similar to each other—one tree is like the image of the other in a mirror. This mirror-tree method goes one step beyond the phylogenetic profiles method: the length of the branches and the structure of the trees are taken into account, whereas in the phylogenetic profiles method, only the pattern of presence/absence (leaves of the tree) is considered. Because of the complexities involved with defining and computing tree similarity directly, the method actually evaluated the similarity indirectly as the element-wise correlation r between the two distance matrices, where each matrix stores the pair-wise distances between homologous proteins for the corresponding protein. High correlation is interpreted as high similarity, and hence high probability that two proteins interact.

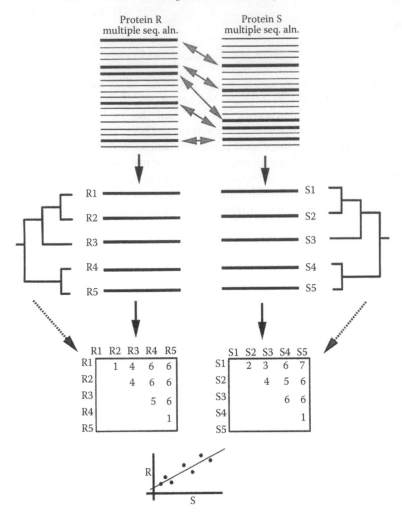

FIGURE 15.4: Similarity of phylogenetic trees (mirrortree). Multiple sequence alignment for homologous proteins of R is used to construct the corresponding intersequence distance matrix. Likewise, such a matrix can be generated for S. The linear correlation between the two distance matrices is used as an indicator: high correlation values are interpreted as high similarity between the phylogenetic trees and hence high probability that R and S interact. (From Pazos, F., and A. Valencia, *Protein Eng.*, 14, 609–614, 2001.)

While it makes sense to improve PFL prediction accuracy by incorporating more information such as phylogenetic trees, one may ask why this would necessarily help make the prediction more specific, since PFL is a very broad relationship. This argument actually touches upon a very fundamental aspect of machine learning, on which many bioinformatics methods of phenomenological nature are based. Certainly, more attributes as input would help differentiate entities, e.g., 2-D gel electrophoresis (*iso-electric point* as one dimension and *molecular weight* as another dimension) can separate molecules that otherwise cannot be separated by 1-D gel electrophoresis (*molecular weight*). However, simply adding more information may not help, and can sometimes complicate learning methods.

Indeed, despite the improved performance, false positives are still abundant in the predictions made by the *mirrortree* method. A main reason is that the extra information (i.e., the phylogenetic trees) is not differentiating; phylogenetic trees, built as shown in Figure 15.4, may bear high similarity due to the resemblance to the underlying species tree, (Sato et al., 2005), regardless if the two proteins interact or not. Techniques need to be developed to incorporate evolutionary information judiciously so as to improve the PFL prediction, and at the same time, to shed light on how PFL and other events such lateral gene transfer that may have intertwined during evolution.

Phylogenetic trees are encoded and concatenated to the phylogenetic profiles for protein classification (Liao, 2006). One key idea in incorporating phylogenetic tree is realizing that not all components in the phylogenetic profiles are equally important—components corresponding to closely related organisms likely contribute differently than those from distantly related organisms in determining PFL for proteins—and therefore the phylogenetic tree can inject some input that can differentiate the profile components. As shown in Figure 15.5, a phylogenetic profile is augmented with extra components representing scores from the interior nodes of the tree. These scores are obtained by letting the zeros and ones in the original profile percolate through the tree from the leaves to the root:

$$S(k) = (1/|C|)\Sigma_{i \in C} S(i) \tag{15.5}$$

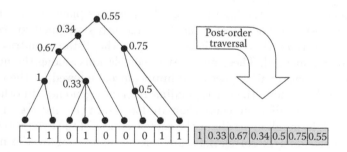

FIGURE 15.5: Extended phylogenetic profiles incorporating phylogenetic tree information. Tree leaves are assigned with scores corresponding to the phylogenetic profile, and each interior node is assigned a score as the average of their descendants'. By a posttraversal order, these scores at the interior nodes including the root node are listed as a vector extending the original phylogenetic profile vector.

where score S(k) at the node k is the arithmetic average of scores at its children nodes, with C standing for the set of children nodes of k and $|C|$ standing for the size of the set. The components in the original profile contribute to the extended part in a way that depends on their positions corresponding to the tree leaves. The original components can be more explicitly differentiated by weights that represent how often proteins tend to be present (or absent) in the corresponding genomes. The score $S(k)$ is now defined as

$$S(k) = (1/|C|)\Sigma_{i \in C} \ S(i)W_{S(i)}(i) \qquad (15.6)$$

where the weights W at the interior nodes are defined as 1, which means no weighting, and the weights $W_{S(i)}(i)$ at a leaf i are collected as the frequency of absence $(S(i) = -1)$ and presence $(S(i) = +1)$ of proteins in genome i in the training data. As the weights at the leaves can be interpreted as probability distribution of proteins presence in a genome, the frequency from counting gives the maximum likelihood estimation of the probability distribution. Since the weighting factor W's reflect how likely proteins may be absent or present at a leaf position in the phylogenetic tree, and such collective information about the proteome, as sampled in the training data, can help distinguish functionally related proteins from non related proteins, it therefore makes intuitive sense to collect weighting factors for the positive training examples (related proteins) and for the negative training examples (non related proteins) separately. This scheme of separate weighting proves to be very effective in training, yet it cannot be directly applied to testing examples; because we do not know if they are positive or negative, we cannot collect separate weights. To overcome this problem, a method of transductive learning coupled with expectation-maximization algorithm was developed by Craig and Liao (2007a) to leverage the hidden information (the W's) from the testing data, which was needed in Equation 15.6 to encode the phylogenetic tree topology as the extended part of the profiles.

In order to reduce the high rate of false positives plaguing the *mirrortree* method, a novel method *TreeSec* has been recently developed to account for some of the intramatrix correlations existing in the distance matrices used in the mirrortree method. These intramatrix correlations among the matrix elements reflect their relative positional importance, endowed by the evolutionary relationship among the corresponding reference genomes. In other words, same values, depending on where they are located in the matrices, may carry different weight in telling us about the coevolvedness of the two proteins. However, such intramatrix correlations are not accounted for by the mirrortree method's linear correlation coefficient, as shown in Figure 15.6, where all elements are treated equally important. This situation is remedied in the new method. Specifically, as shown in Figure 15.6, the species tree of the reference genomes is used as a guide tree for hierarchical clustering of the orthologous proteins. The distances between these clusters, derived from the original pairwise distance matrix using the neighbor joining algorithm, form intermediate distance matrices, which are then transformed and concatenated into a *super-phylogenetic vector*. A support vector machine is trained and tested

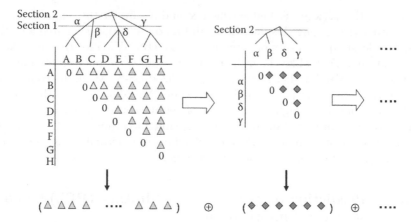

FIGURE 15.6: Schematic illustration of TreeSec method to derive super phylogenetic vector for a given protein from the distance matrix of its orthologous proteins A, B, ..., H. Section 1 across the tree leads to four clusters of the orthologous proteins: $\alpha = \{A,B\}$, $\beta = \{C\}$, $\delta = \{D,E,F\}$, and $\gamma = \{G,H\}$. The distances among these clusters are calculated, resulting in an intermediate matrix. The procedure is repeated for all sections, producing more intermediate matrices. The upper triangles of the matrices are transformed into vectors and concatenated (denoted by the symbol $\oplus$) into a super phylogenetic vector. In this way, the phylogenetic vector used in mirrortree method is extended with extra "bits" that encode the topological information of the protein tree with reference to the species tree.

on pairs of proteins (Cristianini & Shawe-Taylor 2000) and (Scholkopf and Smola, 2001), represented as super phylogenetic vectors, whose interactions are known. The performance, measured as receiver operating characteristic (ROC) score in cross validation experiments, shows significant improvement over the mirrortree method.

The relative positional importance of matrix elements is accounted by the *TreeSec* method, but in a somewhat indirect and subtle way, and the method requires the species tree, which is sometimes either unavailable or controversial when available. Since the components of a phylogenetic vector may contribute differently in determining its similarity with another vector. Therefore, the similarity between m-dimension vectors **x** and **y** can be measured by a weighted dot product defined as follows:

$$K(\mathbf{x}, \mathbf{y}) = \sum_{d=1}^{m} \beta_d \mathbf{x}^d \mathbf{y}^d \tag{15.7}$$

where β_d are the weights. All protein pairs represented as phylogenetic vectors are split into two subsets, each contains both interacting and non interacting pairs, and one subset is used for training and the other for testing. A learning algorithm for this binary classification problem should find the β weights such that the $K(\mathbf{x}, \mathbf{y})$ is higher when $\mathbf{x}$ and $\mathbf{y}$ are in the same class, and lower otherwise. Along this line, Craig and Liao (2007c), formulated a method based on least-squares support vector machine (LS-SVM) for the linear kernel in Equation 15.7. Unlike the ordinary least-squares support vector machines (Suykens and Vandewalle, 1999), this specialized LS-SVM can be trained very efficiently as a linear programming problem. An advantage of this method is that the optimal weights, as a kind of regularization, need no extra grid searching and can be obtained at the same time of training the LS-SVM. The performance, tested in the same cross validation experiments as the *TreeSec* method, shows further significant improvement.

15.3.3   Predicting protein functional linkage (PFL) based on structural information

Structural information such as binding residues has been incorporated into models for predicting PPI for proteins that do not have structural information. Feature selection plays an essential role in identifying structural and functional signatures and for suggesting effective knockout mutations.

Because the important roles that the structure plays in determining functions of the proteins, functional linkage between proteins can in turn also exert pressure for proteins to either keep their individual structure or more importantly maintain the compatibility between relevant structural elements. Such connections are particularly essential for proteins that have direct physical interaction. Proteins interact with one another via some interacting domains, and it has become a common approach for predicting PPI by identifying these domains first. Although the domains responsible for binding two proteins together tend to possess certain biochemical properties that dictate some specific composition of amino acids, such compositional characteristics are typically not unique enough to be solely relied upon for domain identification—some reported that interaction surfaces are almost statistically indistinguishable from other surfaces.

Despite the difficulties considerable progress has been made in identifying interaction interface, also called *protein docking* problem. One class of the current approaches to protein docking consider proteins as rigid bodies (based on X-ray crystallography structure) and use the geometric matching to identify candidate interacting surfaces. Some methods in this class, being the closest to be called *ab initio*, further optimize the matching by allowing interacting surfaces to adapt to each other via conformational changes based on molecular dynamics or *Monte Carlo* simulations, limited to the surface areas. Other approaches try to limit the input to primary sequences. It has been shown by Pazos and Valencia (2001) that the interprotein correlated mutations can be used to point to the residues and regions implicated in the interaction between the two proteins. The advantage of these approaches is their independence of the structural information, which is not often available. Even when the structural information is available, it is often obtained from proteins as a free molecule and therefore may be different from the actual structure when the protein forms a complex with its interacting partner. This is particularly problematic for these structure-based methods without capacity of allowing conformational changes.

A highly relevant problem to PPI in particular is about how to leverage the information about these interaction interfaces to improve PPI prediction. Once the interacting domains are identifies, how can they be characterized and represented in such a way they can be used as pieces to solve the whole PPI puzzle? Particularly, such representations should be able to capture and retain both structural and sequence information, when available. Another desired capability is to be able to transfer knowledge from proteins that have structural information to proteins that do not.

As mentioned above, compositional variations are not uncommon in the multiple sequence alignment of these proteins that contain the same domain. Hidden Markov models are among the most successful efforts in biosequence analysis to capture the commonalities of a given domain while allowing variations (Durbin et al., 1999). A collection of hidden Markov models covering many common protein domains and families is available in the PFAM database—ipfam (http://www.sanger.ac.uk/software/pfam/ipfam).

In Patel et al. (2006), the binding site information is incorporated to improve the accuracy of predicting PPI. Seven attributes including residue propensity, hydrophobicity, accessible surface area, shape index, electrostatic potential, curvedness, and conservation score are previously used for identifying protein binding sites. Principal component analysis (PCA) was utilized to assess the importance of these attributes, indicating that the amino acid propensity and hydrophobicity are the most important while the curvedness and shape index seem to be least important. This finding is interesting, note that the two attributes related to the geometry of a patch (shape index and curvedness) happen to be the two least important attributes for distinguishing a binding patch. Based on the PCA result, the attribute vectors are reduced to the dimensionality, i.e., using only the first and second components, and 85% accuracy is achieved. This is in good standing with the 86% accuracy when seven attributes are used.

Those attributes at the binding sites are then incorporated to predict PPI. The attribute vectors of candidate interacting partners are concatenated, and a support vector machine is trained to predict the interacting partners. This is combined with using the attributes directly derived from the primary sequence at the binding sites. The results from the leave-one-out cross validation experiments show significant improvement in prediction accuracy by incorporating the structural information at the binding sites.

15.3.3.1   Transferring structural information

Despite its utility in enhancing prediction accuracy, structural information is not always available due to the difficulties and high cost of the experimental methods. As a compromise, predicted structure is often used when its quality is acceptable, which is particularly common for protein secondary structure.

Hidden Markov models are a powerful tool for sequence analysis. Recently, an ipHMMs has proposed for identifying interacting domains in proteins whose structural information is available. In addition to the match, insert, and delete states in these ordinary profile hidden Markov models, an extra set of states are introduced to designate whether a residue is a binding residue (see Figure 15.7). Then the model can be used to (i) predict if a query protein contains the interacting domain and, (ii) to mark out these binding residues. These two tasks can be achieved respectively by the Forward algorithm and Viterbi algorithm, only slightly modified from the standard implementation in pHMMs. It is shown in Fredrich et al. (2006) that ipHMMs significantly outperformed pHMMs as implemented in the PFAM database.

To take advantage of ipHMM's improved accuracy in identifying interacting domains, a hybrid method is developed to better identify interacting partners for proteins that do not have X-ray structural information. Since PPI is realized via domain-domain interaction. Therefore, for a pair of interacting domains A and B in the PFAM database, any two member proteins each from one of the two families should interact. However, this conclusion will not be true if the membership is false, as shown in the central panel in Figure 15.8. Given a protein, its membership to a domain family is determined

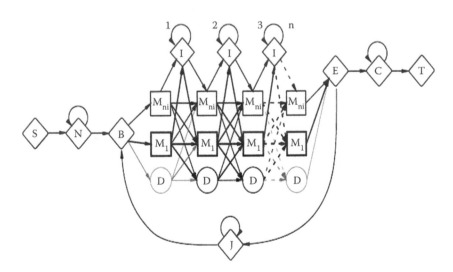

FIGURE 15.7: Topology of the interaction profile hidden Markov model following the restrictions and connectivity of the HMMer architecture. As described in Fredrich et al. (2006), the match states of the classical pHMM are split into a noninteracting (M_{ni}) and an interacting match state (M_i). (From Friedrich, T., B. Pils, T. Dandekar, J. Schltz, and T. Muller, *Bioinformatics*, 22, 2851–2857, 2006.)

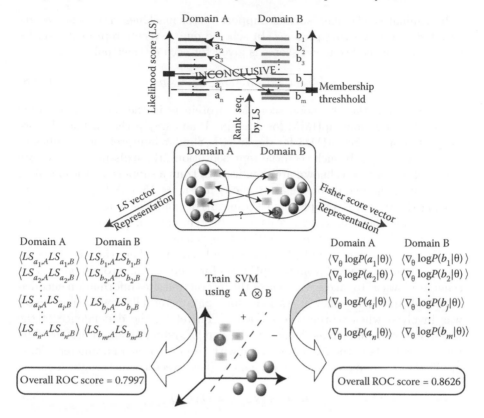

FIGURE 15.8: Transfer domain information. In middle panel, ipHMM for domains A and B with putative members. In upper panel, likelihood score is used to rank potential members. In lower left panel, vectors of cross likelihood scores are used to train the svm. In the lower right panel, fisher score vectors are used to train the svm.

by calculating the likelihood score (using forward algorithm) and comparing it to a preset threshold, i.e., the protein is a member if the likelihood score is higher than the threshold. It is reasonable to expect that the improved accuracy in identifying the interacting domains can translate to improved accuracy in predicting the interacting partners, as shown in the top panel in Figure 15.8. Yet, it is difficult to get rid of false positives and false negatives (FNs) by using a cutoff on a single score. Because membership may not be appropriately established by a clear cut on a single score, false positives and FNs are not uncommon, and get multiplied in numbers when we make prediction on interacting pairs.

In Patel and Liao (2007), a strategy is proposed to combine a generative model with a discriminative classifier to improve the accuracy, since both the negative examples and positive examples are accounted in training a discriminative classifier, such as support vector machines. Two schemes are adopted in representing proteins. In scheme one, a vector representation by concatenating the likelihood scores is given for each distinct pair of proteins:

$$< \mathrm{LS}_{ai,A}, \mathrm{LS}_{ai,B}, \mathrm{LS}_{bj,A}, \mathrm{LS}_{bj,B} > \qquad (15.8)$$

where $\mathrm{LS}_{ai,A}$ is the likelihood score for protein a_i in the interaction profile hidden Markov model ipHMM for domain A, and $\mathrm{LS}_{ai,B}$ the likelihood score for protein a_i in the ipHMM for domain B. Similar interpretation applies to $\mathrm{LS}_{bj,A}$ and $\mathrm{LS}_{bj,B}$. In such vectorial representations, these distinct pairs whose status of interaction is known are then used to train a support vector machine, as shown in the middle of the lower panel in Figure 15.8. A hyperplane separating the positive examples and the negative examples is indicated by the dashed line. The interaction status for a protein pair can be predicted by its location with respect to the separating hyperplane.

Feature selection is applied in the second scheme, where the Fisher scores are selectively extracted from the ipHMMs for representing protein pairs. This scheme allows us to take advantage of the interaction sites information encoded in the ipHMM. The sufficient statistics for a parameter tells how the parameter was involved when scoring the sequence. Therefore, the dependence of the likelihood score on each model parameter can be systematically represented by taking gradients of the likelihood score with respect to each parameter. These gradients are components of the so called Fisher vector which is given as,

$$\mathbf{U}(x) = \nabla_\theta \log P(x|\theta, \pi) \qquad (15.9)$$

where π is the most probable path of hidden states in the model, to which the sequence x is aligned, as identified by the Viterbi algorithm. The use of a probability conditioned on the viterbi path in Equation 15.9 is more intuitive than the original Fischer score, where probability $P(x|\theta, \pi)$ was used. The Fisher scores as defined in Equation 15.9 can be efficiently calculated and has been used to differentiate transmembrane proteins from signal peptide with significant improved accuracy (Kahsay et al., 2005). We have developed an algorithm to calculate these gradients as efficiently as that for the original Fisher score, which uses counts of observations for each parameter in the hidden markov model (HMM) by using the forward-backward algorithm.

By selecting the parameters θ in Equation 15.9 from the states that correspond to the interaction sites, the ipHMM's improved accuracy in predicting the interaction sites can be transferred to high accuracy in predicting the interacting partners. One scheme fisher scores for non-interacting match states (FS_NM), served as a baseline, is to calculate the fisher scores with respect to the non-interacting match states in the ipHMMs. For each sequence in domain A, we compute this fisher vector using Equation 15.9 by choosing M_{ni} states in Figure 15.7. The same applies for domain B. Training set is made by concatenating the fisher vectors of interacting proteins (positive class) and noninteracting proteins (negative class), as shown in the left part of the lower panel in Figure 15.8. In another scheme fisher scores for interacting match states (FS_IM), we compute the fisher scores with respect to the interacting match states. As a baseline, Equation 15.8 is also considered as a scheme (LS) for representing protein pairs as vectors. To take advantage of the SVM's learning power in integrating orthogonal data of different sources (Scholkopf et al., 2004), a combination of schemes (FS_IM) and LS will also be adopted, i.e., $K(x, y) = K1(x, y) + K2(x, y)$, every protein pair is represented as a vector by concatenating the respective vectors from scheme FS_IM and scheme LS.

The other aspect of feature selection is to suggest mutations that would most likely turn on/off the binding affinity and change the binding specificity. This is explored by computing a vector of scores using a formulation conjugate to Equation 15.9:

$$\mathbf{V}(x) = \nabla_x \log P(x|\theta, \pi) \tag{15.10}$$

where π is the viterbi path for the template protein aligned to the ipHMM model.

15.3.4 Predicting protein functional linkage (PFL) by integrating network information

15.3.4.1 Transferring network information

We have seen how knowledge transferring improves the prediction in the case of protein structure. Here we see another example of knowledge transferring in genetic networks, where nodes representing proteins or genes, and edges representing some PFL relationship between proteins, be it physical interaction or regulatory or others. The networks can be conveniently represented as a matrix, in which all rows and columns correspond to proteins and an entry gives the "strength" or certainty of a relationship between the corresponding proteins. This adjacency matrix can be interpreted as a kernel representation of the PFL. A kernel is a real valued function $K(x, y)$ defined on any pair of proteins x and y. The value $K(x, y)$ is thought of as a measure of the relationship between x and y. An example of kernel is given in Equation 15.4 when proteins are represented as vectors.

For a dataset of genetic network, a common and natural choice is the diffusion kernel $K = \exp(\beta H)$, where $\beta > 0$ is a regularization parameter and H=A−D is the opposite Laplacian matrix of the graph with A as the adjacency matrix and D being the diagonal matrix of node connectivity. A typical scenario in genetic networks is, as schematically shown in Figure 15.9, that only part of the network (i.e., part of A) is known. The task for a learning method is to recover the rest of the network, based on some attributes of the proteins, such as phylogenetic profiles. If we design a kernel function $K_a(x, y)$ on the phylogenetic profiles, such as the one we have in Equation 15.7, we would want this $K_a(x, y)$ to be as similar as possible to the diffusion kernel, let's call it $K_d(x, y)$ and at least for the part that is known. Often this may not be the case. Then we may try to redefine our kernel function K_a, which

Adjacency matrix of protein network

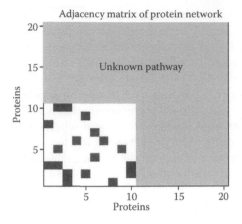

Similarity matrix of the other genomic data

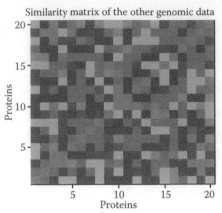

FIGURE 15.9: Transfer network knowledge. In the left panel, the true protein network is only partly known, with the entries in the adjacency matrix dark shaded to indicate pairs of related proteins. In the right panel, similarity for all pairs of proteins is assessed from certain attributes, such as phylogenetic profiles. (From Yamanishi, Y., J.-P. Vert and M. Kanehisa, *Bioinformatics*, 20, i363–i370, 2004.)

is what we have been doing. Another way is, as suggested by Yamanishi et al. (2004), to extract most relevant information out of the phylogenetic profiles by recombining their components in a way that maximizes the correlation between eigenvectors from these two kernels, and then to cluster proteins based on the most relevant information.

The key step of the method is to solve a generalized eigenvalue problem, as defined in the following

$$\begin{pmatrix} 0 & K_a K_d \\ K_d K_a & 0 \end{pmatrix} \begin{pmatrix} \alpha \\ \beta \end{pmatrix} = \rho \begin{pmatrix} (K_{a-}\lambda I)^2 & 0 \\ 0 & (K_{d-}\delta I^2) \end{pmatrix} \begin{pmatrix} \alpha \\ \beta \end{pmatrix} \quad (15.11)$$

where $\mathbf{I}$ is the identity matrix, and $\boldsymbol{\alpha}$ and $\boldsymbol{\beta}$ are the eigenvectors associated with eigenvalue ρ. Much like in the principal component analysis, we can rank all eigenvalues in descending order and focus on the first few top eigenvalues, because the corresponding eigenvectors represent the directions where these two kernels are most similar to each other, in the sense as defined in what is called canonical correlation analysis (CCA). If L is chosen as the number of the top eigenvalues for consideration, any protein x is now represented by L new components by being projected onto the corresponding eigenvectors as follows.

$$f^j(x) = \Sigma_k < \alpha^j, \quad x_k > K_a(x_k, x), \quad \text{for } j = 1 \text{ to } L \quad (15.12)$$

where $\mathbf{x}_k$, are the proteins in the known part of the network, α^j is the eigenvector for the top j-th eigenvalue, and $<,>$ stands for the dot product of two vectors. So, protein **x** is now mapped to a feature space as such an L-dimension vector $\mathbf{u} = (u_1, \ldots, u_L) = [f^1(\boldsymbol{x}), \ldots, f^L(\boldsymbol{x})]$. The rest of the network can be inferred by assessing if any pair of proteins x and y are related based on the correlation coefficient for their corresponding vectors, as defined in Equation 15.12. The method offers a straightforward way to integrate heterogeneous data by creating a new kernel as the sum of the kernels corresponding to different data sources. In Yamanishi et al. (2004), prediction accuracy is shown improved using a kernel $K = K_{\text{ppi}} + K_{\text{exp}} + K_{\text{phy}} + K_{\text{loc}}$ that integrates expression data (K_{exp}), PPI from Y2B (K_{ppi}), phylogenetic profiles (K_{phy}), and protein intracellular localization data (K_{loc}).

Pairwise PPI will make fuller sense when placed in the context of biological networks. It is not uncommon that a pairwise PPI is influenced by, or cannot happen without the presence of, a third protein. For these cases, PPI cannot be accurately and reliably predicted without being put into its context. Also, many high throughput genomic and proteomic data can be more effectively deciphered in a network setting than in a pairwise setting. Some critical biological insights about the cellular behavior as a system, such as relation between the essentiality of a gene and its connectivity and position in biological networks, are difficult to fathom simply from the pairwise PPI perspective.

The concepts and techniques developed for predicting pairwise PFL can be expanded to situations where multiple proteins are involved, in order to accommodate high connectivity observed in the protein networks. The phylogenetic based methods can be generalized to deal with interactions of multiple proteins. The basic concept is presented by Bowers et al. (2004). If a protein's presence in a genome depends on two other proteins presence and/or absence in the same genome, we hypothesize that the protein interacts with these two proteins. Such a hypothesis becomes more plausible if the projected relation among these three proteins holds true in more genomes, namely, we will see the same pattern in all columns in an alignment of the phylogenetic profile of these proteins. In Figure 15.10, the relation $c = a \wedge b$ is true in every column: c

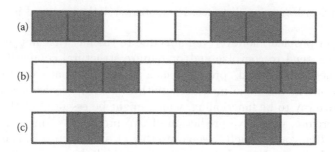

FIGURE 15.10: A row corresponds to a phylogenetic profile of eight genomes. A shaded (open) square indicates binary value 1 (0), which means the presence (absence) of the protein in the corresponding genome. Protein C's presence depends on the presence of both a and b.

is present in a genome if and only if both a and b are present. Even using the binary profiles, such a scheme has already yielded very plausible predictions of protein triplets, some of which have been validated in real networks. Another significant source of data for predicting PPI is genomic data, and approaches based on Bayesian networks have been developed for prediction (Jansen et al., 2003).

15.4 Discussions and Future

Knowledge of PFL plays an important role in functional genomics, proteomics, and systems biology. Predicting protein functional linkage poses special challenges to existing computational approaches. In summary, we discussed some recent developments in tackling the problems via tapping into a variety of rich information beyond protein sequences. Not only have these methods achieved some remarkable progress towards high accuracy prediction, they also represent some very innovative ideas in applying machine learning techniques in this biologically important domain. Specifically involved are the following aspects:

- Learning schemes: unsupervised, supervised, and semisupervised/ transductive

- Learning models: hidden Markov models, support vector machines, and kernel canonical correlation analysis

- Representations: phylogenetic profiles, distance matrices, phylogenetic trees, and heterogeneous data

- Metrics: Hamming distance, Pearson correlation, mutual information

A major problem, however, remains due to the fact that not just interacting proteins tend to coevolve and/or require structural compatibility, so do other "functionally linked" proteins, such as proteins that catalyze biochemical reactions in the same metabolic pathway, their coexistence is required for the pathway to be functioning, which might be essential to the cell's survival. Therefore, a positive prediction of two proteins as functionally linked, although often good enough to separate them from the crowds, is not yet specific enough to pin down their functions, particularly for proteins that have multi functions. Future work would necessarily involve developing computational methods that can make more specific prediction, and can better handle heterogeneous data as new high throughput technologies will certainly create ever more data in systems biology.

References

Bowers, P.M., S.J. Cokus, D. Eisenberg and T.O. Yeates. 2004. Use of logic relationships to decipher protein network organization. *Science*, 306, 2246–2249.

Craig, R and L. Liao. 2007a. Transductive learning with EM algorithm to classify proteins based on phylogenetic profiles. *Int. J. Data Mining Bioinformatics*, 1, 337–351.

Craig, R. and L. Liao. 2007b. Phylogenetic tree information aids supervised learning for predicting protein-protein interaction based on distance matrices. *BMC Bioinformatics*, 8, 6.

Craig, R.A. and L. Liao. 2007c. Improving protein-protein interaction prediction based on phylogenetic information using least-squares SVM. *Ann. New York Acad. Sci.*, 1115(1), 154–167.

Craig, R.A., K. Malaviya, K. Balasubramanian and L. Liao. 2008. Inferring functional linkage from residue level co-evolution information. *The International Conference on Bioinformatics and Computational Biology (Bio-Comp08)*, Las Vegas, NV.

Cristianini, N. and J. Shawe-Taylor. 2000. *An Introduction to Support Vector Machines and Other Kernel-based Learning Methods*. Cambridge University Press, Cambridge, UK.

Durbin, R., S.R. Eddy, A. Krogh and G. Mitchison. 1999. *Biological Sequence Analysis: Probabilistic Models of Proteins and Nucleic Acids*. Cambridge University Press, Cambridge, UK.

Friedrich, T., B. Pils, T. Dandekar, J. Schltz and T. Muller. 2006. Modeling interaction sites in protein domains with interaction profiles hidden Markov models. *Bioinformatics*, 22, 2851–2857.

Jansen, R., H. Yu, D. Greenbaum, Y. Kluger, N.J. Krogan, S. Chung, A. Emili, M. Snyder, J.F. Greenblatt and M. Gerstein. 2003. A Bayesian networks approach for predicting protein-protein interactions from genomic data. *Science*, 302, 449–453.

Kahsay, R., G. Gao and L. Liao. 2005. Discriminating transmembrane proteins from signal peptides using SVM-Fisher approach. *Proceedings of 4th International Conference on Machine Learning and Applications (ICMLA)*, Los Angeles, CA, 151–155.

Kim, Y., M. Koyuturk, U. Topkara, A. Grama and S. Subramaniam. 2006. Inferring functional information from domain co-evolution. *Bioinformatics*, 22, 40–49.

Liao, L. 2006. Hierarchical profiling, scoring and applications in bioinformatics. In *Advanced Data Mining Technologies in Bioinformatics*, edited by Hui-Huang Hsu. Idea Group, Inc., Hershey, USA.

Patel, T. and L. Liao. 2007. Predicting protein-protein interaction using Fisher scores extracted from domain profiles. *Proceedings of IEEE 7th International Symposium for Bioinformatics and Bioengineering (BIBE)*, Boston, MA.

Patel, T., M. Pillay, R. Jawa and L. Liao. 2006. Information of binding sites improves prediction of protein-protein interaction. *Proceedings of the Fifth International Conference on Machine Learning and Applications (ICMLA'06)*, Orlando, FL, 205–210.

Pazos, F. M.H. Citterich, G. Ausiello and A. Valencia. 1997. Correlated mutations contain information about protein-protein interaction. *J. Mol. Biol.*, 271(4), 511–523.

Pazos, F. and A. Valencia. 2001. Similarity of phylogenetic trees as indicator of protein–protein interaction. *Protein Eng.*, 14, 609–614.

Pellegrini, M., E.M. Marcotte, M.J. Thompson, D. Eisenberg and T.O. Yeates. 1999. Assigning protein functions by comparative genome analysis: Protein phylogenetic profiles. *Proc. Natl. Acad. Sci. USA,* 96, 4285–4288.

Sato, T., Y. Yamanishi, M. Kanehisa and H. Toh. 2005. The inference of protein-protein interactions by co-evolutionary analysis is improved by excluding the information about the phylogenetic relationships. *Bioinformatics*, 21, 3482–3489.

Scholkopf, B. and A.J. Smola. 2001. *Learning with Kernels: Support Vector Machines, Regularizaton, Optimization, and Beyond.* The MIT Press, Cambridge, MA.

Scholkopf, B., K. Tsuda and J.P. Vert, editors. 2004. *Kernel Methods in Computational Biology.* The MIT Press, Cambridge, MA.

Suykens, J.A.K. and J. Vandewalle. 1999. Least squares support vector machine classifiers. *Neural Proc. Lett.*, 9, 293–300.

Valencia, A and F. Pazos. 2003. Prediction of protein-protein interactions from evolutionary information. In *Structural Bioinformatics*, edited by P.E. Bourne and H. Weissig. Wiley-Liss, Inc.

Yamanishi, Y., J.-P. Vert and M. Kanehisa. 2004. Protein network inference from multiple genomic data: a supervised approach. *Bioinformatics*, 20, i363–i370.

Chapter 16

Computational Methods for Unraveling Transcriptional Regulatory Networks in Prokaryotes

Dongsheng Che

East Stroudsburg University

Guojun Li

University of Georgia and Shandong University

16.1   Introduction

In the past decade, high-throughput sequence technology has made 700 genomes fully sequenced as of October 2008 (http://www.ncbi.nlm.nih.gov/genomes/lproks.cgi). In order to keep up with the sequencing pace, researchers have annotated coding genes of the genomic data and predicted gene function. This level of genome annotation is known as *one-dimensional genome annotation* [1]. The next grand challenge is to understand how gene products (either RNA genes or proteins) interact with each other and govern the transcriptional rates of genes in the network. Such an annotation, a regulatory network annotation, is also known as *two-dimensional genome annotation* [1].

Our current knowledge of regulatory networks in organisms is limited within a few model organisms, such as *Escherichia coli* and yeast. To bridge this gap, various high-throughput techniques have been employed to elucidate the transcriptional regulation machinery of organisms. For instance, high-throughput microarray experiments in *E. coli* were performed under different conditions, and the mutual information-based algorithm was developed to elucidate the relationships of the co-regulated genes in the network [2]. Such kind of an approach, which relies on expression data to infer regulation interactions, is known as *reverse engineering* [3]. Due to the limited knowledge of what conditions should be tested, the high-throughput dataset may represent only a fraction of all conditions that organisms have to face in the real environment, and thus this approach suffers from low prediction sensitivity [2]. In addition, the expensive labor and time-consuming process limit this approach to be applied in a few organisms. Therefore, the development of computational tools for regulatory network construction has become more attractive.

A straightforward but effective way is to use known regulatory networks of closely related organisms as templates. Babu et al. [4] used the regulatory network of *E. coli* to build networks for those genomes that are less characterized, or newly sequenced genomes. This template-based approach assumes that orthologous *transcription factors* (TFs), which bind DNA segments of regulated genes and control their transcription process, regulate the transcription of orthologous target genes in a similar way. Therefore, in order to build a regulatory network for a target genome, we used a BLAST-based method to find the orthologous genes corresponding to those genes in the known regulatory network. As our knowledge of regulatory networks even for the model organisms is far from complete, the application of this approach to construct regulatory networks is inherently incomplete. In addition, each organism must possess its own unique regulatory machinery to make it different from other organisms. These unique regulatory mechanisms are also beyond the template-based approach.

A more general computational method for regulatory network construction is by identifying *cis*-regulatory elements, which are DNA segments in the promoter regions of genes, and are regulated by TFs. With our partial knowledge of TFs and their corresponding bound DNA elements (also known as binding sites or motifs), computational methods can be used to identify more conserved *cis*-motifs using statistical models, thus enriching the known regulatory network. As more genomes become available, we can use comparative genomic analysis to detect *cis*-regulatory elements even without using any prior knowledge of TFs and motifs. The *cis*-regulatory elements can be computationally detected from the collection of promoter sequences of orthologous genes across multiple organisms, and; this technique is known as "phylogenetic footprinting" [5]. In this chapter, we will focus on the regulatory network construction in prokaryotes using the *cis*-motif identification-based approach.

16.2 Framework for Regulatory Network Construction in Prokaryotes

The computational methods of regulatory network constructions designed for eukaryotes, in general, can be applied into prokaryotes. The Genomic structure differences between eukaryotes and prokaryotes, however, do exist. For example, each gene of a eukaryote has its own promoter and transcription terminator, and a regulatory region, where one or more TFs bind to DNA elements (i.e., TF-binding sites (TFBSs)). In prokaryotes, only about half of all genes in the genome have their own regulatory regions and terminators (Figure 16.1a). The remaining genes share their regulatory regions and terminators, indicating that all neighboring genes in the group are transcriptionally regulated by the same TF (Figure 16.1b). These neighboring genes that have the same regulation mechanism are usually known as *multigene operons*, while a single gene that has its own regulation is known as a *single-gene operon*. The operon structures must be determined in order to elucidate the regulatory networks in prokaryotes.

In addition, evolutionary studies of operon structures across prokaryotic genomes have shown that operons are fully conserved across multiple genomes, while a higher order than operons (i.e., a union of operon structures) does. These collections of operons are known as *uber-operons*. In many cases, operons within an uber-opoeron belong to the same regulatory network. Thus, the uber-operon information in prokaryotes can be complementary to the current computational methods. Considering the special genomic features of prokaryotes and the state-of-the-art computational methods for regulatory network prediction, we outline the computational framework for a regulatory network

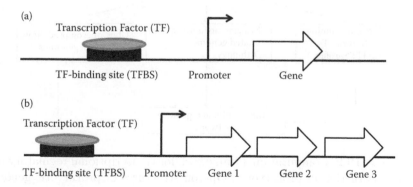

FIGURE 16.1: A schematic representation of the transcriptional regulation in prokaryotes. In (a), a single gene is transcriptionally regulated, while in (b) multiple genes (genes 1, 2, and 3) are regulated by a TF.

of prokaryotes, including: predicting operons using multiple features, scanning the whole genome to identify new regulated genes in the network, collecting *cis*-regulatory motifs and their co-regulated genes using the phylogenetic footprinting technique, and predicting uber-operons (Figure 16.2). The predicted operons are used for the other three tasks in the framework, and the integration of scanning, phylogenetic footprinting and uber-operon prediction makes the constructed regulated networks more complete and reliable. We now describe each of those modules in the remainder of the chapter.

16.3   Operon Structure Prediction

The operon prediction problem can be simply considered as the partitioning of a genome into gene clusters, where all genes within the cluster share a promoter and terminator. Alternatively, it can be treated as a classification problem, i.e., determining whether an adjacent gene pair belongs to the same operon or not.

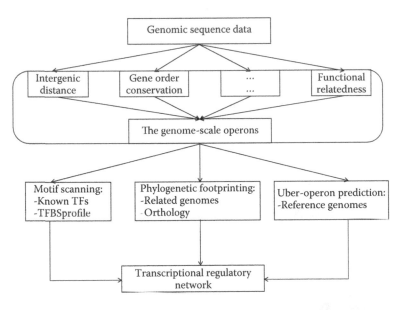

FIGURE 16.2: Computational framework for transcriptional regulatory network construction in prokaryotes. Operon structures of prokaryotic genomes are predicted using multiple genomic features (i.e., intergenic distance, gene order conservation, functional relatedness, etc.), and they are used as transcription units in the computational methods of scanning, phylogenetic footprinting, and uber-operon prediction.

Various features have been studied for the operon prediction problem. Initial attempts were focused on identifying promoter and terminator regions. In general, it is difficult to characterize promoters and terminators, though we could accurately determine the terminators of the phylum of Firmicutes, in which inverted repeats are followed by a stretch of thymine residues [6]. Compared to the boundary signals, the feature of the intergenic distance between two adjacent genes is more effective for operon prediction. Salgado et al. [7] found that adjacent gene pairs within the same operons tend to have shorter intergenic distances, while gene pairs from two consecutive operons tend to have longer distances. When multiple genome sequences are available, the gene order conservation across multiple genomes can imply operon structures [8]. In addition, the functional relatedness of neighboring genes, measured by clusters of orthologous genes (COG) [9], and experimental data (i.e., microarrays and metabolic pathways) are also useful in predicting operons [10].

Numerous computational methods have been developed by using the combined features. We can roughly group them into two major categories, machine-learning-based and graph-based approaches. The basic framework for the machine-learning-based approaches is to construct models based on the training set (i.e., known operon data), and then to predict the remaining unknown genome using the trained models. The graph based approaches use comparative genomic data to identify conserved gene clusters. The identification of conserved gene clusters is implemented by the graph-based algorithms. Interestingly, recent performance evaluations on operon predictions showed that the decision tree approach, a machine learning approach, performed the best [10]. In another recent independent evaluation of programs, Li et al. [11] showed that UNIPOP, a graph-based algorithm, performed the best, without using training data or experimental data. We now describe two representative programs, OperonDT (a decision tree approach) and UNIPOP (a graph-based approach).

16.3.1   OperonDT

OperonDT [12] is a decision tree-based operon predictor that uses the program of J48 implemented in the Weka software package [13]. J48 implements the Quinlan's model of C4.5 [14], where nonterminal nodes are tests on features and terminal nodes are decision outcomes. OperonDT uses five features (including shorter intergenic distance (S_Dist), longer intergenic distance (L_Dist), COG score ($COGN$), gene order conservation (GOC), and transcription direction information (LRS)) to classify whether a gene is an *operonic* gene or *non-operonic* gene. An operonic gene is a gene that shares a promoter and a terminator with its neighbor gene(s), while a non-operonic gene doesn't. For any test gene, two intergenic distances (the one between the test gene and its left neighboring gene, and the one between the test gene and its right neighboring gene) are computed and compared. The longer one is called the *longer intergenic distance*, while the shorter one is called the *shorter intergenic distance.* The feature of transcription direction information is used to record the transcription direction relationship between the test gene and its neighbors.

OperonDT starts with all genes of the training set in the root node, and picks the feature that best differentiates the output attribute value (i.e., *operonic* gene or *nonoperonic* gene) based on the calculated *information gain* (IG), which is defined as

$$\text{IG}(S, A) = E(S) - \sum_{v \in Value(A)} \frac{|S_v|}{|S|} E(S_v) \tag{16.1}$$

where *Value(A)* is the set of all possible values for the feature of A ($A \in \{S\_Dist, L\_Dist, COGN, GOC, LRS\}$). S_v is the subset of S for which feature A has the value of v (i.e., $S_v = \{s \in S | A(s) = v\}$), where v could be an *operonic* gene or a *nonoperonic* gene. $E(S)$ is the entropy of S, which is defined as

$$E(S) = -p_o \log_2 p_o - p_{no} \log_2 p_{no} \tag{16.2}$$

p_o is the probability that the gene is an *operonic* gene (i.e., the percentage of *operonic* genes in S), and p_{no} is the probability that the gene is a *nonoperonic* gene. The algorithm splits the set based on the possible values of the selected feature. If all the instances in a subset have the same output value (i.e., *operonic* or *nonoperonic*), then the process stops and the node is a terminal node. If the subset does contain instances from two classes (*operonic* and *non-operonic*), this process will be repeated until no further distinguishing features can be determined.

Figure 16.3 shows the decision tree model that OperonDT built for the organism *Bacillus subtilis*. For a given gene with its feature values, the trained decision tree model could classify it as an operonic gene or not. For instance, for a gene with *S_Dist*=30, *L_Dist*=100, *GOC*=10, *COGN*=2, *LRS*=0, the decision tree model classifies this gene as a *nonoperonic* gene.

16.3.2   UNIversal Prediction Of oPerons (UNIPOP)

UNIversal Prediction Of oPerons (UNIPOP) [11] is designed to predict operons of any sequenced microbial genome and guarantee high prediction accuracy. Most operon predictors are designed for those organisms that have various kinds of data available. The prediction accuracies of such methods on the model organisms (e.g., *E. coli* or *B. subtilis*) are generally high. In practice, however, almost all new sequenced genomes don't have additional information other than the genome data itself, making these methods inapplicable to new genomes. A few operon predictors were designed by using a minimal amount of information for operon prediction, but they suffered the problem of low prediction accuracy [11]. UNIPOP achieves the goal that uses a minimal amount of information while keeping the accuracy high.

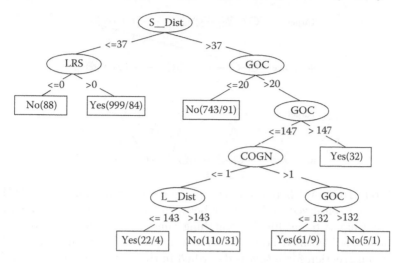

FIGURE 16.3: An example of a decision tree model for the operon prediction of *B. subtilis*. Each interior node is one of the features (i.e., *S_Dist*, *L_Dist*, *COGN*, *GOC*, and *LRS*), while each leaf node is the classification (i.e., "Yes" indicates the test gene is an operonic gene, while "No" indicates it is a non-operonic gene).

The key idea of UNIPOP is to identify conserved gene clusters between two genomes as such gene clusters are operons in real scenarios. The identification of conserved gene clusters is achieved by applying the maximum bipartite cardinality matching based algorithm [15]. In order to make the operon prediction of a target genome reliable, multiple reference genomes are used to predict multiple versions of gene clusters for one target genome, and a consensus gene cluster is adopted. We now describe the procedure of indentifying conserved gene clusters between two genomes (Figure 16.4).

Given two genome genomes (a target genome and a reference genome), and homologous gene relationships between the two genomes, the algorithm looks for the largest conserved cluster with the constraint of a *maximum allowed distance* (*W*) for the cluster in each step. Specifically, each constrained cluster in the first genome is formed based on two rules: (1) all genes within the cluster must have the same transcription direction, either forward or reverse, and (2) the sum of all internode distances must be less than *W*, where each node could be a gene initially, or a gene cluster collapsed in previous steps (see Figure 16.4). The *internode distance* between two individual genes is the intergenic distance between them. The internode distance between a gene and a gene cluster is the intergenic distance between the individual gene and the first gene of the gene cluster if the individual gene is upstream from the gene cluster, or the intergenic distance between the last gene of the gene cluster and the individual gene if the individual gene is downstream from the

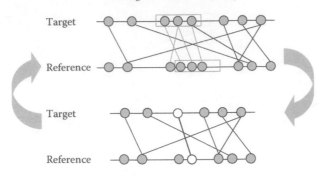

FIGURE 16.4: A schematic diagram of operon prediction using UNIPOP. A maximum conserved node cluster between two genomes is identified using maximum bipartite cardinality matching (*mbcm*) (see the top part) at each step, and this identified node cluster is collapsed into two new nodes. Node cluster construction for *mbcm* is described in the text. This process continues until there is no conserved node cluster whose maximum matching size is greater than 1.

gene cluster. Each constrained clusters in the second genome is formed in a similar way.

For any pair of constrained clusters from two genomes, a bipartite graph $G = (U, V, E)$ is constructed, where nodes U, V represent the set of genes (or could be from a collapsed gene cluster) from the target genome, and the reference genome, respectively, and E represents the homologous gene relationships between the two genomes. A maximum cardinality matching M corresponding to each bipartite graphs is calculated, and the bipartite graph with the largest M is identified as the largest conserved cluster. At this stage, the whole gene cluster is collapsed into one node in each of the two genomes, and, correspondingly, all edges are collapsed into one edge, which connects the two new nodes (Figure 16.4). This iteration continues until no cluster with $M \geq 2$ can be found. Finally, we collect all genes in each contracted node in the genome and consider them as an operon.

16.4   Motif Scanning

In the scanning method, experimentally verified TFBSs are first collected from databases such as regulonDB [16], and all TFBSs of the same TF are aligned (Figure 16.5a). A frequency matrix ($F_{4 \times l}$) corresponding to the aligned matrix is constructed to record the occurrence of base A, C, G, or T in the alignment with the length of l (Figure 16.5b). A position weight matrix (PWM) $M_{4 \times l}$ (Figure 16.5c) is constructed using $F_{4 \times l}$ as follows [17]:

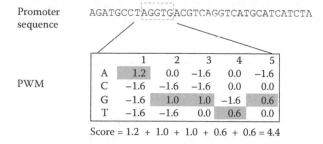

(a) Alignment

	1	2	3	4	5
S1	A	A	G	A	C
S2	A	G	G	C	G
S3	A	G	G	T	T
S4	A	G	T	T	G

(b) Frequency matrix

	1	2	3	4	5
A	4	1	0	1	0
C	0	0	0	1	1
G	0	3	3	0	2
T	0	0	1	2	1

(c) Position weight matrix

	1	2	3	4	5
A	1.2	0.0	−1.6	0.0	−1.6
C	−1.6	−1.6	−1.6	0.0	0.0
G	−1.6	1.0	1.0	−1.6	0.6
T	−1.6	−1.6	0.0	0.6	0.0

(d) Scanning for TFBSs with PWMs

Promoter sequence: AGATGCCTAGGTGACGTCAGGTCATGCATCATCTA

PWM

	1	2	3	4	5
A	1.2	0.0	−1.6	0.0	−1.6
C	−1.6	−1.6	−1.6	0.0	0.0
G	−1.6	1.0	1.0	−1.6	0.6
T	−1.6	−1.6	0.0	0.6	0.0

Score = 1.2 + 1.0 + 1.0 + 0.6 + 0.6 = 4.4

FIGURE 16.5: Motif scanning approach using position weight matrix (PWM). PWM is constructed based on the aligned TFBSs and is used to identify candidate TFBSs in the promoter sequence.

$$M_{i,j} = \ln \frac{(n_{i,j} + p_i)/(N+1)}{p_i}, i \in \{A, C, G, T\}, j \in \{1, 2, \ldots, l\} \qquad (16.3)$$

where p_i is the a priori probability (i.e., the background frequency) for the letter i, N is the total number of aligned sequences, and $n_{i,j}$ is the number of times that letter i is observed at position j in the alignment (Figure 16.5b and c).

The constructed PWM for each TF is used to identify more TFBSs associated with this known TF, based on whether promoter regions of predicted operons in the genome have high scores when compared to the PWM. In particular, for each possible l-mer sequence extracted from the promoter region (usually the upstream sequence up to 400 base-pairs (bp) from the translation start site (TSS)), the total score S is the sum of the weights for the letter aligned at each position. Those sequences whose S values are greater than the predefined threshold value T are considered to be TFBSs. For instance, if T is set to 4.0, and the score for the sequence "AGGTG" is $1.2 + 1.0 + 1.0 + 0.6 + 0.6 = 4.4$, then this sequence is considered to be a TFBS (Figure 16.5d).

This method has been applied in identifying more TFBSs in several studies. For example, Tan et al. [18] identified new regulated genes by the TFs of Cyclic-AMP receptor protein (CRP) and fumarate and nitrate reduction regulatory protein (FNR), based on the number of known TFBSs. Su et al. [19] searched the literature for NtcA-regulated TFBSs, and constructed a PWM to predict more NtcA members. The scanning method usually produces high false positive motifs, due to the nature of short motif length or the degenerated pattern for many motifs. This problem can be resolved by incorporating comparative analysis, where predicted motifs or reliable motifs conserved among orthologous genes are, in general, biological meaningful. For instance, Tan et al. used conserved motifs in the orthologous genes of *E. coli* and *Haemophilus influenzae* to identify new FNR and CRP members [18]. On the other hand, incorporating extra binding site information in the scanning method could also dramatically reduce the high false positives. This has been seen in predicting NtcA-regulated genes by adding the binding site information of σ^{70}, in which case the false positive rate was reduced up to 40-fold [19].

16.5   Phylogenetic Footprinting

High-throughput sequencing technology has made it possible to compare genomic sequence analyses. By collecting the promoter sequences of orthogolous genes (or conserved operons) across multiple species, and analyzing the motif patterns in these regions, we can derive TFBSs in the whole genome and construct conserved regulation networks, without relying on known TFs and their TFBSs. This strategy that, which uses comparative genomic data to derive *cis*-regulatory elements, is also known as *phylogenetic footprinting*. The assumption of this strategy is that orthologous genes should have one (or more) orthologous TFs across multiple species, where TFs regulate orthologous genes (or conserved operons) in a similar fashion. Those TFBSs that are regulated by the same TFs should be much more conserved, compared to those background sequences, and thus can be predicted using statistical models. This strategy can be roughly divided into three major steps: (1) collecting the promoter sequences of orthologous genes (or conserved operons); (2) predicting the TFBSs (or motifs) of all collected sequence sets by using motif-finding programs; and (3) grouping all motifs into clusters based on the similarity of the motif patterns. The corresponding genes (or operons) that share the same *cis*-elements within the same clusters are known as *coregulated genes*, and are thus used for regulatory network construction. We now describe some representative methods for each step in detail.

16.5.1   Sequence collection

Two kinds of approaches of sequence collection have been used: the orthologous-gene-based and the conserved-operon-based approach. In the orthologous-gene-based approach, the orthologous genes of multiple species are first predicted using sequence similarity-based approaches such as BLASTP. The real sequences to be collected are dependent on whether the genes are the first gene of the operon. If the upstream intergenic region of the gene is longer than 100 bp, this gene is simply treated as the first gene of an operon, and the upstream region sequence up to 300 bp from the TSS of the gene is collected. Otherwise, the intergenic sequence is concatenated to the intergenic region of its upstream gene until the first gene of an operon is reached. This kind of strategy includes all sequences of "orthologous" genes by sequence similarity, and it has been used in several studies [20,21]. In the program of "PhyloClus" [22], only those orthologous genes where upstream intergenic regions are longer than 50 bp are included; the remaining ones are discarded. This selection strategy assumes that those orthologous genes that have short upstream regions are not the first genes in the operons, and thus their upstream sequences should not be included.

The rationales of the conserved-operon-based approach are the followings: (1) for any prokaryotic organism, promoter regions exist in the upstream of operon units, not in the intergenic regions within an operon, thus, using operon units for sequence collection simplifies the problem; (2) operon structures of prokaryotic organisms can be accurately predicted, and they are available to use; and (3) the orthologous gene approach using the sequence similarity cannot guarantee the identification of true orthologous relationships in some cases. This problem can be improved by using the conserved operon approach as the cooccurring homologous relationships between two operons usually reflect the true orthology relationships [23].

Sequence collection using the conserved operon approach starts with the identification of conserved operon sets, and then the collection of the upstream sequence (300–400 bp) for each operon of the conserved operon set. OPERMAP [23], a maximum bipartite graph matching based approach, was proposed to identify a conserved operon in a reference genome corresponding to a query operon in a target organism (Figure 16.6). The basic idea of OPERMAP is to identify the conserved operon that shares homolgous gene relationships with the query operon, quantified by the maximum bipartite matching.

Algorithm OPERMAP (U, RO, T)

[**Input**]: A query operon (U) from a target genome, a set of all operons (RO) from a
    reference genome, and a threshold value (T) of the mapping degree

[**Output**]: a pair conserved operon if the threshold requirement is satisfied.

1. Let maxMD = 0, $V^* = \varnothing$
2. for each operon V from RO
3.   Construct the bipartite graph $G = (U, V, E)$, where E represents the homology
      relationship between genes in U and V
4.   Calculate the maximum weighted maximum cardinality bipartite matching M on G
5.   Let the mapping degree MD = $|M|$ /max (opSize (U), opSize (V))
6.   if MD > maxMD
7.   then $V^* = V$ and maxMD = MD
8. return (U, V^*)

FIGURE 16.6: Algorithm for finding a conserved operon for a query operon.

16.5.2   Motif prediction

Given a set of collected sequences of orthologous genes (or conserved oper-ons), the motif-finding problem is to identify motifs (or TFBSs) that are mostly conserved in the sequence set. While different kinds of approaches have been proposed and developed, the basic idea of these approaches is based on the following assumption: each sequence contains zero or more TFBSs bound by one or more TF(s). These aligned TFBSs by the same TF should be much more conserved than those of background sequences. Here, we briefly intro-duce two major categories of methods, enumerative methods and PWM-based methods, with detailed descriptions of several representative programs.

16.5.2.1   Enumerative method

The general idea for the enumerative method is to enumerate all possible words in the sequence set and then check to see which ones are enriched. The estimation of motif enrichment may be employed by statistical analysis like P values. A number of programs, such as Oligodyad [24] and Weeder [25], use the enumerative strategy to identify motifs in the sequence set. In a recent comparison study [26] of 14 motif-finding algorithms, Weeder was ranked the best according to the benchmark evaluation on the datasets of yeasts, humans and flies. We describe here more about the Weeder program to understand why this program performs better than any other program.

Weeder was first developed by Pesole et al. [27]. In order to reduce the search time, the Weeder algorithm preprocesses and organizes the input se-quences as a suffix tree. Like many other enumerative methods, Weeder evalu-ates the motifs based on the statistical point of view. In the recent development of the Weeder Web server [25], however, Weeder analyzes the known transcrip-tional factor binding sites of yeast, and incorporates the prior knowledge into the parameter setting. In addition, the Weeder algorithm analyzes and com-pares the different runs of top-ranked predicted motifs and forms clustering on them to generate a consensus motif. These additional features might explain the superior performance of this algorithm.

The main drawback of the enumerative methods is its time complexity. The enumeration of words is exponential, and thus enumerating all longer consensus motifs is impossible. To address this issue, some enumerative meth-ods allow some numbers of errors for patterns in order to make the searches more computationally feasible [27,28].

16.5.2.2   Position weight matrix (PWM)-based method

The general framework of the PWM approach is to randomly initiate the motif matrix (i.e., the aligned TFBS profile) from the sequence set, refine it iteratively using heuristic algorithms, and reach the stage in which the mo-tif pattern cannot be optimized. In general, the PWM based approaches run fast, but they don't guarantee to identify global optimal results. Different kinds of algorithms have been developed for the PWM optimization problem, including: a greedy based algorithm such as CONSENSUS [17], Gibbs-sampling based algorithms such as Bioprospector [29], and an expectation maximization based approach such as MEME [30]. We describe here two representative algorithms, CONSENSUS and Bioprospector.

CONSENSUS was developed by Hertz and Stormo [17] and has been used widely in motif discovery. The basic idea of the algorithm is to first find the two closest l-mer segments in terms of minimum hamming distance. The algorithm then scans a new sequence in each step, and adds a new l-mer segment in the current PMW that maximizes the score matrix of the PMW, which was evaluated in the information content [17]. In practice, CONSENSUS takes the sequences in a random order for each run so that order bias can be avoided. In addition, CONSENSUS stores a large number of seed matrices in each step and thus avoids losing the optimal solution.

Bioprospector [29] is a Gibbs-sampling based approach, which starts with initial motifs randomly selected from each sequence of a sequence set. For each step, the algorithm chooses a sequence and updates the motif profile by excluding the selected sequence. The algorithm then calculates the probability of each possible l-mer segment of the selected sequence and randomly samples a new site according to the calculated probability distribution. The iteration stops when the motif profile converges. Multiple runs of the algorithm are performed in order to avoid the local optima. Bioprospector adopts a new motif scoring function so that it can accept zero or more sites for each sequence. In addition, gapped motifs (i.e., a two-block motif) and motifs with palindromic patterns are also incorporated in Bioprospector.

While numerous methods have been developed, the benchmark evaluation of several datasets has indicated that no single approach performs better than the others in all cases [26,31,32]. In addition, most of motif-finding programs generate a large number of false positive motifs. The possible reasons are that motifs are usually short, ranging from 6 to 30 nucleotides, and they are usually variable, making it more difficult to assess how sophisticated a computational model is. To alleviate such problems, the prediction results of different programs might be combined by using a consensus ensemble algorithm [32]. Alternatively, the predicted motifs of different algorithms can be re-evaluated and improved by the motif optimization function [33,34].

16.5.3 Motif clustering

The systematic prediction of motif patterns at the whole-genome level generates many similar motif patterns. Such similar motifs generated by the phylogenetic footprinting strategy might be regulated by the same TFs, or the same family of TFs. Those TFs that regulate many genes are called *global regulators*. For example, CPR regulates 199 operons, covering 404 genes, in *E. coli.* The clustering of all motifs predicted by the motif-finding programs can be used to construct conserved regulatory networks. To do so, a motif similarity measurement should be established first.

Different distance metrics have been used for pairwise motif comparisons and alignments, including the Pearson correlation coefficient of nucleotide frequencies [35], the average log likelihood ratio (ALLR) [36], and the sum of squared distances [37]. The alignment algorithms are applied using either local alignments such as Smith–Waterman, or global alignments such as Needleman–Wunsch. The similarity score of the pairwise alignment is then used in the clustering procedure, and various clustering strategies have been employed. We describe here two clustering algorithms: PHYLONET (a graph-based algorithm) [38] and Bayesian motif clustering (BMC, a stochastic sampling algorithm) [39].

The PHYLONET algorithm starts with the identification of high-scoring pairs (HSPs) by calculating their ALLR scores and P values. A motif graph is constructed in such a way that each vertex is a predicted motif profile, and each edge represents an HSP. A maximum clique-based approach is applied in the motif graph in order to identify highly connected subgraphs. The significance of the identifies subgraphs (or clusters) is estimated by the P values, which are approximation of the Poisson distribution.

The PHYLONET algorithm was applied in *Saccharomyces cerevisiae*. Two hundred and ninety-six statistically significant motifs, covering 3315 promoters, were discovered in the regulatory network. Further analysis showed that more than 90% experimentally verified TFBSs were included in the predicted motifs. Most of the predicted gene clusters in the network were supported by microarray experiments and biological function enrichment.

The Gibbs-sampling based approach, widely used in motif-finding programs, can also be used for motif clustering. The BMC algorithm starts with a random partition of motifs into an arbitrary number of clusters. The algorithm iterates a number of times until no improvement can be made. For each iteration, one motif is selected to be reassigned to a new cluster, given the current partition of all the other motifs. The probability of being assigned into one of the existing clusters or a new cluster is calculated, and then the motif is randomly sampled into one of the clusters based on the calculated probabilities. The BMC algorithm was applied in *Escherichia coli*, and it correctly identified many known co-regulated genes, as well as some new members.

16.6   Uber-Operon Prediction

The prediction of uber-operons is completely dependent on the availability of the genome data, as well as on operon structures. Currently, experimentally verified operon data are limited to only in a few genomes, and they are not even complete. The computational advance in operon prediction, however, has made it possible to obtain predicted operon data at the genome scale with high prediction accuracy. The predicted operons can be used to derive uber-operons using the comparative genome analysis. Here, we describe two algorithms for uber-operon prediction.

For the method described in Lathe et al. [40], the orthologous relationships across multiple genomes are assumed to be known. The algorithm starts with identifying a gene whose orthologous gene number is greater than a predefined threshold value. The algorithm then collects all neighbors (within the same operons) of those orthologous genes and considers them as *conserved neighbors* if the total number of ortholog for a neighboring gene is greater than three. This procedure is repeated on the newly identified neighboring genes until there are no such conserved neighbors.

Che et al. [41] developed a graph-based approach for uber-operon prediction without knowing the orthology relationships in advance. The maximum bipartite cardinality matching based algorithm can identify uber-operons, and simultaneously determine the orthology relationships between two genomes. The algorithm is designed to meet two objectives: (a) separate as many operon groups (or *components*) as possible, where operons within the operon group are linked by homologous relationships; and (b) maximize the total number of orthlologys between two genomes (see Figure 16.7).

The input of the algorithm are the operon data of two genomes and the homologous gene relationships predicted by BLAST. The algorithm repeatedly finds two *components* (C_i, C_j) in each step, where the weight $(w(C_i, C_j))$ of the two components is maximum in all components for that step. $w(C_i, C_j)$ is defined as follows:

$$w(C_i, C_j) = |M_{i,j}| - |M_i| - |M_j| \qquad (16.4)$$

where $|M_i|$, $|M_j|$, $|M_{i,j}|$ are the maximum matching sizes of C_i, C_j and the merged component of C_i and C_j. Each component is composed of operons

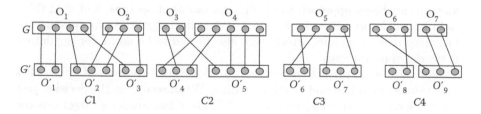

FIGURE 16.7: Example of predicted uber-operons for the genomes G and G'. The genome G contains seven operons $(O_1, O_2, \ldots, O_7)$, while G' contains nine operons $(O'_1, O'_2, \ldots, O'_9)$. The algorithm partitions the whole bipartite graph into a maximum number of connected components (four components here) while keeping the sum of the maximum matchings of all components maximum. The predicted uber-operons for the genome G are O_1O_2, O_3O_4, O_5, and O_6O_7, while the predicted ones for the genome G' are $O'_1O'_2O'_3$, $O'_4O'_5$, $O'_6O'_7$, and $O'_8O'_9$.

from two genomes, and a bipartite graph is built, where the vertices represent the genes of these operons and the edges represent the homologs of these genes. A maximum matching can be calculated [15]. Initially, each individual operon is considered as a component, and the maximum matching size for each component is zero. The two identified components are merged into a new, larger component. This process continues until there is no such pair of components whose weight is greater than zero.

The predicted uber-operons of *E. coli* using this algorithm were compared with experimentally verified regulons, which were sets of operons that were regulated by the same TF. This analysis showed that uber-operons and regulons are highly related, indicating that operon members within an uber-operon are regulated by the same TF. Since uber-operons are sets of operons that are conserved across multiple genomes, they should be inherently related, which could be reflected in the same transcriptional regulation. The prediction of uber-operons for regulatory network construction is complementary to the current mainstream approach, the *cis*-element approach, as we know this approach usually has low prediction sensitivity.

16.7   Conclusions

In this chapter, we began with the introduction of different kinds of approaches to tackle the transcriptional regulatory network problem, and then focused on the *cis*-motif based approach for the regulatory network construction in microbial organisms. Since the basic transcription units for the prokaryotic organisms are operons, we described various methods for operon prediction. We then described the motif scanning method and the phylogenetic footprinting based approach to identify *cis*-motifs of operons, and used them to build regulatory networks. Finally, we described algorithms for finding evolutionary related operons, which were also proved to be useful for regulatory networks in prokaryotes.

For the operon prediction problem, we described two kinds of approaches in detail: decision tree and graph matching. We showed that the decision tree approach could be very effective when all kinds of features for a target genome were available, while the graph matching based approach proved to be very useful when only a minimal amount of information for the target genome was available, but with other genomic data available for comparative analysis.

For the detection of *cis*-motifs using the scanning and phylogenetic footprinting methods, we used the predicted operons as the transcription units, and then collected their promoter regions for motif identification. We introduced different approaches for the motif discovery problem. We also mentioned the challenging issues related to this problem, and made suggestions for using multiple algorithms to improve prediction results. In order to obtain a reliable and complete regulatory network, we might incorporate all sorts of experimental regulation datasets, as well as state-of-the-art computational tools.

Acknowledgments

D. Che's work was supported in part by the Pennsylvania KISZK grant (C000032549). GJ Li's work was supported in part by grants (60873207, 10631070 and 60373025) from the NSFC and the Taishan Scholar Fund from Shandong Province, China. The authors are also grateful to anonymous referees for their helpful insights during the review process.

References

[1] J. L. Reed, I. Famili, I. Thiele et al. Towards multidimensional genome annotation. *Nat Rev Genet,* 7(2), 130–41, 2006.

[2] J. J. Faith, B. Hayete, J. T. Thaden et al. Large-scale mapping and validation of Escherichia coli transcriptional regulation from a compendium of expression profiles. *PLoS Biol,* 5(1), 54–66, 2007.

[3] T. S. Gardner, D. di Bernardo, D. Lorenz et al. Inferring genetic networks and identifying compound mode of action via expression profiling. *Science,* 301(5629), 102–5, 2003.

[4] M. Madan Babu, S. A. Teichmann, and L. Aravind. Evolutionary dynamics of prokaryotic transcriptional regulatory networks. *J Mol Biol,* 358(2), 614–33, 2006.

[5] D. A. Tagle, B. F. Koop, M. Goodman et al. Embryonic epsilon and gamma globin genes of a prosimian primate (*Galago crassicaudatus*). Nucleotide and amino acid sequences, developmental regulation and phylogenetic footprints. *J Mol Biol,* 203(2), 439–55, 1988.

[6] M. J. de Hoon, Y. Makita, K. Nakai et al. Prediction of transcriptional terminators in *Bacillus subtilis* and related species. *PLoS Comput Biol,* 1(3), 212–221, 2005.

[7] H. Salgado, G. Moreno-Hagelsieb, T. F. Smith et al. Operons in Escherichia coli: genomic analyses and predictions. *Proc Natl Acad Sci USA,* 97(12), 6652–57, 2000.

[8] M. D. Ermolaeva, O. White, and S. L. Salzberg. Prediction of operons in microbial genomes. *Nucleic Acids Res,* 29(5), 1216–21, 2001.

[9] R. L. Tatusov, E. V. Koonin, and D. J. Lipman. A genomic perspective on protein families. *Science,* 278(5338), 631–37, 1997.

[10] R. W. Brouwer, O. P. Kuipers, and S. A. van Hijum. The relative value of operon predictions. *Brief Bioinform,* 9(5), 367–75, 2008.

[11] G. Li, D. Che, and Y. Xu. A universal operon predictor for prokaryotic genomes. *J Bioinform Comput Biol,* 7(1), 19–38, 2009.

[12] D. Che, J. Zhao, L. Cai et al. Operon prediction in microbial genomes using decision tree approach. *IEEE Symposium on Computational Intelligence in Bioinformatics and Computational Biology,* 135–42, 2007.

[13] I. H. Witten, and E. Frank. *Data Mining: Practical machine learning tools and techniques,* 2nd ed. San Francisco, CA: Morgan Kaufmann, 2005.

[14] J. R. Quinlan. *C4.5 Programs for machine learning.* San Mateo, CA: Morgan Kaufmann Publishers, 1993.

[15] T. Cormen, C. Leiserson, R. Rivest et al. *Introduction to algorithms.* Cambridge, MA: The MIT Press, 2001.

[16] S. Gama-Castro, V. Jimenez-Jacinto, M. Peralta-Gil et al. RegulonDB (version 6.0): gene regulation model of Escherichia coli K-12 beyond transcription, active (experimental) annotated promoters and Textpresso navigation. *Nucleic Acids Res,* 36, Database issue, D120–24, 2008.

[17] G. Z. Hertz, and G. D. Stormo. Identifying DNA and protein patterns with statistically significant alignments of multiple sequences. *Bioinformatics,* 15(7–8), 563–77, 1999.

[18] K. Tan, G. Moreno-Hagelsieb, J. Collado-Vides et al. A comparative genomics approach to prediction of new members of regulons. *Genome Res,* 11(4), 566–84, 2001.

[19] Z. Su, V. Olman, F. Mao et al. Comparative genomics analysis of NtcA regulons in cyanobacteria: regulation of nitrogen assimilation and its coupling to photosynthesis. *Nucleic Acids Res,* 33(16), 5156–71, 2005.

[20] L. A. McCue, W. Thompson, C. S. Carmack et al. Factors influencing the identification of transcription factor binding sites by cross-species comparison. *Genome Res,* 12(10), 1523–32, 2002.

[21] S. Neph, and M. Tompa. MicroFootPrinter: a tool for phylogenetic footprinting in prokaryotic genomes. *Nucleic Acids Res,* 34, Web Server issue, W366–68, 2006.

[22] S. T. Jensen, L. Shen, and J. S. Liu. Combining phylogenetic motif discovery and motif clustering to predict co-regulated genes. *Bioinformatics,* 21(20), 3832–39, 2005.

[23] D. Che, G. Li, S. Jensen et al. PFP: a computational framework for phylogenetic footprinting in prokaryotic genomes. *Lecture Notes Comput Sci,* 4983, 110–21, 2008.

[24] J. van Helden, A. F. Rios, and J. Collado-Vides. Discovering regulatory elements in non-coding sequences by analysis of spaced dyads. *Nucleic Acids Res,* 28(8), 1808–18, 2000.

[25] G. Pavesi, P. Mereghetti, G. Mauri et al. Weeder Web: discovery of transcription factor binding sites in a set of sequences from co-regulated genes. *Nucleic Acids Res,* 32, Web Server issue, W199–203, 2004.

[26] M. Tompa, N. Li, T. L. Bailey et al. Assessing computational tools for the discovery of transcription factor binding sites. *Nat Biotechnol,* 23(1), 137–44, 2005.

[27] G. Pavesi, G. Mauri, and G. Pesole. An algorithm for finding signals of unknown length in DNA sequences. *Bioinformatics,* 17(Suppl 1), S207–14, 2001.

[28] K. D. MacIsaac, and E. Fraenkel. Practical strategies for discovering regulatory DNA sequence motifs. *PLoS Comput Biol,* 2(4), 201–210, 2006.

[29] X. Liu, D. Brutlag, and J. Liu. BioProspector: discovering conserved DNA motifs in upstream regulatory regions of coexpressed genes. *Pac. Symp. Biocomput.* 127–138, 2001.

[30] T. L. Bailey, and C. Elkan. Fitting a mixture model by expectation maximization to discover motifs in biopolymers. *Proc Int Conf Intell Syst Mol Biol 2,* 28–36, 1994.

[31] C. T. Harbison, D. B. Gordon, T. I. Lee et al. Transcriptional regulatory code of a eukaryotic genome. *Nature,* 431(7004), 99–104, 2004.

[32] J. Hu, B. Li, and D. Kihara. Limitations and potentials of current motif discovery algorithms. *Nucleic Acids Res,* 33(15), 4899–913, 2005.

[33] D. Che, S. Jensen, L. Cai et al. BEST: binding-site estimation suite of tools," *Bioinformatics,* 21(12), 2909–11, 2005.

[34] S. T. Jensen, and J. S. Liu. BioOptimizer: a Bayesian scoring function approach to motif discovery. *Bioinformatics,* 20(10), 1557–64, 2004.

[35] S. Pietrokovski. Searching databases of conserved sequence regions by aligning protein multiple-alignments. *Nucleic Acids Res,* 24(19), 3836–45, 1996.

[36] T. Wang, and G. D. Stormo. Combining phylogenetic data with co-regulated genes to identify regulatory motifs. *Bioinformatics*, 19(18), 2369–80, 2003.

[37] A. Sandelin, and W. W. Wasserman. Constrained binding site diversity within families of transcription factors enhances pattern discovery bioinformatics. *J Mol Biol*, 338(2), 207–15, 2004.

[38] T. Wang, and G. D. Stormo. Identifying the conserved network of cis-regulatory sites of a eukaryotic genome. *Proc Natl Acad Sci USA*, 102(48), 17400–5, 2005.

[39] Z. S. Qin, L. A. McCue, W. Thompson et al. Identification of co-regulated genes through Bayesian clustering of predicted regulatory binding sites. *Nat Biotechnol*, 21(4), 435–39, 2003.

[40] W. C. Lathe, 3rd, B. Snel, and P. Bork. Gene context conservation of a higher order than operons. *Trends Biochem Sci*, 25(10), 474–79, 2000.

[41] D. Che, G. Li, F. Mao et al. Detecting uber-operons in prokaryotic genomes. *Nucleic Acids Res*, 34(8), 2418–27, 2006.

Chapter 17

Computational Methods for Analyzing and Modeling Biological Networks

Nataša Pržulj and Tijana Milenković

University of California, Irvine

17.1   Introduction

Large networks have been used to model and analyze many real world phenomena including biomolecular systems. Although graph theoretic modeling in systems biology is still in its infancy, network-based analyses of cellular systems have already been used to address many important biological questions. We survey methods for analyzing, modeling, and comparing large biological networks that have given insights into biological function and disease.

We focus on protein-protein interaction (PPI) networks, since proteins are important macro-molecules of life and understanding the collective behavior of their interactions is of importance. After discussing the major challenges in the field, we survey network properties, measures used to characterize and compare complex networks. We also give an overview of network models for PPI networks. We discuss to what extent each of the models fits PPI networks and demonstrate that geometric random graphs provide the best fit. We also provide an overview of available network alignment methods and discuss their potential in predicting function of individual proteins, protein complexes, and larger cellular machines. Finally, we present a method that establishes a link between the topological surrounding of a node in a PPI network and its involvement in performing biological functions and in disease.

17.2   Motivation

Recent technological advances in experimental biology have yielded large amounts of biological network data. Many other real-world phenomena have also been described in terms of large *networks* (also called *graphs*), such as various types of social and technological networks. Thus, understanding these complex phenomena has become an important scientific problem that has lead to intensive research in network modeling and analyses. The hope is that utilizing such systems-level approaches to analyzing and modeling complex biological systems will provide insights into biological function, evolution, and disease.

17.2.1   Types of biological networks and availability of data sets

Biological networks come in a variety of forms. They include PPI networks, transcriptional regulation networks, metabolic networks, signal transduction networks, protein structure networks, and networks summarizing neuronal connectivities. These networks differ in whether their nodes represent biomolecules such as genes, proteins, or metabolites, and whether their edges indicate functional, physical, or chemical interactions between the corresponding biomolecules. Studying biological networks at these various granularities could provide valuable insight into the inner working of cells and might lead to important discoveries about complex diseases. However, it is the proteins that execute the genetic program and that carryout almost all biological processes. Thus, we primarily focus on PPI networks, in which nodes correspond to proteins, and undirected edges represent physical interactions amongst them. Nevertheless, methods and tools presented in this chapter can also be easily applied to other types of biological networks.

There exist a variety of methods for obtaining these rich PPI network data, such as yeast 2-hybrid (Y2H) screening, protein complex purification methods using mass-spectrometry (e.g., tandem affinity purification (TAP) and high-throughput mass-spectrometric protein complex identification (HMS-PCI)), correlated messenger RNA (mRNA) expression profiles, genetic interactions, or in silico interaction prediction methods. Y2H and mass spectrometry techniques aim to detect physical binding between proteins, whereas genetic interactions, mRNA coexpression, and in silico methods seek to predict functional associations, for example, between a transcriptional regulator and the pathway it controls. For a more detailed survey of these biochemical methods, see [1].

Numerous datasets resulting from small- and large-scale screens are now publicly available in several databases including: Saccharomyces Genome Database (SGD),* Munich Information Center for Protein Sequences (MIPS),† the Database of Interacting Proteins (DIP),‡ the Molecular Interactions Database (MINT),§ the Online Predicted Human Interaction Database (OPHID),¶ Human Protein Reference Database (HPRD),‖ and the Biological General Repository for Interaction Datasets (BioGRID).** For a more detailed survey of these databases, see [1].

17.2.2 Major challenges

Major challenges when studying biological networks include network analyses, comparisons, and modeling, all aimed at discovering a relationship between network topology on one side and biological function, disease, and evolution on the other.

Since proteins are essential macromolecules of life, we need to understand their function and their role in disease. However, the number of functionally unclassified proteins is large even for simple and well studied organisms such as baker's yeast. Moreover, it is still unclear in what cellular states serious diseases, such as cancer, occur. Methods for determining protein function and identifying disease genes have shifted their focus from targeting specific proteins based solely on sequence homology to analyses of the entire proteome based on PPI networks [2]. Since proteins interact to perform a function, PPI networks by definition reflect the interconnected nature of biological processes. Therefore, analyzing structural properties of PPI networks may provide useful clues about the biological function of individual proteins, protein complexes, pathways they participate in, and larger subcellular machines [2,3].

---

* http://www.yeastgenome.org/
† http://mips.gsf.de/
‡ http://dip. doe-mbi.ucla.edu/
§ http://mint.bio.uniroma2.it/mint/
¶ http://ophid.utoronto.ca/ophid/
‖ http://www. hprd.org/
** http://www.thebiogrid.org/

Additionally, recent studies have been investigating associations between diseases and network topology in PPI networks and have shown that disease genes share common topological properties [3–5]. Finding this relationship between PPI network topology and biological function and disease remains one of the most challenging problems in the post-genomic era.

Analogous to genetic sequence comparison, comparing large cellular networks will revolutionize biological understanding. However, comparing large networks is computationally infeasible due to NP-completeness of the underlying subgraph isomorphism problem. Note that even if the subgraph isomorphism was feasible, it would not find a practical application in biological network comparisons, since biological networks are extremely unlikely to be isomorphic. Thus, large network analyses and comparisons rely on heuristics, commonly called *network properties*. These properties are roughly categorized into *global* and *local*. The most widely used global properties are the *degree distribution*, the *average clustering coefficient*, the *clustering spectrum*, the *average diameter*, and the *spectrum of shortest path lengths* [6]. Local properties include *network motifs*, small over-represented subgraphs [7], and two measures based on *graphlets*, small induced subgraphs of large networks: the *relative graphlet frequency distance (RGF-distance)*, which compares the frequencies of the appearance of graphlets in two networks [8], and the *graphlet degree distribution agreement (GDD-agreement)*, which is a graphlet-based generalization of the degree distribution [9]. These properties have been used to compare biological networks against model networks and to find well-fitting network models for biological networks [8–10], as well as to suggest biological function of proteins in PPI networks [2]. The properties of models have further been exploited to guide biological experiments and discover new biological features, as well as to propose efficient heuristic strategies in many domains, including time- and cost-optimal detection of the human interactome.

Modeling biological networks is of crucial importance for any computational study of these networks. Only a well-fitting network model that precisely reproduces the network structure and laws through which the network has emerged can enable us to understand and replicate the biological processes and the underlying complex evolutionary mechanisms in the cell. Various network models have been proposed for real-world biological networks. Starting with Erdös–Rényi random graphs [11], network models have progressed through a series of versions designed to match certain properties of real-world networks. Examples include random graphs that match the degree distribution of the data [12], network growth models that produce networks with scale-free degree distributions [13] or small network diameters [14], geometric random graphs [15], or networks that reproduce some biological and topological properties of real biological networks (e.g., stickiness model [10]). An open-source software tool called *GraphCrunch* [16] implements the latest research on biological network models and properties and compares real-world networks against a variety of network models with respect to a wide range of network properties.

17.3   Network Properties

Global network properties give an overall view of a network. The most commonly used global network properties are the degree distribution, the average network diameter, the spectrum of shortest path lengths, the average clustering coefficient, and the clustering spectrum. The *degree* of a node is the number of edges incident to the node. The *degree distribution, P(k)*, describes the probability that a node has degree k. The smallest number of links that have to be traversed to get from a node x to a node y in a network is called the *distance* between nodes x and y and a path through the network that achieves this distance is called the *shortest path* between nodes x and y. The average of shortest path lengths over all pairs of nodes in a network is called the *average network diameter*. The *spectrum of shortest path lengths* is the distribution of shortest path lengths between all pairs of nodes in a network. The *clustering coefficient* of a node z in a network, C_z, is defined as the probability that two nodes x and y which are connected to the node z are themselves connected. The average of C_z over all nodes z of a network is the *average clustering coefficient, C*, of the network; it measures the tendency of the network to form highly interconnected regions called clusters. The distribution of the average clustering coefficients of all nodes of degree k in a network is the *clustering spectrum, C(k)*.

However, global network properties are not detailed enough to capture complex topological characteristics of real-world networks. Network properties should encompass large number of constraints, in order to reduce degrees of freedom in which networks being compared can vary. Thus, more constraining measures of local network structure have been introduced. Local properties include *network motifs*, small over-represented subgraphs [7], and highly constraining measures of local structural similarities between two networks: RGF-distance [8] and GDD-agreement [9]. RGF-distance and GDD-agreement are based on *graphlets*, small connected non-isomorphic induced subgraphs of large networks [8]. Note that graphlets are different from network motifs since they must be induced subgraphs (motifs are partial subgraphs) and since they do not need to be over-represented in the data when compared to "randomized" networks. An *induced subgraph* of a graph G on a subset of nodes of G is obtained by taking S and all edges of G having both end-points in S; *partial subgraphs* are obtained by taking S and some of the edges of G having both end-points in S. Thus, graphlets avoid both main criticisms associated with network motifs: (1) they do not depend on the randomization scheme, since they do not need to be over-represented in the data when compared with randomized networks, and (2) they are not susceptible to over-counting as motifs are, since graphlets are induced while motifs are partial subgraphs. Since the number of graphlets on n nodes increases super-exponentially with n, RGF-distance and GDD-agreement computations are currently based on three- to five-node graphlets (presented in Figure 17.1).

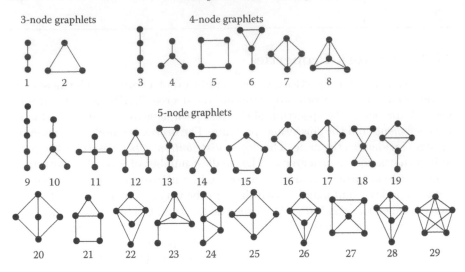

FIGURE 17.1: All three-node, four-node and five-node graphlets. (From Pržulj N, Corneil DG, Jurisica I. *Bioinformatics*, 20(18), 3508–3515, 2004. With permission.)

RGF-distance compares the frequencies of the appearance of all three- to five-node graphlets in two networks (see [8] for details). If networks being compared have the same number of nodes and edges, the frequencies of the occurrence of the only one-node graphlet (a node) and the only two-node graphlet (an edge) are also taken into account by this measure. Thus, RGF-distance encompasses 31 similarity constraints by examining the fit of 31 graphlet frequencies.

GDD-agreement generalizes the notion of the degree distribution to the spectrum of *graphlet degree distributions (GDDs)* in the following way [9]. The degree distribution measures the number of nodes of degree k, i.e., the number of nodes "touching" k edges, for each value of k. Note that an edge is the only graphlet with two nodes (graphlet denoted by G_0 in Figure 17.2). GDDs generalize the degree distribution to other graphlets: they measure for each graphlet $G_i, i \in 0, 1, \ldots, 29$, (illustrated in Figure 17.2) the number of nodes "touching" k graphlets G_i at a particular node. A node at which a graphlet is "touched" is topologically relevant, since it allows us to distinguish between nodes "touching", for example, a copy of graphlet G_1 in Figure 17.2 at an end-node, or at the middle-node. This is summarized by *automorphism orbits* (or just *orbits*, for brevity), as illustrated in Figure 17.2: for graphlets $G_0, G_1, \ldots, G_{29}$, there are 73 different orbits, numerated from 0 to 72 (see [9] for details). For each orbit j, the jth *GDD*, i.e., the distribution of the number of nodes "touching" the corresponding graphlet at orbit j, is measured. Thus, the degree distribution is the 0th GDD. The jth *GDD-agreement* compares the jth GDDs of two networks (see [9] for details). The total GDD-agreement

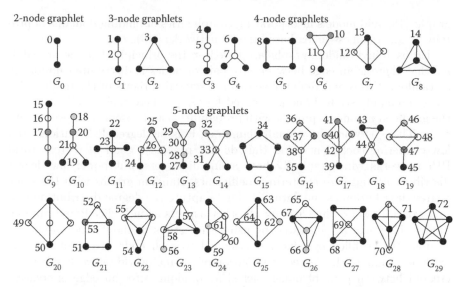

FIGURE 17.2: The 30 two-, three-, four-, and five-node graphlets G_0, $G_1, \ldots, G_{29}$ and their automorphism orbits 0, 1, 2, ..., 72. In a graphlet $G_i, i \in \{0, 1, \ldots, 29\}$, nodes belonging to the same orbit are of the same shade (From Pržulj N., *Bioinformatics*, 23, e177–e183, 2007. With permission.)

between two networks is the arithmetic or the geometric average of the jth GDD-agreements over all j (henceforth arithmetic and geometric averages are denoted by "amean" and "gmean," respectively). GDD-agreement is scaled to always be between 0 and 1, where 1 means that two networks are identical with respect to this property. By calculating the fit of each of the 73 GDDs of the networks being compared, GDD-agreement encompasses 73 similarity constraints. Furthermore, each of these 73 constraints enforces a similarity of two distributions, additionally restricting the ways in which the networks being compared can differ. (Note that the degree distribution is only one of these 73 constraints.) Therefore, GDD-agreement is a very strong measure of structural similarity between two networks. RGF-distance and GDD-agreement were used to discover a new, well-fitting, geometric random graph model of PPI networks [8,9] (see Section 17.3.2).

17.4   Network Models

17.4.1   Survey of network models

The most commonly used network models for PPI networks include: Erdös–Rényi random graphs [11], random graphs with the same degree distribution as the data [12], scale-free networks [13,17], geometric random graphs [15], and models incorporating complexities of binding domains of proteins (e.g., stickiness-index based model networks [10]).

Erdös–Rényi random graphs are based on the principle that the probability that there is an edge between any pair of nodes is distributed uniformly at random. Erdös and Rényi have defined several variants of the model. The most commonly studied one is denoted by $G_{n,p}$, where each possible edge in the graph on n nodes is present with probability p and absent with probability $1-p$. These networks have small diameters, Poisson degree distributions, and low clustering coefficients, and thus do not provide a good fit to real-world PPI networks which typically have small diameters, but power-law degree distributions and high clustering coefficients. Random graphs with the same degree distribution as the data (ER-DD) capture the degree distribution of a real-world network while leaving all other aspects as in Erdös–Rényi random model. They can be generated by using the "stubs method": the number of "stubs" (to be filled by edges) is assigned to each node in the model network according to the degree distribution of the real-world network; edges are created between pairs of nodes picked at random; after an edge is created, the number of "stubs" left available at the corresponding "end-nodes" of the edge is decreased by one. Thus, these networks preserve the degree distributions and small diameters of real-world networks. However, this model also produces networks with low clustering coefficients and thus other network models have been sought. One such example are small-world networks [14]. These networks are created from regular ring lattices by random rewiring of a small percentage of their edges. However, although these networks have high clustering coefficients and small diameters, they fail to reproduce power-law degree distributions of real-world networks.

Scale-free networks are characterized by power-law degree distributions. One such model is the Barabási–Albert preferential attachment model (SF-BA) [13], in which newly added nodes preferentially attach to existing nodes with probability proportional to the degree of the target node. Other variants focused on modeling PPI networks include scale-free network models constructed by mimicking "gene duplications and mutations" [17]. In these model networks, connectivity of some nodes is significantly higher than for the other nodes, resulting in power-law degree distribution. Although the average diameter is small, they typically still have low clustering coefficients.

High clustering coefficients of real-world networks are well reproduced by geometric random graphs (GEO) that are defined as follows: nodes correspond to uniformly randomly distributed points in a metric space and edges are created between pairs of nodes if the corresponding points are close enough in the metric space according to some distance norm [15]. For example, three-dimensional Euclidean boxes (GEO-3D) and the Euclidean distance norm have been used to model PPI networks [8,9]. Although this model creates networks with high clustering coefficients and small diameters, it still fails to reproduce power-law degree distributions of real-world PPI networks. Instead, geometric random graphs have Poisson degree distribution. However, it has been argued that power-law degree distributions in PPI networks are an artifact of noise present in them.

Finally, "stickiness network model" (STICKY) is based on stickiness indices, numbers that summarize node connectivities and thus also the complexities of binding domains of proteins in PPI networks. The probability that there is an edge between two nodes in this network model is directly proportional to the stickiness indices of nodes, i.e., to the degrees of their corresponding proteins in real-world PPI networks (see [10] for details). Networks produced by this model have the expected degree distribution of a real-world network. Additionally, they mimic well the clustering coefficients and the diameters of real-world networks.

17.4.2 An optimized null model for protein–protein interaction (PPI) networks

Modeling bio-chemical networks is a vibrant research area. The choice of an appropriate null model can have important implications for many graph-based analyses of these networks. For example, the use of an adequate null model is vital for structural motif discovery, which requires comparing real-world networks with randomized ones. Using an inappropriate network null model may identify as over-represented (under-represented) subgraphs that otherwise would not have been identified. Another example is that a good null model can be used to guide biological experiments in a time- and cost-optimal way and thus minimize the costs of interactome detection by predicting the behavior of a system. Since incorrect models lead to incorrect predictions, it is vital to have as accurate a model as possible.

17.4.2.1 Geometricity of protein–protein interaction (PPI) networks

Also, as new biological network data becomes available, we must ensure that the theoretical models continue to accurately represent the data. The scale-free model has been assumed to provide such a model for PPI networks. However, in the light of new PPI network data, several studies have started questioning the wellness of fit of scale-free network model. For example, Pržulj et al. [8] and Pržulj [9] have used two highly constraining measures of local network structures to compare real-world PPI networks to various network models and have shown compelling evidence that the structure of yeast PPI networks is closer to the geometric random graph model than to the widely accepted scale-free model. Furthermore, Higham et al. [18] have designed a method for embedding networks into a low-dimensional Euclidean space and demonstrated that PPI networks can be embedded and thus have a geometric graph structure (see below).

In search of a well fitting null model for biological networks, one has to consider biological properties of a system being modeled. Geometric random graph model of PPI networks is biologically motivated. Genes and proteins as their products exist in some highly-dimensional biochemical space. The currently accepted paradigm is that evolution happens through a series of gene duplication and mutation events. Thus, after a parent gene is duplicated, the child gene is at the same position in the biochemical space as the parent and therefore inherits interactions with all interacting partners of the parent. Evolutionary optimization then acts on the child gene to either become obsolete and disappear from the genome, or to mutate distancing itself somewhat from the parent, but preserving some of the parent's interacting partners (due to proximity to the parent in the biochemical space) while also establishing new interactions with other genes (due to the short distance of the mutated child from the parent in the space). Similarly, in geometric random graphs, the closer the nodes are in a metric space, the more interactors they will have in common, and vice-versa. Thus, the superior fit of geometric random graphs to PPI networks over other random models is not surprising.

A well-fitting null model should generate graphs that closely resemble the structure of real networks. This closeness in structure is reflected across a wide range of statistical measures, i.e., network properties. Thus, testing the fit of a model entails comparing model-derived random graphs to real networks according to these measures. Global network properties, such as the degree distribution, may not be detailed enough to capture the complex topological characteristics of PPI networks. The more constraining the measures are, the fewer degrees of freedom exist in which the compared networks can vary. Thus, by using highly constraining measures, such as RGF-distance and GDD-agreement, a better-fitting null model can be found.

RGF-distance was used to compare PPI networks of yeast (*Saccharomyces cereviscae*), and fruitfly (*Drosophila melanogaster*), to a variety of network models and to show the supremacy of the fit of geometric random graph model to these networks over three other random graph models. Pržulj et al. [8] compared the frequencies of the appearance of all three- to five-node graphlets in these PPI networks with the frequencies of their appearance in four different types of random networks of the same size as the data: ER, ER-DD, SF-BA, and GEO. Furthermore, several variants of the geometric random graphs were used, depending on the dimensionality of the Euclidean space chosen to generate them: two-dimensional (GEO-2D), three-dimensional (GEO-3D), and four-dimensional (GEO-4D) geometric random graphs. Four real-world PPI networks were analyzed: high-confidence and lower-confidence yeast PPI networks, and high-confidence and low-confidence fruitfly PPI networks. Pržulj et al. [8] computed RGF-distances between these real-world PPI networks and the corresponding ER, ER-DD, SF-BA and GEO random networks. They found that the GEO random networks fitted the data an order of magnitude better than other network models in the higher-confidence PPI networks, and less so (but still better) in the more noisy PPI networks. The only exception was the noisy fruitfly PPI network which exhibited scale-free behavior. It was hypothesized that this behavior of the graphlet frequency parameter was the consequence of a large amount of noise present in this network.

Since currently available PPI data sets are incomplete, i.e., have a large percentage of false negatives or missing interactions, and thus are expected to have higher edge densities, Pržulj et al. also compared the high-confidence yeast PPI network against three-dimensional geometric random graphs with the same number of nodes, but about three and six times as many edges as the PPI network, respectively. By making the GEO-3D networks corresponding to this PPI network about six times as dense as the PPI network, the closest fit to the PPI network with respect to RGF-distance was observed. Additionally, to address the existence of noise, i.e., false positives, in PPI networks, the high-confidence yeast PPI network was perturbed by randomly adding, deleting, and rewiring 10%, 20% and 30% of edges and RGF-distances between the perturbed networks and the PPI network were computed. The study demonstrated that graphlet frequencies were robust to these random perturbations, thus further increasing the confidence in PPI networks having geometric network structure.

Geometric structure of PPI networks has also been confirmed by GDD-agreements between PPI and model networks drawn from several different random graph models [8]: ER, ER-DD, SF-BA, and GEO-3D. Several PPI networks of each of the following four eukaryotic organisms were examined: yeast (*S. cerevisiae*), frutifly (*D. melanogaster*), nematode worm (*Caenorhabditis elegans*), and human. The total of 14 PPI networks originating from different sources, obtained with different interaction detection techniques (such as Y2H, TAP, or HMS-PCI, as well as human curation), and of different interaction confidence levels were analyzed. GEO-3D network model showed the highest GDD-agreement for all but one of the fourteen PPI networks; for the remaining network, GDD-agreements between GEO-3D, SF, and ER-DD models and the data were about the same (see Figure 17.3).

Additionally, an algorithm that directly tests whether PPI networks are geometric has been proposed [18]. It does so by embedding PPI networks into a low dimensional Euclidean space. If a geometric network model fits the PPI network data, then it is expected that PPI networks can be embedded into some space. Geometric random graphs constructed from Euclidean space are chosen as a proof of concept. These graphs in two-dimensional space are generated by placing N nodes uniformly at random in the unit square, and by connecting two nodes by an edge if they are within a given Euclidean distance. The three- and four-dimensional cases are defined analogously. The task is then to embed the proteins into n-dimensional Euclidean space for $n = 2, 3, 4$, given only their PPI network connectivity information. The algorithm is based on Multi-Dimensional Scaling, with shortest path lengths between protein pairs in a PPI network playing the role of the Euclidean distances between the corresponding nodes embedded in the n-dimensional Euclidian space. After proteins are embedded, a radius r is chosen and each node is connected to all other nodes that are at most at distance r from that node. Each choice of a radius thus corresponds to a different geometric graph. By varying the radius, specificity and sensitivity are measured and the overall goodness of fit

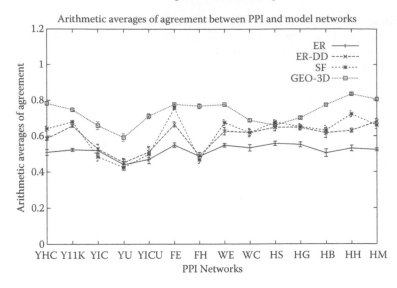

FIGURE 17.3: GDD-agreements between the 14 PPI networks of four organisms and their corresponding model networks. Labels on the horizontal axis denote the 14 PPI networks (see [9] for details). Averages of GDD-agreements between 25 model networks and the corresponding PPI network are presented for each random graph model and each PPI network, i.e., at each point in the figure; the error bar around a point is one standard deviation below and above the point (in some cases, error bars are barely visible, since they are of the size of the point). (From Pržulj N, *Bioinformatics*, 23, e177–e183, 2007. With permission.)

is judged by computing the areas under the receiver operator characteristic (ROC) curve, with higher values indicating a better fit.

The algorithm was applied to 19 real-world PPI networks of yeast, fruit-fly, worm, and human obtained from different sources, as well as to artificial networks generated using seven types of random graph models: ER, ER-DD, GEO (GEO-2D, GEO-3D, and GEO-4D), SF-BA, and STICKY. These networks were embedded into two-dimensional (2D), three-dimensional (3D), and four-dimensional (4D) Euclidean space. The resulting areas under the ROC curves (AUCs) were high for all PPI networks. The highest AUC value was obtained for embedding the high-confidence yeast (YHC) PPI network. The authors focused their further analyses on YIIC network. Random graphs of the same size as YHC network drawn from the seven network models were embedded into 2D, 3D and 4D space. For geometric random networks, AUCs were very high, with values above 0.9. For non-geometric networks AUCs were below 0.78. Since PPI networks are noisy, to test whether PPI networks had a geometric structure, the authors added noise to GEO-3D networks by randomly rewiring 10%, 20%, and 30% of their edges. These rewired networks were then embedded into 2D, 3D and 4D space, and their AUCs were computed. The values of AUCs for the 10% rewired GEO-3D networks were very similar to those for real-world networks. Thus, by embedding networks into low dimensional Euclidean space, the authors performed a direct test of whether PPI networks had a geometric graph structure. The results yielded support to the results of previous studies and to the hypothesis that the structure of currently available PPI networks was consistent with the structure of noisy geometric graphs [18].

17.4.2.2 "Stickiness" of protein–protein interaction (PPI) networks

Another biologically motivated "stickiness index"-based network model has been proposed for PPI networks [10]. It is commonly considered that proteins interact because they share complimentary physical aspects, a concept that is consistent with the underlying biochemistry. These physical aspects are referred to as binding domains [10]. In the stickiness-index-based network model, the stickiness index of a protein is a number based on its normalized degree that summarizes the abundance and popularity of binding domains of the protein. The model assumes that a high degree of a protein implies that the protein has many binding domains and/or its binding domains are commonly involved in interactions. Additionally, the model considers that a pair of proteins is more likely to interact (i.e., share complementary binding domains) if both proteins have high stickiness indices, and less likely to interact if one or both have a low stickiness index. Thus, according to this model, the probability that there exist an edge between two nodes in a random graph is the product of the two stickiness indices of the corresponding proteins in the PPI network (see [10] for details). The resulting model networks are guaranteed to have the expected degree distributions of real-world networks.

To examine the fit of this network model to real-world PPI networks, as well as to compare its fit against the fit of other network models, a variety of global and local network properties were used [10]: the degree distribution, clustering coefficient, network diameter, and RGF-distance. In addition to the stickiness-index-based network model (STICKY), model networks were also drawn from the following network models: ER, ER-DD, SF-BA, and GEO-3D. The fit of fourteen real-world PPI networks of four organisms (yeast, fruitfly, worm, and human) to each of these five network models was evaluated with respect of all above mentioned network properties. With respect to RGF-distance, the stickiness model showed an improved fit over all other network models in ten out of fourteen tested PPI networks. It showed as good results as the GEO-3D model in one and was outperformed by the GEO-3D model in three PPI networks. In addition, this model reproduced well global network properties such as the degree distributions, the clustering coefficients, and the average diameters of PPI networks. Thus, this model that uses biologically motivated assumptions mentioned above clearly outperforms scale-free network models such as SF-BA and ER-DD that also match the degree distribution of a real-world PPI network.

17.5 Network Comparison and Alignment

Just as comparative genomics has led to an explosion of knowledge about evolution, biology, and disease, so will comparative proteomics. As more biological network data is becoming available, comparative analyses of these networks across species are proving to be valuable, since such systems biology types of comparisons may lead to exciting discoveries in evolutionary biology. For example, comparing networks of different organisms might provide deeper insights into conservation of proteins, their function, and PPIs through evolution. Conceptually, network comparison is the process of contrasting two or more interaction networks, representing different species, conditions, interaction types, or time points, aimed at answering some fundamental biological questions [19].

17.5.1 Types of network comparison methods

Three different types of comparative methods exist [19]. The most common methods are *network alignments.* An alignment is achieved by constructing a mapping between the nodes of networks being compared, as well as between the corresponding interactions. In this process, topologically and functionally similar regions of biological networks are discovered. Depending on the properties of mappings, network alignment can be local or global. Most of the research in previous years has been focused on local alignments. With local network alignment algorithms, optimal mappings are chosen independently for each local region of similarity. With global network alignments, one optimal mapping for the entire network is constructed, even though this may imply less perfect alignments in some local regions.

The second type of a method is *network integration,* the process of combining different networks and encompassing interactions of different types over the same set of elements to study their inter-relations. Each type of a network provides an insight into a different slice of biological information, and thus, integrating different network types could provide a more comprehensive picture of the overall biological system under study [19]. The major difference from network alignment is the following. Networks to be integrated are defined over the same set of elements and integration is achieved by merging them into a single network with multiple types of interactions, each drawn from one of the original networks. A fundamental problem is then to identify in the merged network functional modules that are supported by interactions of multiple types.

Finally, the third type of comparison is *network querying,* in which a network is searched for subnetworks that are similar to a subnetwork query of interest. Network alignment and integration are focused on *de novo* discovery of biologically significant regions embedded in a network, based on the assumption that regions supported by multiple networks are functionally important. In contrast, network querying searches for a subnetwork that is previously known to be functional. The goal is to identify subnetworks in a given network that are similar to the query. However, network querying tools are still in their infancy since they are currently limited to sparse topologies, such as paths and trees.

17.5.2 Algorithms for network alignment

Here, we focus on network alignment methods that have been applied to PPI networks. Due to the subgraph isomorphism problem, the problem of network alignment is computationally hard and thus heuristic approaches have been sought.

Conceptually, to perform network alignment, a merged representation of the networks being compared, called a network alignment graph, is created [19]. In a network alignment graph, the nodes represent sets of "similar" molecules, one from each network, and the links represent conserved molecular interactions across the different networks. There are two core challenges involved in performing network alignment and constructing the network alignment graph. First, a scoring framework that captures the "similarities" between nodes originating in different networks must be defined. Then, a way to rapidly identify high-scoring alignments (i.e., conserved functional modules) from among the exponentially large set of possible alignments needs to be specified. Due to the computational complexity, a greedy algorithm needs to be used for this purpose. Methods for network alignment differ in these two challenges, depending how they define the similarity scores between protein pairs and what greedy algorithm they use to identify conserved subnetworks.

The problem of network alignment has been approached in different ways and a variety of algorithms have been developed. Unlike the majority of algorithms focusing on pairwise network alignments, newer approaches have tried to address the problem of aligning networks belonging to multiple organisms; note that multiple network alignment represents a challenge since computational complexity increases dramatically with the number of networks. Additionally, instead of performing local alignments, algorithms for global network alignment have emerged. Finally, whereas previous studies have focused on network alignment based solely on biological functional information such as protein sequence similarity, recent studies have been combining the functional information with network topological information [19].

In the most simple case, the similarity of a protein pair, where one protein originates from each of the networks being aligned, is determined solely by their sequence similarity. Then, the top scoring protein pairs, typically found by applying BLAST to perform all-to-all alignment between sequences of proteins originating in different networks, are aligned between the two networks. The most simple network alignment then identifies pairs of interactions in PPI networks, called interologs, involving two proteins in one species and their best sequence matches in another species. However, beyond this simple identification of conserved protein interactions, it is more interesting to identify network *subgraphs* that might have been conserved across species. Algorithms

Biological Data Mining

for network alignment are still in their early development, since they are currently limited to identifying conserved pathways or complexes. Methods for detecting larger and denser structures are unquestionably needed.

An algorithm called *PathBLAST** searches for high-scoring pathway alignments between two networks under study. Pathway alignments are scored as follows. The likelihood of a pathway match is computed by taking into account both the probabilities of true homology between proteins that are aligned on the path and the probabilities that the PPIs that are present in the path are real, i.e., not false-positive errors. The score of a path is thus a product of independent probabilities for each aligned protein pair and for each protein interaction. The probability of a protein pair is based on the BLAST E-value of aligning sequences of the corresponding proteins, whereas the probability of a protein interaction is based on the false-positive rates associated with interactions. The PathBLAST method has been extended to detect conserved protein clusters rather than paths, by deploying a likelihood-based scoring scheme that weighs the denseness of a given subnetwork versus the chance of observing such network substructure at random. Moreover, it has also been extended to allow for the alignment of more than two networks.

MaWISh† is a method for pairwise local alignment of PPI networks implementing an evolution-based scoring scheme to detect conserved protein clusters. This mathematical model extends the concepts of evolutionary events in sequence alignment to that of duplication, match, and mismatch in network alignment. The method evaluates the similarity between graph structures through a scoring function that accounts for these evolutionary events. Each duplication is associated with a score that reflects the divergence of function between the two proteins. The score is based on the protein sequence similarity and is computed by BLAST. A match corresponds to a conserved interaction between two orthologous protein pairs. A mismatch, on the other hand, is the lack of an interaction in the PPI network of one organism between a pair of proteins whose orthologs interact in the other organism. A mismatch may correspond to the emergence of a new interaction or the elimination of a previously existing interaction in one of the species after the split, or to an experimental error. After each match, mismatch, and duplication is given a score, the optimal alignment is defined as a set of nodes with the maximum score, computed by summing all possible matches, mismatches, and duplications in the given set of nodes.

Graemlin‡ has been introduced as the first method capable of multiple alignment of an arbitrary number of networks, supporting both global and local search, and being capable of searching for dense conserved subnetworks of an arbitrary structure. Graemlin's purpose is to search for evolutionarily conserved functional modules across species. The method supports five types of evolutionary events: protein sequence mutations, protein insertions, protein deletions, protein duplications, and protein divergences (a divergence

---

* http://www. pathblast.org/

† http://www. cs.purdue.edu/homes/koyuturk/mawish/

‡ http://graemlin.stanford.edu/

being inverse of a duplication). The module alignment score is computed by deploying two models that assign probabilities to the evolutionary events: the alignment model that assumes that a module is subject to evolutionary constraint, and the random model that assumes that the proteins are under no constraints. Then, the score of the alignment is the log-ratio of the two probabilities, which is a common method for scoring sequence alignments.

Finally, we conclude with *ISORANK* [20], a method initially designed for pairwise global alignment of PPI networks that has later been extended to allow for multiple local alignment of these networks. The initial version of ISORANK maximizes the overall match between the two networks by using both biological (i.e., BLAST-computed protein sequence similarity) and topological (i.e., protein interaction) information. Given two networks, the output of the algorithm is the maximum common subgraph between the two graphs, i.e., the largest graph that is isomorphic to subgraphs of both networks. The algorithm works in two stages. It first associates a score with each possible match between nodes of the two networks. The scores are computed using the intuition that two nodes, one from each network, are a good match if their respective neighbors also match well with each other. The method captures not only local topology of nodes, but also non local influences on the score of a protein pair: the score of the protein pair depends on the score of the neighbors of the two nodes, and the latter, in turn, depend on the neighbors of their neighbors, and so on. The incorporation of other information, e.g. BLAST scores, into this model is straightforward. The second stage constructs the mapping by extracting from all protein pairs the high-scoring matches by applying the repetitive greedy strategy of identifying and outputting the highest scoring pair and removing all scores involving any of the two identified nodes. The later version of ISORANK is the direct generalization of the algorithm to support multiple networks.

These methods have been used to identify network regions, protein complexes, and functional modules that have been conserved across species. Additionally, since a conserved subnetwork that contains many proteins of the same known function suggests that the remaining proteins also have that function, the network alignment methods have been used to predict new protein functions. Similarly, functions for unannotated proteins in one species have been predicted based on the functions of their aligned annotated partners in the other species. Finally, since proteins that are aligned together are likely to be functional orthologs, orthology relationships across multiple species have been predicted.

17.6 From Structure to Function in Biological Networks

17.6.1 Protein function prediction

We have illustrated how network comparisons across species can help in functional annotation of individual proteins and in identification of network

regions representing functional modules or protein complexes. Similarly, biological function of uncharacterized proteins in a PPI network of a single species
can be determined from the function of other, well described proteins from
the same network. It has been shown that proteins that are closer in a network are more likely to perform the same function [2]. Similarly, proteins with
similar topological neighborhoods show tendency to have similar biological
characteristics [3]. Moreover, cancer genes have been shown to have greater
connectivities and centralities compared to non-cancer genes [5]. Thus, various network structural properties such as the shortest path distances between
proteins, network centralities, or graphlet-based descriptions of protein's topological neighborhoods can suggest their involvement in certain biological functions and disease. Since defining a relationship between PPI network topology
and biological function and disease and inferring protein function from it is
considered to be one of the major challenges in the post-genomic era [2],
various approaches for determining protein function from PPI networks have
been proposed. They can be categorized into two major classes: direct and
cluster-based methods [2].

The methods in the first class infer the function of an individual protein
by exploring the premise that proteins that are closer in the network are more
likely to share a function. The simplest method of this type is the "majority
rule" that investigates only the direct neighborhood of a protein, and an-
notates it with the most common functions among its annotated neighbors.
However, this approach does not assign any significance values to predicted
functions. Additionally, it considers only nodes directly connected to the protein of interest and thus, only very limited topology of a network is used in
the annotation process. Finally, it fails to differentiate between proteins at
different distances from the target protein. Other approaches have tried to
overcome these limitations by observing n-neighborhood of an annotated protein, where n-neighborhood of a protein is defined as a set of proteins that are
at most at distance n from the target protein. Then, the protein of interest is
assigned the most significant function (with the highest χ-square value) among
functions of all n-neighboring proteins. This approach thus covers larger portion of the network compared to the simple majority rule. Additionally, it
assigns confidence scores to the predicted functions. However, it still fails to
distinguish between proteins at different distances from the protein of interest. A forthcoming study has addressed this approach by assigning different
weights to proteins at different distances from the target protein. However,
this method observes only one- and two-neighborhoods of proteins, thus again
covering only their local topologies.

For this reason, several global optimization-based function prediction
strategies have been proposed. For example, any given assignment of functions to the whole set of unclassified proteins in a network is given a score,
counting the number of interacting pairs of nodes with no common function;
the functional assignment with the lowest score maximizes the presence of
the same function among interacting proteins. Since this method again fails

to distinguish between proteins at different distances from the protein of interest, thus not rewarding local proximity, a network flow-based method that considers both local and global effects has been proposed. According to this method, each functionally annotated protein in the network is considered as the source of a "functional flow." Then, the spread of the functional flow through the network is simulated over time, and each unannotated protein is assigned a score for having the function based on the amount of flow it received during the simulation.

Approaches of the second type are exploiting the existence of regions in PPI networks that contain a large number of connections between the constituent proteins. These dense regions are a sign of a common involvement of those proteins in certain biological processes and therefore are feasible candidates for biological complexes or functional modules. Thus, these approaches try to identify clusters in a network and then, instead of predicting functions of individual proteins, they assign an entire cluster with a function based on the functions of its annotated members. Various approaches for identifying these functionally enriched modules solely from PPI network topology have been defined. For example, the highly connected subgraphs and the restricted neighborhood search clustering (RNSC) algorithms have been used to detect complexes in PPI networks [21,22]. However, with some of the cluster-based methods (e.g., MCODE*), the number of clusters or the size of the sought clusters need to be provided as input [2].

Additionally, several iterative hierarchical clustering-based methods that form clusters by computing the similarities between protein pairs have been proposed. Thus, the key decision with these methods is the choice of the appropriate similarity measures between protein pairs. The most intuitive network topology-based measure is to use pairwise distances between proteins in the network [2]: the smaller the distance between the two proteins in the PPI network is, the more "similar" they are, and thus, the more likely they are to belong to the same cluster. In other words, module members are likely to have similar shortest path distance profiles. However, global network properties such as pairwise shortest path distances might not be detailed enough and more constraining measures of node similarities are necessary.

Here, we present a sensitive graph theoretic method for comparing local network structures of protein neighborhoods in PPI networks that demonstrates that in PPI networks, biological function of a protein and its local network structure are closely related [3]. The method summarizes the local network topology around a protein in a PPI network into a vector of "graphlet degrees" (GDs) called the "signature of a node" (i.e., the signature of a protein) and computes the "signature similarities" between all protein pairs (see Section 17.6.1.1 below for details). Proteins with topologically similar network neighborhoods are then grouped together under this measure and the resulting protein groups have been shown to belong to the same protein complexes,

---

* http://baderlab.org/Software/MCODE

perform the same biological functions, are localized in the same subcellular compartments, and have the same tissue expressions. This has been verified for PPI networks of a unicellular and a multicellular eukaryotic organisms of yeast and human, respectively. Thus, it is hypothesized that PPI network structure and biological function are closely related in other eukaryotic organisms as well. Next, since the number of functionally unclassified proteins is large even for simple and well studied organisms such as baker's yeast (*S. cerevisiae*), Milenković and Pržulj [3] have described how to apply their technique to predict a protein's membership in protein complexes, biological functional groups, and subcellular compartments for yet unclassified yeast proteins. Additionally, they have shown how this method can be used for identification of new disease genes, demonstrating that it can provide valuable guidelines for future experimental research.

17.6.1.1   Graphlet degree signatures

Similar to GDDs described in Section 17.2, this node similarity measure generalizes the degree of a node, which counts the number of edges that a node touches, into a vector of *GDs*, counting the number of graphlets that a node touches. The method counts the number of graphlets touching a node for all two- to five-node graphlets (these graphlets are denoted by $G_0, G_1, \ldots, G_{29}$ in Figure 17.2); counts involving larger graphlets become computationally infeasible for large networks. For example, an outer (black) node in graphlet G_9 touches graphlets G_0, G_1, G_3, and G_9 once, and it touches no other graphlets. Clearly, the degree of a node is the first coordinate in this vector, since an edge (graphlet G_0) is the only two-node graphlet. This vector is called the *signature* of a node.

Due to the existence of 73 automorphism orbits for two- to five-node graphlets (described in Section 17.2), the signature vector of a node has 73 coordinates. For example, a node at orbit 15 in graphlet G_9 touches orbits 0, 1, 4, and 15 once, and all other orbits zero times. Thus, its signature will have 1s in the 0th, 1st, 4th, and 15th coordinate, and 0s in the remaining 69 coordinates.

Node *signature similarities* are computed as follows. A 73-dimensional vector W is defined to contain the weights w_i corresponding to orbits $i \in \{0, \ldots, 72\}$. Different weights are assigned to different orbits for the reasons illustrated below. For example, the differences in orbit 0 (i.e., in the degree) of two nodes will automatically imply the differences in all other orbits for these nodes, since all orbits contain, i.e., "depend on," orbit 0. Similarly, the differences in orbit 3 (the triangle) of two nodes will automatically imply the differences in all other orbits of the two nodes that contain orbit 3, such as orbits 14 and 72. This is generalized to all orbits. Thus, higher weights need to be assigned to "important" orbits, those that are not affected by many other orbits, and lower weights need to be assigned to "less important" orbits, those that depend on many other orbits. By doing so, the redundancy of an orbit contained in other orbits is removed. For details on computing orbit weights, see [3].

For a node u, u_i denotes the ith coordinate of its signature vector, i.e., u_i is the number of times node u touches orbit i. The distance $Di(u, v)$ between the ith orbits of nodes u and v is then defined as:

$$D_i(u, v) = w_i \times \frac{|log(u_i + 1)| - log(v_i + 1)|}{log(max\{u_i, v_i\} + 2)}.$$

The authors use *log* in the numerator because the ith coordinates of signature vectors of two nodes can differ by several orders of magnitude and the distance measure should not be entirely dominated by these large values [3]. Also, by using these logarithms, they take into account the relative difference between u_i and v_i instead of the absolute difference. They scale D_i to be in $[0, 1)$ by dividing with the value of the denominator in the formula for $D_i(u, v)$. The total distance $D(u, v)$ between nodes u and v is then defined as:

$$D(u, v) = \frac{\sum_{i=0}^{72} D_i}{\sum_{i=0}^{72} w_i}.$$

Clearly, the distance $D(u, v)$ is in $[0, 1)$, where distance 0 means the identity of signatures of nodes u and v. Finally, the *signature similarity*, $S(u, v)$, between nodes u and v is: $S(u, v) = 1 - D(u, v)$. For example, the two outer (black) nodes at orbit 15 in graphlet G_9 have the same signatures, and thus, their total distance is 0 and their signature similarity is 1.

Clusters in a PPI network are formed as follows: for a node of interest, construct a cluster containing that node and all nodes in a network that have similar signatures to it. This is repeated for each node in the PPI network. Thus, nodes u and v will be in the same cluster if their signature similarity $S(u, v)$ is above a chosen threshold. Thresholds are determined experimentally to be in 0.9–0.95. For thresholds above these values, only a few small clusters are obtained, especially for smaller PPI networks, indicating too high stringency in signature similarities. For thresholds bellow 0.9, the clusters are very large, especially for larger PPI networks, indicating a loss of signature similarity. To illustrate signature similarities and the choices of signature similarity thresholds, Figure 17.4 presents the signature vectors of yeast proteins in a PPI network with signature similarities above 0.90 (Figure 17.4 a) and below 0.40 (Figure 17.4 b); signature vectors of proteins with high signature similarities follow the same pattern, while those of proteins with low signature similarities have very different patterns.

The method described above was applied to six yeast and three human PPI networks of different sizes and different confidence levels [3]. After the creation of clusters in each network, the clusters were searched for common *protein properties*. In yeast PPI networks, protein properties included protein complexes, functional groups, and subcellular localizations described in MIPS*; in human PPI networks, they involved biological processes, cellular components, and tissue expressions described in HPRD.[†] Classification schemes and the

---

* http://mips.gsf.de/

[†] http://www. hprd.org/

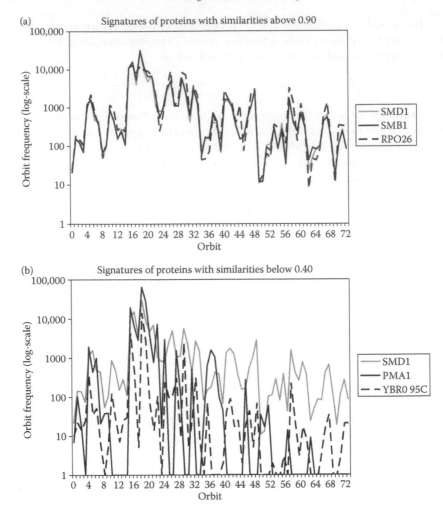

FIGURE 17.4: Signature vectors of proteins with signature similarities: (a) above 0.90; and (b) below 0.40. The 73 orbits are presented on the abscissa and the numbers of times that nodes touch a particular orbit are presented on the ordinate in log scale. In the interest of the aesthetics of the plot, 1 is added to all orbit frequencies to avoid the log-function to go to infinity in the case of orbit frequencies of 0.

data for the protein properties were downloaded from MIPS and HPRD in November 2007. For each of the protein property classification schemes, two levels of strictness were defined: the *strict* scheme that uses the most specific MIPS annotations, and the *flexible* one that uses the least specific ones. For example, for a protein complex category annotated by *510.190.900* in MIPS, the strict scheme returns *510.190.900,* and the flexible one returns 510.

In the clusters, the sizes of the largest common categories for a given protein property were measured as a percentage of the cluster size and this was referred to as the *hit-rate*. Clearly, a protein can belong to more than one protein complex, be involved in more than one biological function, or belong to more than one subcellular compartment. Thus, it is possible to have an overlap between categories, as well as more than one largest category in a cluster for a given protein property. A *miss-rate* was also defined as a percentage of the nodes in a cluster that are not in any common category with other nodes in the cluster. For each of the PPI networks, the corresponding protein properties, and the two schemes (the strict and the flexible), the percentage of clusters having given hit- and miss-rates was measured. The hit- and miss-rates were binned in increments of 10%. The method identified high hit-rates and low miss-rates in a large percentage of clusters in all six yeast and three human PPI networks, for all protein properties (see [3] for details). For example, for protein complexes in the high-confidence yeast PPI network, 44% of clusters had 100% hit-rate according to the flexible scheme. This additionally validated the method, since PPIs in this network have been obtained mainly by TAP and HMS-PCI, which are known to favor protein complexes. Thus, the method established a strong relationship between a biological function of a protein and the network structure around it in a PPI network.

To evaluate the effect of noise in PPI networks on the accuracy of the method, the authors compared the results for the high-confidence and the lower-confidence yeast networks. As expected, clusters in a noisier network had lower hit-rates than the clusters in a high-confidence network. However, low miss-rates were still preserved in clusters of both networks for all three protein properties, indicating the robustness of the method to noise present in PPI networks.

Furthermore, the statistical significance of the results was examined by the following two analyses: (a) random reshuffling of node labels in PPI networks and comparisons of the results with those obtained from such randomized networks; and (b) computing p-values for obtaining the observed homogeneous protein properties within clusters. Hit-rates for all protein properties for the PPI networks were higher than for randomized networks, thus indicating that the signature-based algorithm captured the true biological signal from the topology of PPI networks. Similarly, miss-rates for the data were lower than for randomized networks with respect to all protein properties in PPI networks. This was especially true for protein complexes and biological function in yeast. Only for subcellular localization, miss-rates in the data and in randomized networks were comparable, but hit-rates were still higher in the data than in the randomized networks. The main reason for observing similarity in miss-rates for subcellular localization in the data and randomized networks was the large size of most subcellular localization categories. Therefore, it was expected that most proteins in a cluster would be in a common localization category with at least one other protein in a cluster, which led to low miss-rates.

Additionally, the probability that a given cluster was enriched by a given category merely by chance was computed, following the hypergeometric distribution (as in [22]). The total number of proteins in all clusters was denoted by $|N|$, the size of cluster C by $|C|$, and the number of proteins in cluster C that were enriched by category P by k, where category P contained $|P|$ out of $|N|$ proteins. Thus, the hit-rate of cluster C was $k/|C|$, and the $p-$value for cluster C and category P, i.e., the probability of observing the same or higher hit-rate, was:

$$p\text{-}\mathrm{value} = 1 - \sum_{i=0}^{k-1} \frac{\binom{|P|}{i}\binom{|N|-|P|}{|C|-i}}{\binom{|N|}{|C|}}.$$

The p-value of a cluster was considered to be its smallest p-value over all categories for a given protein property. p-values for all signature-based clusters with hit-rates of at least 50% were computed for all yeast and human PPI networks and for all protein properties. Then, the percentage of clusters (out of the total number of clusters with hit-rates of at least 50%) having a given p-value was found. Depending on a method and its application, sensible cut-offs for p-values were reported to range from 10^{-2} to 10^{-8} [22].

For yeast, with respect to both the strict and the flexible scheme, low p-values of $O(10^{-2})$ or lower were observed for protein complexes and biological functions, whereas for subcellular localizations, a percentage of clusters had higher p-values of $O(10^{-1})$. As explained above, since subcellular localization categories typically contained a large number of proteins, this high p-values were expected. Since the flexible scheme by definition meant that a larger number of proteins was contained within each category, somewhat higher, but still low enough, p-values were observed compared to the strict scheme. For humans, a significant percentage of clusters had high p-values, for all three human protein properties. The reason for this was the same as for subcellular localization in yeast: many proteins existed within each category in all three human protein properties. However, a certain percentage of clusters had low p-values. For example, for tissue expressions, about 50% of clusters for all three human PPI networks had p-values of $O(10^{-2})$ or lower. Therefore, although p-values varied depending on a given protein property and the size of its categories, the algorithm identified clusters in which true biological signal was captured. This was especially true for protein complexes and biological functions in all six yeast PPI networks, with respect to both the strict and the flexible scheme.

This technique can also be applied to predict protein properties of yet unclassified proteins by forming a cluster of proteins that are similar to the unclassified protein of interest and assigning it the most common properties of the classified proteins in the cluster. The authors did this for all 115 functionally unclassified yeast proteins from MIPS that had degrees higher than four in any of the six yeast PPI networks that they analyzed (the list of predictions is in [3]). Note that a yeast protein can belong to more than one yeast PPI network that was analyzed. Thus, biological functions that such proteins perform can be predicted from clusters derived from different yeast PPI networks. An overlap of the predicted protein functions obtained from multiple PPI networks for the same organism was observed, additionally confirming the validity of predictions. Furthermore, there existed an overlap between protein function predictions produced by this method and those of others. Finally, the predictions were verified in the literature.

The graphlet degree signatures-based method has several advantages over direct approaches for protein function prediction (described above). Not only that it assigns a confidence score to each predicted annotation (in terms of hit- and miss-rates), but also for doing that it takes into account up to four-neighborhoods of a node along with all interconnectivities, since it is based on two- to five-node graphlets. Additionally, although the signature of a node describes its "four-deep" local neighborhood, due to typically small diameters of PPI networks, it is possible that two- to five-node-graphlet-based signatures capture the full, or almost full topology of these networks.

Also, the graphlet degree signatures method belongs to the group of clustering-based approaches. However, unlike other methods of this type that define a cluster as a dense interconnected region of a network (typically enriching for a biological function, as described above), this method defines a cluster as a set of nodes with similar topological signatures. Thus, nodes belonging to the same cluster do not need to be connected or belong to the same part of the network. Additionally, whereas other approaches typically assign the function to the entire cluster, since this method forms a cluster for each protein in the PPI network individually, it assigns function(s) to individual proteins. Moreover, the method does not require the number of clusters or their size to be predefined, unlike some of the other above mentioned approaches. Furthermore, to create pairwise similarities between protein pairs, this method uses highly constraining local-topology-based measure of similarity of proteins' signatures, unlike other studies that use only global network properties; additionally, this clustering method is not hierarchical, and it allows for overlap between clusters.

It is difficult to perform direct comparisons of the performance of all methods described above. Attempts to perform a comparison of several cluster-based methods have been made. However, due to different performance measures across different studies, fundamental differences between different annotation types, and a lack of the golden standards for functional annotation, any comprehensive comparison is very difficult [2]. For example, only this study used hit- and miss-rates to measure the success of the results of the method. Additionally, some studies used the MIPS* annotation catalogs, whereas other studies used Gene Ontology[†] as the annotation source, and some annotations that exist in one data source might not exist in the other. Moreover, to our knowledge, the above described study by Milenković and

---

* http://mips.gsf.de/
[†] http://www.geneontology.org/

Pržulj [3] is the only one that related the PPI network structure to all of the following: protein complexes, biological functions and subcellular localizations for yeast and cellular components, tissue expressions, and biological processes for human, thus making it impossible to do comparisons with other methods from this aspect. However, even if they were unable to quantify their results with respect to other studies, Milenković and Pržulj provided other indications of the correctness of their approach (see above).

The graphlet signatures-based method is easily extendible to include larger graphlets, but this would increase the computational complexity. The complexity is currently $O(|V|^5)$ for a graph $G(V, E)$, since the searches for graphlets with up to five nodes are performed. Nonetheless, since the algorithm is "embarrassingly parallel," i.e., can easily be distributed over a cluster of machines, extending it to larger graphlets is feasible. In addition to the design of the signature similarity measure as a number in (0, 1), this makes the technique usable for much larger networks.

17.6.2   Disease gene identification

In addition to protein function prediction, several studies have investigated associations between diseases and PPI network topology. Radivojac et al. [4] have tried to identify candidate disease genes from a human PPI network by encoding each gene in the network based on the distribution of shortest path lengths to all genes associated with disease or having known functional annotation. Additionally, Jonsson and Bates [5] analyzed network properties of cancer genes and demonstrated greater connectivity and centrality of cancer genes compared to non-cancer genes indicating an increased central role of cancer genes within the interactome. However, these studies have been mainly based on global network properties, which might not be detailed enough to encompass complex topological characteristics of disease genes in the context of PPI networks.

Similarly, graphlet degree signature-based method has also been applied to disease genes. A set of genes implicated in genetic diseases available from HPRD* was examined. To increase coverage of PPIs, the human PPI network that was analyzed was the union of the human PPI networks from HPRD, BIOGRID, and Rual et al. [23], which consisted of 41,755 unique interactions amongst 10,488 different proteins. There were 1,491 disease genes in this PPI network out of which 71 were cancer genes. If network topology is related to function, then it is expected that genes implicated in cancer might have similar graphlet degree signatures. To test this hypothesis, Milenković and Pržulj looked for all proteins with a signature similarity of 0.95 or higher with protein TP53. The resulting cluster contained ten proteins, eight of which were disease genes; six of these eight disease genes were cancer genes (TP53, EP300, SRC, BRCA1, EGFR, and AR). The remaining two proteins in the

---

* http://www. hprd.org/

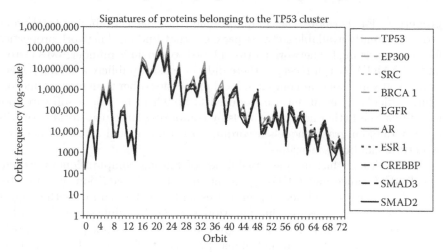

FIGURE 17.5: Signature vectors of proteins belonging to the TP53 cluster. The cluster is formed using the threshold of 0.95. The axes have the same meaning as in Figure 17.4.

cluster were SMAD2 and SMAD3 which are members of TGF-beta signaling pathway whose deregulation contributes to the pathogenesis of many diseases including cancer. The striking signature similarity of this 10-node cluster is depicted in Figure 17.5.

17.7 Software Tools for Network Analyses and Modeling

The recent explosion in biological and other real-world network data has created the need for improved tools for large network analyses. In addition to well established *global* network properties, several new mathematical techniques for analyzing *local* structural properties of large networks have been developed (see Section 17.2). Adequate null-models for biological networks have been sought in many research domains and various network models have been proposed (see Section 17.3.1). Network properties are used to assess the fit of network models to the data.

Computing global network properties is computationally and conceptually easy and various software tools are available for this purpose, such as tYNA* and pajek.† However, none of the tools have built-in capabilities to compare real-world networks against a series of network models based on these

---

* http://tyna.gersteinlab.org/tyna/
† http://vlado.fmf.unilj.si/pub/networks/pajek/

properties. Furthermore, the computational challenge is in finding local properties. Currently available software packages that find local network properties focus on searching for network motifs. These tools include mfinder,* MAVisto,[†] and FANMOD.[‡] Until recently, there did not exist a publicly available, open-source software tool that computed local properties other than network motifs. Thus, Milenković et al. have introduced GraphCrunch[§] [16], a software tool that finds well-fitting network models by comparing large real-world networks against random graph models according to various global and local network structural similarity measures.

GraphCrunch has unique capabilities of finding computationally expensive RGF-distance and GDD-agreement measures. In addition, it computes several standard global network measures and therefore supports the largest variety of network measures thus far. More specifically, while some of the other above mentioned software tools compute only local network properties, others that analyze both local and global properties offer fewer functions than GraphCrunch does. The main purpose of mfinder, MAVisto, and FANMOD is motif search; they do not compute global network properties. On the other hand, pajek focuses on global network properties and has very limited local network analysis capabilities; its search for subgraphs is limited to three- to four-node rings. tYNA's global and local network analyses are limited: it calculates the statistics of global network properties and focuses on three network motif types only. Unlike any of these software packages, GraphCrunch uses all of the two- to five-node graphlets for computing its two highly constraining graphlet-based local network properties, GDD-agreement [9] and RGF-distance [8], along with five standard global properties. The properties currently supported by GraphCrunch are presented in Table 17.1 and described in Section 17.2.

Furthermore, GraphCrunch uses all of these properties for comparing real-world networks against a series of network models. Five network models are currently supported by GraphCrunch. They are presented in Table 17.2 and explained in Section 17.3.1. Although mfinder, FANMOD and pajek offer more than one network model (MAVisto does not), none of these tools supports a variety of network models as GraphCrunch does. Note that tYNA does not generate random models at all and it searches for subgraphs in real-world networks only. Furthermore, GraphCrunch determines the fit of various network models to real-world networks with respect to an array of global and local network properties; none of the other currently available network analysis software tools have this functionality.

Finally, although mfinder and FANMOD both include an option of using random subgraph sampling heuristics for speeding up the computing time,

---

* http://www.weizmann.ac.il/mcb/UriAlon/groupNetworkMotifSW.html
† http://mavisto.ipk-gatersleben.de/
‡ http://www.minet.uni-jena.de/~wernicke/motifs/index.html
§ http://www.ics.uc.edu/~bio-nets/graphcrunch/. Labeled as "highly accessed" by *BMC Bioinformatics*.

TABLE 17.1: Network properties currently supported by GraphCrunch.

Global properties:

Degree distribution
Clustering coefficient
Clustering spectrum
Average diameter
Spectrum of shortest path lengths

Local properties:

Relative graphlet frequency distance (RGF-distance)
Graphlet degree distribution agreement (GDD-agreement)

TABLE 17.2: Network models currently supported by GraphCrunch.

Models:

Erdös–Rényi random graphs (ER)
Random graphs with the same degree distribution as the data (ER-DD)
Scale-free Barabási-Albert model graphs (SF-BA)
N-dimensional geometric random graphs (GEO-nd; default: GEO-3d)
Stickiness model graphs (STICKY)

---

a feature that GraphCrunch currently does not support, GraphCrunch's exhaustive graphlet counting is very competitive. Moreover, GraphCrunch is easily extendible to include additional network measures and models and it has built-in parallel computing capabilities allowing for a user specified list of machines on which to perform compute intensive searches for local network properties. This feature is not supported by mfinder, FANMOD, MAVisto, tYNA, or pajek. This functionality will become crucial as biological network data sets grow.

Many network analysis and modeling software tools are already available. However, as biological data becomes larger and more complete, the need for improving these tools and the algorithms that they implement will continue to rise.

17.8 Concluding Remarks

Biological networks research is still in its infancy, but has already become a vibrant research area that is likely to have impacts onto biological understanding and therapeutics. As such, it is rich in open research problems that we are currently only scratching a surface of. Many new, unforeseen problems will keep emerging. Thus, the field is likely to stay at the top of scientific endeavor in the years to come.

References

[1] Pržulj N. Graph theory analysis of protein-protein interactions. In *Knowledge Discovery in Proteomics*. Edited by Jurisica I, Wigle D. CRC Press, Boca Raton, FL, 2005:73–128.

[2] Sharan R, Ulitsky I, Ideker T. Network-based prediction of protein function. *Molecular Systems Biology* 2007:3(88).

[3] Milenković T, Pržulj N. Uncovering biological network function via graphlet degree signatures. *Cancer Informatics* 2008:6:257–273.

[4] Radivojac P, Peng K, Clark WT, Peters BJ, Mohan A, Boyle SM, Mooney SD. An inte-grated approach to inferring gene-disease associations in humans. *Proteins* 2008:72(3):1030–1037.

[5] Jonsson P, Bates P. Global topological features of cancer proteins in the human interactome. *Bioinformatics* 2006:22(18):2291–2297.

[6] Newman MEJ. The structure and function of complex networks. *SIAM Review* 2003:45(2):167–256.

[7] Milo R, Shen-Orr SS, Itzkovitz S, Kashtan N, Chklovskii D, Alon U. Network motifs: simple building blocks of complex networks. *Science* 2002:298:824–827.

[8] Pržulj N, Corneil DG, Jurisica I. Modeling interactome: scale-free or geometric? *Bioinformatics* 2004:20(18):3508–3515.

[9] Pržulj N. Biological network comparison using graphlet degree distribution. *Bioinformatics* 2007:23:e177–e183.

[10] Pržulj N, Higham D. Modelling protein-protein interaction networks via a stickiness index. *Journal of the Royal Society Interface* 2006:3(10):711–716.

[11] Erdös P, Rényi A. On random graphs. *Publicationes Mathematicae* 1959:6:290–297.

[12] Molloy M, Reed B. A critical point for random graphs with a given degree sequence. *Random Structures and Algorithms* 1995:6:161–179.

[13] Barabási AL, Albert R. Emergence of scaling in random networks. *Science* 1999:286(5439):509–512.

[14] Watts DJ, Strogatz SH. Collective dynamics of 'small-world' networks. *Nature* 1998:393:440–442.

[15] Penrose M. *Geometric Random Graphs*. Oxford University Press, USA, 2003.

[16] Milenković T, Lai J, Pržulj N. GraphCrunch: a tool for large network analyses. *BMC Bioinformatics* 2008:9(70).

[17] Vazquez A, Flammini A, Maritan A, Vespignani A. Modeling of protein interaction networks. *ComPlexUs* 2003:1:38–44.

[18] Higham D, Rašajski M, Pržulj N. Fitting a geometric graph to a protein-protein interaction network. *Bioinformatics* 2008:24:1093–1099.

[19] Sharan R, Ideker T. Modeling cellular machinery through biological network comparison. *Nature Biotechnology* 2006:24(4):427–433.

[20] Singh R, Xu J, Berger B. Pairwise global alignment of protein interaction networks by matching neighborhood topology. *RECOMB 2007, LNBI* 2007:4453:16–31.

[21] Pržulj N, Wigle D, Jurisica I. Functional topology in a network of protein inter-actions. *Bioinformatics* 2004:20(3):340–348.

[22] King AD, Pržulj N, Jurisica I. Protein complex prediction via cost-based clus-tering. *Bioinformatics* 2004:20(17):3013–3020.

[23] Rual JF, Venkatesan K, Hao T, Hirozane-Kishikawa T, Dricot A, Li N, Berriz GF, et al. Towards a proteome-scale map of the human protein-protein interaction network. *Nature* 2005:437:1173–1178.

Chapter 18

Statistical Analysis of Biomolecular Networks

Jing-Dong J. Han

Chinese Academy of Sciences

Chris J. Needham

University of Leeds

18.1   Brief Introduction to Biomolecular Networks

In the cell biomolecules, proteins, genes, metabolites, and miRNAs are interconnected with each other in a biomolecular network. In networks, nodes are biomolecules, and edges are functional relationships among the nodes. Functional relationships can be transcriptional and translational regulation, protein interactions, gene modifications, protein modifications, metabolic reactions, and indirect interactions like genetic interactions.

Based on the node and edge type, the networks can be classified into a few large categories: (1) protein–protein interaction (PPI) networks, where nodes are proteins and edges are direct PPIs, such as binding, complex formation, phosphorylation, dephosphorylation, ubiquitination, deubiquitination, and so on, with the first two types of interactions as undirected edges; (2) regulatory networks, where the nodes are of two types, transcription factors and their target genes or binding sequences, and the edges are directed from transcription factors to their targets; (3) genetic interaction networks, where the nodes are genes and the edges are epistatic relationship (directed) or synthetic interactions (undirected); (4) metabolic networks, where nodes are metabolites, or small chemical molecules, and the edges are the reactions connecting them catalyzed by enzymes; (5) small molecule-target networks and microRNA-target networks are also emerging. The biomolecular networks are encoded by the genetic or gene networks as determined by the genomic sequences. However, "molecular networks," "genetic networks" or "gene networks" are sometimes used interchangeably for practical reasons.

A network can be characterized by many topological and dynamic features, and many of them specifically reveal the function of the network. The topological and dynamic features of the network can be found through statistical analysis. Therefore, statistical analyses of the networks have been increasingly used to discern the function of the biomolecular networks and to predict genes that contribute to complex phenotypes or complex diseases (Han, 2008). In the sections below, we will go through frequently used network analyses, the contexts and goals of using them, as well as the biological feature they might reveal. We provide a high level overview of the most important concepts and the most common methodologies in the field to introduce the statistical analysis of biomolecular networks.

18.2 Network Topologies

Network topology measurements can be roughly divided into global and local measurements. Although interlinked, the value of former measurements depends on the whole network, where as that of the latter depends on small regions of a network, or a subnetwork. The former includes characteristic path length (CPL), degree distribution, betweenness of nodes or edges, and the latter includes network motifs, clusters, etc.

18.2.1 Characteristic path length (CPL)

Between any pair of nodes that are connected through either direct or indirect edges, there may be multiple paths to reach each other. The path length is the number of edges within a path. Among any pair of connected nodes, at least one path has the shortest path length. In a PPI network, where each edge represents a direct PPI, a shortest path between a pair of nodes is then the minimal number of interactions needed to connect the two nodes (proteins) given they are in the same network component (a connected subnetwork). This is also referred to as the degree of separation between nodes. A six-degree separation indicates that the two nodes are six direct interactions away from each other. The shorter the path, the more closely related the two nodes are in terms of biological functions. For example, the phenotypic and gene expression similarities are both inversely correlated with the shortest path length of gene pairs in the PPI network (Gunsalus et al., 2005).

The average of all the shortest path lengths of all connected node pairs is also called the CPL of the network. The shorter the CPL, the more densely connected the network is. Therefore, CPL has been used as a good indicator of the density of network connectivity. A change of CPL upon removal of node(s) hence signifies the impact of the removed nodes on network connectivity, or the network structural stability.

Biological networks usually have very short CPLs, which can mainly be attributed to the over-representation of densely connected regions (such as cliques) and highly connected nodes, also known as hubs (see below).

18.2.2 Degree distributions

The number of edges attached to a node is also called the "degree" of the node (often denoted as k). The degree distribution of a network shows the proportion of nodes having a particular number of interactions within a network. Since it is a property of the whole network, it is usually referred to as a global property of a network. But when we focus on an individual node, the degree (although based on a relative distribution) is a local property of the node.

Biological networks are frequently scale-free, that is, the degree distribution is linear when plotted on log-log scale ($log(p(k)) \sim log(k)$), where k is degree and $p(k)$ is the fraction of nodes having degree k. Such a distribution implies that the majority of the nodes have only one or two interactions, whereas a small, but significant number of nodes have a large number of interactions, which are also called hubs of a scale-free network.

Derived from the degree distribution is the concept of degree centrality. Simulation of node removal in a scale-free network reveals that removals of hubs have a stronger impact on the connectivity of the network than non-hubs, as measured by the CPL. In yeast PPI networks, the higher interaction degree a protein has, the more likely the gene encoding the protein is essential to the survival of the yeast (Jeong et al., 2001). Therefore, high degree proteins, or hubs, are also regarded as more central to the network, and node degree is sometimes referred to as the "degree centrality." Power law degree distribution-related topics have been extensively described previously (Chung and Lu, 2006).

18.2.3   Betweenness

Mirroring degree centrality, betweenness centrality is also frequently used to determine the importance of a node to the network structural stability. Betweenness can be applicable to both nodes and edges. It is defined as the number of shortest paths passing through a node ("node betweenness") or an edge ("edge betweenness"). Edge betweenness has been used to find the network communities through sequentially removing the edges of highest betweenness, based on the assumption that network communities are connected through sparse connections, and hence heavier traffic per edge when the edge bridges between communities. A high betweenness value indicates the edges or the nodes are bottlenecks for node to node communication within a network. They are sometimes directly called "bottlenecks." This feature also gives rise to the concept of betweenness centrality, where a bottleneck has central position in the network. Betweenness centrality of yeast proteins in yeast PPI networks correlates better to the essentiality of the gene/protein than degree centrality although node betweenness is highly correlated with node degree (Yu et al., 2007).

18.2.4   Network motifs

Network motifs were first proposed by Alon and coworkers as statistically over-represented small subnetworks (usually up to four nodes because of computational intensity) compared to randomized networks of the same degree distribution (Milo et al., 2002). In yeast and *E. coli* regulatory networks, small circuits of three or four nodes have been surveyed, and structures such the feedback and feed-forward configurations (Figure 18.1) have been shown to appear more frequently than random expectations, and are therefore called the motifs of the networks (Milo et al., 2002). The motifs in a regulatory network, consist of all directed edges, and often have unique functions on network dynamics. For example, a negative feedback loop usually stabilizes the signal passing through it, and can serve as a barrier to the random noise accidentally input into the system, whereas a positive feedback loop may potentiate the signal and drive the system toward a new stable state. Feed-forward loops, on the other hand can create combinatory effect or form redundancy. These circuits are also frequently seen in signaling networks.

The overrepresented motifs in undirected PPI networks are however often full or partial cliques within the networks (also referred to as network clusters, see below) corresponding to protein complexes (Wuchty et al., 2003).

18.3   Network Modules

A network module can be defined as a subnetwork meeting three criteria: precisely defined inputs and outputs, comparable timescale, and conditionally

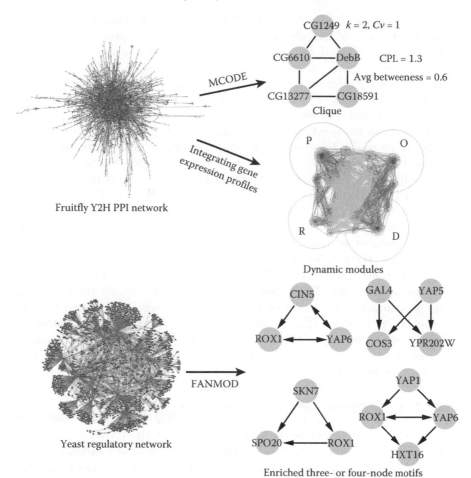

FIGURE 18.1: Network measurements and basic structural and functional units. Measurements include degree (k) and clustering coefficient ($C(v)$) illustrated on the graph. The calculation of CPL, betweenness and others are described in the text. Functional units range from small motifs (found through the FANMOD program, http://www.minet.uni-jena.de/~wernicke/motifs/), to clusters/cliques (found through the MCODE program, http://baderlab.org/Software/MCODE), to large size modules, including but not limited to those dissected based on edge density (structural modules), edge type consistency (epistatic modules), or expression profiles (dynamic profiles). The fruitfly aging network modules were used as examples (Xue et al., 2007).

spatial colocalization (Papin et al., 2005). Such a definition emphasizes the context-coherence of modules and is applicable to both static and dynamic network modules. Modules can also be classified into structural, epistatic, dynamic modules and so on by the procedures applied to the network to find the modules. However, for any subnetworks to be recognized as modules, the nodes (genes or proteins) within it must be functionally homogenous or coherent. Functional coherency of a module can be measured by various metrics, such as the function diversity, function category entropy, function enrichment and so on. Function diversity of a module is defined as the number of unique function categories associated with the nodes within the module divided by the total number of nodes in the module. It is inversely correlated with functional homogeneity or uniformity (Han et al., 2004). Function category entropy of a module is expressed as $-\sum_i F_i \log F_i$, where F_i is the frequency of appearance of function category i. $F_i = T / \sum_i^n T_i$, where T_i is the number of times the function category i appears in the subnetwork and n is the number of distinct function categories present in the module (Snel et al., 2002). Lower function category entropy indicates greater function homogeneity. These metrics have been successfully used for yeast gene/protein networks, based on the functional categories assigned to genes by the MIPS database (http://mips.gsf.de/genre/proj/yeast/). In other organisms, gene ontology (GO) terms (http://www.geneontology.org/) can be used as function categories. However, as GO terms are organized in hierarchical tree-like structures, the extent of GO term enrichment might be a better metric to accommodate redundancy in GO terms. The enrichment can be evaluated as fold enriched over the average, or a P-value of enrichment given by hypergeometric or Fisher exact test. Many software packages, for example, BINGO (a plug-in of the Cytoscape package, http://www.cytoscape.org), Vampire (http://genome.ucsd.edu/microarray/) and Ontologizer (http://compbio.charite.de/index.php/ontologizer2.html) among many others are widely used programs to judge GO term enrichment in subnetworks.

18.3.1 Structural modules

Modules can be defined based on network configurations alone without specifically restricting the temporal or spatial context of the modules, which are referred to here as structural modules or network structure-based modules. Such a module is a subnetwork where nodes are more densely connected within a module than toward the outside of the subnetwork, or more than random expectation. The modularity metric Q is one of the metrics designed to describe such modules. For a network divided into a number of modules, the fraction of edges that link a node in module i to a node in module j can be represented as e_{ij}. The modularity metric is defined as $Q = \sum_i (e_{ii} - (\sum_j e_{ij})^2)$, which describes the fraction of edges which link vertices within a module minus the expected value of the same quantity of edges in a network with the same module divisions but with random connections between the vertices. The Q value varies from 0 to 1, with 0 being no modularity, and 1 being maximally modular for the network (Newman and Girvan, 2004). Although defined based on network structures only, these modules may correspond to protein complexes, pathways or subnetwork that are often temporally, spatially or functionally coherent.

Network clusters are a special type of structure modules that mostly but not completely consist of cliques. A full clique is a subgraph in which all nodes are fully connected to each other, that is, all the possible edges among the nodes are present. The clustering coefficient can be used to measure the level of clustering of the neighborhood of a node. It is defined as the ratio of observed edges and possible edges among the interactors of a node, which is expressed as $C(v) = 2E_{obs}/((N-1)N)$, where N is the number of direct interactors of a node, E_{obs} is the total number of observed edges among the N interactors (Watts and Strogatz, 1998). Clustering can also be measured for an edge using, for example, the Jaccard Index, which is defined for an edge between nodes v and w as $JI = (N_v \cap N_w)/(N_v \cup N_w)$, where N_v and N_w are the numbers of interactors of node v and w, respectively. The MCODE algorithm is based on $C(v)$ to find small neighborhood in a network where nodes have high $C(v)$. Such network clusters often correspond to full or partial cliques and are denser than those derived based on the modularity metric. In PPI networks they often correspond to protein complexes (Bader and Hogue, 2003). Many other algorithms have been developed to identify structural modules or in particular network clusters, such as those that based on similarity of shortest distances between node pairs (Rives and Galitski, 2003), or the enrichment or conservation of subgraphs or certain network configuration compared to random expectation (in this latter case, the modules can be also regarded as network motifs) (Sharan et al., 2005), and so on. Some of the other algorithms are covered in other chapter(s) and reference (Han, 2008).

18.3.2 Epistatic modules

When edges in a network are directed and represent hierarchical or epistatic relationships, modules can be dissected from a network based on the consistency of interaction types of nodes inside a module and towards other modules. For example, epistatic interactions can be classified as "buffering" or "aggravating" interactions, referring to that the effect of the double mutant of a pair of genes is more or less severe than the sum of the two single mutants of the genes. Then using an algorithm to hierarchically cluster genes that have the same type of epistatic relationships, the network can be partitioned into epistatic modules, where all the nodes within a module have the same type of relationship toward other modules. This then converts a gene network into a module network, with either buffering or aggravating interactions between modules (Segre et al., 2005). Such a network structure indicates that epistatic networks are also modular.

18.3.3   Dynamic modules

Dynamic modules are subnetworks whose nodes have similar dynamic behaviors under a specific biological condition or context. For example, when we select a subnetwork that consists of only PPIs between genes that have similar or opposite expression profiles during the human brain or fruitfly aging, then through hierarchical clustering of nodes according to their expression similarity and separating transcriptionally anticorrelated PPI gene pairs into different modules, we identified modules that are transcriptionally anticorrelated and correspond to cellular switches, such as the proliferation to differentiation switch or the oxidative to reductive metabolism switch. These modules also assume opposite expression changes during aging. Furthermore, the PPIs connecting the transcriptionally anticorrelated modules play important regulatory roles in controlling the temporal switches as well as the aging process (Xue et al., 2007).

18.4   Testing the Significance of a Network Feature

Figure 18.1 briefly summarizes and illustrates some of the concepts we described in Sections 18.2 and 18.3. Now that we have all the measurements to describe the properties of a network, how do we know which property is statistically and/or biologically meaningful? To test the statistical significance of a network feature, we must compare it to random expectation. To derive random or background distributions for degree distribution, random networks or networks generated by the "Erdős–Rényi" network model are usually used. These networks are constructed based on the same number of nodes and edges as the test network, except the edges are randomly allocated among nodes. The degree distributions for such networks follow Poisson distributions. For most other features, the degree distribution of a test network is usually controlled for when generating random background models, or randomized networks. This is because degree distributions may be biased by nonbiologically relevant factors, such as limited sampling during network mapping (Han et al., 2005), and that some network features, such as appearance frequency of subgraphs, can be biased by degree distributions alone (Itzkovitz et al., 2003). Randomized networks can be constructed by simply permuting the node labels if only the nonrandomness of the node grouping is concerned. Or when the nonrandomness of edge configuration or arrangements is also concerned, the randomized networks can be constructed by randomly allocating edges among nodes with the restriction that the frequencies of nodes with various degrees remain exactly the same as the testing network. For example, when testing the significance of a configuration of a feedback loop or the size of a subnetwork, the P values or Z-scores of the candidate motif's frequency or the examined subnetwork size among those of randomly generated networks can be used to evaluate the significance of the motifs and subnetworks (Milo et al., 2002; Pujana et al., 2007; Wernicke and Rasche, 2006).

Statistical significance can be also evaluated through model based approaches. In a pioneering work, Itzkovitz et al. (2003) proposed a mathematical model for calculating the expected number of a particular subgraph based on the configuration of the subgraph and the average in-degree, out-degree, and mutual edge degrees (number of incoming, outgoing and mutual edges per node, respectively) of nodes in the networks with random or scale-free degree distributions. Later He and Singh proposed that a frequent subgraph or candidate motif's statistical significance can be estimated as the probability of a collection of features of the subgraph to appear in a random feature vector, where the features can be nodes, edges and small graphs, and that the probability distribution of a feature vector is modeled as a multinomial distribution (He and Singh, 2006). Koyutürk (2007) developed a method to estimate significance of the network clusters by comparing their sizes to the estimated critical size of the largest subgraph with the same density in a random graph generation model, whose degree distribution matches the biological network. Here density is defined as the ratio of the number of edges between all the nodes within a subgraph versus the number of all possible edges if the subgraph is complete.

To evaluate the biological meaning of a reasonable size (usually >5 nodes) node group, the metrics measuring functional coherence among nodes (described in Section 18.3) can be used. For smaller subnetworks, an annotated database or even literature can be referenced. For example, the overlap to MIPS complex list can be used to score the clusters found in yeast PPI networks. To evaluate the biological relevance of an edge group, pair-wise similarity of nodes in function annotations or gene expression profiles can be used.

In the sections above, we discussed methods for analyzing the properties of a preexisting network. Some of the networks can be constructed through experimentally measuring the interactions between nodes, while others can be inferred through related but indirect data. In the next section, we will examine some of the methods used to reverse engineer or infer a network from large datasets that do not directly measure the inter-relationships among nodes.

18.5 Network Structure Inference

With the increasing quantities of high throughput data it is possible to infer biological networks from these measurements. First, we consider how complex a network is, and then look at the combinatorics of how quickly the number of possible networks grows as more nodes are added to the network.

For networks with n nodes, there are $n(n-1)/2$ possible edges between nodes. For undirected networks it is possible for each edge to be either present or absent, giving $2^{(n(n-1)/2)}$ possible networks. For directed networks it is possible for an edge to point one way, the other way, or to be absent, giving $3^{(n(n-1)/2)}$ possible networks. This is illustrated in Table 18.1. Thus, for networks with ten nodes, there are more than 10^7 times more possibilities and for 50 nodes more than 10^{215} times more possibilities.

As a consequence, building an undirected network involves determining $n(n-1)/2$ edges; for example, identifying the presence or absence of an edge by mining PPI data, or calculating if the correlation between gene expression profiles is greater than a threshold. (Some tools also do some pruning of these networks to remove some extra rogue edges—due to the large number of edges, the false positive rate for incorporating edges into an association network is often high). Whereas, more sophisticated techniques need to be used in order to infer network structures from data due to the combinatorics of the possible number of network graphs.

The analysis and inference of directed networks is a somewhat harder task than undirected networks. Rather than finding associations or correlations between variables (be these genes or other biological entities) directed networks capture the directionality of the relationships between variables. Causal relationships between the variables can be captured by such models. Thus a model A→B, may indicate that gene A promotes gene B, whereas this information is not present in an undirected model (A–B). Thus, directed network models are suitable for modeling dynamic (temporal) processes, such as gene regulation. In addition to adding directionality to edges, models also exist which do not treat all the edges as independent, but also allow for more complex

TABLE 18.1:   Details of the numbers of possible networks for a given number of nodes (excluding self-loops).

# Nodes	# Possible edges	# Undirected graphs	# Directed graphs	# Directed acyclic graphs
1	0	1	1	1
2	1	2	3	3
3	3	8	27	25
4	6	64	729	543
5	10	1024	59,049	29,281
6	15	32,768	14,348,907	3,781,503
7	21	2,097,152	104,603,553,203	1,138,779,265
8	28	268,435,456	2.288e+13	783,702,329,343
9	36	6.872e+10	1.501e+17	1.213e+15
10	45	3.518e+13	2.954e+21	4.175e+18
n	$n(n-1)/2$	$2^{(n(n-1)/2)}$	$3^{(n(n-1)/2)}$	a_n

Note: The number of directed acyclic graphs on n nodes is defined by the following recurrence relation with $a_0 = 1$, $a_n = \Sigma_{k=1}^{n} (-1)^{k-1} \, C(n,k) 2^{k(n-k)} \, a_{n-k}$.

relationships to be represented—modeling the state of a biomolecule as a Boolean function of a number of other biomolecules' states in the case of Boolean networks, or as a conditional probability distribution (CPD) in the case of Bayesian networks. This adds to the complexity of structural inference of the networks. We will now introduce a number of structural inference methods, beginning with methods for undirected graphs, then for directed graphs.

18.5.1 Inference of nondirectional relationships

Nondirectional relationships are often derived by correlations between nodes throughout various events, such as gene expression, evolution or even literature citations.

18.5.1.1 Mutual information or correlation

Mutual information of two discrete random variables X and Y is defined as

$$I(X;Y) = \sum_{x \in X} \sum_{y \in Y} p(x,y) \log \left(\frac{p(x,y)}{p_1(x)p_2(y)} \right)$$

where $p(x,y)$ is the joint probability of X and Y for different combinations of x and y values, and $p_1(x)$ and $p_2(y)$ are the marginal probabilities of X and Y, respectively at the same x and y values. A correlation measure, such as Pearson correlation coefficient (PCC), between two variables X and Y is defined as

$$\text{PCC} = \frac{\sum_{i=1}^{n} (Xi - \overline{X})(Yi - \overline{Y})}{(n-1)S_x S_y},$$

where S_x and S_y are the standard deviations of variables X and Y, respectively, n is the length of the vector, or the number of data points shared between X and Y.

Using mutual information, the PCC or other correlation measures between a pair of nodes (genes or proteins) based on, for instance, gene expression profiles, followed by a cutoff on the calculated values is the simplest way to form a network among genes or proteins. However, the network thus built must be filtered with multiple datasets of the same type or various other evidences to be useful for deriving biological hypothesis, because of high false positive rate and extensive transitive indirect relationships. That is, when gene A is highly correlated with B and B with C, A, and C are very likely to be also correlated, resulting in a full clique or a fully connected network, which is unlikely to occur for a real biomolecular network.

18.5.1.2 Naïve Bayesian inference

Building a network by the naïve Bayes model is essentially generating probabilities of the existence of an edge (or functional interaction) between two nodes (genes/proteins) using the naïve Bayes rule based on a set of gold standard positive and negative interactions and multiple independent evidence data.

For example, we can define an edge as positive when two proteins interact and as negative when they do not. Considering the total number of positive edges among all the possible node pairs, the prior odds of finding a positive pair is

$$O\text{prior} = \frac{P(\text{positive})}{P(\text{negative})},$$

where $P(\text{positive})$ is the possibility of getting a positive edge among all the possible edges, while $P(\text{negative})$ stands for the possibility of getting a negative edge. The posterior odds are the odds of getting a positive edge after considering the N given evidence

$$O\text{posterior} = \frac{P(\text{positive}|\text{evidence}1, \ldots, \text{evidence}N)}{P(\text{negative}|\text{evidence}1, \ldots, \text{evidence}N)}.$$

For example, the evidences can include gene expression correlations, protein domain-domain interactions, synthetic lethal interactions, etc. The terms "prior" and "posterior" refer to the condition before and after considering the evidences. Then a likelihood ratio (L) is defined as

$$L(\text{evidence}1, \ldots, \text{evidence}N) = \frac{P(\text{evidence}1, \ldots, \text{evidence}N|\text{positive})}{P(\text{evidence}1, \ldots, \text{evidence}N|\text{negative})},$$

which relates prior and posterior odds according to the Bayes rule $O\text{posterior} = L(\text{evidence}1, \ldots, \text{evidence}N) * O\text{prior}$. When all the evidences are independently derived, the Bayes rule can be simplified to naïve Bayes rule and L can be simplified as

$$L(\text{evidence}1, \ldots, \text{evidence}N) = \prod_{i=1}^{N} L(\text{evidence}i)$$

$$= \prod_{i=1}^{N} \frac{P(\text{evidence}i|\text{positive})}{P(\text{evidence}i|\text{negative})},$$

that is, L equals the product of the likelihood ratios of the individual evidences. The likelihood ratio L of evidence i can be calculated from the positive and negative hits by binning the evidence into contiguous intervals, for instance intervals of expression profile correlations. Then based on a cutoff of L, a network consisting of edges weighted by their probability can be derived.

Such networks can form the basis for predicting functional associations, as well as further structural and dynamic analyses of the networks.

18.5.1.3 Regression models

Similar to naïve Bayes predictors, the logistic regression model also assumes independency among different evidences, but unlike naïve Bayes predictors, it also assumes the log likelihood ratio L of each evidence i contributes linearly to the logit of the probability (p) of an edge

$$\text{logit}(p) = \log\left(\frac{p}{1-p}\right) = c + \sum_{i=1}^{n} ai \log(Li).$$

Linear regression models have also been used to infer the effect of one node to another based on the linear regression slope for each variable, with a positive slope signifying activation and a negative slope for repression (Gardner et al., 2003). In this case, the network is directed.

18.5.2 Inference of directed networks with Bayesian networks (BNs)

Bayesian networks are a subset of probabilistic graphical models, for which efficient inference and learning algorithms exist. BNs aim to capture the joint probability distribution over a set of variables. The representation and use of probability theory makes BNs suitable for combining domain knowledge and data, expressing causal relationships, avoiding over-fitting a model to training data, and learning from incomplete data sets (Heckerman, 1998; Needham et al., 2007; Yu et al., 2004).

A Bayesian network (BN) represents the joint probability distribution over the set of variables as a product of conditional probability distributions (CPDs), exploiting the independence between some of the variables. The network structure encodes the dependencies between variables and takes the form of a directed acyclic graph (DAG), with edges pointing from the parent to child nodes. The fact that the graph is directed allows for causal relationships to be expressed, and being acyclic is important for exact inference algorithms to be employed. The CPD for each node depends only on its parents. As an example, BNs are often used to represent biological networks and pathways, particularly relating gene expression levels. In this case, the variables are gene expression levels, and so the nodes are the genes, the network structure encodes the dependencies between genes, and for each gene, there is a CPD representing the probability of that gene's expression level given the expression level of those genes it depends upon (its parents). These expression levels can be continuous, or discretized into classes. An example CPD is illustrated in Figure 18.2.

More formally, if we let $\mathbf{x} = \{x_1, \ldots, x_n\}$ denote the variables (nodes in the BN), and θ_s denote the model parameters for structure S^h, then the full Bayesian posterior—marginal likelihood—for the predictive distribution for the state of x given a set of training examples D is given by:

$$p(\mathbf{x}|D) = \sum_{S^h} p(S^h|D) \int p(x|\theta_s, S^h) p(\theta_s|D, S^h) d\theta_s$$

So, the equation reads that the probability of x given the training data D is the sum over all possible models of the product of the probability

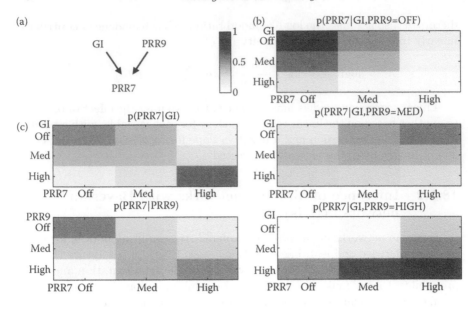

FIGURE 18.2: This figure illustrates the model parameters in a simple example network for three *Arabidopsis thaliana* genes. The parameters are learned from a set of over 2500 gene expression microarrays, quantized into three classes denoted OFF, MED, and HIGH. (a) A network structure for three genes showing that the conditional probability distribution for the gene PRR7 depends on the expression levels of GI and PRR9. (b) The CPD for PRR7. Black squares represent likely configurations; when GI and PRR9 are both OFF, it is likely that GI is also OFF, and when GI and PRR9 are both HIGH, so is PRR7. (c) Marginal probabilities for PRR7. The first table shows the likely states of PRR7 given the state of GI, when PRR9 is marginalized over. Likewise the second table shows p(PRR7|PRR9). Both illustrate the positive relationship between GI/PRR9 and PRR7.

of the model structure given the data and the probability of x given the model structure (which in turn is calculated by integrating over all possible model parameters θ_s for that structure—thus within the integral we average over all possible model parameters the product of the probability of x given the parameters and structure, and the likelihood of those model parameters given the examples D and the structure S^h). However, this calculation is intractable for all practical applications, and some approximations can made, such as sampling a set of good structures as there are infeasibly too many to consider exhaustively (see Table 18.1) or just to take a single "best" structure, and also approximating the marginalization over the parameters, through taking maximum likelihood (ML) or maximum a posteriori (MAP) estimates.

However, here we are interested in inferring networks from data, rather than averaging over networks in order to predict or classify data. Thus, the elucidation of networks is our focus, and we will discuss approaches to learning network structures (or distributions of network structures).

Importantly, rather than building a network from data based on correlations or other *pair-wise* measures of associations between nodes (genes) the CPD for each node is able to capture a potentially complex CPD for that variable dependent upon the state of a number of other variables. Thus relationships which are strong and positive, weaker and positive, negative, or nonlinear can be captured for multinomial data, and the CPDs can be examined in order to determine the type of interaction inferred from the data. Often data is discretized into three classes representing over-, under- and non-differentially expressed gene expression levels. With three classes, the model size remains manageable, whilst being robust, with more classes the models would soon grow large in size, prohibiting CPDs with dependencies on many other nodes.

Algorithms for reverse engineering networks tend to incorporate two features, a method for scoring structures and a method for searching for structure hypotheses. The scoring of a network structure can be done by calculating the full Bayesian posterior—integrating over the model parameters for that structure—or more commonly due to computational cost, using an approximation to the marginal likelihood. One such example of a scoring function is the Bayesian information criterion which contains a term for the likelihood that the data was generated by the model, and a term to penalize complex models (those with many parameters, resulting from nodes with many parents). There is a super-exponential number of possible network structures (see Table 18.1 for the number of directed acyclic graphs (DAGs) on n nodes), a common method is to use a greedy search which starts from an initial network, and iteratively an edge is added, reversed or deleted, until the score for the network reaches a local optimum. This is then done with a number of restarts from different random starting DAGs in order to explore more of the search space, as it is susceptible to only find local optima. Alternatively, a distribution of good model structures may be formed using Markov chain Monte Carlo methods to sample from the posterior distribution of models.

Interventional data such as perturbations can provide powerful information for identifying causal networks. This has been demonstrated in the reverse engineering of a protein signaling network from thousands of observations of multiple signaling molecules measured by flow cytometry in the presence of stimulatory and inhibitory perturbations (Sachs et al., 2005).

Time series data can be particularly useful, and *dynamic BNs* are particularly promising for this sort of data, as they can exploit the learning mechanisms of BNs, but have the additional temporal information which aids in interpreting causal relationships. A network can be unrolled in time, so that the state of a gene at time $t + \Delta t$ depends only on the state of genes at time t (a first order Markov process). Such models allow feedback and cyclic behavior observed in biological systems, and can be implemented using continuous or discretized variables to represent gene expression.

Gene regulatory networks can not only be reverse engineered from gene expression data, the enormous amount of genomic DNA-binding profiles for transcription regulatory factors generated by recent chromatin-immunoprecipitation followed by microarray (ChIP-chip) or by deep sequencing (ChIP-seq) technologies are especially suited for using BN to reconstruct the dependency or causal relationships among these regulatory factors (Yu et al., 2008). The probabilistic formalism provides a natural treatment for the stochastic nature of biological systems and measurements.

18.6 Summary

In this chapter, we briefly introduced the frequently used network metrics to statistically analyze the characteristics of a biomolecular network, and the biological meanings of the network features, when and why we examine them, and how to determine the statistical relevance of a network feature. Then we described common approaches to infer or reverse engineer network structures from data. With more and more large-scale data generated for biomolecules and their relationships genome-wide, the statistical network analysis methods discussed here will be increasingly more useful and necessary when analyzing the sea of data to generate biological hypotheses and insights.

References

Bader, G.D., and Hogue, C.W. 2003. An automated method for finding molecular complexes in large protein interaction networks. *BMC Bioinformatics* 4, 2.

Chung, F., and Lu, L. 2006. *Complex Graphs and Networks*. Providence, RI: American Mathematical Society.

Gardner, T.S., di Bernardo, D., Lorenz, D., and Collins, J.J. 2003. Inferring genetic networks and identifying compound mode of action via expression profiling. *Science* 301, 102–105.

Gunsalus, K.C., Ge, H., Schetter, A.J., Goldberg, D.S., Han, J.D., Hao, T., Berriz, G.F., Bertin, N., Huang, J., Chuang, L.S., et al. 2005. Predictive models of molecular machines involved in *Caenorhabditis elegans* early embryogenesis. *Nature* 436, 861–865.

Han, J.D. 2008. Understanding biological functions through molecular networks. *Cell Res* 18, 224–237.

Han, J.D., Bertin, N., Hao, T., Goldberg, D.S., Berriz, G.F., Zhang, L.V., Dupuy, D., Walhout, A.J., Cusick, M.E., Roth, F.P., et al. 2004. Evidence for dynamically organized modularity in the yeast protein–protein interaction network. *Nature* 430, 88–93.

Han, J.D., Dupuy, D., Bertin, N., Cusick, M.E., and Vidal, M. 2005. Effect of sampling on topology predictions of protein–protein interaction networks. *Nat Biotechnol* 23, 839–844.

He, H., and Singh, A.K. 2006. GraphRank: statistical modeling and mining of significant subgraphs in the feature space. The 2006 IEEE-WIC-ACM International Conference on Date Mining (ICDM 2006), Hong Kong, Published by IEEE Computer Society Press, 45–59.

Heckerman, D. 1998. A tutorial on learning with Bayesian networks. In: Jordan MI, editor. *Learning in Graphical Models.* Kluwer Academic, Boston, 301–354.

Itzkovitz, S., Milo, R., Kashtan, N., Ziv, G., and Alon, U. 2003. Subgraphs in random networks. *Phys Rev E Stat Nonlin Soft Matter Phys* 68, 026127.

Jeong, H., Mason, S.P., Barabasi, A.L., and Oltvai, Z.N. 2001. Lethality and centrality in protein networks. *Nature* 411, 41–42.

Koyutürk, M., Szpankowski, W., and Grama, A. 2007. Assessing significance of connectivity and conservation in protein interaction networks. *J Comput Biol*, 14(6): 747–764.

Milo, R., Shen-Orr, S., Itzkovitz, S., Kashtan, N., Chklovskii, D., and Alon, U. 2002. Network motifs: simple building blocks of complex networks. *Science* 298, 824–827.

Needham, C.J., Bradford, J.R., Bulpitt, A.J., and Westhead, D.R. 2007. A primer on learning in Bayesian networks for computational biology. *PLoS Comput Biol* 3, e129.

Newman, M.E., and Girvan, M. 2004. Finding and evaluating community structure in networks. *Phys Rev E Stat Nonlin Soft Matter Phys* 69, 026113.

Papin, J.A., Hunter, T., Palsson, B.O., and Subramaniam, S. 2005. Reconstruction of cellular signalling networks and analysis of their properties. *Nat Rev Mol Cell Biol* 6, 99–111.

Pujana, M.A., Han, J.D., Starita, L.M., Stevens, K.N., Tewari, M., Ahn, J.S., Rennert, G., Moreno, V., Kirchhoff, T., Gold, B., et al. 2007. Network modeling links breast cancer susceptibility and centrosome dysfunction. *Nat Genet* 39, 1338–1349.

Rives, A.W., and Galitski, T. 2003. Modular organization of cellular networks. *Proc Natl Acad Sci USA* 100, 1128–1133.

Sachs, K., Perez, O., Pe'er, D., Lauffenburger, D.A., and Nolan, G.P. 2005. Causal protein-signaling networks derived from multiparameter single-cell data. *Science* 308, 523–529.

Segre, D., Deluna, A., Church, G.M., and Kishony, R. 2005. Modular epistasis in yeast metabolism. *Nat Genet* 37, 77–83.

Sharan, R., Suthram, S., Kelley, R.M., Kuhn, T., McCuine, S., Uetz, P., Sittler, T., Karp, R.M., and Ideker, T. 2005. Conserved patterns of protein interaction in multiple species. *Proc Natl Acad Sci USA* 102, 1974–1979.

Snel, B., Bork, P., and Huynen, M.A. 2002. The identification of functional modules from the genomic association of genes. *Proc Natl Acad Sci USA* 99, 5890–5895.

Watts, D.J., and Strogatz, S.H. 1998. Collective dynamics of 'small-world' networks. *Nature* 393, 440–442.

Wernicke, S., and Rasche, F. 2006. FANMOD: a tool for fast network motif detection. *Bioinformatics* 22, 1152–1153.

Wuchty, S., Oltvai, Z.N., and Barabasi, A.L. 2003. Evolutionary conservation of motif constituents in the yeast protein interaction network. *Nat Genet* 35, 176–179.

Xue, H., Xian, B., Dong, D., Xia, K., Zhu, S., Zhang, Z., Hou, L., Zhang, Q., Zhang, Y., and Han, J.D. 2007. A modular network model of aging. *Mol Syst Biol* 3, 147.

Yu, H., Kim, P.M., Sprecher, E., Trifonov, V., and Gerstein, M. 2007. The importance of bottlenecks in protein networks: correlation with gene essentiality and expression dynamics. *PLoS Comput Biol* 3, e59.

Yu, H., Zhu, S., Zhou, B., Xue, H., and Han, J.D. 2008. Inferring causal relationships among different histone modifications and gene expression. *Genome Res* 18, 1314–1324.

Yu, J., Smith, V.A., Wang, P.P., Hartemink, A.J., and Jarvis, E.D. 2004. Advances to Bayesian network inference for generating causal networks from observational biological data. *Bioinformatics* 20, 3594–3603.

Part IV

Literature, Ontology, and Knowledge Integration

Chapter 19

Beyond Information Retrieval: Literature Mining for Biomedical Knowledge Discovery

Javed Mostafa

University of North Carolina

Kazuhiro Seki

Kobe University

Weimao Ke

University of North Carolina

19.1 Motivations for Mining Literature

The growth rate and production of biomedical information far exceeds the capacity of individual researchers to keep up with developments in the field. There is a strong interest in biology now for a more integrative and "systems" approach whereby characterization for specific genes or proteins are sought in terms of a broader context which includes other genes or proteins, requiring biomedical researchers to be aware of findings that go beyond their individual focus. Furthermore, there is a growing interest in linking outcomes of basic biomedical research to evidence produced in clinical or operational settings, with an eye toward achieving improved translation of research, which is also pushing the demand for efficient means for integrating evidence in published literature with other biomedical resources.

Most common information retrieval tools today only produce document surrogates or pointers to documents. Significant progress has been made, however, based on statistical natural language processing (NLP) and machine learning (ML) techniques which allows a researcher to probe the retrieved content of a typical information retrieval (IR) system automatically. That is, there are new task-centric functions available that are capable of generating useful information quickly to address the query that motivated the original search instead of a path involving printing or reading numerous documents. Additionally, new functions exist that permit a researcher to conduct exploratory analysis on literature content to identify relationships among critical biomedical entities or produce hypotheses about biomedical phenomena that may be subjected to subsequent experiments or trials. The focus of this chapter is on these new analytical functions that are likely to complement the current state of the art IR applications.

The chapter will begin with a brief overview of IR systems and how they may form the building blocks for next generation mining functions. Next, the challenges associated with identifying critical biomedical entities accurately and the related problem of generating or updating biomedical ontologies will be discussed along with successful mining approaches. This area, known as named entity recognition (NER), may appear to be simple on the surface, but in fact is extremely challenging due to inconsistent use of language and the highly dynamic nature of the associated fields in biomedicine. Along with presentation of approaches developed for NER, we will also discuss related techniques for abbreviation and synonym detection.

Moving to the individual corpus level the focus transitions to mining for key assertions or facts in published documents. In this section of the chapter, we will cover automatic summarization techniques and fact retrieval. These techniques primarily rely on statistical language models and less so on classic NLP approaches. But their use range from generating digests of facts for quick review to an association network among key entities. One step beyond detecting entity relationships is detection of indirect relationships among entities motivated by an a priori hypothesis or a hypothesis to be discovered post facto. Certain manually intensive techniques have proven to be useful in hypothesis validation (for example, the well-known ARROWSMITH approach), however, techniques that can produce new hypotheses (i.e., not start with one asserted by the researcher) and produce evidence of causality are in much demand and progress in this area will also be discussed in the chapter.

Finally, shifting to the corpora level, we will discuss mining functions that allow the researcher to detect patterns related to certain critical research topics and also identify influential scholars in those areas. The patterns may include waning or rising trends in associated topics or collaboration networks among researchers. Staying at the collection level, we will then describe new developments in data integration based on automatic linking of data across diverse resources such as document corpora, sequence sets, molecular structure databases, and image libraries. Increasingly, various specialized applications are becoming available under open source licenses that allow manipulation of data associated with published literature (for example, viewers or simulation programs) that expose their functions through programmable interfaces or services. Both automatic liking of content across disparate resources (for example as currently conducted in Entrez-NCBI) or linking of content to publicly available services depend on classification algorithms with roots in ML. These new classification approaches will be discussed in the chapter. In the concluding section, we will cover well-known metrics and evaluation methodologies for analyzing the utility of literature mining techniques and user interface features that have the potential to seamlessly integrate the new mining functions with core IR functions.

19.2   Literature Mining and its Roots

The origins of modern computer-based information retrieval systems can be traced back at least half a century to the period close to World War II when there was a realization that advances in computer technologies could perhaps be harnessed to improve access to scientific and technical information. Serious research on improving system design and evaluation of IR systems soon followed. For example, the classic experiments on retrieval evaluation known as the Cranfield Project was conducted during the period between the 1950s and 1960s.

Originally the aim of improving access to scientific and technical information primarily concerned itself with representation of bibliographic information with a strong focus on representing terms appearing in document titles and end users' queries. The representation challenge and the associated challenge of term-matching were handled using a variety of strategies applied at various levels, ranging from a basic level involving term stemming up to more sophisticated levels including relevance-feedback methods. A full range of these representation (or re-representation) techniques, sometimes referred to as text analysis and retrieval refinement methods, were implemented and thoroughly studied in Gerard Salton's SMART project (Salton and McGill, 1983). As the emphasis of IR systems gradually shifted to full-text content, some of the same techniques are still applied, albeit with some tuning to handle larger volumes of information. The roots of current literature mining methods actually are these text analysis and refinement methods, some of which we will discuss next.

A key initial challenge in text mining is extraction of selective words or terms from text. Many methods have been developed for this, but the most popular ones rely on statistical techniques for identifying significant or distinctive terms based on inter- and intradocument frequency distribution patterns. An example of a well-known method is known as *tf.idf* (Salton and McGill, 1983). The first component in the formula, *tf* is a frequency count of a term in a document and the second component of the formula, *idf*, is equivalent to $\log(N/n)$, where the N is total number of documents in the collection and the n is total number of documents where the term appears. The *idf* part is used to derive a more appropriate representation of term distribution by adjusting the raw frequency per document to account for collection-wide distribution of the same term. Upon calculation of all the term weights using the same technique, it is then possible to rank the terms per document and select the most distinctive terms based on maximum number of occurrences in the ranked lists where the length of the lists is determined based on some preset threshold value.

Upon extraction of tokens a logical next step is establishing if any semantic associations exist among these tokens. Such associations can be derived in many ways and has obvious implications for text mining applications as well. A typical derivative of a statistically driven token extraction process (as described above) is that each token can be represented as a numerical weight relative to the document they appear in. Hence, the tokens can be transformed to weighted vectors where each vector corresponds to a document. The ultimate outcome of the process is a *document×token* matrix of weights. A few basic matrix manipulation steps applied on the latter structure can yield results that are critical to both retrieval and mining. A simple one which is frequently conducted is calculation of "association" matrices, such as a *document.document* matrix and a *term.term* matrix.

In retrieval systems, a common need is to refine search results (i.e., hit set) by identifying additional documents that are closely related to relevant documents or pruning the hit set by removing documents that are not related to relevant documents. Such a need can be addressed efficiently if similarity values across all documents in the collection exist and if the documents are organized in terms of their similarity values so that closely related documents can be easily located. A scheme which permits this is hierarchical clustering of documents based on document similarity, which requires as its input a document association matrix as calculated and described above. Such hierarchical structure has other benefits such as permitting interactive browsing of large hit sets—such as a shallow visual text analysis approach similar to the classic scatter-gather interaction modality.[*]

A refinement technique which has its roots in retrieval research and has influenced text mining is term association discovery. A common approach to establish associations is to apply a clustering scheme starting with the association matrix as described above as input. For example, a hierarchical dendrodgram outcome from such a scheme can be used to expand an initial term set by identifying closely related terms in the same branch or nearby branches. The "distance" among terms determined according to cluster centroids after applying a clustering scheme is often used as a surrogate for "semantic distance" among the terms and applied to prune or supplement query terms. Automatically generated term clusters based on similarity may also be treated as a rudimentary thesaurus and this thesaurus then may support functions for term set expansion or pruning depending on the needs of the user.

Before the *term.term* matrix is used to generate term clusters, to improve the likelihood of discovering term association, sometimes a technique known as latent semantic analysis[†] is applied (Landauer et al., 1988). To illustrate the value of applying such a technique, let us consider a simple example.[‡] In Table 19.1, we have a tiny test set of documents with nine items. Based on selection of a dozen key terms,[§] the tiny test set can be converted to a *document.term* matrix. In Table 19.2, we show the *document.term* matrix with term frequencies. We also show in Table 19.2, the Pearson correlation coefficient calculated for two pairs of term vectors, namely *human.user* and *human.minors*.

The term "minors" as used in the documents in this particular set has its origins in graph theory and has little to do with humans, hence it makes sense that we have a negative correlation coefficient for the pair *human.minors*. However, for the pair *human.user* the negative correlation is hard to justify, as the terms appear to share the same context in the document set (i.e., they are part of a group of documents that are related) and they obviously are

---

[*] Originally developed by Cutting et al. in 1990s, this approach starts with top level clusters of documents generated from a collection and allows the user to select a subset of these clusters (gather step) based on key terms displayed as representative terms for each cluster. For refinement the user then can request a reclustering of documents applied only on the documents belonging to the chosen clusters (scatter step), which produces a new and more narrowly focused set of document clusters. The user may continue iterating between these gather and scatter steps until a desired set of documents are identified.

[†] Based on the statistical technique known as singular value decomposition, which detects and aids in dimensionality reduction to raise the correlation values of individual terms based on latent cooccurrence patterns in the original association matrix.

[‡] We are grateful to Landauer, Foltz, and Laham for the example they described in their 1998 paper which we are using in this chapter to illustrate the application of LSA for matrix refinement.

[§] Key term selection can be conducted manually or it can be fully automated by applying a method which is exclusively statistical as described earlier or a hybrid one which combines a classic NLP technique with a statistical method (see discussions on NER in this chapter).

TABLE 19.1: A sample document set.

Doc1	Human machine interface for PARC computer application
Doc2	A survey of user opinion of computer system response time
Doc3	The EPS user interface management system
Doc4	System and human system engineering testing of EPS
Doc5	Relation of user perceived response time to error measurement
Doc6	The generation of random, binary ordered trees
Doc7	The intersection graph of paths in trees
Doc8	Graph minors IV: Widths of trees and well-quasi-ordering
Doc9	Graph minors: A survey

TABLE 19.2: Document × term matrix with raw term frequency.

	Doc1	Doc2	Doc3	Doc4	Doc5	Doc6	Doc7	Doc8	Doc9
HUMAN	1	0	0	1	0	0	0	0	0
INTERFACE	1	0	1	0	0	0	0	0	0
COMPUTER	1	1	0	0	0	0	0	0	0
USER	0	1	1	0	1	0	0	0	0
SYSTEM	0	1	1	2	0	0	0	0	0
RESPONSE	0	1	0	0	1	0	0	0	0
TIME	0	1	0	0	1	0	0	0	0
EPS	0	0	1	1	0	0	0	0	0
SURVEY	0	1	0	0	0	0	0	0	1
TREES	0	0	0	0	0	1	1	1	0
GRAPH	0	0	0	0	0	0	1	1	1
MINORS	0	0	0	0	0	0	0	1	1
r(human.user)					-0.38				
r(human.minors)					-0.29				

TABLE 19.3: Document × term matrix after applying LSA.

	Doc1	Doc2	Doc3	Doc4	Doc5	Doc6	Doc7	Doc8	Doc9
HUMAN	0.16	0.4	0.38	0.47	0.18	-0.05	-0.12	-0.16	-0.09
INTERFACE	0.14	0.37	0.33	0.4	0.16	-0.03	-0.07	-0.1	-0.04
COMPUTER	0.15	0.51	0.36	0.41	0.24	0.02	0.06	0.09	0.12
USER	0.26	0.84	0.61	0.7	0.39	0.03	0.08	0.12	0.19
SYSTEM	0.45	1.23	1.05	1.27	0.56	-0.07	-0.15	-0.21	-0.05
RESPONSE	0.16	0.58	0.38	0.42	0.28	0.06	0.13	0.19	0.22
TIME	0.16	0.58	0.38	0.42	0.28	0.06	0.13	0.19	0.22
EPS	0.22	0.55	0.51	0.63	0.24	-0.07	-0.14	-0.2	-0.11
SURVEY	0.1	0.53	0.23	0.21	0.27	0.14	0.31	0.44	0.42
TREES	-0.06	0.23	-0.14	-0.27	0.14	0.24	0.55	0.77	0.66
GRAPH	-0.06	0.34	-0.15	-0.3	0.2	0.31	0.69	0.98	0.85
MINORS	-0.04	0.25	-0.1	-0.21	0.15	0.22	0.5	0.71	0.62
r(human.user)					0.94				
r(human .minors)					-0.83				

semantically related. Here is where a latent semantic analysis (LSA) technique is helpful by detecting the contextual overlap based on other terms that the two terms, "human" and "user," cooccur with. In Table 19.3, we show the outcome after the LSA technique is applied on the original matrix as shown in Table 19.2. Upon recalculation of the Pearson correlation coefficients for the two pairs, based on the values of the new matrix, we find that the coefficient of the *human.user* actually switches to a strong positive value (0.94) and the coefficient of the second pair *human.minors* becomes even more negative (−0.83), reinforcing further the difference between these two terms.

LSA as demonstrated above, therefore, can be used as a matrix refinement technique for improving the outcome of association discovery among key tokens/terms. To sum up, in this section we provided a brief but detailed discussion on some of the fundamental IR concepts and techniques. As we present basic literature mining techniques in the subsequent sections, it will become clearer how the IR techniques discussed in this section are in fact critical building blocks for mining techniques.

19.3 Named Entity Recognition (NER)

NER in biomedicine generally refers to the process of spotting individual occurrences of biomedical entities, such as genes, proteins, and diseases, in text. NER is the first step toward knowledge extraction from natural language text and is critical for successful literature mining. At first glance, it may seem trivial to identify such entities by, for example, simple dictionary lookup, which is unfortunately not the case for gene/protein names.* The difficulty mainly comes from two sources. One is associated with the flexible terminology, including the inconsistent use of capital and small letters, word inflections, extra/lack of tokens, and different word orders (Cohen et al., 2002; Fang et al., 2006). Therefore, even entities existing in a dictionary may be overlooked due to the way they are actually written. The other is the dynamic nature of the biomedical domain. For instance, new genes/proteins continue to be discovered, where a predefined, static dictionary is useless for identifying new names not contained in the dictionary.

To deal with those problems, much research has been conducted. The approaches proposed in the literature roughly fall into three categories: dictionary-based, heuristic rule-based, and statistical. The following sections introduce each type of the approaches.

19.3.1 Dictionary-based approaches

Dictionary-based approaches (Hanisch et al., 2003; Krauthammer et al., 2001) utilize a predefined name dictionary and identify the target named entities by comparing the dictionary entries and input text. A problem associated with this type of approaches is that biomedical terms have many variants as to how they are actually written due to the flexible terminology as described

---

*Disease names are more standardized and used more consistently than gene and protein names (Jimeno et al., 2008).

above. Therefore, dictionary lookup needs to be somehow "relaxed" so that minor differences can be tolerated.

Hanisch et al. (2003) proposed a dictionary-based approach combined with synonym (or variant) generation. They used HUGO, OMIM, SWISS-PROT, and TrEMBL databases to compile an initial name dictionary. Then, the dictionary was semiautomatically curated to improve the quality. The curation process consisted of expansion and pruning phases. The expansion phase added name variants according to some heuristic rules and an acronym list to increase the coverage of the dictionary. For instance, "IL1" generates "IL 1" by inserting a space between different character types (e.g., alphabets and numerals) and "interleukin 1" by replacing "IL" with "interleukin" based on the acronym list. The pruning phase removed general or ambiguous names from the dictionary in order to prevent false positives. After curation, the dictionary contained approximately 38,200 unique names and 151,700 synonyms. To further improve recall, they used approximate word match with different term weights (penalties) for different term classes. For example, precursor is categorized into "nondescriptive," which has little influence on gene name match, and thus a small weight is given. As a result, the existence or absence of precursor is virtually ignored in dictionary lookup.

In general, dictionary-based approaches are easy to implement when existing databases are available. However, it is not easy to achieve high performance. First, automatically compiled dictionaries are usually not very accurate due to inconsistent formats in the databases. To obtain a high-quality dictionary, manual curation is often needed. Second, the coverage of the dictionary may be limited due to various name variants, resulting in low recall. Third, simple dictionary-lookup cannot disambiguate multisense names. Lastly, new names cannot be identified by this type of approaches unless they are very similar to existing dictionary entries.

19.3.2   Heuristic rule-based approaches

As discussed in the above sections, biomedical named entities have rather flexible forms. However, there is still some regularity that indicates their occurrences. Heuristic rule-based approaches take advantage of such surface clues. The following summarizes major features commonly used in the literature (for example, Fukuda et al., 1998; Franzén et al., 2002) for detecting gene/protein names.

- Capital letters (e.g., ADA, CMS)

- Arabic numerals (e.g., ATF-2, CIN85)

- Greek alphabets (e.g., Fc alpha receptor, 17beta-estradiol dehydrogenate)

- Roman numerals (e.g., dipeptidylpeptidase IV, factor XIII)

- Suffixes frequently appearing with protein name (e.g., -nogen, -ase, and -in)

Because simply relying on the above clues would misdetect irrelevant terms as well, postprocessing is usually needed to rule out those false positives. Then, multiword names are recognized with the help of morphological information, such as affixes and parts of speech. For instance, one of the heuristic rules used by Fukuda et al. (1998) is "*Connect nonadjacent annotations if every word between them are either noun, adjective, or numeral.*" Franzén et al. (2002) proposed an approach similar to Fukuda et al. (1998) except that they used more comprehensive filters to rule out false positives and a syntactic parser to find name boundaries of multiword names.

The advantages of rule-based approaches are that (i) they are intuitive and easily understood by humans, (ii) they can be flexibly defined and extended as needed, and that (iii) the rules may be able to recognize even unknown, newly created names which may not be detected by a static dictionary if they possess some features observed in the target name type. On the other hand, their disadvantages are that (i) it is laborious and time-consuming to create a comprehensive and consistent set of rules, and that (ii) in contrast to dictionary-based approaches, recognized named entities are not automatically linked to existing database entries.

19.3.3 Statistical approaches

Statistical (or ML) approaches have been successfully applied to natural language processing tasks such as part-of-speech tagging and parsing. In the biomedical domain, examples applied for NER include decision trees (Hatzivassiloglou et al., 2001), naïve Bayes classifiers (Hatzivassiloglou et al., 2001), hidden Markov models (HMM) (Zhou et al., 2004), support vector machines (SVMs) (Mika and Rost, 2004), and conditional random fields (CRF) (Hsu et al., 2008). It is worth mentioning that CRF-based approaches are becoming the most popular choice, and are generally effective, as witnessed at the second BioCreative challenge evaluation workshop (Wilbur et al., 2007).

Among others, Hsu et al.'s (2008) method for recognizing gene names appears to be one of the best. They adopted CRF, where NER was treated as a task to assign a tag sequence $Y = y_1, y_2 \ldots, y_n$ to a given word (or token) sequence $X = x_1, x_2, \ldots, x_n$. There were three possible tags considered in Hsu et al.'s work, namely, B (beginning of a gene name), I (inside a gene name), and O (outside a gene name). The CRF tagger chooses Y which maximizes the conditional probability:

$$P(Y|X) = \frac{\exp\left((\Theta \cdot F(X, Y)\right)}{Z_\Theta(X)} \tag{19.1}$$

where $F(X,Y) = (f_1, f_2, \ldots, f_d)$ is a vector of binary features with dimension d, $\Theta \in \mathbf{R}^d$ is a vector of feature weights, and $Z_\Theta(X)$ is a normalizing factor.

While the feature set needs to be manually designed as described shortly, the feature weights Θ are learned from training data annotated with gene names. Each feature $f_i, i = 1, 2, \ldots, d$, is defined as factored representation

$$f_i = P_j(x_t)\Lambda q_k(y_{t-1}, y_t) \tag{19.2}$$

where $P_j(x_t)$ denotes a predicate (Boolean function) for token x_t and $q_k(y_{t-1}, y_t)$ is a predicate for the pair of tags. The former, $P_j(x_t)$, examines if x_t belongs to a certain class (e.g., *"Is x_t a singular common noun?"*). In total, they used 1,686,456 such predicates. The latter, $q_k(y_{t-1}, y_t)$, returns true if and only if y_{t-1} and y_t match a combination of tags specified by predicate q_k. Because there are three tags (i.e., B, I, and O), there are nine possible combinations (predicates) in total. The total number of features, d, is a product of the numbers of the two types of predicates (i.e., $1, 686, 456 \times 9$), which amounts to over 10 million.

As compared to dictionary- and heuristic rule-based approaches, statistical approaches are more principled and well grounded on a mathematical foundation. Applying statistical techniques to a given task is straightforward with the availability of an appropriate model and a feature set, which, however, are often not easy to identify. In addition, statistical approaches usually require a large amount of training data in order to train the model, which is time-consuming and laborious to create. One way to mitigate the burden is to employ semi-supervised approaches using both annotated and unannotated corpora as explored by Ando (2007). Another is to develop a framework to automatically create *weakly* labeled data. Although the quality of the data may be inferior to manually labeled data, the automatic framework can easily produce a large-scale training data set, which may compensate for the low quality as we will see in the next section.

19.3.4   Hybrid approaches

Dictionary-based, rule-based, and statistical approaches are not necessarily exclusive. Seki and Mostafa (2005) proposed a cascaded framework for identifying protein names by sequentially applying heuristics rules, probabilistic models, and a name dictionary. The heuristic rules, similar to Fukuda et al. (1998) and Franzén et al. (2002), first detect protein name fragments, then the probabilistic models are applied for determining name boundaries based on word and word class transition probabilities. Finally, the dictionary is consulted to detect protein names that are not detected by the rules and the probabilistic models. Since the rules and the dictionary are not essentially different from the others described in the previous sections, the following focuses on their proposed probabilistic models.

The probabilistic models aim to determine protein name boundaries by iteratively expanding each name fragment (e.g., peptidase) detected by heuristic rules in order to locate complete protein names (e.g., dipeptidyl peptidase IV). Let w_i denote a protein name fragment detected by the rules. Given w_i, the

TABLE 19.4: Examples of word classes.

Class	Examples
suffix_in	protein, oncoprotein, lactoferrin
suffix_ase	kinase, transferase, peptidase
word	the, a, an
acronym	CN, TrkA, USF
arabic_num	2, 3, 18, 76
roman_num	I, II, III, IV
roman_alpha	alpha, beta, gamma
punctuation	comma (,), period (.)
symbol	), (, %, +

probability that word w_{i-1} immediately preceding w_i is also a part of the protein name can be expressed as a conditional probability $P_P(w_{i-1}|w_i)$ assuming a first-order Markov process. Likewise, the probability that w_{i-1} is *not* a protein name fragment can be expressed as $P_N(w_{i-1}|w_i)$. Then, in the case where there is no name boundary between w_{i-1} and w_i (i.e., w_{i-1} is also a protein name fragment), $P_P(w_{i-1}|w_i)$ is expected to be greater than $P_N(w_{i-1}|w_i)$. Thus, we regard w_{i-1} as a protein name fragment if the following condition holds:

$$P_P(w_{i-1}|w_i) > P_N(w_{i-1}|w_i). \qquad (19.3)$$

However, directly estimating these parameters would require a large amount of training data annotated with protein names. Also, simply using a large-scale corpus cannot be an absolute solution due to the nature of protein names, that is, new proteins continue to be discovered. Previously unseen data are fatal for probability estimation such as maximum likelihood estimation. To remedy the data sparseness problem, Seki and Mostafa generalize words to word classes based solely on surface clues. Table 19.4 shows some examples of word classes.

By incorporating the word classes to the probabilistic models, the following bigram class models are formalized, where c_i denotes the word class of w_i.

$$\begin{aligned} P_P(w_{i-1}|w_i) &= P_P(w_{i-1}|c_{i-1}).P_P(c_{i-1}|c_i) \\ P_N(w_{i-1}|w_i) &= P_N(w_{i-1}|c_{i-1}).P_N(c_{i-1}|c_i) \end{aligned} \qquad (19.4)$$

The probabilistic parameters for the above models can be estimated based on a corpus annotated with protein names, which generally requires labor-intensive work to create. Instead, Seki and Mostafa proposed to leverage database references to automatically create weakly annotated corpus. Their basic idea is to annotate text (a Medline abstract) if a certain protein name (which is already known to be associated with the text) literally appears in the text. Such known associations between Medline abstracts and proteins can be easily obtained from a number of databases. For example, according to the Protein Information Resource Nonredundant REFerence protein database, the Medline article with PubMed ID 10719003 is associated with maturase-like protein. It does not mean that the exact name appears in the Medline abstract due to many variations of the protein names. However, if it does appear, it is likely to be an occurrence of the protein and can be automatically annotated. Following the idea, they annotated 32,328 abstracts with 91,773 protein mentions, which were then used to learn the probabilistic models. On a comparative experiment with a heuristic rule-based approach (Franzén et al., 2002), Seki and Mostafa's approach was reported especially effective for recognizing multiword protein names.

19.4   Text Summarization and Fact Extraction

19.4.1   Automatic information organization and concept extraction

Text clustering and classification, as information organization mechanisms, involve the aggregation of like-entities (Sebastiani, 2002). While clustering can be used to group similar or related features together and derive concepts from text, text categorization, or classification, is potentially useful in finding associations of documents and concepts from free text. Automatic clustering and classification, as applied in automatic information extraction and knowledge discovery, have been important research topics in ML and IR (Sebastiani, 2002).

An automatic text classifier, e.g., a neural-network (NN)-based method, can be used to improve text representation given that key tokens can be properly extracted from free text based on, e.g., frequency distributions. Empirical studies in text classification have shown that the effectiveness of NN classifiers are comparable to benchmark methods such as kNN and support vector machine (SVM) (Sebastiani, 2002). As compared to linear NN classifiers, multilayer nonlinear neural networks (NNs) have the potential to learn from higher-order interactions or dependencies between features (terms).

To use classification for concept identification and better text representation, it requires a method be operating on multiple labels (concepts). One approach to multilabel classification is simply using multiple binary classifiers. However, this type of method assumes the independence of concepts and ignores their potential associations. A nonlinear multilayer NN has the advantage of detecting multiple concepts simultaneously within one model (Sebastiani, 2002). Hence, a multilayer NN is potentially useful for multiconcept identification/discovery. In terms of NN training for text classification, a comparative study found that a three-layer NN of about 1000 input and 100 output nodes produced satisfying results with only 8000 training samples (Yang and Liu, 1999), much smaller compared to other nontext applications.

Automatic text clustering and classification have been applied to various aspects of biomedical text mining. For example, inspired by the fact that medical specialties have individual sublanguages, Bernhardt et al. (2005) proposed the use of these sublanguages for classification of text and identification of prominent subdomains in medicine. The work demonstrated the utility of the journal descriptor index (JDI) tool for automatically indexing biomedical documents and mapping them to specialties or subdomains.

Our previous work employed a word distribution weighting scheme and some heuristics to extract terms from medical documents and applied automatic clustering for concept discovery. The function included two operations, namely, term extraction and term clustering, and is available through online Web services. It was shown that concepts discovered through automatic clustering facilitated text representation and improved information filtering effectiveness (converged normalized precision 0.8) (Mostafa et al., 1997). In addition, the proper use of classification reduced the memory size for information representation but maintained a level of granularity that could be accurately mapped to the query space.

In another research, we explored a hybrid method for key entity (protein name) identification in biomedical texts. The research used a set of heuristics for the initial extraction of key entities and employed a probabilistic model for expanding the concepts. This method, together with a pre-compiled dictionary, produced superior results (Seki and Mostafa, 2005). For example, on compound concept identification, it achieved precision 0.74 and recall 0.79 as compared to a gold-standard method with precision 0.69 and recall 0.73.

19.5   Hypothesis Validation and Discovery

Beyond extracting NEs and factoids, literature-based hypothesis discovery aims at discovering unknown implicit knowledge that has not been recognized in the literature based on an analysis of already known, explicit knowledge. This field of research was initiated by Swanson in the 1980ss and typically follows his proposed framework based on a chain of reasoning or syllogism.

19.5.1   Chain of reasoning

Over two decades, Swanson has advocated the potential use of the biomedical literature as a source of new knowledge (Swanson, 1986b). The underlying idea is a chain of reasoning or syllogism, that is, discovering an implicit connection between two concepts, such as "A causes C," where it is well acknowledged that "A causes B" and "B causes C" while A and C do not have a relationship explicitly reported in the literature. It is important to note that the discovered connection is not (yet) new knowledge but a "hypothesis" that has to be put into a direct test to gain general acceptance.

The idea of the ABC syllogism originally came from Swanson's two independent observations regarding fish oil and Raynaud's syndrome (a peripheral circulatory disorder) in the literature (Swanson, 1986a). He noticed that dietary fish oil was shown in many experiments to lead to reductions in blood lipids, platelet affregability, blood viscosity, and vascular reactivity, whereas Raynaud's syndrome was associated with high platelet affregability, high blood viscosity, and vaso-constriction. Taken together, these observations suggest that fish oil may be beneficial for Raynaud patients, although they had never been brought together in print until Swanson made the hypothesis. (The hypothesis was later corroborated by experimental evidence.) In addition, his follow-up study revealed that two literatures representing Raynaud's syndrome and fish oil had few bibliographic connections (Swanson, 1987), which endorses the separation of the literatures. Based on the same reasoning framework, Swanson and his colleagues proposed several other hypotheses, including the associations between magnesium deficiency and migraine, indomethacin and Alzheimer's disease, and estrogen and Alzheimer's disease. The association between magnesium and migraine was also validated by Castelli et al. (1993) and others.

Most of the hypotheses proposed by Swanson were obtained by a scheme called "closed discovery" which starts with both concepts A (e.g., fish oil) and C (e.g., Raynaud's syndrome) and identifies linking concepts B's (e.g., blood lipids and platelet affregability) which support the implicit association between A and C (Figure 19.1(a)). To initiate the discovery, we need to know beforehand a pair of concepts, A and C, that might have hidden connections. Therefore, the emphasis of closed discovery is not on finding the association

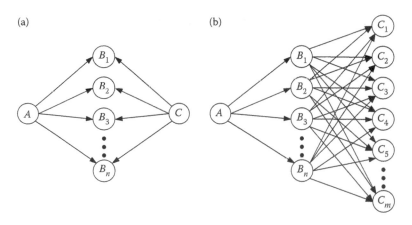

FIGURE 19.1: Closed and open discovery. (a) Closed discovery is bidirectional and starts with both concepts A and C. (b) Open discovery starts with a single concept A.

between A and C but on finding plausible pathways linking these two concepts. On the other hand, another scheme called "open discovery" starts with only concept A, and explores concepts Bs associated with A, then Cs associated with any B (Figure 19.1(b)). In contrast to closed discovery, the emphasis of open discovery is on discovering a logical yet unnoticed connection from concept A to concept C by way of B.

A challenge associated with hypothesis discovery is that the number of the potential intermediate concepts Bs or the target concepts Cs becomes prohibitively large, especially for open discovery due to a combinatorial explosion. This makes it difficult to identify truly important connections worth further investigation. A general approach to deal with the problem is to rank the output concepts based on some criteria (Swanson et al., 2006) and/or to group them to facilitate better understanding of the output (Srinivasan, 2004). Another challenge is to validate the discovered potential connections or hypotheses, which requires substantial expertise, intensive manual assessment of potential pathways, and/or even experimental verification possibly involving human subjects.

While Swanson's early works were mostly carried out by thorough manual investigation, several others as well as Swanson himself later developed computer systems to aid literature-based hypothesis discovery (Gordon and Lindsay, 1996; Lindsay and Gordon, 1999; Srinivasan, 2004; Weeber et al., 2003). The following briefly introduces a few of the representative works.

Swanson's system, called ARROWSMITH (Swanson and Smalheiser, 1997), implements closed discovery and is available for use on a web-based interface.* To use the system, a user first conducts a Medline title-word search for a concept of interest (A) and then conduct another search for concept C that s/he would like to investigate whether any potential association with A exists. The two sets of search results may contain articles mentioning relationships between A and B and those between B and C, where B represents the intermediate concepts in the ABC syllogism. ARROWSMITH takes the two search results and automatically identifies all the words or phrases (up to six contiguous words) that are common to the two sets of titles. These words or phrases are then filtered using an extensive stopword list containing approximately 5000 words, including nontopic words (e.g., able, about, and above) and general words (e.g., clinical, comparative, and drugs) in the domain. Lastly, the user, typically a domain expert, removes redundancies and useless terms. The remaining words or phrases are the candidates of the B concepts bridging A and C. For each concept B, ARROWSMITH juxtaposes two sets of titles, one containing A and B and the other containing B and C, to help the user assess the proposed pathways via B. In the output, B concepts are ranked according to the number of common Medical Subject Heading (MeSH) terms between the set of articles containing $A - B$ connections and the other set containing $B - C$ connections.

---

* http://kiwi.uchicago.edu/

Srinivasan (2004) developed another system, called Manjal,* for literature-based discovery. In contrast to the previous work which mainly uses the textual portion of Medline records, i.e., titles and abstracts, she focused solely on MeSH terms assigned to Medline records in conjunction with the UMLS semantic types and investigated their effectiveness for discovering implicit associations. Given a starting concept A, Manjal conducts a Medline search for A and extracts MeSH terms from the retrieved Medline records as B concepts. Then, the B concepts are grouped into the corresponding UMLS semantic types according to predefined mapping. By introducing UMLS semantic types, the subsequent processes can be limited only to the concepts under specific semantic types of interest to narrow down the potential pathways, and also related concepts can be grouped to help the user browse the system output. Also, Manjal incorporates the TFIDF term weighting scheme (Sparck Jones, 1972) to rank B and C concepts to make potentially significant connections more noticeable to the user. Srinivasan tested the system on several hypotheses proposed by Swanson, specifically, the associations between fish oil and Raynaud's syndrome, migraine and magnesium, indomethacin and Alzheimer's disease, somatomedin C and arginine, and Schizophrenia and calcium-independent phospholipase A2, and showed that Manjal successfully ranked in most cases the key concepts (B or C) within top ten concepts under respective semantic types.

A criticism against Swanson's framework is that many real-world phenomena would involve more complex mechanisms than the simple ABC syllogism. For instance, the risk of many common diseases is influenced by multiple genes as well as environmental factors, which may not be properly modeled by the syllogism. Another criticism against the previous work is that the proposed approaches have underused textual information, relying only on word cooccurrences at best. Moreover, although the previous work has shown the potential of literature-based discovery, it is difficult to determine what strategies would guide successful discovery of hypotheses from the small number of examples examined in the evaluation.

The next section looks at another work related to hypothesis discovery but targeting more specific types of entities, namely, genes and diseases.

19.5.2 An extension of an information retrieval (IR) model for hypothesis discovery

Focusing on implicit associations between hereditary diseases and their causative genes, Seki and Mostafa (2007) proposed a new discovery framework adapting an existing IR model, specifically, the inference network (Turtle and Croft, 1991), for hypothesis discovery. Since IR models are not designed for finding implicit associations, they need to be modified to enable hypothesis

---

* http://sulu.info-science.uiowa.edu/Manjal.html

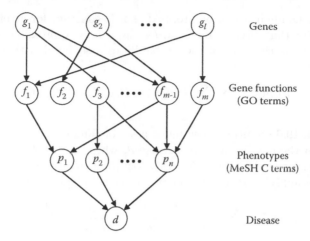

FIGURE 19.2: An inference network to model gene-disease associations.

discovery. The following summarizes their proposed framework and estimation methods for probabilistic parameters involved in the model.

In the original inference network model, a user query and documents are represented as nodes in a network and are connected via intermediate nodes representing keywords that compose the query and documents. To adapt the IR model to represent gene-disease associations, Seki and Mostafa treated a disease as a query and genes as documents and used two types of intermediate nodes: gene functions and phenotypes which characterize genes and disease, respectively. Figure 19.2 shows the network model, which consists of four types of nodes: genes (g), gene functions (f) represented by gene ontology (GO) terms,[*] phenotypes (p) represented by MeSH terms under the C category (hereafter referred to as MeSH C terms for short),[†] and disease (d). Each gene node g represents a gene and corresponds to the event that the gene is found in search for the causative genes underlying d. Each gene function node f represents a function of gene products. There are directed arcs from genes to functions, representing that instantiating a gene increases the belief in its functions. Likewise, each phenotype node p represents a phenotype of d and corresponds to the event that the phenotype is observed. The belief in p is dependent on the belief in f's since phenotypes are (partly) determined by gene functions. Finally, observing certain phenotypes increases the belief in d.

Given the inference network model, causative gene set G for given disease d can be predicted by the probability:

$$P(G|d) = \frac{P(d|G) \times P(G)}{P(d)} \tag{19.5}$$

---

[*] http://www.geneontology.org
[†] http://www.nlm.nih.gov/mesh

where the denominator can be dropped as it is constant for given d. In addition, assuming that $P(G)$ is uniform, $P(G|d)$ can be approximated to $P(d|G)$ defined below as the sum of the probabilities associated with all the paths from G to d.

$$P(d|G) = \sum_i \sum_j P(d|\vec{p}_i) \times P(\vec{p}_i|\vec{f}_j) \times P(\vec{f}_j|G). \tag{19.6}$$

Equation 19.6 quantifies how much a set of candidate genes, G, increases the belief in the development of disease d, where $\vec{p}_i$ (or $\vec{f}_j$) is defined as a vector of binary random variables with i-th (or j-th) element being 1 and all others 0. By applying Bayes' theorem and some independence assumptions,

$$P(d|G) \propto \sum_i \sum_j \left(\frac{P(p_i|d)}{P(\bar{p}_i|d)} \times \frac{P(f_j|p_i)P(\bar{f}_j|\bar{p}_i)}{P(\bar{f}_j|p_i)P(f_j|\bar{p}_i)} \right.$$

$$\left. \times F(p_i) \times F(f_j) \times P(f_j|G) \right) \tag{19.7}$$

is derived, where p and $\bar{p}$ (f and $\bar{f}$) are used as the shorthand of $p = 1$ and $p = 0$ ($f = 1$ and $f = 0$), respectively, and

$$F(p_i) = \prod_{h=1}^{m} \frac{P(\bar{f}_h|p_i)}{P(\bar{f}_h|\bar{p}_i)}, \quad F(f_i) = \prod_{k=1}^{n} \frac{P(\bar{f}_j)P(f_j|\bar{p}_k)}{P(f_j)P(\bar{f}_j|\bar{p}_k)} \tag{19.8}$$

The first factor of the right-hand side of Equation 19.7 represents the interaction between disease d and phenotype p_i, and the second factor represents the interaction between p_i and gene function f_j, which is equivalent to the odds ratio of $P(f_j|p_i)$ and $P(f_j|\bar{p}_i)$. The third and fourth factors are functions of p_i and f_j, respectively, representing their main effects. The last factor takes either 0 or 1, indicating whether f_j is a function of any gene in G under consideration. Only the estimation method for $P(f|p)$ is summarized below.

The probability $P(f|p)$ indicates the degree of belief that gene function f underlies phenotype p. Seki and Mostafa take advantage of literature data and domain ontologies to estimate the probabilities. To be precise, they used Medline records that are assigned any MeSH C terms and are referenced from any gene entry in the Entrez Gene database. Given such record, one can obtain a set of phenotypes (MeSH C terms assigned to the record) and a set of gene functions (GO terms) associated with the Entrez Gene entry. The fact that the phenotypes and gene functions are associated with the same Medline record implies that some of the phenotypes and gene functions are possibly associated. To quantify the possible associations, Seki and Mostafa sought for evidence in the textual portion of the Medline record, i.e., title and abstract. To be more precise, they searched for cooccurrences of gene functions (GO terms) and phenotypes (MeSH C terms) in a sliding, fixed-size window, assuming that associated concepts tend to cooccur more often in proximity than unassociated ones. A possible shortcoming of this approach is that gene functions or phenotypes can be described in many different forms and are not often expressed in concise GO and MeSH terms (Camon et al., 2005; Schuemie et al., 2004). To deal with the problem, Seki and Mostafa proposed to use a set of "proxy" terms to represent GO and MeSH terms. The proxy terms were automatically extracted from their definitions (or scope notes). Each cooccurrence of the terms from those two proxy term sets (one representing a gene function and the other representing a phenotype) can be seen as evidence that supports the association between the gene function and phenotype, increasing the estimated strength of their association. Based on the idea, the association between gene function f and phenotype p was defined as a sum of the products of the term weights for cooccurring proxy terms. As term weights, the TFIDF weighting scheme was utilized.

Seki and Mostafa conducted evaluative experiments on 212 known gene-disease associations extracted from the Genetic Association Database (GAD). To simulate scientific discovery, they estimated probabilistic parameters based only on Medline records published before the target associations were reported in the literature. As compared with the previous work in hypothesis discovery, this methodology allowed them to assess their discovery framework on a large number of "undiscovered" associations. The results showed that their framework achieved an AUC of 0.723, which was significantly greater than the case where textual information was not utilized. On a different set of (monogenic) diseases from the Online Mendelian Inheritance in Man database,* it was also shown that their framework compared favorably with the state-of-the-art gene association discovery system incorporating DNA sequence comparison (Perez-Iratxeta et al., 2005) despite the fact that they relied only on knowledge buried in the literature (Seki and Mostafa, 2009).

19.6 Mining Across Collections: Data Integration

19.6.1 Deriving meaning from text and links

Information is captured not only in text but also in structures, e.g., connections between data points. Link and network analyses have become popular in information retrieval and text mining for the identification of authoritative and accurate sources from different and distributed resources.

The World Wide Web is a highly distributed collection of information resources where it is challenging to retrieve orderly and quality information. The linked structure of the Web provides some potential. Link analysis has been widely used in bibliometrics and, in recent years, adopted by researchers to measure hyper-link based similarities and authorities on the Web

---

* http://www.ncbi.nlm.nih.gov/sites/entrez?db=omim

(Kleinberg, 1999). Various link-based algorithms have been implemented for finding related pages on the World Wide Web and demonstrated promising results (Kleinberg, 1999; Page et al., 1998).

Kleinberg (1999) defined *authoritative* Web pages as those being frequently pointed to by other pages and *hubs* as those that have significant concentration of links to the *authoritative* pages on particular search topics. Employing an iterative algorithm to weigh Web pages and to identify hubs and authorities, the research demonstrated the effectiveness of using links for locating high-quality information.

In a similar manner, Page et al. (1998) proposed and implemented Page-Rank to evaluate information items by analyzing collective votes through hyperlinks. PageRank effectively supported the identification of authoritative information resources on the Web and has enabled Google, one of the most popular search engines today, for ranking searched items.

Related techniques have been applied to and studied in biomedical literature mining to identify important publications, scholars, communities, and conceptual associations (Newman, 2001). Link analysis based on citations, coauthorships, and textual associations provides a promising means to discover relations and meanings embedded in the structures, which we continue to discuss in the following sections.

19.6.2 Spotting research trends and influencers

To understand a field, we usually ask questions about the major research areas of the field and how they connect to each other. We also want to know who the key players (e.g., scholars and institutes) are, their impact on research trends, and how all evolves as a field. Bibliometrics based on network analysis provides a means to answer some of these questions. Bibliometrics was defined as dealing with the application of mathematics and statistical methods to books and other media of written communication (Nicolaisen, 2007). The quantitative methods offered by bibliometrics have enabled some degree of objective evaluations of scientific publications.

19.6.2.1 Citation analysis

Among measures used to analyze various facets of written communication, citations are considered good indicators of the use and usefulness of publications (Nicolaisen, 2007). Citation analysis has been widely used to assess research productivity and scholarly impact. Researchers also employed citation and cocitation analysis and visualization techniques to identify topical communities in various domains.

However, it is still challenging to propose a theory to explain why scholars cite each other and when citations are effective indicators of impact (Nicolaisen, 2007). Today's digital infrastructure not only facilitates the dissemination of medical publications but also introduces a lot more unpredictable factors that complicate the analysis. In addition, citation analysis requires a scholarly reference dataset containing comprehensive variables and a long period of publications. Without the availability of comprehensive data, such analysis will only show a partial picture.

One shall not treat citations as if they are objective indicators of scholarly impact or quality because bias does exist. Citing a previous work does not necessarily convey credit to the work given different motivations involved (Nicolaisen, 2007). Sometimes scholars cite another paper because they are not in agreement with points expressed in it. In addition, previous research on network science and citation analysis proposed that citation networks are scale-free networks subject to *preferential attachment*, that is, they have a distribution of connectivities that decays with a power law function (Albert and Barabasi, 2002). In other words, the citation score distribution is approximately linear on log-log coordinates.

The majority of publications are rarely cited and extremely "poor" whereas highly cited nodes, i.e., influential publications or scholars, tend to be cited even more—demonstrating the rich get richer effect. This, as other cumulative advantage processes, indicates that success tends to breed success and it is challenging to single out impact or quality from the preferential attachment tendency. Research also found associations between authors' institutional affiliations and what they tend to cite. Particularly, in various domains, geographic distance strongly correlates with the frequency of two institutions citing each other.

Another bias involves the age of citation that follows some common patterns. Gupta (1997) studied distributions of citation age and found that a citation decay curve consists of two parts: (1) an increase of citations during first couple of years, followed by (2) gradual decline of citations when the paper gets older. Similar patterns were found by other researchers.

The impact of this age distribution is obvious—older papers tend to be forgotten and replaced by new ones in terms of attracting additional citations. This favoritism might be a reasonable nature of the scholarly communication; but it invites cautions when we are to treat citations as an objective measure.

Regardless of all the bias involved in scholarly references, the use of citation data has proved effective in identifying important works and influential scholars in various domains. Its application in biomedical literature mining produced promising results. For example, Bernstam et al. (2006) defined importance as an article's influence on the scientific discipline and used citation analysis to operationalize it for biomedical information retrieval. They found that citation-based methods, namely Citation Count and PageRank, are significantly more effective at identifying important articles from Medline.

Besides direct citation counting, other forms of citation analysis involve the methods *bibliographic coupling* (or coreference) and *cocitation*. While bibliographic coupling examines potentially associated papers that refer to a common literature, cocitation analysis aims to identify important and related papers that have been cited together in a later literature. These techniques have been extended to identify key scholars, groups, and topics in some fields.

19.6.2.2   Collaboration analysis and others

Coauthorship analysis is another widely used method for identifying important scholars, scientific collaborations, and clusters of research topics of a field. Research investigated coauthorship networks in various scientific domains such as neurosciences and found interesting common characteristics, e.g., power-law distributions of connectivities (Albert and Barabasi, 2002). Yet, disciplines have different patterns in terms of, e.g., the degree scholars collaborated with each other and how isolated research groups are from each other (Newman, 2001).

Newman (2001) studied four different scholarly collaboration networks from bibliographic data during 1995–1999. Using a measure called clustering coefficient, or the fraction of connected triples of vertices (authors) that form triangles, he found that collaborations of three or more scholars are common in science. As compared to the others, the coauthorship network in biomedical research had a smaller clustering coefficient value, demonstrating a laboratory-based collaboration pattern where students, postdocs, and technicians worked with a supervising principal investigator and formed a treelike structure. The study also showed that probability distributions of # papers and # collaborators per author follow roughly a power-law form, i.e., a roughly linear function on log–log coordinates.

One difficulty involved in the coauthorship network analysis is the disambiguation of author names (Newman, 2001). While two authors may have the same name, a single author may use different forms of his/her name. The use of additional information such as affiliations helps distinguish and/or unify author names in some cases but fails in others. A single scholar may have multiple affiliations and move from one place to another.

As is in PubMed/Medline databases, the use of first initials further complicates the problem. Torvik et al. (2005) hypothesized that different papers written by the same author would have common features and proposed a model that estimated the probability of two identical names (in terms of last name and first initial) actually referred to the same person. The study showed that affiliation words, title words, and MESH are useful in measuring pair-wise similarities of papers for the disambiguation of author names.

Other approaches to intelligently identifying unique scholars have been pursued. For example, cocitation analysis was used to bring together related authors, sometimes identical scholars, by examining citation patterns of their authored works.

19.6.2.3   Information visualization for network analysis

The citation and collaboration analysis techniques discussed earlier usually result in large complex networks that cannot be easily captured and understood. Information visualization, which attempts to facilitate information communication through visual representations, is potentially useful in the discovery of citation and collaboration patterns in a domain.

Clustering, sometimes in the form of information visualization, was often performed on coauthorship networks to discover major research groups/ collaborations where important research topics can be identified. Coauthorship and citation analysis are often used together to discover influential collaborations in a field. Figure 19.3 shows a visualization of clusters in a collaboration network, on which major research institutions and topics can be labeled.

Börner et al. (2005) applied entropy-based measures to the analysis and visualization of a weighted coauthorship network in an emerging science domain over 31 years. They found an increasingly interconnected scholar communities and a heterogeneous distribution of high impact collaborations. Visualizations of the coauthorship network over the years clearly presented the evolution of the clusters of research groups/topics in the field.

Link analysis and visualization tools abound and many are applicable to biomedical text mining. Bhavnani et al. (2007) employed network analysis tools to examine associations between toxic chemicals and symptoms. They presented the distribution of the number of matched chemicals over the number of random symptoms and visualized how the two types of nodes were interconnected. Then the bipartite network (of chemicals and symptoms) was projected to a one-node visualization on which intersymptom links and regularities were shown. This work helped explain, based on the visualizations, why current toxic identification systems required a large number of symptom inputs and proposed how first-responder systems could be improved to identify chemicals more rapidly. It demonstrated the usefulness of network analysis and information visualization in related domains.

Mane and Börner (2004) applied the burst detection algorithm, proposed by Kleinberg for detecting *hot* research topics (bursts), to the analysis of a literature across multiple domains over a 20-year period. By extracting keywords and analyzing their frequencies of use over time, the algorithm enabled the discovery of significant concepts that experienced a sudden increase in their usage. The results were visualized on a concept map, which conveys information about key concepts, associations of the concepts, when they burst (suddenly became popular), and the degree of the bursts. The visualization reveals a clear research trend from investigations of biological entities in 1980s to molecular sequence data in 1990s and then to protein research toward the end of 20th century.

19.6.3 Linking services to texts and data

While biomedical research domains and subdomains become increasingly specialized, some of them also come to intersections. Linking research corpus from different specialties are critical for providing a bigger picture for newcomers to understand the field and for directions on continued research/investigations. As Cohen and Hersh (2005) pointed out, integration of multiple databases and greater access to the full collections are very

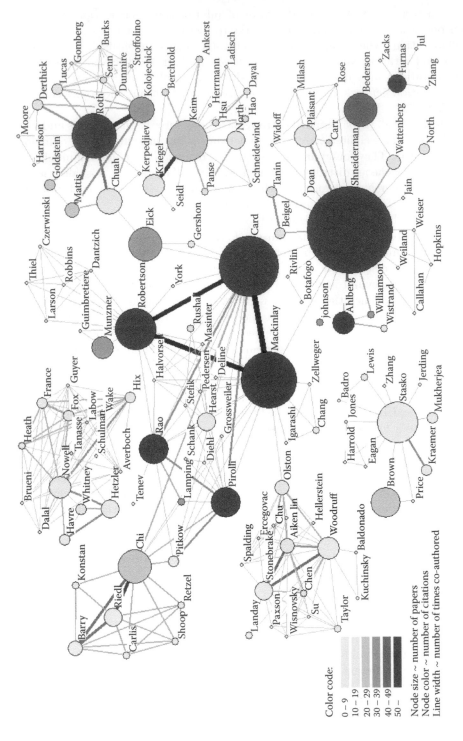

Color code:

0 – 9
10 – 19
20 – 29
30 – 39
40 – 49
50 –

Node size ~ number of papers
Node color ~ number of citations
Line width ~ number of times co-authored

FIGURE 19.3: A sample collaboration network in information visualization (1974–2004).

important in biomedical text mining. However, the lack of a globally standardized and structured framework for concept representation is making the integration of biomedical data sources extremely challenging.

The availability of free text in biomedical literature provides the potential for association discovery based on concept extraction; it also adds another layer of difficulties. Cantor and Lussier (2003) hypothesized that the use of existing biomedical thesauri, with a large number of associated concepts and their relations, would be useful for linking data sources semantically and logically. After terms were extracted from NCBI OMIM, related concepts in UMLS Metathesaurus and SNOMED CT were matched and their relations verified. It was found that a considerable portion of the concepts were properly mapped to both vocabulary sets. This provides a means to extract a semantic/conceptual structure from existing databases and to apply the semantic structure back to the free text corpora. Potentially, preestablished relationships are useful for researchers and clinicians to discover/infer unknown associations across collections.

To take advantage of the UMLS Metathesaurus for the identification of biomedical concepts from free text, the National Library of Medicine designed and implemented a software package called MetaMap (Aronson, 2001). The tool had various modules for text parsing, (concept) variant generation, candidate (concept) retrieval and evaluation, and mapping construction. MetaMap showed significant improvements over existing methods and has been widely used in biomedical text mining applications.

Bashyam et al. (2007) observed the popularity and effectiveness of MetaMap in various settings but realized its efficiency was not tuned due in part to the complexity of the modules/methods. Reasoning that the process could be accelerated by ignoring modules that are less useful in common applications, the authors proposed a new method that tokenized sentences into stemmed words without part-of-speech (POS) taggers and matched the longest spanning words with individual preindexed UMLS words. Comparative experiments showed an 86% improvement of efficiency over MetaMap with an about 9% loss in effectiveness. This type of method will have its application in situations where real-time results are desirable.

19.7 Conclusion: Evaluation and Moving Forward

19.7.1 Metrics and methodologies for evaluation

The standard methods for evaluating literature mining results have strong overlap with metrics applied in traditional IR. The two established metrics are known as specificity and sensitivity and they can be directly derived from two well-known IR metrics known as recall and precision. The best way to explain

TABLE 19.5:   Contingency
matrix.

	True	False
Classified yes	TP	FP
Classified no	FN	TN

these metrics and show their relationships is to describe them in terms of a contingency matrix as in Table 19.5.

Assume that the task for the literature miner is one involving extracting a certain set of tokens that represent a disease class. We can treat the task as a retrieval or classification involving each token identified by the literature miner. Assume further that each time a token is classified an assessment routine or a human judge may consider it and generate a verdict of true or false. Now, we can explain the content of the matrix in the above table (Table 19.5). Upon completion of the classification task, we can generate a sum for the number of tokens that are correctly classified for the disease class (TP), those that belong but were not classified to be in the class (FN), those that do not belong but were wrongly classified to be in the class (FP), and finally those do not belong and were correctly excluded from the class (TN). With the accumulated statistics it then becomes possible to express the two basic metrics, namely *sensitivity* $= TP/(TP + FN)$ and *specificity* $= TP/(TP + FP)$. Sometimes, the two metrics are combined into a harmonic mean score, typically calculated as $F = (2 \times sensitivity \times specificity)/(sensitivity + specificity)$.

Increasingly, in addition to the use of the above metrics for evaluation purposes, there is a preference for using a visual analysis approach for understanding the relationship between TP (defined as the benefits derived) and FP (cost of using a less than effective approach). The method known as receiver operating characteristics (ROC) curves allow one to map the results of several classifiers (or several classification approaches) into a single 2-D space and determine their superiority in terms of the cost versus benefit factors (see Figure 19.4).*

For example from Figure 19.4 we can easily pick out the system A as superior than system B or system B as superior than system E. The diagonal line represents results produced randomly with equal chance of being correct and wrong. For example, the system C can be considered to be unreliable as it has produced near random results. Our inherent goal is to expand the benefits achieved by increasing the number of true positives while reduce the cost by decreasing the number of false positives—this is the basic rationale used to determine relative superiority of systems once their results are plotted. According to this rationale, the systems (or their classification results) that fall

---

* This figure is adapted from Fawcett (2003). The paper is a nice tutorial on the application of ROC graphs for evaluation and provides guidelines for its appropriate usage.

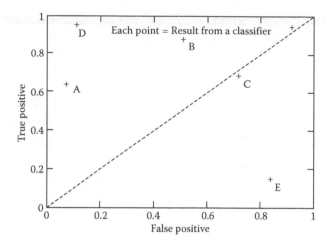

FIGURE 19.4: An ROC graph showing the results from several different extractor/classifier systems. (Adapted from Fawcett, T. Technical Report HPL-2003-4, HP Laboratories Palo Alto, CA, 2003.)

approximately in the top left hand corner zone (or near it) would always be superior compared to those that are not near this zone, or conversely the systems that fall near the bottom right hand corner are always going to be less effective than those that do not fall near this zone. One issue that requires closer examination in the evaluation realm is the distribution of positive examples in the test corpus and how it influences the standard metrics (i.e., specificity and sensitivity). It has been noted correctly that skewed distribution of positive examples in a corpus would have a significant impact on the results achieved, independent of the actual performance of the system (Baldi et al., 2000; Jensen et al., 2006). Hence, it is important to demonstrate the performance of systems across corpora that vary both in size and content.

19.7.2 User interface features for expanding information retrieval (IR) functions

An important trend in IR systems, particularly in the biomedical area, has been the incorporation of new functions that supplement the core retrieval operations. The new functions vary greatly in sophistication and usefulness, but their main aim is to assist the user in attaining additional insights or understanding. Such functions are typically based on the "text/s in hand" (i.e., the article/s) retrieved as a response to the most recent query. A basic set of such functions can be seen in the Oxford University Press journal, *Bioinformatics* which upon display of an individual article offers the user the option of conducting a relevance-feedback search whereby the user may continue the

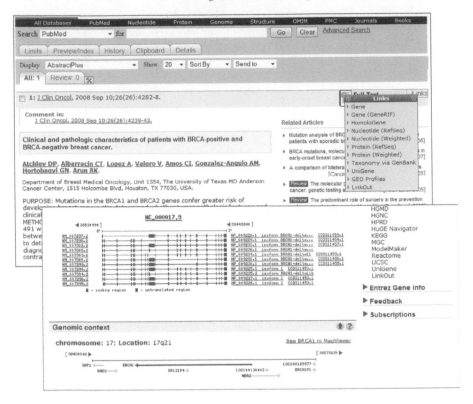

FIGURE 19.5: Starting with a basic search and ending with detailed information on a particular gene.

search to identify additional relevant articles by using the current article as a "query." The search in fact can be extended to other online reference sources such as Pubmed and Google Scholar. In the NCBI Pubmed system, upon retrieval of individual articles, the user is offered the option of "linking-out" from a specific article to associated protein sequences, structures, genes, etc. Here, the retrieval operation is no longer limited to finding articles but is extended to identifying key biological entities that are relevant to the article. It is possible in NCBI system to continue pursuing the search by focusing on a single protein sequence to identify other proteins or genes that are highly correlated. Here, the line between classic retrieval and mining functions begins to blur.

In Figure 19.5, we show a sequence of two screen shots from the NCBI Pubmed system which starts with a simple query to identify genes that are known to have a strong role in certain types of cancer (neoplasm is the mesh term for cancer). As can be seen in Figure 19.5, the user may retrieve additional information on specific genes mentioned in an article by following the "gene" link-out option to the location-based information for the BRCA1

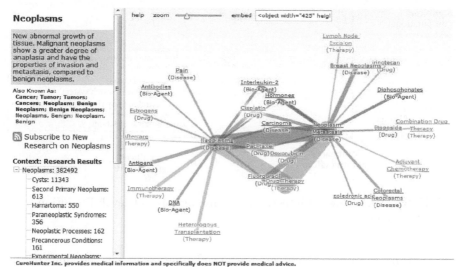

FIGURE 19.6: Interactive visual analysis of data in CureHunter.

gene (coding regions versus noncoding regions). The search operation here was supplemented with preprocessed information on specific gene occurrences in the literature and integration of such information with other data sets in the NCBI system (in this case genomics data).

Another example of a more advanced function can be seen in the Cure-Hunter system.* Here, a user may access Pubmed data but in a retrieval modality that permits interactive analysis and visual mining of the data to reveal subtle relationships among key biomedical concepts and terminologies. In Figure 19.6, we show an example where the search was initiated on the term "neoplasm" leading to extraction of other critical data associated with drugs, therapy, biological agents, and disease subclasses.

Before wrapping up this section, we would like to present a conceptual framework to explain the rapid change in IR systems which we believe is currently in progress. An earlier version of Figure 19.7 was presented in a *Scientific American* article by the first author (Mostafa, 2004). As the figure shows, the integration of text mining functions with IR systems is a logical advance—going beyond "access" to "analysis." It is worth pointing out several key characteristics of this new class of systems which has three fundamental dimensions, namely: (1) they are integrative (searching plus mining), (2) they are intelligent because they are capable of taking proactive actions based on deep analysis of content and predicting steps where user participation is essential, and (3) they support interleaving of user efforts and system operations because they assume both the user and the system are critical resources, thus

---

* A demonstration of this system can be found here: http://www.curehunter.com

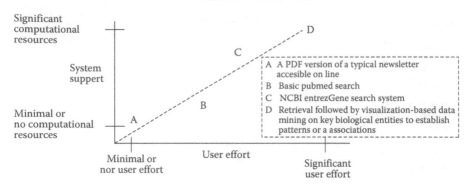

FIGURE 19.7: Toward integration of mining into retrieval: Evolution path for IR systems.

maximizing the ultimate outcome achieved (one could say the sum, of the system and the user influence, is greater than the simple aggregation of their dual input into the workflow). As a shorthand, we refer to these systems supporting the new interaction modality as I^3 for: integrative, intelligent, and interleaving.

As can be seen in Figure 19.7, the ideal system demands a large and approximately equal level of participation by the user and the system. The new I^3 modality involves the user driving the direction of the inquiry and the system executing critical functions and revealing associated data along the way. It should also be noted that the addition of text mining functions *does not coopt the core information retrieval functions*. The I^3 modality, if successfully implemented, would provide a seamless iterative path. A typical path may include the following: (a) start with search operations that may reveal a small subset of relevant document, (b) move to mining of biological entities and patterns associated with them, and (c) return to searching of related documents. These systems will not distinguish between a focus on documents versus a focus on biological entities and would allow the user to perform inquiry at any desired level of abstraction. The bottom-line is that the main goal of these new systems, supporting the I^3 modality, is not to bring voluminous information closer to the user but to facilitate *understanding*.

19.7.3 Frontiers in literature mining

We believe the future of literature mining is bright indeed. There are several likely areas of growth.

One potentially fruitful area is integration of text mining with data mining, particularly with the goal of increasing the confidence in potential associations among biological entities by combining evidence from multiple sources. For example, if a micro array analysis reveals interesting relationship among some genes, such evidence may be combined with association information among genes discovered through literature mining and vice versa. Such integrative mining, conducted on different resources, has critical implications for advancing translational science as well. There is a growing trend in the creation of clinical data warehouses that aim to make health records aggregated over a period of time available for research purposes. Interesting patterns discovered through mining operations on such records may potentially be validated through analysis of published literature. Going in the other direction, it may also offer the possibility of evaluating key associations that are recently published* against "realistic" associations found through mining of clinical records.

Another area of potential growth is development of more powerful interactive literature mining tools that can be seamlessly combined with current retrieval systems. We have discussed a few of these systems in the last section, using the I^3 framework. It should be noted here that increasing the intensity and level of involvement by both the system and the user are not "free," rather they introduce associated costs. On the system side, we know that smooth execution of many of the new mining functions requires dedication of significant computational resources. Some of the computations can be conducted in an "offline" manner, for example, establishing association between critical items discussed in the literature and sequence information of biological entities for the provision of services such as "link-out" in the NCBI system. However, increasingly, users will demand on-the-fly and real-time discovery of such associations which are driven by the users' specific search contexts or their desire for exhaustive coverage of all relevant items in the database including those that have been addeded in the recent past (for example, in the last 30 minutes). The shift toward the requirement for real-time mining operation has significant implications for the design of new mining algorithms as they most likely will have to take advantage of multicore or parallel architectures. On the human side, we must point out that the increasing engagement must be preceded by education or training in the domain, as a deeper and nuanced knowledge base is required for users to drive the mining operations in rationale directions and to accurately interpret the information the system presents along the way.

Acknowledgments

We thank the UNC biomedical research imaging center (BRIC) and Translational and Clinical Sciences (TraCS) Institute for support.

---

* Actually, it will be possible to conduct such integrative analysis before publication as well.

References

Albert, R., and Barabasi, A.-L. 2002. Statistical mechanics of complex networks. *Reviews of Modern Physics*, 74(1):47–97.

Ando, R. K. 2007. BioCreative II gene mention tagging system at IBM watson. In *Proceedings the Second BioCreative Challenge Evaluation Workshop*, Madrid, Spain, 101–103.

Aronson, A. R. 2001. Effective mapping of biomedical text to the umls metathesaurus: the metamap program. *Proc AMIA Symp*, Washington, DC, 17–21.

Baldi, P., Brunak, S., Chauvin, Y., Andersen, C., and Nielsen, H. 2000. Assessing the accuracy of prediction algorithms for classification: An overview. *Bioinformatics*, 16:412–424.

Bashyam, V., Divita, G., Bennett, D., Browne, A., and Taira, R. 2007. A normalized lexical lookup approach to identifying umls concepts in free text. In *Proceedings of the 12th World Congress on Health Informatics MEDINFO (2007)*, Brisbane, Australia, 545–549.

Bernhardt, P. J., Humphrey, S. M., and Rindflesch, T. C. 2005. Determining prominent subdomains in medicine. In *AMIA Annual Symposium Proceeding 2005*. American Medical Informatics Association, Washington, DC, 46–50.

Bernstam, E. V., Herskovic, J. R., Aphinyanaphongs, Y., Aliferis, C. F., Sriram, M. G., and Hersh, W. R. 2006. Using citation data to improve retrieval from medline. *Journal of the American Medical Informatics Association*, 13(1):96–105.

Bhavnani, S., Abraham, A., Demeniuk, C., Gebrekristos, M., Gong, A., Nainwal, S., Vallabha, G., and Richardson, R. 2007. Network analysis of toxic chemicals and symptoms: Implications for designing first-responder systems. In *Proceedings of AMIA '07*, Chicago, IL, 51–55.

Börner, K., Dall' Asta, L., Ke, W., and Vespignani, A. 2005. Studying the emerging global brain: Analyzing and visualizing the impact of co-authorship teams: Research articles. *Complexity*, 10(4):57–67.

Camon, E., Barrell, D., Dimmer, E., Lee, V., Magrane, M., Maslen, J., Binns, D., and Apweiler, R. 2005. An evaluation of GO annotation retrieval for BioCreAtIvE and GOA. *BMC Bioinformatics*, 6(Suppl 1):S17.

Cantor, M. N., and Lussier, Y. A. 2003. Putting data integration into practice: Using biomedical terminologies to add structure to existing data sources. In *AMIA Annual Symposium Proceedings*, Washington, DC, 125–129.

Castelli, S., Meossi, C., Domenici, R., Fontana, F., and Stefani, G. 1993. Magnesium in the prophylaxis of primary headache and other periodic disorders in children. *Pediatria Medica e Chirurgica*, 15(5):481–488.

Cohen, A. M., and Hersh, W. R. 2005. A survey of current work in biomedical text mining. *Brief Bioinformatics*, 6(1):57–71.

Cohen, K. B., Acquaah-Mensah, G. K., Dolbey, A. E., and Hunter, L. 2002. Contrast and variability in gene names. In *Proceedings of the ACL-02 Workshop on Natural Language Processing in the Biomedical Domain*. Association for Computational Linguistics, Morristown, NJ, 14–20.

Cutting, D. R., Karger, D., Pedersen, J. O., and Tukey, J. W. 1992. Scatter/Gather: A cluster-based approach to browsing large document collections. In *SIGIR '92. Proceedings of the 15th Annual International ACM SIGIR Conference on Research and Development in Information Retrieval*. ACM, Copenhagen, Denmark, 318–329.

Fang, H. R., Murphy, K., Jin, Y., Kim, J., and White, P. 2006. Human gene name normalization using text matching with automatically extracted synonym dictionaries. In *Proceedings of the HLT-NAACL BioNLP Workshop on Linking Natural Language Processing and Biology*. Association for Computational Linguistics, Morristown, NJ, 41–48.

Fawcett, T. 2003. Roc graphs: Notes on practical considerations for data mining researchers. Technical Report HPL-2003-4, HP Laboratories, Palo Alto, CA.

Franzén, K., Eriksson, G., Olsson, F., Asker, L., and Lidén, P. 2002. Exploiting syntax when detecting protein names in text. In *Proceedings of Workshop on Natural Language Processing in Biomedical Applications NLPBA 2002*. Nicosia, Cyprus.

Fukuda, K., Tsunoda, T., Tamura, A., and Takagi, T. 1998. Toward information extraction: Identifying protein names from biological papers. In *Proceedings of the Pacific Symposium on Biocomputing*, Hawaii, 705–716.

Gordon, M. D. and Lindsay, R. K. 1996. Toward discovery support systems: a replication, re-examination, and extension of Swanson's work on literature-based discovery of a connection between Raynaud's and fish oil. *Journal of the American Society for Information Science*, 47(2):116–128.

Gupta, B. M. 1997. Analysis of distribution of the age of citations in theoretical population genetics. *Scientometrics*, 40(1):139–162.

Hanisch, D., Fluck, J., Mevissen, H., and Zimmer, R. 2003. Playing biology's name game: Identifying protein names in scientific text. In *Proceedings of the Pacific Symposium on Biocomputing*, Hawaii, 403–414.

Hatzivassiloglou, V., Duboué, P. A., and Rzhetsky, A. 2001. Disambiguating proteins, genes, and RNA in text: A machine learning approach. *Bioinformatics*, 17(Suppl 1):s97–s106.

Hsu, C.-N., Chang, Y.-M., Kuo, C.-J., Lin, Y.-S., Huang, H.-S., and Chung, I.-F. 2008. Integrating high dimensional bi-directional parsing models for gene mention tagging. *Bioinformatics*, 24(13):i286–i294.

Jensen, L., Saric, J., and Bork, P. 2006. Literature mining for the biologist: From information retrieval to biological discovery. *Nature Reviews Genetics*, 7:119–129.

Jimeno, A., Jimenez-Ruiz, E., Lee, V., Gaudan, S., Berlanga, R., and Rebholz-Schuhmann, D. 2008. Assessment of disease named entity recognition on a corpus of annotated sentences. *BMC Bioinformatics*, 9(Suppl 3):S3.

Kleinberg, J. M. 1999. Authoritative sources in a hyperlinked environment. *Journal of the ACM*, 46(5):604–632.

Krauthammer, M., Rzhetsky, A., Morozov, P., and Friedman, C. 2001. Using BLAST for identifying gene and protein names in journal articles. *GENE*, (259):245–252.

Landauer, T., Foltz, P., and Laham, D. 1988. Introduction to latent semantic analysis. *Discourse Processes*, 25:259–284.

Lindsay, R. K., and Gordon, M. D. 1999. Literature-based discovery by lexical statistics. *Journal of the American Society for Information Science*, 50(7):574–587.

Mane, K., and Börner, K. 2004. Mapping topics and topic bursts in PNAS. *Proceedings of the National Academy of Science*, 101:5287–5290.

Mika, S., and Rost, B. 2004. Protein names precisely peeled off free text. *Bioinformatics*, 20(Suppl 1):i241–i247.

Mostafa, J. 2004. Seeking better web searches. *Scientific American*, 292(2):51–71.

Mostafa, J., Mukhopadhyay, S., Lam, W., and Palakal, M. 1997. A multilevel approach to intelligent information filtering: Model, system, and evaluation. *ACM Transactions on Information Systems*, 368–399.

Newman, M. E. J. 2001. Scientific collaboration networks. i. network construction and fundamental results. *Physical Review E*, 64(1):016131.

Nicolaisen, J. 2007. Citation analysis. *Annual Review of Information Science and Technology,* 43:609–641.

Page, L., Brin, S., Motwani, R., and Winograd, T. 1998. The PageRank citation ranking: Bringing order to the Web. Technical report, Stanford Digital Library Technologies Project.

Perez-Iratxeta, C., Wjst, M., Bork, P., and Andrade, M. 2005. G2D: a tool for mining genes associated with disease. *BMC Genetics,* 6(1):45.

Salton, G., and McGill, M. 1983. *Introduction to Modern Information Retrieval,* 1st edition. McGraw-Hill, Ohio.

Schuemie, M. J., Weeber, M., Schijvenaars, B. J. A., van Mulligen, E. M., van der Eijk, C. C., Jelier, R., Mons, B., and Kors, J. A. 2004. Distribution of information in biomedical abstracts and full-text publications. *Bioinformatics,* 20(16):2597–2604.

Sebastiani, F. 2002. Machine learning in automated text categorization. *ACM Computing Survery,* 34(1): 1–47.

Seki, K., and Mostafa, J. 2005. A hybrid approach to protein name identification in biomedical texts. *Information Processing and Management,* 41(4):723–743.

Seki, K., and Mostafa, J. 2007. Discovering implicit associations between genes and hereditary diseases. *The Pacific Symposium on Biocomputing,* 12:316–327.

Seki, K. and Mostafa, J. 2009. Discovering implicit associations among critical biological entities. *International Journal of Data Mining and Bioinformatics,* 3(2), in press.

Sparck Jones, K. 1972. Statistical interpretation of term specificity and its application in retrieval. *Journal of Documentation,* 28(1):11–20.

Srinivasan, P. 2004. Text mining: generating hypotheses from Medline. *Journal of the American Society for Information Science and Technology,* 55(5):396–413.

Swanson, D. R. 1986a. Fish oil, Raynaud's syndrome, and undiscovered public knowledge. *Perspectives in Biology and Medicine,* 30(1):7–18.

Swanson, D. R. 1986b. Undiscovered public knowledge. *Library Quarterly,* 56(2):103–118.

Swanson, D. R. 1987. Two medical literatures that are logically but not bibliographically connected. *Journal of the American Society for Information Science,* 38(4):228–233.

Swanson, D. R., and Smalheiser, N. R. 1997. An interactive system for finding complementary literatures: a stimulus to scientific discovery. *Artificial Intelligence,* 91(2):183–203.

Swanson, D. R., Smalheiser, N. R., and Torvik, V. I. 2006. Ranking indirect connections in literature-based discovery: The role of medical subject headings. *Journal of the American Society for Information Science and Technology,* 57(11): 1427–1439.

Torvik, V. I., Weeber, M., Swanson, D. R., and Smalheiser, N. R. 2005. A probabilistic similarity metric for medline records: A model for author name disambiguation. *Journal of the American Society for Information Science and Technology,* 56(2):140–158.

Turtle, H., and Croft, W. B. 1991. Evaluation of an inference network-based retrieval model. *ACM Transactions on Information Systems,* 9(3):187–222.

Weeber, M., Vos, R., Klein, H., de Jong-van den Berg, L. T., Aronson, A. R., and Molema, G. 2003. Generating hypotheses by discovering implicit associations in the literature: A case report of a search for new potential therapeutic uses for thalidomide. *Journal of the American Medical Informatics Association,* 10(3):252–259.

Wilbur, J., Larry, S., and Lorrie, T. 2007. BioCreative 2. Gene mention task. In *Proceedings the Second BioCreative Challenge Evaluation Workshop,* Madrid, Spain, 7–16.

Yang, Y., and Liu, X. 1999. A re-examination of text categorization methods. In *SIGIR '99: Proceedings of the 22nd Annual International ACM SIGIR Conference on Research and Development in Information Retrieval.* ACM, New York, NY, 42–49.

Zhou, G., Zhang, J., Su, J., Shen, D., and Tan, C. 2004. Recognizing names in biomedical texts: A machine learning approach. *Bioinformatics,* 20(7):1178–1190.

Chapter 20

Mining Biological Interactions from Biomedical Texts for Efficient Query Answering

Muhammad Abulaish

Jamia Millia Islamia

Lipika Dey

Tata Consultancy Services

Jahiruddin

Jamia Millia Islamia

20.1 Introduction

The information age has made the electronic storage of large amounts of data effortless. The proliferation of documents available on the Internet, corporate Intranets, news wires, and elsewhere is overwhelming. Technological

advances and professional competition have contributed to the large volume
of scientific articles, making it impossible for researchers to keep up with the
literature. This information overload also exists in the biomedical field, where
scientific publications and other forms of text-based data are produced at an
unprecedented rate due to growing research activities in the recent past. The
number of text documents disseminating knowledge in this area has gone up
manifolds. Most scientific knowledge is registered in publications and other
unstructured representations that make it difficult to use and to integrate the
information with other sources (e.g., biological databases).

Given that almost all current biomedical knowledge is published in scientific articles, researchers trying to make use of this information and consequently there is an increasing demand for automatic curation schemes to
extract knowledge from scientific documents and store them in a structured
form without which the assimilation of knowledge from this vast repository
is becoming practically impossible. While search engines provide an efficient
way of accessing relevant information, the sheer volume of the information
repository on the Web makes assimilation of this information a potential bottleneck in the way its consumption. One approach to overcome this difficulty
could be to use intelligent techniques to collate the information extracted from
various sources into a semantically related structure which can aid the user
for visualization of the content at multiple levels of complexity. Such a visualizer provides a semantically integrated view of the underlying text repository
in the form of a consolidated view of the concepts that are present in the
collection, and their inter-relationships as derived from the collection along
with their sources. The semantic net thus built can be presented to the user
at arbitrary levels of depth as desired.

Several disciplines including *information retrieval* (IR), *natural language
processing* (NLP), and *information extraction* (IE) involve the automated handling of text. IR deals mostly with finding documents that satisfy a particular
information need (e.g., all the documents relevant to a certain protein or disease) within a large database of documents [3]. In the biomedical domain,
IR technologies are in widespread use. Most experimental biologists take advantage of the PubMed IR system available at the NCBI, which run on the
PubMed database [5]. NLP is a broad discipline concerned with all aspects of
automatically processing both written and spoken language [4]. IE is a sub-
field of NLP, which generally uses pattern matching to find explicit entities
and facts in free text [11,12]. An example of IE applications in biomedical
domain is the identification of protein interactions. Techniques such as simple pattern matching can highlight relevant text passages from large abstract
collection. However, generating new insights to future research is far more
complex.

Text mining is a knowledge discovery process that attempts to find hidden
information in the literature by exploring the internal structure of the knowledge network created by the textual information. Text mining is the combined, automated process of analyzing unstructured, natural language text to

discover information and knowledge that is typically difficult to retrieve [15]. Text mining in molecular biology is defined as the automatic extraction of information about genes, proteins and their functional relationships from text documents. Text mining has emerged as a hybrid discipline on the edges of the fields of information science, bioinformatics, and computational linguistics. A range of text-mining applications have been developed recently that will improve access to knowledge for biologists and database annotators [1,2]. Knowledge discovery could be of major help in the discovery of indirect relationships, which might imply new scientific discoveries. Such new discoveries might provide hints for experts working on specific biological processes. A survey of current work in biomedical text mining can be found in [10].

However, while dealing with unstructured text documents domain knowledge plays an important role to resolve different cases of ambiguity that can occur during syntactic and semantic analysis of text [6]. Consider the following examples from biomedical documents:

- These data also suggest that AFB1 binds preferentially to DNA with an alternating G-C sequence compared to DNA with a sequence of contiguous Gs or Cs.

- GMPPCP binds to tubulin with a low affinity relative to GTP or GDP.

In the first sentence, the prepositional phrase introduced by "with" should be attached to the previous noun phrase "DNA" while in the second sentence, the prepositional phrase should be attached to the noun phrase "CMPPCP" before the verb "bind," and not to "tubulin." This ambiguity can only be resolved by knowing that G-C sequences are DNA characteristics and that low affinity is an attribute of a binding event.

Ontologies, which store domain knowledge in a structured machine-interpretable format, can play a very important role in text IR to resolve such ambiguities by providing additional domain knowledge to the search engine. Ontology represents domain knowledge in a structured form and is increasingly being accepted as the key technology wherein key concepts and their interrelationships are stored to provide a shared and common understanding of a domain across applications [7]. In the biomedical field, ontologies have been widely used for the structured organization of domain specific knowledge.

Though there has been a lot of work on entity extraction from text documents, relation mining has received far less attention. In this chapter, we have proposed the design of an intelligent biomedical text mining system, *BioTextMiner*, which uses linguistic and semantic analysis of text to identify key information components (ICs) from biomedical text documents and stores them in a structured knowledge base over which user queries are processed. The ICs are centered on domain entities and their relationships, which are extracted using NLP techniques and cooccurrence-based analysis. The system is also integrated with a biomedical named entity recognizer, ABNER [27], to identify a subset of GENIA ontology concepts (DNA, RNA, protein, cell-line, and cell-type) and tag them accordingly. This helps in answering biological queries based on biological concepts rather than on particular entities only. The process of extracting relevant ICs from text documents and automatic construction of structured knowledge bases is termed as curation [29]. Curation is very effective in managing online journal collections. Given a query, BioTextMiner aims at retrieving all relevant sentences that contain a set of biological concepts stated in a query, in the same context as specified in the query, from the curated knowledge base. Schutz and Buitelaar [28] state that verbs play an important role in defining the context of concepts in a document. BioTextMiner is designed to locate and characterize verbs within the vicinity of biological entities in a text, since these can represent biological relations that can help in establishing query context better. The verbs thus mined from documents are subjected to feasibility analysis and then characterized at concept level. We have shown that relation mining can yield significant ICs from text whose information content is much more than entities. We have also presented a scheme for semantic integration of information extracted from text documents. The semantic net highlights the role of a single entity in various contexts, which are useful both for a researcher as well as a layman. The network provides distinct user perspectives and allows navigation over documents with similar ICs and is also used to provide a comprehensive view of the collection. It is possible to slice and dice or aggregate to get more detailed or more consolidated view as desired. We have also presented methodologies for collating information from multiple sources and present them in an integrated fashion that can help in easy comprehension and smooth navigation through the pile of information. The biological relations extracted from biomedical text along with other relevant ICs like biological entities occurring within a relation, are stored in a database. The database is integrated with a query-answering module. The query-answering module has an interface, which guides users to formulate queries at different levels of specificity. The efficacy of BioTextMiner is established through experiments on top 500 abstracts retrieved by PubMed search engine for the query term "breast cancer."

Unlike most of the related works [16–19] on biological relation extraction, which have described methods for mining a fixed set of biological relations occurring with a set of predefined tags, the proposed system identifies all verbs in a document, and then identifies the feasible biological relational verbs using contextual analysis. While mining biological relations the associated prepositions are also considered which very often changes the nature of the verb. For example, the relation "activates in" denotes a significant class of biological reactions. Thus, we also consider the biological relations, which are combinations of *root verbs or their morphological variants, and prepositions* that follow these. Typical examples of biological relations identified in this category include "activated in," "binds to," "stimulated with" etc. Besides mining relational verbs and associated entities, the novelty of the system lies in extracting entities termed as *catalyst*, whose presence or absence validates a particular biological interaction. For example, consider the following sentence:

"Diallyl sulfite (DAS) prevents cancer in animals (PMID: 8636192)."

In the above sentence, "prevents" is identified as relational verb relating the biological entities "Diallyl sulfite (DAS)" and "cancer" while "animals" is identified as catalyst. Last but not least, our system also extracts the adverbs associated with relational verbs, which plays a very important role especially to identify the negation in sentences.

The remaining chapter is structured as follows. Section 20.2 presents a brief introduction to ontologies with a focus on biological ontologies. A review of related works on biomedical text mining is presented in Section 20.3. The architectural detail of BioTextMiner is given in Section 20.4. Sections 20.5 through 20.9 present the functioning details of different modules of BioTextMiner. Section 20.10 validates the efficacy of the proposed relation extraction process. Finally, in Section 20.11 we conclude with a summary and direction for possible enhancements to the proposed system.

20.2 Brief Introduction to Ontologies

The word ontology has been borrowed from philosophy, where it means "*a systematic explanation of being*" and has been used in the field of NLP for quite sometime now to represent and manipulate meanings of texts. Gruber [14] defined an ontology as "*an explicit specification of a conceptualization,*" which has become one of the most acceptable definitions to the ontology community. Usually a concept in ontology is defined in terms of its mandatory and optional properties along with the value restrictions on those properties. A typical ontology consists of a set of classes, a set of properties to describe these classes, a taxonomy defining the hierarchical organization of these classes and a set of semantic and structural relations defined among these classes. Ontology is also accompanied by a set of inference rules implementing functions for reasoning with ontology concepts.

20.2.1 Biological ontologies

Biology is one of the fields, where knowledge representation has always followed a structured mechanism and hence lends itself to machine-interpretable ontology building readily. Biological concepts are well defined, and biological ontologies usually store the taxonomical and partonomical relations among these concepts. A large number of medically oriented ontologies are summarized in the Unified Medical Language System, maintained at the US National Library of Medicine (NLM). Two of the well-known ontologies in the field of molecular biology domain are gene ontology (GO [9]) and GENIA ontology [13].

The GO is mainly used for database annotation. GO has grown up from within a group of databases, rather than being proposed from outside. It seeks to capture information about the role of gene products within an organism. GO lacks any upper-level organizing ontology and organizes processes in three levels of hierarchy, representing the function of a gene product, the process in which it takes place and cellular location and structure.

The GENIA ontology provides the base collection of molecular biological entity types and relationships among them. It is a taxonomy of 47 biologically relevant nominal categories. The top three concepts in GENIA ontology are *biological source*, *biological substance*, and *other_name*. The other_name does not refer to any fixed biological concept but is used for describing the terms that are regarded as biological concepts but are not identified with any other known concept in the ontology. The GENIA ontology concepts are related through "is-a" and "part-of" semantic relations. In GENIA ontology substances are classified according to their chemical characteristics rather than their biological roles, since chemical classification of a substance is quite independent of the biological context in which it appears, and therefore more stably defined. The biological role of a substance may vary depending on the biological context. Therefore, in this model substances are classified as proteins, DNA, RNA etc. They are further classified into families, complexes, individual molecules, subunits, domains, and regions. The sources are classified into natural and cultured sources that are further classified as an organism (human), a tissue (liver), a cell (leukocyte), a sublocation of a cell (membrane or a cultured cellline (HeLa)). Organisms are further classified into multicell organisms, virus, and mono-cell organisms other than virus. Our BioTextMiner uses a subset of GENIA ontology concepts shown as a shaded box in Figure 20.1 as these are used by biomedical named entity recognizer, ABNER [27] to mark biological entities in text documents.

20.3   Related Work on Biomedical Text Mining

In this section we present an overview of some of the recent research efforts that have been directed towards the problems of extraction of biological entities, biological interactions between them from unstructured biological documents. A brief description of the biological knowledge visualization and ontology-based biomedical text processing will be also a part of this section.

20.3.1   Biological entity recognition

The goal of biological entity recognition is to identify, within a collection of biological text, all of the instances of a name for a specific type of biological object, e.g., all of the gene names and symbols within a collection of MEDLINE

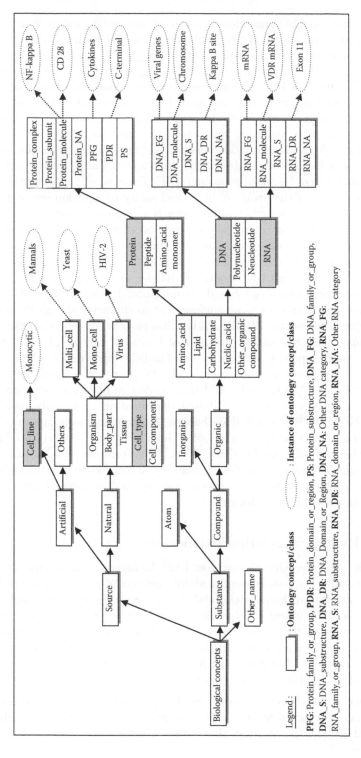

FIGURE 20.1: GENIA ontology with partial instances. Shaded rectangles represent the concepts used in ABNER system.

abstracts. The unambiguous identification of biological entities is a fundamental requirement for IE in the biomedical domain. Some of the common challenges associated with biological entity recognition concern the handling of unknown words (e.g., new gene names), multiple names for the same entity (e.g., *PTEN* and *MMAC1* refer to the same gene) and the identification of entities composed of more than one word (e.g., carotid artery). In addition, same word or phrase can refer to different things depending on the context (e.g., ferritin can be a biological substance or a laboratory test).

Complex IE from biological documents is largely dependent on the correct identification of biological entities in the documents. These entities are then tagged or annotated for more meaningful IE. Because of the potential utility and complexity of the problem, biological entity recognition has attracted the interest of many researchers, and there is a tremendous amount of published research in this topic. Initially the process of named entity recognition and their tagging were done manually. But the sheer volume of texts arriving everyday has initiated a significant research effort towards automated identification of biological entities in journal articles and tagging them. Fukuda et al. [21] have proposed a rule-based system called PROPER, to extract material names from sentences using surface clue on character strings in medical and biological documents. PROPER identifies protein names in articles with a recall value of 98.8% and a precision of 94.7%. Proux et al. [22] used a dictionary-based approach to identify non-English words in a document and identify them as gene terms. Hanisch et al. [23] proposed a hybrid approach including the use of dictionaries and hand-coded rules in combination with robust linear programming (RLP)-based optimization to identify biological entities. The ABNER developed by Settles [27] uses linear-chain conditional random fields (CRFs) with a variety of orthographic and contextual features to identify biomedical named entities (DNA, RNA, proteins, cell-line, and cell-type) in biomedical texts with F-score value of 70.5%. Machine learning-based techniques [25,26] have been also successfully applied to identify and classify gene/protein names in text documents.

20.3.2 Biological relation extraction

Though, named-entity recognition from biological text documents has gained reasonable success, reasoning about contents of a text document however needs more than identification of the entities present in it. Context of the entities in a document can be inferred from an analysis of the interentity relations present in the document. Hence, it is important that the relationships among the biological entities present in a text are also extracted and interpreted correctly. Related work in biological relation extraction can be classified into the following three categories:

- Co-occurrence-based approach: in this approach, after the biological entities are extracted from a document, relations among them are inferred based on the assumption that two entities in the same sentence or abstract are related. Negation in the text is not taken into account. Jenssen et al. [37] collected a set of almost 14,000 gene names from publicly available databases and used them to search MEDLINE abstracts. Two genes were assumed to be linked if they appeared in the same abstract; the relation received a higher weight if the gene pair appeared in multiple abstracts. For the pairs with high weights i.e., with five or more occurrences of the pair, it was reported that 71% of the gene pairs were indeed related. However, the primary focus of the work is to extract related gene pairs rather than studying the nature of these relations.

- Linguistics-based approach: in this approach, usually shallow parsing techniques are employed to locate a set of handpicked verbs or nouns. Shallow parsing is the process of identifying syntactical phrases (noun phrases, verb phrases etc.) in natural language sentences on the basic of the parts of speech of the words but does not specify their internal structure, nor their role in the main sentence. Rules are specifically developed to extract the surrounding words of these predefined terms and to format them as relations. As with the co-occurrence-based approach, negation in sentences is usually ignored. Sekimizu et al. [16] collected the most frequently occurring verbs in a collection of abstracts and developed partial and shallow parsing techniques to find the verb's subject and objects. The estimated precision of inferring relations is about 71%. Thomas et al. [17] modified a pre-existing parser based on cascaded finite state machines to fill templates with information on protein interactions for three verbs—*interact with, associate with, bind to*. They calculated recall and precision in four different manners for three samples of abstracts. The recall values ranged from 24 to 63% and precision from 60 to 81%. The PASTA system is a more comprehensive system that extracts relations between proteins, species and residues [38]. Text documents are mined to instantiate templates representing relations among these three types of elements. This work reports precision of 82% and a recall value of 84% for recognition and classification of the terms, and 68% recall and 65% precision for completion of templates. Craven and Kumlien [39] have proposed identification of possible drug-interaction relations between protein and chemicals using a bag of words approach applied at the sentence level. This produces inferences of the type: *drug-interactions* (protein, pharmacologic-agent), which specify an interaction between an agent and a protein. Ono et al. [18] reports a method for extraction of *protein-protein interactions* based on a combination of syntactic patterns. They employ a dictionary look-up approach to identify proteins in the document. Sentences that contain at least two proteins are selected and parsed with parts-of-speech (POS) matching rules. The rules are triggered by a set of keywords, which are frequently used to name protein interactions (e.g., "associate," "bind" etc.). Rinaldi et al. [19] have proposed an approach towards automatic extraction of a predefined set of seven relations in the domain of molecular biology, based on a complete syntactic analysis of an existing corpus. They extract relevant relations from a domain corpus based on full parsing of the documents and a set of rules that map syntactic structures into the relevant relations. Friedman et al. [40] have developed a NLP system, GENIES, for the extraction of molecular pathways from journal articles. GENIES uses the MedLEE parser to retrieve target structures from full-text articles. GENIES identifies a predefined set of verbs using templates for each one of these, which are encoded as a set of rules. This work [40] reports a precision of 96% for identifying relations between biological molecules from full-text articles.

- Mixed approach: Ciaramita et al. [34] report an unsupervised learning mechanism for extracting semantic relations between molecular biology concepts from tagged MEDLINE abstracts which are a part of the GENIA corpus. For each sentence containing two biological entities, a dependency graph highlighting the dependency between the entities is generated based on linguistic analysis. A relation $\Re$ between two entities C_i and C_j is extracted as the shortest path between the pair following the dependency relations. The approach used is that of hypothesis testing, where the null hypothesis H_0 is formulated as concepts A and B do not cooccur more frequently than expected at chance. The probability of cooccurrence of A and B denoted by $P(AB)$ is constructed using corpus statistics and H_0 is rejected if $P(AB)$ is less than the significance level. The major emphasis of this work is to determine the role of a concept in a significant relation and enhance the biological ontology to include these roles and relations. Sentences that contained complex embedded conjunctions/disjunctions or contained more than 100 words were not used for relation extraction. In the presence of nested tags, the system considers only the innermost tag.

It can be observed that most of the systems other than Ciaramita et al. [34] have been developed to extract a prespecified set of relations. The relation set is manually chosen to include a set of frequently occurring relations. Each system is tuned to work with a predetermined set of relations and does not address the problem of relation extraction in a generic way. For example the method of identification of interaction between genes and gene products cannot work for extraction of enzyme interactions from journal articles, or for automatic extraction of protein interactions from scientific abstracts. On the way of Ciaramita et al. [34], the proposed BioTextMiner attempts to extract generic biological relations along with the associated entities and store them in a structured repository. While mining biological relations the associated prepositions are also considered which very often changes the nature of the verb. Unlike most of the systems mentioned above, BioTextMiner also identify the negations in sentences and store them along with the relational verbs. Besides mining relational verbs and the associated entities, the entities linked with conjunctional-prepositions, which presence or absence validates a particular biological interaction, are also identified and stored in the knowledge repository.

20.3.3 Biological knowledge visualization

The powerful combination of precise analysis of the biomedical documents with a set of visualization tools enables the user to navigate and use easily the abundance of biomedical document collection. Visualization is a key element for effective consumption of information. Semantic nets provide a consolidated view of domain concepts and can aid in the process. Wagner et al. [30] have suggested building a semantic net using the Wiki technology for making e-governance easier through easy visualization of information. In [31], similar approaches have been proposed for integrating and annotating multimedia information. In [32] a soft computing-based technique is proposed to integrate information mined from biological text documents with the help of biological databases. Castro et al. [33] propose building a semantic net for visualization of relevant information with respect to use cases like the nutrigenomics use case, wherein the relevant entities around which the semantic net is built are predefined.

The proposed method differs from all these approaches predominantly in its use of pure linguistic techniques rather than use of any preexisting collection of entities and relations. Moreover, the knowledge visualizer component of BioTextMiner is also integrated with the document collection, which on selecting a particular entity or relation in the graph displays the relevant documents with highlighting the snippet in which the target knowledge is embedded. Though biological relation mining [34] have gained attention of researchers for unraveling the mysteries of biological reactions—their use in biological information visualization is still limited.

20.3.4 Ontology-based biomedical text processing

The growth of several biological ontologies mentioned in the earlier section has prompted a lot of attention towards ontology-based processing of biomedical texts. The aim is to provide intelligent search mechanisms for extracting relevant ICs from a vast collection.

Textpresso [2] is an ontology-based biological IR and extraction system which analyzes tagged biological documents. Two types of tags are used for tagging text elements manually. The first set of tags defines a collection of biological concepts and the second set of tags defines a set of relations that can relate two categories of biological concepts. A tag is defined by a collection of terms including *nouns, verbs, etc.*, that can be commonly associated to the concept. Portions of the document containing a relevant subset of terms are marked by the corresponding biological concept or relation tag. The search engine allows the user to search for combinations of concepts, keywords and relations. With specific relations like commonly occurring gene-gene interactions etc., encoded as a relation tag, Textpresso assists the user to formulate semantic queries. The recall value of the system is reported to vary from 45 to 95%, depending on whether the search is conducted over abstracts or full text documents.

Uramoto et al. [1] have proposed a text-mining system, MedTAKMI, for knowledge discovery from biomedical documents. The system dynamically and interactively mines a large collection of documents with biomedically motivated categories to obtain characteristic information from them. The MedTAKMI system performs entity extraction using dictionary lookup from a collection of two million biomedical entities, which are then used along with their associated category names to search for documents that contain keywords belonging to specific categories. Users can submit a query and receive a document collection in which each document contains the query keywords or their synonyms. The system also uses syntactic information with a shallow parser to extract binary (a verb and a noun) and ternary (two nouns and a verb) relations that are used as keywords by various MedTAKMI mining functions like dictionary-based full text searching, hierarchical category viewer, chronological viewer, etc.

20.4   Architecture of BioTextMiner

In this section, we present the complete architectural detail of the proposed biomedical text mining system—BioTextMiner, which is specially designed to accept PubMed abstracts as input and extract from these biological relations. The proposed approach to relation extraction traverses the phrase structure tree created by Stanford parser* and analyzes the linguistic dependencies in order to trace the relational verbs and consequently the ICs centered around it. The extracted ICs are later used for knowledge visualization and to assist users in extracting information from text documents in a more efficient way. An information component IC can be defined formally as follows:

Definition 1: IC is a seven tuple of the form $< \mathcal{E}_i, \mathcal{A}, \mathcal{V}, \mathcal{P}_v, \mathcal{E}_j, \mathcal{P}_c, \mathcal{E}_k >$ where, $\mathcal{E}_i, \mathcal{E}_j$, and $\mathcal{E}_k$ are biological entities identified as noun phrases; $\mathcal{E}_i$ and $\mathcal{E}_j$ forms the subject and object, respectively for $\mathcal{V}$, $\mathcal{A}$ is adverb; $\mathcal{V}$ is relational verb, $\mathcal{P}_v$ is verbal-preposition associated with $\mathcal{V}$; $\mathcal{P}_c$ is conjunctional-preposition linking $\mathcal{E}_j$ and $\mathcal{E}_k$.

Figure 20.2 presents the complete architecture of the system, which comprises of the following five main modules.

- Document processor. This module accepts PubMed abstracts as input and parses them using Stanford parser after removal of unwanted snippets of text like author name and affiliation. The output of this

---

* http://nlp.stanford.edu/software/lex-parser.shtml

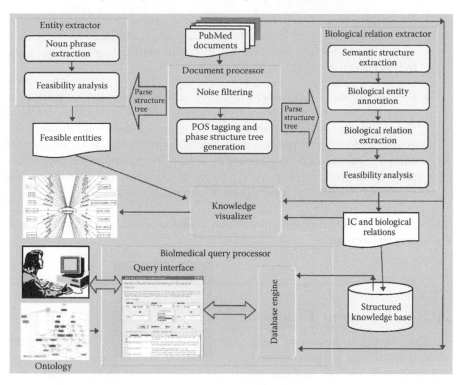

FIGURE 20.2: Architecture of BioTextMiner.

module is *phrase structure tree*, which encodes different types of phrases and their hierarchical organization in a tree form.

- Entity extractor. This module accepts phrase structure tree created by *document processor* module as input and extracts entity names as noun phrase from the text. The feasibility of the extracted noun phrases is analyzed as function of their term frequency (TF) and inverse document frequency (IDF).

- Biological relation extractor. This module also uses the phrase structure tree generated by Stanford parser as input and employs rule-based techniques to extract ICs from them. The ICs are later analyzed to identify feasible biological relations from them. This module employs a biomedical entity recognizer, ABNER [27], which marks the biological entities in the associated noun phrases through a relational verb. This process helps in identifying the valid biological relations relating two biological entities. The feasibility of the biological relations is analyzed as a function of the relative frequency of occurrences of the relations.

- Knowledge visualizer. This module presents the extracted biological knowledge in an integrated fashion with the help of semantic net.

The major idea of generating a semantic net is to highlight the role of a concept in a text corpus by eliciting its relationship to other concepts. The nodes in a semantic net represent entities/concepts and links indicate relationships.

- Biomedical query processor. The query processor provides an intelligent query interface to allow users to formulate queries at multiple levels of specificity. Queries are analyzed by the query analyzer and answers are generated for correct queries. User queries may contain both simple and complex information requirements ranging from requirement specification about presence of an entity name, to presence of a pair of entities or concepts bound by a particular biological relation.

The design and working principles of the different modules are explained in the following sections.

20.5 Document Processor

The document processor cleans the PubMed abstracts by filtering unwanted texts, e.g., authors name, their affiliations etc. The cleaned sentences are subjected to POS analysis which assigns POS tags to every word in a sentence, where a tag reflects the syntactic category of the word [4]. Since our intention is to identify relevant action verbs along with accompanying prepositions and associated entities, POS analysis plays an important role in text IE. The POS tags are also useful to identify the grammatical structure of the sentences like nouns and verb phrases and their interrelationships. For POS analysis and phrase structure tree creation we have used the Stanford parser, which is a statistical parser. The Stanford parser receives documents as input and works out the grammatical structure of sentences to convert them into equivalent phrase structure tree. A list of POS tags used by Stanford parser and their descriptions is shown in Table 20.1. The phrase structure tree output by this module is used by entity extractor and biological relation extrator modules for further analysis. A list of sample sentences and their corresponding phrase structure tree generated by Stanford parser is shown in Table 20.2. These sentences are also refered in rest of the paper to explain the functioning details of other modules.

20.6 Entity Extractor

This module is responsible to extract feasible entities from a corpus of PubMed abstracts. The entity extractor module takes the phrase structure

TABLE 20.1:   POS tags and their descriptions used by Stanford parser.

Tag	Description	Tag	Description
NN	Noun singular or mass	NP	Noun phrase
NNS	Noun plural	VP	Verb phrase
NNP	Proper noun singular	MD	Modal
NNPS	Proper noun plural	PDT	Predeterminer
JJ	Adjective	POS	Possessive ending
JJR	Adjective comparative	PRP	Personal pronoun
JJS	Adjective superlative	PRP	Possessive pronoun
CC	Coordinating conjunction	RB	Adverb
IN	Preposition or subordinating conjunction	RBR	Adverb comparative
TO	to	RBS	Adverb superlative
CD	Cardinal number	RP	Particle
DT	Determiner	SYM	Symbol
EX	Existential there	VB	Verb base form
FW	Foreign word	VBD	Verb past tense
LS	List item marker	VBG	Verb gerund or present participle

tree generated by document processor as input and output the feasible entities after analyzing the noun phrases. In order to avoid long noun phrases, which generally contain conjunctional words, we have considered only those noun phrases that appear at the lowest level nodes in the phrase structure tree. In case, more than one noun phrase appears at the lowest level node as siblings, the string concatenation function is applied to club them into a single noun phrase. While clubbing, only those words tagged as NN* (noun), where * is a wildcard character, or JJ (adjective) are considered. For example, consider the following two subtrees of a phrase structure tree corresponding to a document (PMID: 18636275).

- (NP (NP (DT the) (NN protein) (NNS expressions)) (PP))

- (NP (NP (DT the) (JJ clinical) (NN stage)) (PP (IN of) (NP (NN breast) (NN cancer))))

From first subtree the extracted noun phrase is "protein expression" while from the second one the extracted phrases are "clinical stage" and "breast cancer."

After extracting all noun phrases from all text documents in the corpus their weights are calculate as a function of TF and IDF by using Equation 20.1.

$$\omega(P) = TF \times Log\left(\frac{N}{n}\right) \tag{20.1}$$

TABLE 20.2: A list of PubMed abstracts and corresponding phrase structure tree generated through Stanford parser.

PMID	Sentence	Phrase structure tree
18627608	ISG15 protein expression was analysed in two independent cohorts on tissue microarrays (TMAs), an initial evaluation set, consisting of 179 breast carcinomas and 51 normal breast tissues, and a second large validation set, consisting of 646 breast carcinomas and 10 normal breast tissues.	(ROOT (S (S (NP (CD ISG15) (NN protein) (NN expression)) (VP (VBD was) (VP (VBN analyzed) (PP (IN in) (NP (NP (CD two) (JJ independent) (NNS cohorts)) (PP (IN on) (NP (NP (NN tissue) (NNS microarrays)) (PRN (-LRB- -LRB-) (NP (NNP TMAs)) (-RRB- -RRB-)))))) (PRN (, ,) (S (NP (DT an) (JJ initial) (NN evaluation)) (VP (VBD set) (, ,) (S (VP (VBG consisting) (NP (NP (QP (IN of) (CD 179)) (NN breast) (NNS carcinomas)) (CC and) (NP (CD 51) (JJ normal) (NN breast) (NNS tissues))))))) (, ,)) (CC and) (S (NP (DT a) (JJ second) (JJ large) (NN validation)) (VP (VBD set) (, ,) (S (VP (VBG consisting) (PP (IN of) (NP (NP (CD 646) (NN breast) (NNS carcinomas)) (CC and) (NP (CD 10) (JJ normal) (NN breast) (NNS tissues)))))))) (. .)))
18636156	The results prove that, in cancer cells, some mutagenic or nonmutagenic carcinogens can epigenetically influence the transcription levels of imprinted genes and also suggest the possibility that some chemical carcinogens may have epigenetic carcinogenic effects in human cells.	(ROOT (S (NP (DT The) (NNS results)) (VP (VBP prove) (SBAR (IN that) (S (, ,) (PP (IN in) (NP (NN cancer) (NNS cells))) (, ,) (NP (DT some) (JJ mutagenic) (CC or) (JJ non-mutagenic) (NNS carcinogens)) (VP (MD can) (ADVP (RB epigenetically)) (VP (VP (VB influence) (NP (NP (DT the) (NN transcription) (NNS levels)) (PP (IN of) (NP (JJ imprinted) (NNS genes))))) (CC and) (VP (ADVP (RB also)) (VB suggest) (NP (DT the) (NN possibility)) (SBAR (IN that) (S (NP (DT some) (NN chemical) (NNS carcinogens)) (VP (MD may) (VP (VB have) (NP (NP (ADJP (JJ epigenetic) (JJ carcinogenic) (NNS effects)) (PP (IN in) (NP (JJ human) (NNS cells)))))))))))))) (. .)))

18634034	Common single-nucleotide polymorphisms (SNPs) in miRNAs may change their property through altering miRNA expression and/or maturation, and thus they may have an effect on thousands of target mRNAs, resulting in diverse functional consequences.	(ROOT (S (S (NP (NP (NP (JJ Common) (JJ single-nucleotide) (NNS polymorphisms)) (PRN (-LRB- -LRB-) (NP (NNP SNPs)) (-RRB- -RRB-))) (PP (IN in) (NP (NNS miR-NAs)))) (VP (VP (MD may) (VP (VB change) (NP (PRP$ their) (NN property)) (PP (IN through) (S (VP (VBG altering) (NP (NN miRNA) (NN expression)))))))) (CC and\/or) (VP (NN maturation)))) (, ,) (CC and) (S (ADVP (RB thus)) (NP (PRP they)) (VP (MD may) (VP (VB have) (NP (DT an) (NN effect)) (PP (IN on) (NP (NP (NNS thousands)) (PP (IN of) (NP (NN target) (NNS mRNAs))))) (, ,) (VP (VBG resulting) (PP (IN in) (NP (ADJP (JJ diverse) (JJ functional)) (NNS consequences))))))) (. .)))
18636564	Moreover, high-fat diets rich in omega-6 polyunsaturated fatty acids (PUFAs) have produced higher frequencies of HA-induced mammary gland tumors in rats compared to those fed low-fat diets.	(ROOT (S (ADVP (RB Moreover)) (, ,) (NP (NP (JJ high-fat) (NNS diets)) (ADJP (JJ rich) (PP (IN in) (NP (NP (JJ omega-6) (VBN polyunsaturated) (NN fatty) (NNS acids)) (PRN (-LRB- -LRB-) (NP (NNP PUFAs)) (-RRB- -RRB-)))))) (VP (VBP have) (VP (VBN produced) (NP (NP (JJR higher) (NNS frequencies)) (PP (IN of) (NP (JJ HA-induced) (ADJP (RB mammary) (JJ gland)) (NNS tumors)))) (PP (IN in) (NP (NNS rats))) (PP (VBN compared) (PP (TO to) (NP (DT those) (VBN fed) (JJ low-fat) (NNS diets)))))) (. .)))
18636564	The aim was to evaluate prospectively if intake of HAs is associated with breast cancer incidence, and if the association is independent of omega-6 PUFA intakes.	(ROOT (S (NP (DT The) (NN aim)) (VP (VBD was) (S (VP (TO to) (VP (VB evaluate) (ADVP (RB prospectively)) (SBAR (SBAR (IN if) (S (NP (NP (NN intake)) (PP (IN of) (NP (NNP HAs)))) (VP (VBZ is) (VP (VBN associated) (PP (IN with) (NP (NN breast) (NN cancer) (NN incidence))))))) (, ,) (CC and) (SBAR (IN if) (S (NP (DT the) (NN association)) (VP (VBZ is) (ADJP (JJ independent) (PP (IN of) (NP (JJ omega-6) (NNP PUFA) (NNS intakes))))))))))))) (. .)))
18636415	Suppression of AO enzymes associated with breast cancer and aging is most likely the cause of increased levels of reactive oxygen species (ROS).	(ROOT (S (NP (NP (NN Suppression)) (PP (IN of) (NP (NP (NNP AO) (NNS enzymes)) (VP (VBN associated) (PP (IN

(Continued)

TABLE 20.2: (Continued)

		with) (NP (NN breast) (NN cancer) (CC and) (NN aging))))))) (VP (VBZ is) (NP (NP (ADJP (RBS most) (JJ likely)) (DT the) (NN cause) (PP (IN of) (NP (NP (VBN increased) (NNS levels) (PP (IN of) (NP (NP (JJ reactive) (NN oxygen) (NNS species) (PRN (-LRB- -LRB-) (NNP ROS) (-RRB- -RRB-))))))) (. .)))
18635967	Moreover, the levels of MDC1 and BRIT1 inversely correlated with centrosome amplification, defective mitosis and cancer metastasis in human breast cancer.	(ROOT (S (ADVP (RB Moreover)) (, ,) (NP (NP (DT the) (NNS levels)) (PP (IN of) (NP (NNP MDC1) (CC and) (NNP BRIT1)))) (ADVP (RB inversely)) (VP (VBN correlated) (PP (IN with) (NP (NP (JJ centrosome) (NN amplification)) (, ,) (NP (JJ defective) (NNS mitosis)) (CC and) (NP (NN cancer) (NNS metastasis)))) (PP (IN in) (NP (JJ human) (NN breast) (NN cancer))))) (. .))
18635390	Our previous work on implantable and 7,12-dimethylbenz[a]anthracene (DMBA)-induced breast cancer showed that oral glutamine (GLN) inhibited tumor growth possibly through stimulation of host- and selective inhibition of tumor glutathione (GSH).	(ROOT (NP (NP (NP (PRP$ Our) (JJ previous (NN work)) (PP (IN on) (NP (ADJP (JJ implantable) (CC and) (JJ 7,12-dimethylbenz)) (PRN (-LRB- -LRB-) (X (SYM a)) (-RRB- -RRB-)) (NNS anthracene))) (PRN (-LRB- -LRB-) (NP (NNP DMBA)) (-RRB- -RRB-))) (: -) (NP (NP (NP (JJ induced) (NN breast) (NN cancer) (SBAR (S (VP (VBD showed) (SBAR (IN that) (S (NP (NP (JJ oral) (NN glutamine)) (PRN (-LRB- -LRB-) (NNP GLN) (-RRB- -RRB-))) (VP (VBD inhibited) (NP (NN tumor) (NN growth)) (ADVP (RB possibly)) (PP (IN through) (NP (NP (NN stimulation)) (PP (IN of) (CC and) (NP (NP (JJ selective) (NN inhibition)) (PP (IN of) (NP (NN tumor) (NN glutathione) (PRN (-LRB- -LRB-) (NNP GSH) (-RRB- -RRB-)))))) (. .))
18628456	Conclusions: these results indicate that RSK4 expression may limit the oncogenic, invasive, and metastatic potential of breast cancer cells.	(ROOT (NP (NP (NNS CONCLUSIONS)) (: :) (S (NP (DT These) (NNS results)) (VP (VBP indicate) (SBAR (IN that)

(S (NP (CD RSK4) (NN expression)) (VP (MD may) (VP (VB limit) (NP (NP (DT the) (ADJP (JJ oncogenic) (, ,) (JJ invasive) (, ,) (CC and) (JJ metastatic) (NN potential)) (PP (IN of) (NP (NN breast) (NN cancer) (NNS cells)))))))) (. .)))

18627387	Tufted hair folliculitis in a woman treated with lapatinib for breast cancer.	(ROOT (S (NP (NP (VBN Tufted) (NN hair) (NNS folliculitis)) (PP (IN in) (NP (DT a) (NN woman)))) (VP (VBN treated) (PP (IN with) (NP (NP (NN lapatinib)) (PP (IN for) (NP (NN breast) (NN cancer))))))) (. .)))
18627377	Cucurbitacin B has a potent antiproliferative effect on breast cancer cells *in vitro* and *in vivo*.	(ROOT (S (NP (NNP Cucurbitacin) (NNP B)) (VP (VBZ has) (NP (NP (DT a) (JJ potent) (NN antiproliferative) (NN effect)) (PP (IN on) (NP (NN breast) (NN cancer) (NNS cells))))) (PP (IN in) (UCP (ADJP (NN vitro)) (CC and) (PP (IN in) (NP (NN vivo)))))) (. .)))
18632656	Fifteen COMT single nucleotide polymorphisms (SNPs) selected on the basis of in-depth resequencing of the COMT gene were genotyped in 1,482 DNA samples from a Mayo Clinic breast cancer case control study.	(ROOT (S (NP (NNP Fifteen)) (VP (VBP COMT) (SBAR (S (NP (NP (JJ single) (NN nucleotide) (NNS polymorphisms)) (PRN (-LRB- -LRB-) (NP (NNP SNPs)) (-RRB- -RRB-))) (VP (VBN selected) (PP (IN on) (NP (NP (DT the) (NN basis)) (PP (IN of) (NP (NP (JJ in-depth) (NN resequencing)) (PP (IN of) (NP (DT the) (JJ COMT) (NN gene))))))) (VP (VBD were) (VP (VBN genotyped) (PP (IN in) (NP (CD 1,482) (NN DNA) (NNS samples))) (PP (IN from) (NP (DT a) (NNP Mayo) (NNP Clinic) (NN breast) (NN cancer) (NN case) (NN control) (NN study)))))))))) (. .)))

Where $\omega(P)$ represents the weight of the noun phrase P, TF denotes the total number of times the phrase P occurs in the corpus, $Log(N/n)$ is the IDF of P, N is the total number of documents in the corpus, and n is the number of documents that contain the phrase P.

The algorithm *feasible_noun_phrase_extraction* given in Table 20.3 summarizes the noun phrase extraction and feasibility analysis process formally. A partial list of noun phrases and their weights extracted from a corpus of PubMed abstracts on "breast cancer" is shown in Table 20.4.

20.7 Biological Relation Extractor

A biological relation is assumed to be binary in nature, which defines a specific association between an ordered pair of biological entities. The process of identifying biological relations is accomplished in two stages. During the first stage, prospective ICs (Definition 1) which might embed biological relations within them are identified from the sentences. During the second stage, a feasibility analysis is employed to identify correct biological relations. These stages are elaborated further in the following subsections.

20.7.1 Extraction of information components (ICs)

An IC is usually manifested in a document centered around relational verb. The proposed approach to IC extraction traverses the phrase structure tree and analyzes the phrases and their linguistic dependencies in order to trace relational verbs and other constituents (Definition 1). Since the entities are marked as noun phrases in the phrase structure tree, this module exploits the phrase boundary and proximitivity, to identify relevant ICs. Initially all tuples of the form $< \mathcal{E}_i, \mathcal{A}, \mathcal{V}, \mathcal{P}_v, \mathcal{E}_j, \mathcal{P}_c, \mathcal{E}_k >$ where, $\mathcal{E}_i$, $\mathcal{E}_j$, and $\mathcal{E}_k$ are entities identified as noun phrases; $\mathcal{A}$ is adverb; $\mathcal{V}$ is relational verb, $\mathcal{P}_v$ is verbal-preposition associated with $\mathcal{V}$; $\mathcal{P}_c$ is conjunction-preposition linking $\mathcal{E}_j$ and $\mathcal{E}_k$, are retrieved from the documents.

A verb may occur in a sentence in its root form or as a variant of it. Different classes of variants of a relational verb are recognized by our system. The first of this class comprises of morphological variants of the root verb, which are essentially modifications of the root verb itself. In the English language the word *morphology* is usually categorized into "inflectional" and "derivational" morphology. Inflectional morphology studies the transformation of words for which the root form only changes, keeping the syntactic constraints invariable. For example, the root verb "activate," has three inflectional verb forms—"activates," "activated" and "activating." Derivational morphology on the other hand deals with the transformation of the stem of a word to generate other words that retain the same concept but may

TABLE 20.3:   Algorithm to extract feasible noun phrases.

Algorithm: feasible_noun_phrase_extraction(F)

Input: A forest F of phrase structure trees
Output: A list L_{feas} of feasible noun phrases
Steps:
1. $L \leftarrow \varphi$ // list of noun phrases
2. $L_{feas} \leftarrow \varphi$ // list of feasible noun phrases
3. for each $T \in F$ do
4. for each node $\lambda \in T$ do
5. flag←false
6. if λ = "NP" then
7. flag←true
8. for each node $\xi \in$ child[λ] do
9. if $\xi \neq$ leaf node then
10. flag←false
11. go to step 15
12. end if
13. end for
14. end if
15. if flag=true then
16. np="" // Initialize np with null string
17. for each node $\xi \in$ child[λ] do
18. if (tag(ξ) = NN* OR tag(ξ) = JJ) then // * is wildcard character
19. np $\leftarrow$ np+ξ
20. end if
21. end for
22. $L \leftarrow L \cup$ np
23. end if
24. end for
25. end for
26. for each np $\in L$ do
27. calculate weight value using equation 20.1
28. if weight(np) $\geq \theta$ then // θ is a threshold value
29. $L_{feas} \leftarrow L_{feas} \cup$ np
30. end if
31. end for
32. return L_{feas}
33. stop

---

have different syntactic roles. Thus, "activate" and "activation" refer to the concept of "making active," but one is a verb and the other one a noun. Similarly, inactivate, transactivate, deactivate etc., are derived morphological variants created with addition of prefixes. Presently the system does not

TABLE 20.4:   A partial list of noun phrases and their weights extracted
by entity extractor from a corpus containing PubMed abstracts on "breast
cancer."

Noun phrase (NP)	ω(NP)	Noun phrase (NP)	ω(NP)	Noun phrase (NP)	ω(NP)
Breast cancer	610.14	Surgery	09.75	Prostate cancer	136.72
Women	379.64	Estrogen	92.65	Postmenopausal women	135.19
Cancer	857.61	Proliferation	80.22	Colorectal cancer	133.01
Treatment	639.49	Prognosis	74.01	Zoledronic acid	130.50
Risk	540.77	Tamoxifen	74.00	Breast tumor	124.30
Tumor	528.24	Protein	61.57	Metastatic breast cancer	124.29
Breast	466.09	Breast carcinoma	55.36	Tumor cells	111.86
Cell	360.44	Cancer cells	55.36	CYP genes	18.64
Breast cancer patient	354.23	Breast cancer cells	49.15	Tumor suppressor genes	12.42
Breast cancer risk	236.15	Cancer patient	49.15	Apoptotic stress response genes	6.21
Disease	226.37	Doxorubicin	36.73	Breast cancer susceptibility genes BRCA1 BRCA2	6.21
Chemotherapy	220.85	Estrogen receptor	36.72	E2-responsive genes	6.21

consider derivational morphology, and only inflectional variants of a root verb
are recognized.

In the context of biological relations, we also observe that the occurrence of
a verb in conjunction with a preposition very often changes the nature of the
verb. For example, the functions associated to the verb activates may be quite
different from the ones that can be associated to the verb form activates in,in
which the verb activates is followed by the preposition in. Thus our system
also considers biological relations represented by a combination of root verbs
or their morphological variants, and prepositions that follow these. Typical
examples of biological relations identified in this category include "activated

in," "binds to," "stimulated with" etc., which denotes a significant class of biological reactions. Besides mining relational verbs with accompanying prepositions and associated entities, the entities associated with object entity through conjunctional prepositions are also extracted and termed as catalyst, which presence or absence validates a particular biological interaction.

IC extraction process is implemented as a rule-based system. Dependencies output by the parser are analyzed to identify subject, object, verb, preposition, and various other relationships among elements in a sentence. Some sample rules are presented in Table 20.5 to highlight the functioning of the system. Table 20.6 presents an algorithm, *information_component_extraction*, to convert these rules into working module. A partial list of ICs extracted from PubMed documents (Table 20.2) by applying these rules is presented in Table 20.7. The biological entities appearing in ICs are marked with a biological entity recognizer that helps in identifying valid biological relations and answering user queries based on biological concepts. For this purpose, the Bio-TextMiner is integrated with a biological entity recognizer, ABNER v1.5 [27], which is a molecular biology text analysis tool. ABNER employs statistical machine learning using linear-chain CRFs with a variety of orthographic and contextual features and it is trained on both the NLPBA and BioCreative corpora. A sample output generated by ABNER for the first three sample sentences in Table 20.2 is shown in Figure 20.3.

20.7.2 Identifying feasible biological relations

A biological relation is usually manifested in a document as a relational verb associating two or more biological entities. The biological actors associated to a relation can be inferred from the biological entities located in the proximity of the relational verb. At present, we have considered only binary relations. In order to compile biological relations from ICs, we consider only those tuples in which either subject or object field has at least one biological entity. This consideration deals with the cases in which pronouns are used to refer the biological entities appearing in previous sentences. In this way, a large number of irrelevant verbs are eliminated from being considered as biological relations. Since, our aim is not just to identify possible relational verbs but to identify feasible biological relation. Hence we engage in statistical analysis to identify feasible biological relations. To consolidate the final list of feasible biological relations we take care of two things. Firstly, since various forms of the same verb represent a basic biological relation in different forms, the feasible collection is extracted by considering only the unique root forms after analyzing the complete list of ICs. The root verb having frequency count greater than or equal to a threshold value is retained as root biological relations. There after, ICs are again analyzed to identify and extract the morphological variants of the retained root verbs.

The core functionalities of the biological relation and morphological variants finding module is summed up in the following steps.

TABLE 20.5: Rules to extract ICs from phrase structure tree.

Rule no.	Rules
1	$[\mathcal{C}(\mathcal{R},\mathcal{E}_i) \wedge \mathcal{C}(\mathcal{R},\mathcal{VP}) \wedge \mathcal{L}(\mathcal{VP},\mathcal{E}_i) \wedge \mathcal{C}_l(\mathcal{VP},\mathcal{V}) \wedge$ $\mathcal{S}(\mathcal{V},\mathcal{E}_j)] \Rightarrow \mathcal{E}_i \to \mathcal{V} \leftarrow \mathcal{E}_j$
2	$[\mathcal{C}(\mathcal{R},\mathcal{E}_i) \wedge \mathcal{C}(\mathcal{R},\mathcal{VP}) \wedge \mathcal{L}(\mathcal{VP},\mathcal{E}_i) \wedge \mathcal{C}_l(\mathcal{VP},\mathcal{V}) \wedge$ $\mathcal{S}(\mathcal{V},\mathcal{E}_j) \wedge \mathcal{S}(\mathcal{V},\mathcal{PP}) \wedge \mathcal{C}_l(\mathcal{PP},p) \wedge \mathcal{S}(p,\mathcal{E}_k)] \Rightarrow \mathcal{E}_i \to$ $\mathcal{V} \leftarrow \mathcal{E}_j \to p \leftarrow \mathcal{E}_k$
3	$[\mathcal{C}(\mathcal{R},\mathcal{E}_i) \wedge \mathcal{C}(\mathcal{R},\mathcal{VP}) \wedge \mathcal{L}(\mathcal{VP},\mathcal{E}_i) \wedge \mathcal{C}_l(\mathcal{VP},\mathcal{V}) \wedge$ $\mathcal{S}(\mathcal{V},\mathcal{NP}) \wedge \mathcal{C}(\mathcal{NP},\mathcal{E}_j) \wedge \mathcal{C}(\mathcal{NP},\mathcal{PP}) \wedge \mathcal{L}(\mathcal{PP},\mathcal{E}_j) \wedge$ $\mathcal{C}_l(\mathcal{PP},p) \wedge \mathcal{S}(p,\mathcal{E}_k)] \Rightarrow \mathcal{E}_i \to \mathcal{V} \leftarrow \mathcal{E}_j \to p \leftarrow \mathcal{E}_k$
4	$[\mathcal{C}(\mathcal{R},\mathcal{E}_i) \wedge \mathcal{C}(\mathcal{R},\mathcal{VP}_1) \wedge \mathcal{L}(\mathcal{VP}_1,\mathcal{E}_i) \wedge \mathcal{C}(\mathcal{VP}_1,\mathcal{VP}_2) \wedge$ $\mathcal{C}_l(\mathcal{VP}_2,\mathcal{V}) \wedge \mathcal{S}(\mathcal{V},\mathcal{E}_j)] \Rightarrow \mathcal{E}_i \to \mathcal{V} \leftarrow \mathcal{E}_j$
5	$[\mathcal{C}(\mathcal{R},\mathcal{E}_i) \wedge \mathcal{C}(\mathcal{R},\mathcal{VP}_1) \wedge \mathcal{L}(\mathcal{VP}_1,\mathcal{E}_i) \wedge \mathcal{C}(\mathcal{VP}_1,\mathcal{VP}_2) \wedge$ $\mathcal{C}_l(\mathcal{VP}_2,\mathcal{V}) \wedge \mathcal{S}(\mathcal{V},\mathcal{E}_j) \wedge \mathcal{S}(\mathcal{V},\mathcal{PP}) \wedge \mathcal{L}(\mathcal{PP},\mathcal{E}_j) \wedge$ $\mathcal{C}_l(\mathcal{PP},p) \wedge \mathcal{S}(p,\mathcal{E}_k)] \Rightarrow \mathcal{E}_i \to \mathcal{V} \leftarrow \mathcal{E}_j \to p \leftarrow \mathcal{E}_k$
6	$[\mathcal{C}(\mathcal{R},\mathcal{E}_i) \wedge \mathcal{C}(\mathcal{R},\mathcal{VP}_1) \wedge \mathcal{L}(\mathcal{VP}_1,\mathcal{E}_i) \wedge \mathcal{C}(\mathcal{VP}_1,\mathcal{VP}_2) \wedge$ $\mathcal{C}_l(\mathcal{VP}_2,\mathcal{V}) \wedge \mathcal{S}(\mathcal{V},\mathcal{NP}) \wedge \mathcal{C}(\mathcal{NP},\mathcal{E}_j) \wedge \mathcal{C}(\mathcal{NP},\mathcal{PP}) \wedge$ $\mathcal{L}(\mathcal{PP},\mathcal{E}_j) \wedge \mathcal{C}_l(\mathcal{PP},p) \wedge \mathcal{S}(p,\mathcal{E}_k)] \Rightarrow \mathcal{E}_i \to \mathcal{V} \leftarrow \mathcal{E}_j \to$ $p \leftarrow \mathcal{E}_k$
7	$[\mathcal{C}(\mathcal{R},\mathcal{E}_i) \wedge \mathcal{C}(\mathcal{R},\mathcal{VP}) \wedge \mathcal{L}(\mathcal{VP},\mathcal{E}_i) \wedge \mathcal{C}_l(\mathcal{VP},\mathcal{V}) \wedge$ $\mathcal{S}(\mathcal{V},\mathcal{PP}) \wedge \mathcal{C}_l(\mathcal{PP},p) \wedge \mathcal{S}(p,\mathcal{E}_j)] \Rightarrow \mathcal{E}_i \to \mathcal{V} - p \leftarrow \mathcal{E}_j$
8	$[\mathcal{C}(\mathcal{R},\mathcal{E}_i) \wedge \mathcal{C}(\mathcal{R},\mathcal{VP}) \wedge \mathcal{L}(\mathcal{VP},\mathcal{E}_i) \wedge \mathcal{C}_l(\mathcal{VP},\mathcal{V}) \wedge$ $\mathcal{S}(\mathcal{V},\mathcal{PP}_1) \wedge \mathcal{C}_l(\mathcal{PP}_1,p_1) \wedge \mathcal{S}(p_1,\mathcal{E}_j) \wedge \mathcal{S}(\mathcal{V},\mathcal{PP}_2) \wedge$ $\mathcal{L}(\mathcal{PP}_2,\mathcal{PP}_1) \wedge \mathcal{C}_l(\mathcal{PP}_2,p_2) \wedge \mathcal{S}(p_2,\mathcal{E}_k)] \Rightarrow \mathcal{E}_i \to$ $\mathcal{V} - p_1 \leftarrow \mathcal{E}_j \to p_2 \leftarrow \mathcal{E}_k$
9	$[\mathcal{C}(\mathcal{R},\mathcal{E}_i) \wedge \mathcal{C}(\mathcal{R},\mathcal{VP}) \wedge \mathcal{L}(\mathcal{VP},\mathcal{E}_i) \wedge \mathcal{C}_l(\mathcal{VP},\mathcal{V}) \wedge$ $\mathcal{S}(\mathcal{V},\mathcal{PP}_1) \wedge \mathcal{C}_l(\mathcal{PP}_1,p_1) \wedge \mathcal{S}(p_1,\mathcal{NP}) \wedge \mathcal{C}(\mathcal{NP},\mathcal{E}_j) \wedge$ $\mathcal{C}(\mathcal{NP},\mathcal{PP}_2) \wedge \mathcal{L}(\mathcal{PP}_2,\mathcal{E}_j) \wedge \mathcal{C}_l(\mathcal{PP}_2,p_2) \wedge \mathcal{S}(p_2,\mathcal{E}_k)] \Rightarrow$ $\mathcal{E}_i \to \mathcal{V} - p_1 \leftarrow \mathcal{E}_j \to p_2 \leftarrow \mathcal{E}_k$
10	$[\mathcal{C}(\mathcal{R},\mathcal{E}_i) \wedge \mathcal{C}(\mathcal{R},\mathcal{VP}_1) \wedge \mathcal{L}(\mathcal{VP}_1,\mathcal{E}_i) \wedge \mathcal{C}(\mathcal{VP}_1,\mathcal{VP}_2) \wedge$ $\mathcal{C}_l(\mathcal{VP}_2,\mathcal{V}) \wedge \mathcal{S}(\mathcal{V},\mathcal{PP}) \wedge \mathcal{C}_l(\mathcal{PP},p) \wedge \mathcal{S}(p,\mathcal{E}_j)] \Rightarrow$ $\mathcal{E}_i \to \mathcal{V} - p \leftarrow \mathcal{E}_j$
11	$[\mathcal{C}(\mathcal{R},\mathcal{E}_i) \wedge \mathcal{C}(\mathcal{R},\mathcal{VP}_1) \wedge \mathcal{L}(\mathcal{VP}_1,\mathcal{E}_i) \wedge \mathcal{C}(\mathcal{VP}_1,\mathcal{VP}_2) \wedge$ $\mathcal{C}_l(\mathcal{VP}_2,\mathcal{V}) \wedge \mathcal{S}(\mathcal{V},\mathcal{PP}_1) \wedge \mathcal{C}_l(\mathcal{PP}_1,p_1) \wedge \mathcal{S}(p_1,\mathcal{E}_j) \wedge$ $\mathcal{S}(\mathcal{V},\mathcal{PP}_2) \wedge \mathcal{L}(\mathcal{PP}_2,\mathcal{PP}_1) \wedge \mathcal{C}_l(\mathcal{PP}_2,p_2) \wedge \mathcal{S}(p_2,\mathcal{E}_k)] \Rightarrow$ $\mathcal{E}_i \to \mathcal{V} - p_1 \leftarrow \mathcal{E}_j \to p_2 \to \mathcal{E}_k$
12	$[\mathcal{C}(\mathcal{R},\mathcal{E}_i) \wedge \mathcal{C}(\mathcal{R},\mathcal{VP}_1) \wedge \mathcal{L}(\mathcal{VP}_1,\mathcal{E}_i) \wedge \mathcal{C}(\mathcal{VP}_1,\mathcal{VP}_2) \wedge$ $\mathcal{C}_l(\mathcal{VP}_2,\mathcal{V}) \wedge \mathcal{S}(\mathcal{V},\mathcal{PP}_1) \wedge \mathcal{C}_l(\mathcal{PP}_1,p_1) \wedge \mathcal{S}(p_1,\mathcal{NP}) \wedge$ $\mathcal{C}(\mathcal{NP},\mathcal{E}_j) \wedge \mathcal{C}(\mathcal{NP},\mathcal{PP}_2) \wedge \mathcal{L}(\mathcal{PP}_2,\mathcal{E}_j) \wedge \mathcal{C}_l(\mathcal{PP}_2,p_2) \wedge$ $\mathcal{S}(p_2,\mathcal{E}_k)] \Rightarrow \mathcal{E}_i \to \mathcal{V} - p_1 \leftarrow \mathcal{E}_j \to p_2 \to \mathcal{E}_k$
13	$[\mathcal{C}(\mathcal{R},\mathcal{E}_i) \wedge \mathcal{C}(\mathcal{R},\mathcal{VP}_1) \wedge \mathcal{L}(\mathcal{VP}_1,\mathcal{E}_i) \wedge \mathcal{C}(\mathcal{VP}_1,\mathcal{VP}_2) \wedge$ $\mathcal{C}(\mathcal{VP}_2,\mathcal{VP}_3) \wedge \mathcal{C}_l(\mathcal{VP}_3,\mathcal{V}) \wedge \mathcal{S}(\mathcal{V},\mathcal{E}_j)] \Rightarrow \mathcal{E}_i \to \mathcal{V} \leftarrow \mathcal{E}_j$

(Continued)

TABLE 20.5: (Continued)

14	$[\mathcal{C}(\mathcal{R},\mathcal{E}_i) \wedge \mathcal{C}(\mathcal{R},\mathcal{VP}_1) \wedge \mathcal{L}(\mathcal{VP}_1,\mathcal{E}_i) \wedge \mathcal{C}(\mathcal{VP}_1,\mathcal{VP}_2) \wedge$ $\mathcal{C}(\mathcal{VP}_2,\mathcal{VP}_3) \wedge \mathcal{C}_l(\mathcal{VP}_3,\mathcal{V}) \wedge \mathcal{S}(\mathcal{V},\mathcal{E}_j) \wedge \mathcal{S}(\mathcal{V},\mathcal{PP}) \wedge$ $\mathcal{L}(\mathcal{PP},\mathcal{E}_j) \wedge \mathcal{C}_l(\mathcal{PP},p) \wedge \mathcal{S}(p,\mathcal{E}_k)] \Rightarrow \mathcal{E}_i \rightarrow \mathcal{V} \leftarrow \mathcal{E}_j \rightarrow$ $p \leftarrow \mathcal{E}_k$
15	$[\mathcal{C}(\mathcal{R},\mathcal{E}_i) \wedge \mathcal{C}(\mathcal{R},\mathcal{VP}_1) \wedge \mathcal{L}(\mathcal{VP}_1,\mathcal{E}_i) \wedge \mathcal{C}(\mathcal{VP}_1,\mathcal{VP}_2) \wedge$ $\mathcal{C}(\mathcal{VP}_2,\mathcal{VP}_3) \wedge \mathcal{C}_l(\mathcal{VP}_3,\mathcal{V}) \wedge \mathcal{S}(\mathcal{V},\mathcal{NP}) \wedge$ $\mathcal{C}(\mathcal{NP},\mathcal{E}_j) \wedge \mathcal{C}(\mathcal{NP},\mathcal{PP}) \wedge \mathcal{C}_l(\mathcal{PP},p) \wedge \mathcal{S}(p,\mathcal{E}_k)] \Rightarrow$ $\mathcal{E}_i \rightarrow \mathcal{V} \leftarrow \mathcal{E}_j \rightarrow p \leftarrow \mathcal{E}_k$
16	$[\mathcal{C}(\mathcal{R},\mathcal{E}_i) \wedge \mathcal{C}(\mathcal{R},\mathcal{VP}) \wedge \mathcal{L}(\mathcal{VP},\mathcal{E}_i) \wedge \mathcal{C}_l(\mathcal{VP},\mathcal{V}) \wedge$ $\mathcal{S}(\mathcal{V},\mathcal{ADVP}) \wedge \mathcal{C}(\mathcal{ADVP},\mathcal{PP}) \wedge \mathcal{C}_l(\mathcal{PP},p) \wedge$ $\mathcal{S}(p,\mathcal{E}_j)] \Rightarrow \mathcal{E}_i \rightarrow \mathcal{V} - p \leftarrow \mathcal{E}_j$
17	$[\mathcal{C}(\mathcal{R},\mathcal{E}_i) \wedge \mathcal{C}(\mathcal{R},\mathcal{VP}) \wedge \mathcal{L}(\mathcal{VP},\mathcal{E}_i) \wedge \mathcal{C}_l(\mathcal{VP},\mathcal{V}) \wedge$ $\mathcal{S}(\mathcal{V},\mathcal{ADVP}) \wedge \mathcal{C}(\mathcal{ADVP},\mathcal{PP}_1) \wedge \mathcal{C}_l(\mathcal{PP}_1,p_1) \wedge$ $\mathcal{S}(p_1,\mathcal{E}_j) \wedge \mathcal{S}(\mathcal{V},\mathcal{PP}_2) \wedge \mathcal{L}(\mathcal{PP}_2,\mathcal{ADVP}) \wedge$ $\mathcal{C}_l(\mathcal{PP}_2,p_2) \wedge \mathcal{S}(p_2,\mathcal{E}_k)] \Rightarrow \mathcal{E}_i \rightarrow \mathcal{V} - p_1 \leftarrow \mathcal{E}_j \rightarrow$ $p_2 \rightarrow \mathcal{E}_k$
18	$[\mathcal{C}(\mathcal{R},\mathcal{E}_i) \wedge \mathcal{C}(\mathcal{R},\mathcal{VP}) \wedge \mathcal{L}(\mathcal{VP},\mathcal{E}_i) \wedge \mathcal{C}_l(\mathcal{VP},\mathcal{V}) \wedge$ $\mathcal{S}(\mathcal{V},\mathcal{ADVP}) \wedge \mathcal{C}(\mathcal{ADVP},\mathcal{PP}_1) \wedge \mathcal{C}_l(\mathcal{PP}_1,p_1) \wedge$ $\mathcal{S}(p_1,\mathcal{NP}) \wedge \mathcal{C}(\mathcal{NP},\mathcal{E}_j) \wedge \mathcal{C}(\mathcal{NP},\mathcal{PP}_2) \wedge$ $\mathcal{L}(\mathcal{PP}_2,\mathcal{E}_j) \wedge \mathcal{C}_l(\mathcal{PP}_2,p_2) \wedge \mathcal{S}(p_2,\mathcal{E}_k)] \Rightarrow \mathcal{E}_i \rightarrow$ $\mathcal{V} - p_1 \leftarrow \mathcal{E}_j \rightarrow p_2 \rightarrow \mathcal{E}_k$

Legend:

$\mathcal{T}$: Phrase structure tree

$\mathcal{R}$: The root of a subtree of $\mathcal{T}$

$\mathcal{E}_i, \mathcal{E}_j, \mathcal{E}_k$: Entities appearing as noun phrases at the leaf nodes $\mathcal{T}$

$\mathcal{C}(\mathcal{X},\mathcal{Y})$: $\mathcal{Y}$ is child of $\mathcal{X}$ or $\mathcal{X}$ is parent of $\mathcal{Y}$

$\mathcal{C}_l(\mathcal{X},\mathcal{Y})$: $\mathcal{Y}$ is leftmost child of $\mathcal{X}$

$\mathcal{S}(\mathcal{X},\mathcal{Y})$: $\mathcal{X}$ and $\mathcal{Y}$ are siblings

$\mathcal{L}(\mathcal{X},\mathcal{Y})$: $\mathcal{Y}$ is left to $\mathcal{X}$

- Let $\mathcal{L}_\mathcal{V}$ be the collection of verbs or verb-preposition pairs, which are extracted as part of ICs. Each verb can occur in more than one form in the list $\mathcal{L}_\mathcal{V}$. For example, the verb *activate* may occur in the form of *activate, activates, activated* or *activated in* etc., all of them essentially representing the biological interaction *"activation"* in some form. The list $\mathcal{L}_\mathcal{V}$ is analyzed to determine the set of unique root forms. The frequency of occurrence of each root verb is the sum-total of its occurrence frequencies in each form. All root verbs with frequency less than a user-given threshold are eliminated from further consideration. The surviving verbs are stored in $\mathcal{L}_{\Re\mathcal{V}}$ and termed as *most-frequently occurring* root verbs representing important *biological relations.*

TABLE 20.6: Algorithm to extract ICs from phrase structure tree.

Algorithm: information_component_extraction(T)

Input: A parse tree T, created though Stanford parser
Output: A list of Information Components L_{IC}
Steps:

1. $L_{IC} \leftarrow \varphi$
2. for each node $N \in T$ do
3. $ic \leftarrow \varphi$
4. for each child $\eta_i \in N$ do
5. If η_i=NP AND η_j =VP AND $i < j$ then
6. If $\lambda_0 \in$child$[\eta_j]$=V AND $\lambda_i \in$child$[\eta_j]$ =NP AND $\lambda_j \in$child$[\eta_j]$ =PP AND i$\neq$0, j$\neq$0, $i < j$ AND $\xi_0 \in$child$[\lambda_j]$ =p AND $\xi_i \in$child$[\lambda_j]$ =NP AND i$\neq$0 then
7. ic=E(η_i) $\rightarrow$V$\leftarrow$E(λ_i) $\rightarrow$p$\leftarrow$E$_k$(ξ_i) // Rule 2
8. else if $\lambda_0 \in$child$[\eta_j]$=V AND $\lambda_i \in$child$[\eta_j]$=NP AND i$\neq$0 then
9. if $\xi_i \in$child$[\lambda_i]$=NP AND $\xi_j \in$child$[\lambda_i]$=PP AND $i < j$ AND $\tau_0 \in$child$[\xi_j]$=p AND $\tau_i \in$child$[\xi_j]$=NP AND i$\neq$0 then
10. ic=E(η_i) $\rightarrow$V$\leftarrow$E(ξ_i) $\rightarrow$p$\leftarrow$E(τ_i) // Rule 3
11. else
12. ic=E(η_i) $\rightarrow$V$\leftarrow$E(λ_i) // Rule 1
13. end if
14. else if $\eta_k \in$child$[\eta_j]$=VP AND $\lambda_0 \in$child$[\eta_k]$=V AND $\lambda_i \in$child$[\eta_k]$=NP AND $\lambda_j \in$child$[\eta_k]$=PP AND i$\neq$0, j$\neq$0, $i < j$ AND $\xi_0 \in$child$[\lambda_j]$=p AND $\xi_i \in$child$[\lambda_j]$=NP AND i$\neq$0 then
15. ic=E(η_i) $\rightarrow$V$\leftarrow$E(λ_i) $\rightarrow$p$\leftarrow$E(ξ_i) // Rule 5
16. else if $\eta_k \in$child$[\eta_j]$=VP AND $\lambda_0 \in$child$[\eta_k]$=V AND $\lambda_i \in$child$[\eta_k]$=NP AND i$\neq$0 then
17. if $\xi_i \in$child$[\lambda_i]$=NP AND $\xi_j \in$child$[\lambda_i]$= PP AND $i < j$ AND $\tau_0 \in$child$[\xi_j]$=p AND $\tau_i \in$child$[\xi_j]$=NP AND i$\neq$0 then
18. ic=E(η_i) $\rightarrow$V$\leftarrow$E(ξ_i) $\rightarrow$p$\leftarrow$E(τ_i) // Rule 6
19. else
20. ic=E(η_i) $\rightarrow$V$\leftarrow$E(λ_i) // Rule 4
21. end if
22. else if $\lambda_0 \in$child$[\eta_j]$=V AND $\lambda_i \in$child$[\eta_j]$=PP AND $\lambda_j \in$child$[\eta_j]$=PP AND i$\neq$0, j$\neq$0, $i < j$ AND $£_0 \in$ child$[\lambda_i]$=p$_1$ AND $£_m \in$child$[\lambda_i]$=NP AND m$\neq$0 AND $\xi_0 \in$child$[\lambda_j]$=p$_2$ AND $\xi_i \in$child$[\lambda_j]$=NP AND i$\neq$0 then
23. ic=E(η_i) $\rightarrow$V-p$_1$ $\leftarrow$E($£_m$) $\rightarrow$p$\leftarrow$E(ξ_i) // Rule 8
24. else if $\lambda_0 \in$child$[\eta_j]$=V AND $\lambda_i \in$child$[\eta_j]$=PP AND $£_0 \in$child$[\lambda_i]$=p$_1$ AND $£_m \in$child$[\lambda_i]$=NP AND m$\neq$0 then
25. if $\xi_i \in$child$[£_m]$=NP AND $\xi_j \in$child$[£_m]$=PP AND $i < j$ AND $\tau_0 \in$child$[\xi_j]$=p$_2$ AND $\tau_i \in$child$[\xi_j]$=NP AND i$\neq$0 then
26. ic=E(η_i) $\rightarrow$V-p$_1$ $\leftarrow$E(ξ_i) $\rightarrow$p$_2$ $\leftarrow$E(τ_i) // Rule 9

(Continued)

TABLE 20.6:   (Continued)

27.	else
28.	ic=$E(\eta_i)$ →V-p_1 ←$E(\pounds_m)$ Rule 7
29.	end if
30.	else if η_k ∈child$[\eta_j]$=VP AND λ_0 ∈child$[\eta_k]$=V AND λ_i ∈child$[\eta_k]$=PP AND λ_j ∈child$[\eta_k]$=PP AND i≠0, j≠0, $i < j$ AND $\pounds_0$ ∈child$[\lambda_i]$=p_1 AND $\pounds_m$ ∈child$[\lambda_i]$=NP AND m≠0 AND ξ_0 ∈child$[\lambda_j]$=p_2 AND ξ_i ∈child$[\lambda_j]$=NP AND i≠0 then
31.	ic=$E(\eta_i)$ →V-p_1 ←$E(\pounds_m)$ →p_2 ←$E(\xi_i)$ // Rule 11
32.	else if η_k ∈child$[\eta_j]$=VP AND λ_0 ∈child$[\eta_k]$=V AND λ_i ∈child$[\eta_k]$=PP AND i≠0 AND $\pounds_0$ ∈child$[\lambda_i]$=p_1 AND $\pounds_m$ ∈child$[\lambda_i]$=NP AND m≠0 then
33.	if ξ_i ∈child$[\pounds_m]$=NP AND ξ_j ∈child$[\pounds_m]$=PP AND $i < j$ AND τ_0 ∈child$[\xi_j]$=p_2 AND τ_i ∈child$[\xi_j]$=NP AND i≠0 then
34.	ic=$E(\eta_i)$ →V-p_1 ←$E(\xi_i)$ →p_2 ←$E(\tau_i)$ // Rule 12
35.	else
36.	ic=$E(\eta_i)$ →V←$E(\pounds_m)$ // Rule 10
37.	end if
38.	else if η_k ∈child$[\eta_j]$=VP AND η_l ∈child$[\eta_k]$=VP AND λ_0 ∈child$[\eta_l]$=V AND λ_i ∈child$[\eta_l]$=NP AND λ_j ∈ child$[\eta_l]$=PP AND i≠0, j≠0, $i < j$ AND $\pounds_0$ ∈child$[\lambda_i]$=p_1 AND ξ_0 ∈child$[\lambda_j]$=p AND ξ_i ∈child$[\lambda_j]$=NP AND i≠0 then
39.	ic=$E(\eta_i)$ →V←$E(\lambda_i)$ →p←$E(\xi_i)$ // Rule 14
40.	else if η_k ∈child$[\eta_j]$=VP AND η_l ∈ child$[\eta_k]$=VP AND λ_0 ∈child$[\eta_l]$=V AND λ_i ∈child$[\eta_l]$=NP AND i≠0 then
41.	if ξ_i ∈child$[\lambda_i]$=NP AND ξ_j ∈child$[\lambda_i]$=PP AND $i < j$ AND τ_0 ∈child$[\xi_j]$=p AND τ_i ∈child$[\xi_j]$=NP AND i≠0 then
42.	ic=$E(\eta_i)$ →V←$E(\xi_i)$ →p←$E(\tau_i)$ // Rule 15
43.	else
44.	ic=$E(\eta_i)$ →V←$E(\lambda_i)$ // Rule 13
45.	end if
46.	else if λ_0 ∈child$[\eta_j]$=V AND λ_i ∈child$[\eta_j]$=ADVP AND λ_j ∈child$[\eta_j]$=PP AND i≠0, j≠0, $i < j$ AND λ_k ∈child$[\lambda_i]$=PP AND $\pounds_0$ ∈child$[\lambda_k]$=p_1 AND $\pounds_m$ ∈child$[\lambda_k]$=NP AND m≠0 AND ξ_0 ∈child$[\lambda_j]$=p_2 AND ξ_i ∈child$[\lambda_j]$=NP AND i≠0 then
47.	ic=$E(\eta_i)$ →V-p_1 ←$E(\pounds_m)$ →p←$E(\xi_i)$ // Rule 17
48.	else if λ_0 ∈child$[\eta_j]$=V AND λ_i ∈child$[\eta_j]$=ADVP AND i≠0 AND λ_k ∈child$[\lambda_i]$=PP AND $\pounds_0$ ∈child$[\lambda_k]$=p_1 AND $\pounds_m$ ∈child$[\lambda_k]$=NP AND m≠0 then
49.	if ξ_i ∈child$[\pounds_m]$=NP AND ξ_j ∈child$[\pounds_m]$= PP AND $i < j$ AND τ_0 ∈child$[\xi_j]$=p_2 AND τ_i ∈child$[\xi_j]$=NP AND i≠0 then

(Continued)

TABLE 20.6:   (Continued)

50.	ic=E(η_i) →V-p$_1$ ←E(ξ_i) →p$_2$ ←E(τ_i) // Rule 18
51.	else
52.	ic=E(η_i) →V-p$_1$ ←E($\pounds_m$) // Rule 16
53.	end if
54.	end if
55.	end if
56.	end for
57.	if ic $\neq \varphi$ then
58.	L$_{IC}$ ← L$_{IC}$∪ ic
59.	end if
60.	end for
61.	return L$_{IC}$

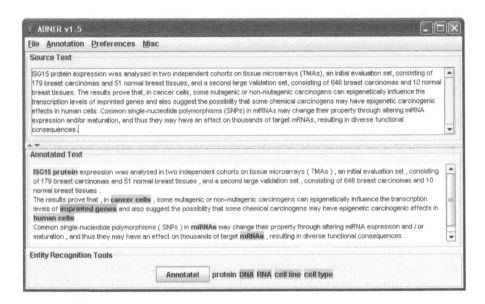

FIGURE 20.3: Tagged version of the first three sample sentences given in Table 20.2 by ABNER.

- Once the frequent root verb list is determined, a pattern matching technique is applied on $\mathcal{L}_V$ to identify and extract the morphological variants of all root verbs in $\mathcal{L}_{\mathfrak{R}V}$.

Algorithm *biological_relation_extraction*, given in Table 20.8, defines this process formally. A partial list of feasible biological relations and their morphological variants extracted from a corpus of 500 PubMed abstracts related to "breast cancer" is shown in Table 20.9.

TABLE 20.7: A partial list of ICs extracted from PubMed documents (given in Table 20.2) by applying the rules given in Table 20.5.

	Information components					PubMed ID	Applied rule no.
Left entity	Relational verb	Verbal prep.	Right entity	Conjunction preposition	Catalyst		
High-fat diets rich in omega-6 polyunsaturated fatty acids (PUFAs)	Produced	—	Higher frequencies of HA-induced mammary gland tumors	In	Rats	18636564	5
Intake of HAs	Associated	With	Breast cancer incidence	—	—	18636564	10
Suppression of AO enzymes associated with breast cancer and aging	Is	—	Most likely the cause of increased levels of reactive oxygen species (ROS)	—	—	18636415	1
AO enzymes	Associated	With	Breast cancer and aging	—	—	18636415	7
Some mutagenic or nonmutagenic carcinogens	Influence	—	The transcription levels of imprinted genes	—	—	18636156	13
Some chemical carcinogens	Have	—	Epigenetic carcinogenic effects	In	Human cells	18636156	6
The levels of MDC1 and BRIT1	Correlated	With	Centrosome amplification, defective mitosis and cancer metastasis	In	Human breast cancer	18635967	8
Oral glutamine (GLN)	Inhibited	—	Tumor growth	Through	Stimulation of host	18635390	2

(Continued)

TABLE 20.7: (Continued)

| Left entity | Information components | | | | | | PubMed ID | Applied rule no. |
	Relational verb	Verbal prep.	Right entity	Conjunction preposition	Catalyst			
RSK4 expression	Limit	—	The oncogenic, invasive, and metastatic potential of breast cancer cells	—	—		18628456	4
Common single-nucleotide polymorphisms (SNPs) in miRNAs	Change	—	Their property	Through	Altering miRNA expression		18634034	14
ISG15 protein expression	Analyzed	In	Two independent cohorts	On	Tissue microarrays (TMAs)		18627608	12
Tufted hair folliculitis in a woman	Treated	With	Lapatinib	For	Breast cancer		18627387	9
Cucurbitacin B	Has	—	A potent antiproliferative effect	On	Breast cancer cells		18627377	3
Single nucleotide polymorphisms (SNPs) selected on the basis of in-depth resequencing of the COMT gene	Genotyped	In	1482 DNA samples	From	A Mayo Clinic breast cancer case control study		18632656	11

TABLE 20.8: Algorithm to extract biological relations.

Algorithm: biological_relation_extraction($\mathbf{L}_{\mathcal{IC}}$)

Input: $\mathcal{L}_{\mathcal{IC}}$ - A list of information components
Output: A set $\mathcal{R}$ of feasible biological relations and their morphological variants
Steps:

1   $\mathcal{L}_{\mathcal{V}} \leftarrow \varphi$, $\mathcal{L}_{\mathcal{UV}} \leftarrow \quad \varphi$, $\mathcal{L}_{\mathcal{RV}} \leftarrow \varphi$
2   for all $\mathcal{IC} \in \mathcal{L}_{\mathcal{IC}}$ do
3   if $\mathrm{E}_i \in \mathcal{IC}.subject$ OR Ei $\in \mathcal{IC}.object$ then   // $\mathcal{E}_i$is a biological entity
4   $\mathcal{L}_{\mathcal{V}} \leftarrow \mathcal{L}_{\mathcal{V}} \cup \mathcal{IC}.verb+\mathcal{IC}.preposition$
5   end if
6   end for
7   $\mathcal{L}_{\mathcal{UV}} \leftarrow \mathcal{UNIQUE}(\mathcal{L}_{\mathcal{V}})$
8   Filter out verbs from $\mathcal{L}_{\mathcal{UV}}$ with a prefix as ξ, where $\xi \in$ {cross-, extra-, hydro-, micro-, milli-, multi-, photo-, super-, trans-, anti-, down-, half-, hypo-, mono-, omni-, over-, poly-, self-, semi-, tele-, dis-, epi-, mis-, non-, pre-, sub-, de-, di-, il-, im-, ir-, un-, up-}
9   Filter out verbs from $\mathcal{L}_{\mathcal{UV}}$ with a suffix as $\Im$, where $\Im \in$ {-able, -tion, -ness, -less, -ment, -ally, -ity, -ism, -ous, -ing, -er, -or, -al, -ly, -ed, -es, -ts, -gs, -ys, -ds, -ws, -ls, -rs, -ks, -en}
10   for all $\mathcal{V} \in \mathcal{IC}$ do
11   $\mathcal{N} =FreqCount(\mathcal{V})$
12   if $\mathcal{N} \geq \theta$ { threshold value} then
13   $\mathcal{L}_{\mathcal{RV}} \leftarrow \mathcal{L}_{\mathcal{RV}} \cup \mathcal{V}$
14   end if
15   end for
16   $\mathrm{R} \leftarrow \mathcal{L}_{\mathcal{RV}}$
17   for all $\mathcal{V}_i \in \mathcal{L}_{\mathcal{RV}}$ do
18   for all $\mathcal{V}_j \in \mathcal{L}_{\mathcal{UV}}$ do
19   if $\mathcal{V}_i \in SubString(\mathcal{V}_j)$ then
20   $\mathcal{R} \leftarrow \mathcal{R} \cup \mathcal{V}_j$
21   end if
22   end for
23   end for
24   return $\mathcal{R}$
25   stop

20.8   Biological Knowledge Visualizer

One of the crucial requirements when developing a text mining system is the ability to browse through the document collection and be able to *visualize* various elements within the collection. This type of interactive exploration

TABLE 20.9: A partial list of feasible biological relations and their morphological variants extracted from a corpus of 500 PubMed abstracts related to "breast cancer."

Biological relations	Morphological variants
Activate	Activates, activated, activated by, activated in, activating
Stimulate	Stimulated, stimulates, stimulated by, stimulated with
Encode	Encodes, encodes for, encoded
Bind	Binds, bind to, binds to, binds with
Associate	Associates, associated, associated with, associated to
Treat	Treat, treated, treated with, treated for, treated by, treated at, treat with, treated after
Correlate	Correlated, correlated with, correlated to
Inhibit	Inhibits, inhibited, inhibited by, inhibited on, inhibited in, inhibiting
Induce	Induced, induces, induced by
Detect	Detected, detected in, detected by, detected on, detected with, detected at
Express	Expressed, expressed in, expressing, expressed at, expresses, expressed on
Affect	Affects, affected, affected by, affected in, affecting
Cause	Caused, caused by, causes
Exhibit	Exhibited, exhibiting, exhibited by
Signal	Signaling, signaling in, signaling through, signaling via,signals from, signaling with
Enhance	Enhanced, enhances, enhanced by, enhanced during, enhanced in
Mediate	Mediated by, mediates, mediated in
Regulate	Regulates, regulated by, regulated in
Initiate	Initiated, initiated in, initiated with, initiated on
Suppress	Suppressed, suppressed by, suppressed in

The relations are ordered in descending order of their frequency count.

enables the identification of new types of entities and relationships that can be extracted, and better exploration of the results from the IE phase [35]. Semantic net created as relationship maps provide a visual means for concise representation of the relationships among many terms in a given context.

The major idea of generating a semantic net is to highlight the role of a concept in a text corpus by eliciting its relationship to other concepts. The nodes in a semantic net represent entities/concepts and links indicate relationships. While concept ontologies are specialized types of semantic net, which also highlight the taxonomical and partonomical relations among concepts, the proposed semantic net is designed only to represent the biological relations mined from the text corpus. Hence, a subset of an IC, termed a relation triplet, is used for this purpose. The relation triplet can be defined formally as follows:

Definition 2: Relation triplet is a subset of an IC and can be defined as a triplet of the form $< \mathcal{E}_i, \mathcal{V}, \mathcal{E}_j >$ where, $\mathcal{V}$ is relational verb, and $\mathcal{E}_i$ and $\mathcal{E}_j$ are biological entities appearing as subject and object respectively for $\mathcal{V}$.

The whole graph is centered around a selected entity belonging to the list of feasible entities recognized by the entity extractor module. For a relation triplet $< \mathcal{E}_i, \mathcal{R}_a, \mathcal{E}_j >$, the entities $\mathcal{E}_i$ and $\mathcal{E}_j$ are used to define classes and $\mathcal{R}_a$ is used to define relationships between them. Biological entities can be complex in nature which includes relations among atomic entities. For example, in the current scenario a noun phrase *"interaction between the ADAM12 and SH3MD1 genes"* represents a complex entity, which contains atomic entities like ADAM12 and SH3MD1. Complex entities are divided into atomic entities identified by the biological entity recognizer and a separate node is created for each atomic entity.

To define a relationship map, the user selects an entity, say ξ, around which the graph is to be created. The selected entity ξ is used to extract all those relation triplets which contains ξ either as a part of *subject* or *object* or both. Hence for a relation triplet $< \mathcal{E}_i, \mathcal{R}_a, \mathcal{E}_j >$ three cases may arise:

Case 1: ξ appears as a part of $\mathcal{E}_i$. In this case, $\mathcal{E}_j$ is divided into a set of atomic biological entities, identified by biological entity recognizer. For each atomic entity, a separate node is created which is linked with a directed edge originating from ξ and labeled with $\mathcal{R}_a$.

Case 2: ξ appears as a part of $\mathcal{E}_j$. In this case, $\mathcal{E}_i$ is divided into a set of atomic biological entities, identified by biological entity recognizer. For each atomic entity, a separate node is created which is linked with a directed edge terminating at ξ and labeled with $\mathcal{R}_a$.

Case 3: ξ appears as a part of both $\mathcal{E}_i$ and $\mathcal{E}_j$. This combines both case 1 and case 2.

Algorithm *semantic_net_generation* shown in Table 20.10 is used to convert the semantic net generation process into a working module. A snapshot of the semantic net generated around the biological entity "breast cancer" is shown in Figure 20.4. The left pan of Figure 20.4 shows the list of all feasible entities identified by entity extractor module around which a semantic net can be generated. The user selected entity is displayed in oval at the centre position and all related entities are displayed around it in rectangles. Since, the ABNER recognizes only a subset of GENIA ontology concepts—protein, DNA, RNA,

TABLE 20.10:   Algorithm for semantic net generation.

Algorithm: *semantic_net_generation*($\mathcal{L}_{\mathcal{IC}}$, ξ)

Input: Information components ($\mathcal{L}_{\mathcal{IC}}$) and a key entity ($\xi$) around which the graph is to be created

Output: Semantic Net – A directed graph $\mathcal{G} = (\mathcal{V}, \mathcal{E})$

Steps:

1 $\mathcal{V} \leftarrow \quad \xi$
2 $\mathcal{E} \leftarrow \quad \phi$
3 for all $< n_i, r, n_j > \in \mathcal{L}_{\mathcal{IC}}$ do
4 if $\xi \in Substring(n_i)$ then
5 Divide n_j into a set $\mathcal{S} = \{\mathcal{E}_1, \mathcal{E}_2, \ldots, \mathcal{E}_n\}$ of biological entities
6 for each $\mathcal{E}_i \in \mathcal{S}$ do
7 if $\mathcal{E}_i \notin \mathcal{V}$ then
8 $\mathcal{V} \leftarrow \mathcal{V} \cup \mathcal{E}_i$
9 $\mathcal{E} \leftarrow \mathcal{E} \cup < \xi, \mathcal{E}_i >$
10 end if
11 end for
12 end if
13 if $\xi = substring(n_j)$ then
14 Divide n_i into a set $\mathcal{S} = \{\mathcal{E}_1, \mathcal{E}_2, \ldots, \mathcal{E}_m\}$ of biological entities
15 for each $\mathcal{E}_i \in \mathcal{S}$ do
16 if $\mathcal{E}_i \notin \mathcal{V}$ then
17 $\mathcal{V} \leftarrow \mathcal{V} \cup \mathcal{E}_i$
18 $\mathcal{E} \leftarrow \mathcal{E} \cup < \mathcal{E}_i, \xi >$
19 end if
20 end for
21 end if
22 end for
23 return $\mathcal{G}$
24 stop

---

cell-line and cell-type, at present the system recognizes the class of only these entities appearing in the text.

The semantic net also facilitates the users to navigate through the pile of documents in an efficient way. While double-clicking on a node, all the ICs in which the entity, contained in the node, appears either as a part of subject or object are selected. Thereafter, the PubMed documents containing these ICs are displayed in which the relevant parts of the sentences are highlighted. The ICs are also displayed separately in the bottom pan of the same window. Figure 20.5 presents a snapshot of the window containing PubMed documents and ICs centered around the entity "stem cell-like" when it was double-clicked in Figure 20.4.

Similarly, on double-clicking an edge, all the ICs centered around the biological relation appearing as the edge label are selected. Thereafter, the

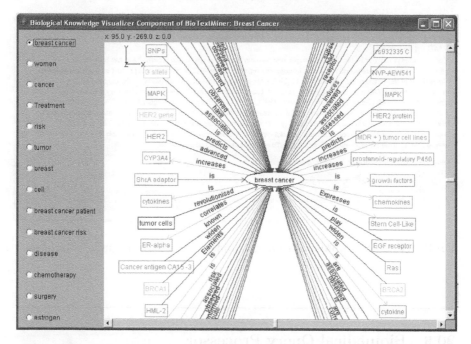

FIGURE 20.4: Semantic net created by BioTextMiner around "breast cancer."

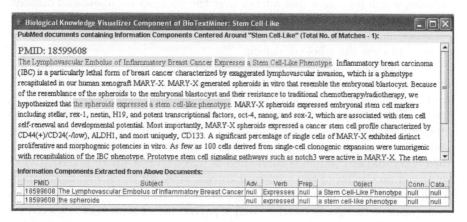

FIGURE 20.5: PubMed documents and ICs centered around "stem cell-like."

PubMed documents containing these ICs are displayed with properly highlighting the relevant snippet of text. The ICs are also displayed separately in the bottom pan of the same window. Figure 20.6 presents a snapshot of the window containing PubMed documents and ICs centered around the biological relation "correlates" when it was double-clicked in Figure 20.4.

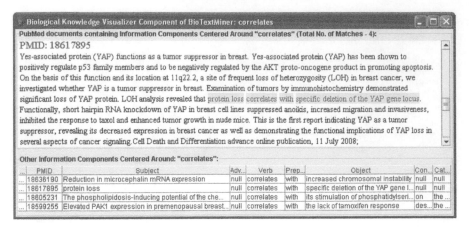

FIGURE 20.6: PubMed documents and ICs centered around "correlates."

20.9 Biomedical Query Processor

In this section we will present the design of the biomedical query processor module, which processes user queries over the abstract database and displays relevant ICs and thereby the sentences retrieved, along with their PubMed reference numbers. A link to the PubMed abstract is also displayed that can be used to view the whole document.

Query processing is a two-step process—acceptance and analysis of the user query and finding the relevant answers from the structured knowledge base. A query is represented by a template <Subject_Entity/Concept/*, Relation/*, Object_Entity/Concept/*> which allows the user to formulate feasible queries at multiple levels of specificity. A query can contain a mixture of concepts and entity names and/or a specific relation. A * in any field represents a wild-card entry and any match is considered as successful. A query is restricted to contain a maximum of two wild-card entries, since all three wild-card entries would be similar to retrieving all documents in the database. Figure 20.7 shows a snapshot of the query interface and a partial list of sentences retrieved for the query <*, associated with, breast cancer> in which the first component is * and can be interpreted as to extract all those sentences and thereby PubMed documents that contains information about genes, proteins, treatment, etc., associated with breast cancer.

By default the feasible relation list that is displayed to the user initially contains all root relations. Similarly default subject and object concept lists initially display all biological concepts present in the underlying ontology. After selecting a root relation, the user may further refine the query by choosing morphological variants of the relations. The frequencies of the biological relations are used by the system to assist the users during query formulation.

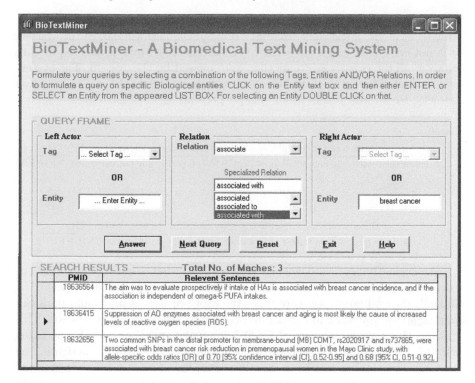

FIGURE 20.7: User interface.

When the user selects a specific concept or relation name, only the feasible elements for the remaining part of query triplet are displayed by the system in decreasing order of strength. Thus guided query formulation allows users to specify only meaningful queries with respect to the underlying corpus.

One of the main uses of the extracted relation instances is in answering user queries intelligently. The most important aspects of querying is that the proposed mechanism queries can accommodate concept names specified at multiple levels of generality and also have relation names contained in them. Moreover, specific biological entities identified by the entity extractor module and stored in the structured knowledge base can also be used to formulate the queries.

20.10   Evaluation of Biological Relation Extraction Process

The performance of the whole system is analyzed by taking into account the performance of the biological relation extraction process. A relation-triplet is said to be "correctly identified" if its occurrence within a sentence along with its left and right entities is grammatically correct, and the system has been able to locate it in the right context. To judge the performance of the system, it is not enough to judge the extracted relations only, but it is also required to analyze all the correct relations that were missed by the system. The system was evaluated for its recall and precision values for each of the selected relation triplets. Recall and precision for this purpose are defined as follows:

Recall

$$= \frac{\text{\# times a relation triplet is correctly identified by the system (TP)}}{\text{\# times the relation triplet actually occurs in the corpus in the correct sense (TP + FN)}}$$

Precision

$$= \frac{\text{\# times a relation triplet is correctly identified by the system (TP)}}{\text{\# times the relation triplet is identified by the system as correct (TP + FP)}}$$

An evaluation software was written in VC++ that exhaustively checks the corpus for possible occurrences of the required relation. For each relation to be judged, the evaluation software takes the root relation as input and performs partial string matching to extract all possible occurrences of the relation. This ensures that various nuances of English language grammar can also be taken care of. For example, if the root relation used in any query is "activate," all sentences containing activates, inactivate, activated by, activated in etc., are extracted. Each sentence containing an instance of the pattern is presented to the human evaluator after its appropriate tagging through ABNER. The sentence without ABNER tags is also presented to the evaluator. This makes it easier for the evaluator to judge the grammatical correctness of the relation in association to the concepts or entities around it. Each occurrence of the relation is judged for correctness by the evaluator, and the correct instances are marked. The marked instances are stored by the evaluation software and later used for computing the precision and recall values. The precision and recall values calculated over 100 randomly selected text documents is shown in Table 20.11.

The precision value of the system reflects its capability to identify a relational verb along with the correct pair of concepts/entities within which it is occurring. The precision of the proposed system is found to be 91.3%. Recall value reflects the capability of the system to locate all instances of a relation within the corpus. The recall value of this module is 87.46%, which can be improved. On analysis, it was found that most of the incorrect identifications and misses occur when the semantic structure of a sentence is wrongly interpreted by the parser. For example, consider the following sentence (PMID: 18636275) and its corresponding phrase structure tree, shown in brackets, generated by the parser.

Expression of Indian Hedgehog signaling molecules in breast cancer [(ROOT (NP (NP (NP (NNP Expression)) (PP (IN of) (NP (NNP

TABLE 20.11: Precision and recall values of the biological relation extraction process.

	# Positive instances present in sample documents	# Negative instances present in the sample documents	Precision	Recall
# Extracted positive instances	418 (TP)	40 (FP)		
			91.3%	87.45%
# Extracted negative instances	60 (FN)	78 (TN)		

Indian) (NNP Hedgehog)))) (VP (VBG signaling) (NP (NP (NNS molecules)) (PP (IN in) (NP (NN breast) (NN cancer))))) (. .)))]

In the above sentence, the word signaling is tagged as verb due to which the phrase "Indian Hedgehog signaling molecules" could not be identified as noun phrase; rather, the sentence is breaked on "signaling" to find the other constituents of the IC. Moreover, it is not possible to interpret that the phrases "Indian Hedgehog signaling molecules" and "breast cancer" are associated with the verb "expression" which is one of the morphological variants of the relational verb "express"—as the word "expression" is tagged as noun.

20.11   Conclusions and Future Work

In this chapter we have proposed the design of an intelligent biomedical text mining system, BioTextMiner, which uses linguistic and semantic analysis of text to identify key ICs from biomedical text documents and stores them in a structured knowledge base over which biomedical queries are processed. The ICs are centered on domain entities and their relationships, which are extracted using NLP techniques and cooccurrence-based analysis. The system is also integrated with a biomedical entity recognizer, ABNER [27], to identify a subset of GENIA ontology concepts (DNA, RNA, protein, cell-line, and cell-type) in the texts and tag them accordingly. This helps in answering biological queries based on biological concepts rather than on particular entities only.

We have also proposed a method for collating information extracted from multiple sources and present them in an integrated fashion with the help of semantic net. The semantic net of the extracted ICs highlights the role of a single entity in various contexts which are useful both for a researcher as well as a layman. The unique aspect of our system lies in its capability to mine and extract information about generic biological relations and the associated prepositions from biomedical text documents. The system also extracts the entities associated with a relation triplet with conjunctional prepositions. The presence or absence of such entities validates a biological interaction which is also a unique aspect of the system over other existing approaches. The system is integrated with a query-answering module. The query-answering module has an interface, which guides users to formulate queries at different levels of specificity.

Since the system advocates using biological relations in queries, the information overload on the users can be substantially reduced.

Right now, the system uses only a subset of GENIA ontology concepts. In future, we are planning to train the biological entity recognizer, ABNER, on GENIA corpus to make it capable to recognize all GENIA ontology concepts in a plain text. The relation extraction rules are also being refined to improve the precision and recall values of the system. Moreover, the query processing module is being redesigned to handle biomedical queries with negations.

References

[1] Uramoto, N., Matsuzawa, H., Nagano, T., Murakami, A., Takeuchi, H. and Takeda, K. A text-mining system for knowledge discovery from biomedical documents. *IBM Systems Journal*, 43(3), 516–533, 2004.

[2] Muller, H. M., Kenny, E. E. and Strenber, P. W. Textpresso: An ontology-based information retrieval and extraction system for biological literature. *PloS Biology* 2(11), e309, URL: http://www.plosbiology.org, 1984–1998, 2004.

[3] Jensen, L. J., Saric, J. and Bork, P. Literature mining for the biologist: From information retrieval to biological discovery. *Nature Reviews Genetics*, 7(2), 119–129, 2006.

[4] Allen, J. *Natural Language Understanding*, 2nd edition. Pearson Education (Singapore) Pte. Ltd., India, 2004.

[5] Schuler, G.D. et al. Entrez: Molecular biology database and retrieval system. *Methods in Enzymology*, 266, 141–162, 1996.

[6] Erhardt, R. A-A., Scheider, R. and Blaschke, C. Status of text-mining techniques applied to biomedical text. *Drug Discovery Today*, 11(7/8), 315–324, 2006.

[7] Fensel, D., Horrocks, I., Harmelen, F. van, McGuinness, D. L. and Patel-Schneider, P. OIL: Ontology infrastructure to enable the semantic web. *IEEE Intelligent Systems*, 16(2), 38–45, 2001.

[8] Bada, M. et al. A short study on the success of the gene ontology. *Journal of Web Semantics,* 1, 235–240, 2004.

[9] Ashburner, M. et al. Gene ontology: Tool for the unification of biology, the gene ontology consortium. *Nature Genetics,* 25, 25–29, 2000.

[10] Cohen, A. M. and Hersh, W. R. A survey of current work in biomedical text mining. *Briefings in Bioinformatics,* 6(1), 57–71, 2005.

[11] McNaught, J. and Black, W. Information extraction. In *Text Mining for Biology and Biomedicine,* Ananiadou S. and McNaught J. (eds). Artech House, Norwood, USA. 143–178, 2006.

[12] Riloff, E. and Lehnert, W. Information extraction as a basis for high-precision text classification. *ACM Transactions on Information Systems,* 12, 296–333, 1994.

[13] Tateisi, Y., Ohta, T., Collier, N., Nobata, C. and Tsujii, J. Building annotated corpus in the molecular-biology domain. In *Proceedings of the COLING 2000 Workshop on Semantic Annotation and Intelligent Content,* Morgan Kaufmann Publisher, San Francisco, USA, 28–34, 2000.

[14] Gruber, T. R. A translation approach to portable ontology specification. *Knowledge Acquisition,* 5(2), 199–220, 1993.

[15] Ding, J. et al. Mining medline: Abstracts, sentences or phrases. In *Pacific Symposium on Biocomputing,* World Scientific, ISBN-10:981024777X, 326–337, 2002.

[16] Sekimizu, T., Park, H. S. and Tsujii, J. Identifying the interaction between genes and genes products based on frequently seen verbs in Medline abstract. *Genome Informatics,* 9, 62–71, 1998.

[17] Thomas, J., Milward, D., Ouzounis, C., Pulman, S. and Carroll, M. Automatic extraction of protein interactions from scientific abstracts. In *Pacific Symposium on Biocomputing,* World Scientific, ISBN-10:9810241887, 538–549, 2000.

[18] Ono, T., Hishigaki, H., Tanigami, A. and Takagi, T. Automated extraction of information on protein-protein interactions from the biological literature. *Bioinformatics,* 17(2), 155–161, 2001.

[19] Rinaldi, F., Scheider, G., Andronis, C., Persidis, A. and Konstani, O. Mining relations in the GENIA corpus. In *Proceedings of the 2nd European Workshop on Data Mining and Text Mining for Bioinformatics,* Pisa, Italy, 2004.

[20] Cohen, A. M. and Hersh, W. R. A survey of current work in biomedical text mining. *Briefings in Bioinformatics,* 6(1),57–71, 2005.

[21] Fukuda, K., Tsunoda, T., Tamura, A. and Takagi, T. Toward information extraction: Identifying protein names from biological papers. In *Pacific Symposium on Biocomputing*, World Scientific, ISBN-10:9810225784, 707–718, 1998.

[22] Proux, D., Rechenmann, F., Julliard, L., Pillet, V. and Jacq, B. Detecting gene symbols and names in biological texts: A first step toward pertinent information extraction. In *Proceedings of the 9th workshop on Genome Informatics*, Universal Academy Press, Japan, 72–80, 1998.

[23] Hanisch, D., Fluck, J., Mevissen, H. T. and Zimmer, R. Playing biology's name game: Identifying protein names in scientific text. In *Pacific Symposium on Biocomputing*, World Scientific, ISBN-10:9812382178, 403–414, 2003.

[24] Rindflesch, T. C., Hunter, L. and Aronson, A. R. Mining molecular binding terminology from biomedical text. In *Proceedings of the AMIA Symposium*, Washington, American Medical Information Association, USA, 127–131, 1999.

[25] Collier, N., Nobata, C. and Tsujii, J. Extracting the names of genes and gene products with a hidden Markov model. In *Proceedings of the 18th International Conference on Computational Linguistics (COLING'2000)*, Morgan Kaufmann Publisher, San Francisco, USA, 201–207, 2000.

[26] Nobata, C., Collier, N., and Tsujii, J. Automatic term identification and classification in biology texts. In *Proceedings of the Natural Language Pacific Rim Symposium*, Beijing, China, 369–375, 1999.

[27] Settles, B. ABNER: An open source tool for automatically tagging genes, proteins, and other entity names in text. *Bioinformatics*, 21(14), 3191–3192, 2005.

[28] Schutz, A. and Buitelaar, P. RelExt: A tool for relation extraction from text in ontology extension. In *Proceedings of the 4th International Semantic Web Conference (ISWC), Galway. Ireland LNCS-3729*, Springer, Berlin, 593–606, 2005.

[29] Yakushiji, A., Teteisi, Y., Miyao, Y. and Tsujii, J. Event extraction from biomedical papers using a full parser. In *Pacific Symposium on Biocomputing*, World Scientific, ISBN-10:9810245157, 408–419, 2001.

[30] Wagner, C., Cheung, K. S. K. and Rachael, K.F. Building semantic webs for e-government with wiki technology. *Electronic Government*, 3(1), 36–55, 2006.

[31] García, R. and Celma, O. Semantic integration and retrieval of multimedia metadata. In *Proceedings of the 5th International Workshop on Knowledge Markup and Semantic Annotation*, Galway, Ireland, 69–80, 2005.

[32] Cox, E. A hybrid technology approach to free-form text data mining. URL: http://scianta.com/pubs/AR-PA-007.htm

[33] Castro, A. G., Rocca-Serra, P., Stevens, R., Taylor, C., Nashar, K., Ragan, M. A. and Sansone, S-A. The use of concept maps during knowledge elicitation in ontology development processes—the nutrigenomics use case. *BMC Bioinformatics*, 7:267, Published online May 25, 2006.

[34] Ciaramita, M., Gangemi, A., Ratsch, E., Saric, J. and Rojas, I. Unsupervised learning of semantic relations between concepts of a molecular biology ontology. In *Proceedings of the 19th International Joint Conference on Artificial Intelligence (IJCAI'05)*, Edinburgh, Scotland, UK, Professional Book Center, ISBN-0938075934, 659–664, 2005.

[35] Aumann, Y. et al. Circle graphs: New visualization tools for text mining. In *Proceedings of the 3rd European Conference on Principles of Data Mining and Knowledge Discovery*. LNCS-1704, Springer-Verlag, UK, 277–282, 1999.

[36] Abulaish, M. and Dey, L. Biological relation extraction and query answering from Medline abstracts using ontology-based text mining. *Data & Knowledge Engineering*, Vol. 61(2). Elsevier Science Publishers, Amsterdam, Netherlands, 228–262, 2007.

[37] Jenssen, T-K., Laegreid, A., Komorowski, J. and Hovig, E. A literature network of human genes for high-throughput analysis of gene expression. *Nature Genetics*, 28, 21–28, 2001.

[38] Gaizauskas, R., Demetriou, G., Artymiuk, P. J. and Willett, P. Protein structures and information extraction from biological texts: the PASTA system. *Bioinformatics*, 19(1), 135–143, 2003.

[39] Craven, M. and Kumlien, J. Constructing biological knowledge bases by extracting information from text sources. In *Proceedings of the 7th International Conference on Intelligent Systems for Molecular Biology (ISMB'99)*, Heidelberg Germany, AAAI Press, ISBN:1-57735-083-9, 77–86, 1999.

[40] Friedman, C., Kra, P., Yu, H., Krauthammer, M. and Rzhetsky, A. GENIES: A natural-language processing system for the extraction of molecular pathways from journal articles. *Bioinformatics*, 17(Suppl. 1), s74–s82, 2001.

Chapter 21

Ontology-Based Knowledge Representation of Experiment Metadata in Biological Data Mining

Richard H. Scheuermann and Megan Kong

University of Texas Southwestern Medical Center

Carl Dahlke

Northrop Grumman, Inc.

Jennifer Cai, Jamie Lee, Yu Qian, and Burke Squires

University of Texas Southwestern Medical Center

Patrick Dunn and Jeff Wiser

Northrop Grumman, Inc.

Herb Hagler

University of Texas Southwestern Medical Center

Barry Smith

University at Buffalo

David Karp

University of Texas Southwestern Medical Center

21.1   General Overview

21.1.1   Paradigm shift in biomedical investigation

The advance of science depends on the ability to build upon information gathered and ideas formulated through prior investigator-driven research and observation. Traditionally, the output of the international research enterprise has been reported in print format—scientific journal articles and books. The intended audiences for these scientific reports are other scientists who carefully read through the text in order to understand the rationale behind the arguments and experimental designs and thereby to gauge the merits of results obtained in addressing the proposed hypothesis. This approach has worked well thus far; it worked well during the advent of molecular biology, when many of the fundamental principles in biology were defined.

However, the last two decades have witnessed a paradigm shift in biomedical investigation, in which reductionistic approaches to investigation in which single functions of single molecules studied using tightly controlled experiments are being replaced by high throughput experimental technologies in which the functions of large numbers of biological entities are evaluated simultaneously. This shift in experimental paradigm was largely initiated when the U.S. Department of Energy, the U.S. National Institutes of Health (NIH) and the European Molecular Biology Laboratory committed to the sequencing of the human genome. In addition to the information derived from the genome sequence itself, the human genome project spawned the development of new research technologies for high throughput investigation that rely on automation and miniaturization to rapidly process large numbers of samples and to simultaneously interrogate large numbers of analytes. For example, microarrays of probes for all known and predicted genes in the human genome are now commercially available to enable simultaneous measurement and comparison of the mRNA levels of all genes in biological samples of interest. The output of these high throughput methodologies is massive amounts of data about the biological systems being investigated, and this has lead to two challenges – how do we analyze and interpret these data, and how do we disseminate the resultant information in such a way as to make it available to (and thus discoverable by) the broader scientific community?

21.1.2 Data sharing standards

In order to maximize its return on investment, NIH established a policy for data sharing in 2003 (http://grants.nih.gov/grants/policy/data_sharing/), to the effect that for any project receiving more than $500,000 per year in NIH funding, the investigators must make their primary data freely available to the scientific community for reuse and meta-analysis. Although most journals now provide electronic versions of their print articles that also include supplemental files of supporting data, these data files are not always available through open access, nor is it easy to find relevant data sets through these sources. Thus, the U.S. National Center for Biomedical Informatics (NCBI), the European Bioinformatics Institute (EBI), and the Stanford microarray community have each established archives for gene expression microarray data—the Gene Expression Omnibus (www.ncbi.nih.gov/geo) (Barrett, 2007), ArrayExpress (www.ebi.uk/arrayexpress) (Parkinson, 2007), and the Stanford Microarray Database (http://genome-www5.stanford.edu/) (Demeter, 2007), respectively. Several other institutes at NIH are supporting projects to develop more comprehensive data sharing infrastructures. The National Cancer Institute's caBIG project is working toward the development of vocabulary standards and software applications that will support data sharing using a distributed grid approach. The Division of Allergy, Immunology and Transplantation of the National Institute of Allergy and Infectious Disease (NIAID) is supporting the development of the Immunology Database and Analysis Portal (ImmPort) to serve as a sustainable archive for research data generated by its funded investigators (www.immport.org). The Division of Microbiology and Infectious Disease also of the NIAID is supporting the development of Bioinformatics Resource Centers for Biodefense and Emerging/Re-emerging Infectious Disease to assemble research data related to selected human pathogens (www.brccentral.org) (Greene, 2007; Squires 2008). The goal of each of these projects is to make primary research data freely available to investigators in a format that will facilitate the incorporation of these data and the information derived there from into new research studies designed to extend these previous findings.

Three new interrelated biological disciplines have emerged to address the challenges of data management and analysis—bioinformatics, computational biology, and systems biology. While there is some debate about whether these are really distinct disciplines of biology, for the purposes of this chapter we will include in the domain of bioinformatics those studies related to defining how laboratory data and biological knowledge relate to each other and how approaches to knowledge representation can aid in data mining for the discovery of new knowledge. We will also make the distinction between data retrieval and data mining, with the former being focused on identifying relevant data sets based on defined characteristics of the experiment (e.g., finding all experiments involving research participants with type 1 diabetes) and the latter being focused on identifying patterns in data sets (e.g., which single nucleotide polymorphisms correlate with the development of type 1 diabetes).

Effective data retrieval requires the accurate and standardized representation of information about each experiment in an easily accessible format (e.g., in one or other standard relational database format). While data mining is also dependent on accurate and standardized representation of data, it is also further enhanced when the information incorporates previous knowledge in such a way as to enable identification of relevant patterns (for example through the use of Gene Ontology (GO) annotation to interpret gene expression patterns in microarray data).

Discussions of data mining tend to focus on the algorithmic portions of the technique. In this chapter we will focus the discussion on how data standards can help support more effective data mining by providing common data structures to support interoperability between data sources and to provide consistent points of integration between disparate data set types. When sharing data between individuals, the use of standards ensures an unambiguous understanding of the information conveyed. For computer programming, the use of standards is essential. In order to accomplish these objectives, data standards should be:

- Useful—provide an aid to storing, sharing, and reuse of data

- Consensual—agreed upon by a plurality of users

- Flexible and evolvable—accommodate all forms of current and future data types

- Comprehensible—understandable by users (and computers)

- Easy to implement—straightforward to use in software development

- Widely adopted—they have to be used

Four related standards will be discussed that, taken together, are necessary for unambiguous and consistent knowledge representation:

- Proposals for the collection of *minimum data elements* necessary to describe an experiment (what information should be provided)

- Common *ontologies* for the vocabulary of data values that will populate these data elements (how that information should be described)

- *Data models* that describe the semantics of how the data elements and values relate to each other (how the information relates to each other)

- Standards that describe the common syntax (format) for *data exchange* (how the information should be transferred between information technology resources)

The chapter will end with an example of how these standards support a type of data mining that we term meta-mining, in which biological knowledge is integrated with primary experimental results for the development of novel hypotheses about the biological systems under evaluation.

21.2   Minimum Data Elements

21.2.1   The MIAME paradigm

Reports of experimental findings and their interpretations published in scientific journals routinely contain specific sections in which certain types of content are provided. The Methods section includes details about how specific assays and other procedures were performed and the materials to which those procedures were applied. The Results section contains information about the design of individual experiments and the data derived. The Introduction section sets the stage by summarizing the current state of the field, and setting out the issues that remain unresolved and that will be addressed in the studies described. The Discussion section provides an interpretation of the experimental findings in the context of the body of knowledge outlined in the Introduction. The Abstract section summarizes the key points of the other sections. While this framework provides some general guidance as to what kind of information should be included in a scientific report, the details concerning what is to be included in each section are left to the authors to decide, thus resulting in considerable variability in the content, structure and level of detail of the information reported. Since there is general agreement that sufficient information should be provided to allow other investigators to reproduce the reported findings, the problem is not so much that important information is missing from scientific publications, but rather that the key information is provided in haphazard, unstructured, and inconsistent ways, requiring readers to distill and organize the relevant content of interest to them.

While this approach to knowledge representation still has its place in the body of scientific investigation, the advent and widespread use of high throughput experiment methodologies has lead to the need to both capture the experimental results in archives and to describe the components of experiments in a standardized way in order to make the data more easily accessible. The importance of these kinds of minimum information check lists for describing the experiment metadata was recognized by the gene expression microarray community, and formalized in the Minimum Information About a Microarray Experiment (MIAME) recommendations (Brazma, 2001). The MIAME recommendations have since been adopted by many scientific journals and microarray archive databases as the de facto standard for reporting the experiment metadata associated with microarray results. Since then a variety of different communities have proposed similar minimum information check lists to capture the unique aspects of their favorite methodologies (e.g., MIFlowCyt for flow cytometry—http://flowcyt.sourceforge.net/) (Lee, et al., 2008).

21.2.2   MIBBI and MIFlowCyt

In order to coordinate efforts in the development of these minimum information checklists, the Minimum Information for Biological and Biomedi-

1. → Experiment overview*
   1.1. → Purpose*
   1.2. → Keywords*
   1.3. → Experiment variables*
   1.4. → Organization*
   1.5. → Primary contact*
   1.6. → Date*
   1.7. → Conclusions*
   1.8. → Quality control measures*
2. → Flow sample/specimen details*
   2.1. → Sample/specimen material description*
      2.1.1. → Biological samples*
         2.1.1.1. → Biological sample description*
         2.1.1.2. → Biological sample source description*
         2.1.1.3. → Biological sample source organism description*
            2.1.1.3.1. → Taxonomy*
            2.1.1.3.2. → Age*
            2.1.1.3.3. → Gender*
            2.1.1.3.4. → Phenotype*
            2.1.1.3.5. → Genotype*
            2.1.1.3.6. → Treatment*
            2.1.1.3.7. → Other relevant information*
      2.1.2. → Environmental samples*
      2.1.3. → Other samples*
   2.2. → Sample treatment(s) description*

2.3. → Fluorescence reagent(s) description*
   2.3.1. → Characteristics(s) being measured*
   2.3.2. → Analyte*
   2.3.3. → Analyte detector[†]
   2.3.4. → Reporter (Fluorochrome)[†]
   2.3.5. → Clone name or number[†]
   2.3.6. → Reagent manufacturer name*
   2.3.7. → Reagent catalog number*
3. → Instrument details*
   3.1. → Instrument manufacturer*
   3.2. → Instrument model*
   3.3. → Instrument configuration and settings*
      3.3.1. → Flow cell and fluidics*
      3.3.2. → Light source(s)[†]
      3.3.3. → Excitation optics configuration[†]
      3.3.4. → Optical filters[†]
      3.3.5. → Optical detectors[†]
      3.3.6. → Optical paths[†]
4. → Data analysis details*
   4.1. → List-mode data file[†]
   4.2. → Compensation details[†]
   4.3. → Data transformation details*
   4.4. → Gating (Data filtering) details[†]

FIGURE 21.1: Minimum information about a flow cytometry experiment (MIFlowCyt). Version 07.09.13 of the MIFlowCyt standard. *Those data elements that are common to most, if not all, MIBBI minimum information standards. [†]Those data elements that are relatively unique to the flow cytometry methodology.

cal Investigations (MIBBI) project (http://www.mibbi.org/) was established by a consortium of investigators from various research communities in order to standardize the content of these data standards and to encourage the reuse of common information elements across methodologies where appropriate (Taylor, 2008). Figure 21.1 shows the required minimum data elements identified in the MIFlowCyt standard. Many of the data elements included correspond to basic elements of experimental design (e.g., experiment purpose and dependent and independent variables) and assay procedures (e.g., biological sample source and treatment details); these kinds of data elements are included in most MIBBI standards as they serve as a common core for biological experiment descriptions. The MIBBI consortium is currently identifying these core data elements that should be consistently represented in all MIBBI-compliant minimum information standards. Other data elements that relate to the kinds of reagents that are used to measure analytes (e.g., fluorochrome reporters and antibody clone names), measurement instrument configuration details (e.g., flow cell and optical filters), and data analysis details (e.g., compensation and gating details) are more technology specific and may only be found as extensions to the common MIBBI core in a few related methodology standards.

    Two important points about these kinds of minimum information standards are worth noting. First, there is a distinction between what information is necessary to reproduce an experiment, as detailed in the MIBBI standards, and the information that would be relevant to capture and support in a database archive. The latter correspond to a subset of MIBBI data elements that might specifically be represented in database tables and would be used to query the database, especially the dependent and independent variables of the experiment, the characteristics of the biological samples and their sources used in the experiment, and the analytes being measured. While the other information is equally important to reproduce the experimental findings, they may not play important roles in the conduct of data meta-analysis. For example, while the instrument configuration details may be necessary to reproduce a particular data set, it is unlikely that one would search for data sets in the database archive based on the details of the optical path. Rather than capturing these details in specific database table columns, this information can be included in text documents that describe all of the protocol details.

The second important point about these minimum information standards relates to who should be responsible for providing the information. In the case of MIFlowCyt, some of the information is derived from the configuration of the instrument and analytical software used in the capture of the resulting data. This information would more appropriately be provided directly by the instrument and software packages themselves, rather than expecting the investigator to have to determine these details for every experiment they run. Thus, in the development of the MIFlowCyt standard, it was important to engage stakeholders from the instrument manufacturer and software developer communities so that they would agree to provide this information as a matter of course in the resulting output files.

Thus, by formalizing the details for the minimum data elements that should be included in the description of an experiment, a higher level of consistency in how to describe experiments can be obtained, both within and between different experimental methodologies. Consistent representation frameworks will facilitate the identification of related experiments in terms of health conditions being investigated, treatment approaches being varied, and responding variable being tested in order to support meta-analysis and reuse of related data sets.

21.2.3   Information artifacts and MIRIAM

Although the biomedical minimum information standards were initially developed to support the description of wet lab experimentation, it became apparent that similar standards would be useful to support the work coming out of the bioinformatics research community in terms of the description of system models and data mining analysis. The BioPax (Luciano, 2005), SBML (Hucka et al., 2003) and CellML (Lloyd et al., 2004) standards have been developed to provide the syntactic standards necessary to exchange biological pathway and systems descriptions. The Minimum Information Requested in the Annotation of biochemical Models (MIRIAM) (Le Novere et al., 2005) was developed to ensure that sufficient information is included in the description of any biological model such that any results obtained from modeling could be reproduced by outside investigators. The Minimum Information About a Simulation Experiment (MIASE) extends MIRIAM to support model simulation. Recently, the European Bioinformatics Institute has established a set of resources based on the MIRIAM standards (Laibe and Le Novere, 2007), including:

- MIRIAM Database—containing information about the MIRIAM data types and their associated attributes;

- MIRIAM Web Services—a SOAP-based application programming interface (API) for querying the MIRIAM Database;

- MIRIAM Library—a library instruction set for the use of MIRIAM Web Services;

- MIRIAM Web Application—an interactive Web interface for browsing and querying MIRIAM Database, and for the submission and editing of MIRIAM data types.

21.3 Ontologies

21.3.1 Ontologies versus controlled vocabularies

While the minimum data standards describe the types of data elements to be captured, the use of standard vocabularies as values to populate the information about these data elements is also important to support interoperability. In many cases, groups develop term lists (controlled vocabularies) that describe what kinds of words and word phrases should be used to describe the values for a given data element. In the ideal case each term is accompanied by a textual definition that describes what the term means in order to support consistency in term use. However, recently many bioinformaticians have begun to develop and adopt ontologies that can serve in place of vocabularies for use as these allowed term lists. As with a specific vocabulary, an ontology is a domain-specific dictionary of terms and definitions. But an ontology also captures the semantic relationships between the terms, thus allowing logical inferencing about the entities represented by the ontology and by the data annotated using the ontology's terms. The semantic relationships incorporated into the ontology represent universal relations between the classes represented by its terms based on knowledge about the entities described by the terms established previously. For example, if we use a disease ontology to annotate gene function that explicitly states that type 1 diabetes and Hashimoto's disease are both types of autoimmune diseases of endocrine glands, then we can infer that gene A, annotated as being associated with type 1 diabetes, and gene B, annotated as being associated with Hashimoto's disease, are related to each other even though this is not explicitly stated in the annotation, through our previous knowledge of disease relationships captured in the structure of the disease ontology used for annotation. Thus, an ontology, in the sense here intended, is a representation of the types of entities existing in the corresponding domain of reality and of the relations between them. This representation has certain formal properties, enabling it to serve the needs of computers. It also employs for its representations the terms used and accepted by the relevant scientific community, enabling it to serve the needs of human beings. To support both humans and computers the terms used are explicitly defined using some standard, shared syntax. Through these definitions and through the relations asserted to obtain between its terms the ontology captures consensual knowledge accepted by the relevant communities of domain experts. Finally, an ontology is a representation of universals; it describes what is general in reality, not what is particular. Thus, ontologies describe classes of entities where databases tend to describe instances of entities.

In recognition of the value that ontologies can add to knowledge representation, several groups have developed ontologies that cover specific domains of biology and medicine. The open biomedical ontology (OBO) library was established in 2001 as a repository of ontologies developed for use by the biomedical research community (http://sourceforge.net/projects/obo). As of August 2008 the OBO library includes 70 ontologies that cover a wide variety of different domains. In some cases, the ontology is composed of a highly focused set of terms to support the data annotation needs of a specific model organism community (e.g., the Plasmodium Life Cycle Ontology). In other cases, the ontology covers a broader set of terms that is intended to provide comprehensive coverage of an entire life science domain (e.g., the Cell Type Ontology). Since 2006, it has become possible to access the OBO library through their BioPortal (http://www. bioontology.org/tools/portal/bioportal.html) of the National Center for Biomedical Ontology (NCBO) project, which also provides a number of associated software services and access to a number of additional ontologies of biomedical relevance. The European Bioinformatics Institute has also developed the Ontology Lookup Service (OLS) that provides a web service interface to query multiple OBO ontologies from a single location with a unified output format (http://www.ebi.ac.uk/ontology-lookup/). Both the BioPortal and the OLS permit users to browse individual ontologies and search for terms across ontologies according to term name and certain associated attributes.

21.3.2 OBO Foundry

While the development of ontologies was originally intended to advance consistency and interoperability in knowledge representation, the recent explosion of new ontologies has threatened to undermine these goals. For example, in some cases multiple ontologies that have been developed independently cover overlapping domains, thus leading to different terms or single terms with different definitions being used to describe the same entity by different communities. While this problem can be partly resolved by efforts to map between terms in different ontologies, in many cases the lack of a one-to-one mapping makes this problematic. Moreover, mappings themselves are difficult to construct, and even more difficult to maintain as the mapped ontologies change independently through time. A second problem is that in many cases the relationships between terms that are used to assemble a single ontology hierarchy are not described explicitly and are not used consistently. It is impossible to support inferencing based on the ontology if it is unclear how adjacent terms in the hierarchy relate to each other. Finally, some ontologies have been developed by small groups from what may be a highly idiosyncratic perspectives and thus may not represent the current consensual understanding of the domain in question.

In order to overcome these and other problems with the current collection of biomedical ontologies, several groups have proposed frameworks for disciplined ontology development (e.g., Aranguren et al., 2008). The OBO Foundry initiative (http://www.obofoundry.org/) was established in 2005 as a collaborative experiment designed to enhance the quality and interoperability of life science ontologies, with respect to both biological content and logical structure (Smith et al., 2007). The initiative is based on the voluntary acceptance by its participant ontology development communities of an evolving set of design and development principles designed to maximize the degree to which their ontologies can support the broader needs of scientists. These best-practice design principles can be roughly subdivided into three broad categories—technical, scientific and societal (Table 21.1). Technical principles include requirements for the inclusion of a common set of meta-data (in a manner similar to the MIBBI-type specifications described above), and the use of a common shared syntax. Scientific principles include the requirement of orthogonality to the effect that the content of each Foundry ontology should be clearly specified and delineated so as not to overlap with other Foundry ontologies, and the requirement for a consistent semantic framework for defining the relations used in each ontology. (These scientific principles are described in more detail below.) Societal principles would include the requirement that the ontology be developed collaboratively, and that the resulting ontology artifact is open and freely available.

As argued above, the great value in using ontologies to represent a particular data set lies in the background knowledge embedded in the relationships that link the terms together in the ontology. OBO Foundry ontologies are expected to utilize relations defined in the Relation Ontology (RO) (Smith et al., 2005) to describe these relations (the first set of relations is depicted in Table 21.2).

The foundational relation that is used primarily in the assembly of OBO Foundry ontologies is the is_a relation that links parent and child terms in the ontology hierarchy. Parent and terms in the is_a hierarchy can be thought of

TABLE 21.1:   OBO Foundry principles as of April 2006.

Technical
The ontology is expressed in a common shared syntax for ontologies (e.g., OBO format, Ontology Web Language (OWL) format)
The ontology possesses a unique identifier space.
The ontology provider has procedures for identifying distinct successive versions.
The ontology includes textual definitions for all terms.
The ontology is well documented.

Scientific
The ontology has a clearly specified and clearly delineated content.
The ontology uses relations that are unambiguously defined following the pattern of definitions laid down in the OBO Relation Ontology.

Societal
The ontology is open and available to be used by all without any constraint.
The ontology has been developed collaboratively with other OBO Foundry members.
The ontology has a plurality of independent users.

TABLE 21.2:   OBO relation ontology.

Foundational	Temporal	Participation
is_a	transformation_of	has_participant
part_of	derives_from	has_agent
Spatial	preceded_by	
located_in		
contained_in		
adjacent_to		

as standing in type-subtype relations similar to the genus-species relations in the species taxonomy. Several advantages arise out of building the ontology structure based on an is_a hierarchy of type-subtype relations, including:

- First, definitions of terms can be constructed using the genus differentia approach proposed by Aristotle, so that the definition of the term "A" will take the form: "An A is_a B that C's" in which A is a subtype (child) of (parent) type B with the special characteristic C that distinguishes instances of A from other instances of B. For example, "type 1 diabetes is_a autoimmune disease of endocrine glands that involves the endocrine pancreas as the primary disease target."

- Second, terms in a well-formed is_a hierarchy inherit characteristics from their parents through the property of transitivity. By defining type 1 diabetes as a subtype of autoimmune disease of endocrine glands, the term inherits characteristics that define all such autoimmune diseases, as well as characteristics of all diseases in general through the definitions of and other attributes associated with these parent terms. Indeed, adherence to the property of transitivity can be a good test for correct positioning of terms in the ontology hierarchy.

21.3.3 Role, quality, function, and type

During ontology development, there is often difficulty in trying to define which sort of characteristic should be used as the primary differentia between sibling subtypes. For example, let's consider the type: old, faded blue Dodge Caravan^{TM} minivan airport shuttle that we might want to represent in a vehicle ontology. Should it be considered to be a subtype of old vehicles, a subtype of faded blue vehicles, a subtype of Dodge vehicles, a subtype of Dodge Caravan^{TM} vehicles, a subtype of minivan vehicles, a subtype of airport shuttle vehicles, or a subtype of all of these parent terms?

In order to address this issue, we need to discuss the distinctions between roles, qualities, functions and types. A role is a special attribute that an entity can be made to play by societal choice. The "airport shuttle" attribute is an example of a role that the driver has assigned to the vehicle. It is not an inherent property of the vehicle that distinguishes it from other vehicles. Indeed, any vehicle could be used as an airport shuttle. The role is also frequently a transient attribute. For example, at the end of the driver's shift the "airport shuttle" might transform into a "soccer practice shuttle," and then back into an "airport shuttle" the next day. The transient, subjective nature of roles makes them a poor choice for primary differentia in ontology hierarchies.

In the example, "old" and "faded blue" describe *qualities* of the vehicle. Qualities are not acquired by choice in the natural world. We cannot choose our age or the color of our skin. And yet qualities are frequently transient in nature. At one point the vehicle was a new, bright blue Dodge Caravan^{TM}. Thus, basing annotations on terms distinguished on the basis of quality characteristics would mean that the annotation would not be invariant and would have to be updated continually to deal with these changes over time. In addition, entities can be described on the basis of a whole range of different quality characteristics—height, width, length, volume, shape, mass, color, age, smell, etc. Without selecting a single defining characteristic, the ontology would explode into a hierarchy of multiple inheritance, in which specific terms would have parents like: old things, blue things, large things, oily-smelling things, and so on.

We conclude that type-subtype is_a relations should be based on properties of the entity that are invariant. A Dodge Caravan^{TM} will always be a minivan regardless of whether it is used as an airport shuttle, whether it is old or new, whether it is painted red or blue, and so forth. It will never be a sports car.

For a more biological example, consider the use of the EcoR1 enzyme in the construction of a recombinant plasmid in a genetic engineering experiment. We would consider EcoR1 to be a type of protein with a molecular function restriction endonuclease activity (GO:0015666). In its normal context, EcoR1 plays a role in the DNA restriction-modification system (GO:0009307) that protects an organism from invading foreign DNA by nucleolytic cleavage of unmethylated foreign DNA. However, EcoR1 can also play another role in an experimental context, its use to open up a double-stranded circular plasmid to accept the insertion of a DNA fragment in a cloning experiment. At the end of the cloning experiment we may want to change the quality of the enzyme from active to inactive through a denaturation process in order to prevent it from realizing its function any further. Thus, while the type of protein and its designed function haven't changed, its role can change based on the process it is involved in and its quality can change dependent on its physical structure state in this case. By precisely distinguishing between roles, functions, qualities and types we can support the accurate representation of entities in their normal states and in artificial experiment contexts, and thereby faster accurate reasoning about these entities in these different contexts.

21.3.4   Orthogonality

The second major scientific OBO Foundry principle relates to the concept of orthogonality. The Foundry is striving toward complete coverage of the entire biological and biomedical domain using one and only one term for a given entity. However, it would be virtually impossible to build a single ontology that covers the entire scope of this domain in a reasonable amount of time with a manageable group of developers who have the requisite expertise in all disciplines of biology. For these reasons, the Foundry has adopted a modular, iterative approach in which smaller subdomains are defined that then become the focus of activity for groups of ontology developers with appropriate expertise. For example, the Chemical Entities of Biological Interest (ChEBI) ontology is being developed by biochemists and organic chemist with knowledge of chemical structure and function in order to cover the small molecules (e.g., drugs, metabolites, peptides, etc.) of interest to biologists. While this modular approach addresses the challenges of biology domain coverage, it brings the problem of potential overlap between subdomains being developed by different groups. Thus, ontologies that are part of the OBO Foundry must submit to the principle of orthogonality in which a given biological entity is covered by one, and only one, ontology. In cases where potential overlap exists, negotiation and consensus building is used to assign terms to a given ontology.

In order to bring some level of consistency in the definition of a subdomain module, the biology domain can be divided into partitions based on two axes (Figure 21.2). The first axis relates to size/granularity, e.g., from molecules to organelles to cells to tissues to organisms to populations to ecosystems. The second axis reflects the general types of entities found in reality as represented in the Basic Formal Ontology (BFO). At the highest level, these entities can be broken down into continuants and occurrents. Continuants are further subdivided into dependent and independent continuants. Continuants

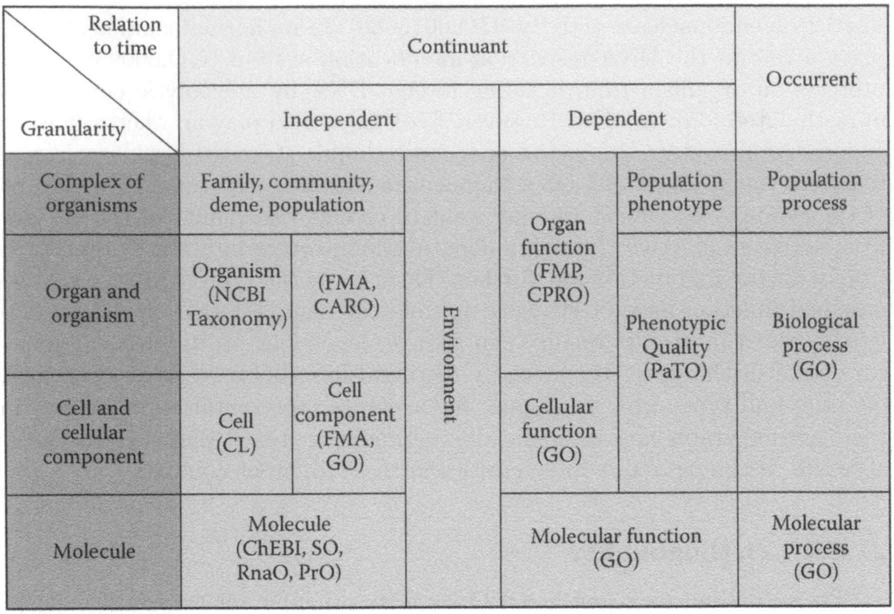

Relation to time / Granularity	Continuant				Occurrent
	Independent			Dependent	
Complex of organisms	Family, community, deme, population			Population phenotype	Population process
Organ and organism	Organism (NCBI Taxonomy)	(FMA, CARO)	Environment	Organ function (FMP, CPRO)	
				Phenotypic Quality (PaTO)	Biological process (GO)
Cell and cellular component	Cell (CL)	Cell component (FMA, GO)		Cellular function (GO)	
Molecule	Molecule (ChEBI, SO, RnaO, PrO)			Molecular function (GO)	Molecular process (GO)

FIGURE 21.2: OBO Foundry candidate ontologies. Domains of biological reality (internal grid) are defined based on the intersection between entity types (columns) as defined by their relationship with time as proposed in the Basic Formal Ontology (BFO) and granularity (rows). Candidate OBO Foundry ontologies for each domain are listed. Foundational Model of Anatomy (FMA), Common Anatomy Reference Ontology (CARO), Cell Ontology (CL), Gene Ontology (GO), Chemical Entities of Biological Interest (ChEBI), Sequence Ontology (SO), RNA Ontology (RnaO), Protein Ontology (PrO), Common Physiology Reference Ontology (CPRO), and Phenotypic Qualities Ontology (PaTO).

exist identically through time. An independent continuant exists on its own without any dependence on another entity; these are the physical objects like tables, cups, proteins, organs, etc. A dependent continuant exists throughout time, but requires adherence in an independent continuant to exist; these are the qualities, roles and functions like the color blue, which only exists in the context of a physical entity. Occurrents are processes, like driving and replication, which exist in a defined time period with a start point and end point. Thus, we can subdivide the biology domain based on a grid in which one axis corresponds to size/granularity and the other to time and entity dependencies (Figure 21.2).

The long-term goal of the OBO Foundry initiative is to achieve, complete coverage of the entire biological domain with single terms for each entity of interest, in which terms are initially linked together using foundational relations (is_a and part_of) into a single hierarchy in a given ontology, which is developed by groups of subdomain experts and reflects the consensual knowledge of the general relations of types of entites in that domain. These terms can then be used to annotate database records in an unambiguous way that supports inference based on the consensual knowledge incorporated into the ontology structure and thus supports database interoperability.

21.4   Data Models

The Immunology Database and Analysis Portal (ImmPort) is being developed to serve as a long-term sustainable archive for data being generated by investigators funded by the Division of Allergy, Immunology, and Transplantation (DAIT) of the U.S. National Institute of Allergy and Infectious Diseases (NIAID). DAIT funds a wide range of basic laboratory and clinical research studies and so the ImmPort system must be able to handle everything from genotyping and gene expression microarray data to clinical trials of new vaccines and therapeutic strategies. As such, ImmPort must be able to manage data associated with a variety of different experiment methodologies and must be able to effectively integrate these data through common metadata features and/or results characteristics. For example, investigators may want to aggregate data from any experiment in which type 1 diabetes is being investigated, as well as experiments in which some characteristic of a particular gene (e.g., TLR4) has been found to be significantly associated with the independent variable of the experiment. Thus, many of the requirements for knowledge representation associated with ImmPort are distinct from those associated with other database archives focused on data from single experiment approaches, like dbGAP for human genetic association data or GEO and ArrayExpress for microarray data. The challenge to support such a wide range of research data lead us to implement a database structure that would reflect the common features of a biomedical investigation.

Several different data model frameworks have been developed over the years—the hierarchical model (used in IBM's IMS database management system), the network model (used in IDS and IDMS), the relational data model (used in IBM DB2, Oracle DBMS, Sybase, Microsoft Access), the object-oriented model (used in Objectstore and Versant) and hybrid object-relational models (Ramakrishnan and Gehrke, 2003). Depending on the application, each of these frameworks has its advantages and disadvantages, but the relational data model has become widely adopted for databases such as ImmPort in that it can handle complex interrelated data in an efficient manner.

The development of a database involves six major steps:

- Requirements analysis—in which an understanding of what the data include and how they will be used are defined;

- Conceptual database design—in which the entities and their relationships are defined;

- Logical database design—in which the conceptual design is converted into a logical model based on the data model framework chosen;

- Schema refinement—in which the logical model is analyzed to identify potential problems and refined accordingly;

- Physical database design—in which the logical model is converted into a physical database schema that incorporate design criteria to optimize system performance;

- Application and security design—integration of the database with other software applications that need to access the data.

Because the uses of biomedical data vary by data type and the specific requirements of the user communities involved, it is virtually impossible to develop a single physical database design that will efficiently meet all user requirements. And yet, in order to support interoperability between database resources, a common approaches for data representation is essential. For these reasons several groups have worked on the development of data models that capture the kinds of entities and relationships in biomedical data that are universal and application independent at the conceptual level. These conceptual models can then be refined in their conversion into physical database schemas optimized to support specific use case applications while still incorporating the common entity types and relationships that allow effective data sharing between data users and data providers (Bornberg-Bauer and Paton, 2002).

21.4.1 BRIDG

Several groups have attempted to develop a data model based on this principle. In the clinical domain, the Biomedical Research Integrated Domain Group (BRIDG) project is a collaborative initiative between the National Cancer Institute (NCI), the Clinical Data Interchange Standards Consortium (CDISC), the Regulated Clinical Research Information Management Technical Committee (RCRIM TC) of Health Level 7 (HL7), and the Food and Drug Administration (FDA) to develop a semantic model of protocol-driven clinical research (Fridsma et al., 2008). It was developed to provide an overarching model that could be used to harmonize between various standards in the clinical research and healthcare domains. It includes representation of "noun things" like organizations, participants, investigators, drugs, and devices, "measurement things" like physical exam assessments, and "interpretation things" like adverse event determination.

21.4.2 FuGE

In the basic research domain, the Functional Genomics Experiment Model (FuGE) is a generic data model to facilitate convergence of data standards for describing high-throughput biological experiments (Jones et al., 2007). Development of FuGE was initially motivated by analysis of a well-established microarray data model: MAGE-OM. The initial goal of FuGE was to deliver a more general model than MAGE-OM by removing concepts that were specific to microarray technology. After receiving a wide range of use cases from different communities on describing experimental designs across "multiomics" and conventional technologies, FuGE developers further generalized the model and added more placeholders in order to broaden the application scope of the model. The current FuGE v1 model aims at not only a generic database schema but also a data exchange standard. In order to capture not only common but also domain-specific experimental information, FuGE is designed to be extensible. Its core model consists of a set of generic object classes to represent the common information in different laboratory workflows and experimental pipelines used in high-throughput biological investigations, while domain-specific extensions of the FuGE core classes are needed to capture specific information requirements in the domain. Due to its extensible characteristic in providing formal and generic data representations, FuGE has been adopted by the Microarray and Gene Expression Data (MGED) Society, the Human Proteome Organization–Proteomics Standards Initiative (PSI), the Genomics Standards Consortium (GSC), and the Metabolomics Standards Initiative (MSI).

Object classes in FuGE v1 are organized under two namespaces: common and bio. There are six packages under common namespace: audit, description, measurement, ontology, protocol, and reference. The bio namespace consists of four packages: conceptualMolecule, data, investigation, and material. Each package has a set of predefined classes that can be either used to describe experimental information directly or reused in domain-specific extensions through inheritance, depending on the nature of the information and the class to be used. Not only entities like instruments, samples, and data files but also relationships among the entities can be described. For example, sample and data can be linked together in FuGE through protocol and protocolApplication, the latter providing a flexible binding between input and output as well as related parameters and the use of instruments. To facilitate data sharing, ontology terms and external resources can be referenced in FuGE to annotate or directly describe the objects. Another benefit of using FuGE is if an extension of FuGE follows the MDA (model driven architecture) standard, a FuGE-compliant database schema (XSD) as well as software code can be automatically generated. A typical extension example is FuGEFlow (http://wiki.ficcs.org/ficcs/FuGEFlow), a recent extension of FuGE to capture MIFlowCyt-compliant information for flow cytometry experiments (Qian, 2009). The generated data schema, called Flow-ML, is planned to be used to help build flow cytometry databases and exchange flow cytometry experimental information among different labs and institutions.

21.4.3 Ontology-based eXtensible Data Model (OBX)

In order to leverage the efforts of the MIBBI and OBO Foundry communities, we have recently developed a data model—the Ontology-Based eXtensible Data Model (OBX) that reflects many of the design principles incorporated in these data standards. Of particular importance to ImmPort database development, a consortium of different research communities has been working on the development of an ontology of terms needed to represent experiment metadata—the Ontology for Biomedical Investigation (OBI; http://purl.obofoundry.org/obo/obi/). The OBI ontology is focused on those aspects of biology that are directly related to scientific investigations—experiment design, protocol specification, biomaterial isolation/purification, assays for the measurement of specific analytes in specimens, instruments and reagents used in these measurement assays, etc. The OBI ontology hierarchical structure has been built on a BFO framework and points to, but does not include, terms from other OBO Foundry ontologies. Several important concepts have been incorporated into the OBI structure, including the typing of assay, biomaterial transformation and data transformation processes based on the kinds of entities that serve as inputs and outputs to those processes, the typing of investigation specifications based on objectives, and the importance of precisely defining the roles played by the various different process components (e.g., specimen, reagent, principle investigator). While the OBI ontology is focused on representing general features of entity classes used in investigations, OBX is focused on capturing instance level information about specific investigations.

A UML diagram of the high level entities represented in OBX is provided in Figure 21.3. The major axis in the model includes objects, events, and qualities, which reflects the major branches of the BFO, namely independent continuants, occurrents, and dependent continuants (Figure 21.3a). Objects include entities like biological specimens, laboratory equipment, assay reagents, chemicals, etc. Procedures are types of events, and are defined based on their inputs and outputs. For example, biomaterial transformation procedures have biomaterial objects as inputs and outputs, whereas assay procedures have biomaterial object inputs and quality outputs (i.e., the assay results). In addition to describing the specific input entity to a given procedure, the role that the entity plays in the procedure is also captured. This allows for the distinction between the use of an antibody as an assay reagent versus the role of an antibody as an analyte to be measured in a given specimen. Every event occurs in a spatial-temporal context and so the event table is linked to the context table to capture these event attributes. In the case of OBX, all events also occur during the conduct of a research study, which

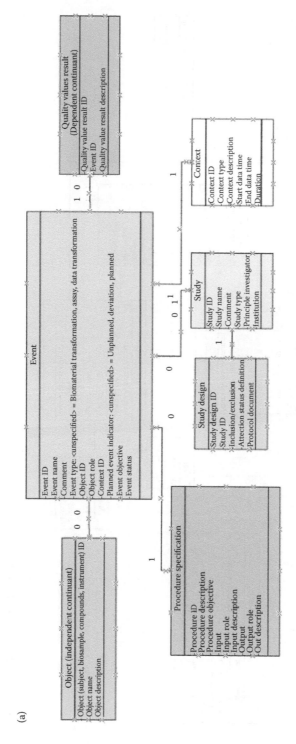

FIGURE 21.3: The Ontology-Based eXtensible Data Model (OBX). See text for a more detailed description. (a) The high-level general entity tables of the OBX model including the individual table attributes. (b) Low-level entity tables of specific object types with subtype-specific attributes.

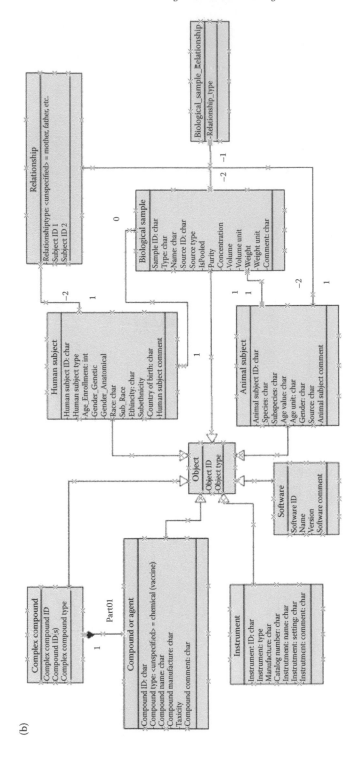

FIGURE 21.3: (*Continued*)

is specified through a study design; the study and the study specification are kept distinct to accommodate deviations from the original design.

Thus, the high-level core set of entities includes processes/events, process specifications, objects and qualities. Each of these high-level tables is then connected to a series of lower level, related tables. The high-level tables include attributes common to all related entities, whereas the low-level tables include attributes that are specific to the related entity types. For example, the objects table is connected to a series of subtables for human subjects, organ/tissue, instrument, compounds/agents, etc. (Figure 21.3b), which include attributes specific for the given entity sub type (e.g., compound manufacturer for the compound/agent entities). Assays include both clinical assessments/physical exams as well as laboratory tests. Therapeutic intervention is a type of material transformation in which a input is a human subject and a compound formulation and the output is a treated subject. Diagnosis is a type of data transformation in which a variety of data inputs from laboratory test results and clinical assessments are processed (i.e., transformed) into a diagnosis data output by the clinician/diagnostician. A detailed description of the OBX UML model can be found at http://purl.oclc.org/NET/bisc/Ontology-Based_eXtensible_Data_Model_(OBX).

An example of how the OBX framework can be used to represent a laboratory experiment is shown in Figure 21.4. The experiment comes from a published study in which the immune response to influenza virus infection is assessed by measuring the levels of interferon gamma in the lungs of infected mice. The first protocol application is the generation of a lung homogenate from infected mice, which can be thought of as three ordered biomaterial transformations—the infection of the mouse, the removal of the lung specimen and the generation of the lung homogenate. In each case the output from the previous process serves as the input for the subsequent process. The process inputs are described as playing specific roles, e.g., host and infectious agent. The first process is a merging (mixing) types of biomaterial transformation, whereas the second two are partitioning (enriching) types of biomaterial transformations. The second protocol application is the measurement of interferon gamma levels in the lung homogenate, which is composed of three ordered subprocesses—the ELISA assay used to derive output data (OD590) as a surrogate of the analyte (interferon gamma) concentration in the specimen, the standard curve interpolation data transformation in which the OD590 value is transformed into interferon gamma mass amount, and finally a simple mathematical data transformation to convert the mass amount into an analyte concentration for the original input specimen. Again the input components are described to play specific roles in the processes, including evaluant, analyte, analyte detector, reagent reporter, comparator, etc. The final protocol application includes a series of data transformation subprocesses to determine if the concentrations of interferon gamma in the lung are significantly different between uninfected and infected mice.

Protocol application name	Sub-process name	Sub-process type	Sub-process step	Input object	Input object type	Input object role	Output object	Output object type	Output object role
Virus-infected lung preparation	Mouse infection	Biomaterial transformation (mixture)	1/3	Mouse	Biomaterial	Host	Virus-infected mouse	Biomaterial	Infected host
				Virus	Biomaterial	Infectious agent			
				Infection/incubation time period	Time	Context			
	Organ harvesting	Biomaterial transformation (enrichment partitioning)	2/3	Infected mouse	Biomaterial	Organ source	Infected lung	Biomaterial	Organ ex vivo
	Organ homogenate preparation	Biomaterial transformation (enrichment partitioning)	3/3	Infected lung	Biomaterial	Transformation subject	Infected lung homogenate	Biomaterial	Enriched franction
Cytokine quantification	Analyte ELISA	Assay (analyte enumeration)	1/3	Infected lung homogenate	Biomaterial	Evaluant	OD590	Data	Amount quality
				IFN-gamma	Biomaterial	Analyte			
				Anti-IFN-gamma	Biomaterial	Analyte detector			
				HRP + substrate	Biomaterial	Reagent reporter			
	Analyte amount determination	Data transformation (standard curve)	2/3	OD590	Data	Value	IFN-gamma amount	Data	Amount quality
				Standard curve	Data	Comparator			
	IFN-gamma concentration calculation	Data transformation (coacentration calculation)	3/3	IFN-gamma amount	Data	Numerator	IFN-gamma concentration	Data	Concentration
				Volume number	Data	Denominator			
IFN-gamma statistical difference evaluation	IFN-gamma concentration average	Data transformation (mean calculation)	1/4	IFN-gamma concentration values 1–n	Data	Numerator	Average IFN-gamma concentration	Data	Mean
				Number count of values	Data	Denominator			
	IFN-gamma concentration variability	Data transformation (standard deviation calculation)	2/4	IFN-gamma concentration values 1–n	Data		IFN-gamma concentration variability	Data	Standard deviation
	T statistic calculation	Data transformation (t statistic)	3/4	Mean	Data		t statistic value	Data	t statistic
				Standard deviation	Data				
	IFN-gamma concentration statistical difference	Data transformation (P value determination)	4/4	t statistic	Data	Value	Significance of difference	Data	P value
				t statistic table	Data	Comparator			

*N.B. All biomaterial objects have associated mass amounts, volume amounts and purity values that should be supported in the database design.
**All assays should have evaluants. The output data from an assay is a quality of the evaluant, which frequency corresponds to the amount of the analyte being measured in the assay.

FIGURE 21.4: OBX-based representation of a virus infection experiment. An OBX-based representation of an experiment from a recent publication (Conenello et al., 2007) in which mice are infected with influenza virus and the amount of interferon gamma in the lung is assessed as a measure of the host immune response to viral infection. Individual subprocesses are defined based on the specific inputs and outputs of the subprocesses together with the roles that each component plays. The three major types of subprocesses—biomaterial transformation, assay, and data transformation—are described. The ordered set of subprocesses forms a specific protocol application defined by its objective (e.g., cytokine quantification).

Several features of this approach to the representation of this experiment are worth noting:

- First, this approach emphasizes the role that the experiment processes play in linking entities together. For example, if a different type of assay were used to measure the interferon gamma analytes or a different standard curve were used for the interpolation step, different concentrations of interferon gamma would be obtained. Thus, in order understand the result it is critical to know the processes that were used to derive it.

- Second, the approach shows how specimen qualities are determined—through combinations of assays and data transformations.

- Third, the relationships between biomaterial objects are captured and can be used to transfer quality information up the biomaterial chain. Thus, the concentration of interferon gamma is both a quality of the lung specimen as well as the mouse source.

- Fourth, the model is focused on capturing the key entities necessary to identify relevant data sets based on the structured meta-data and on common approaches for reanalysis, namely to search for patterns in the experiment results (assay output qualities) based on differences in the input assay variables.

- Fifth, one of the big advantages of taking this approach for database representation is that it naturally interoperates with the ontology terms from the OBO Foundry, which can be used to describe the types of specimens (FMA), chemical therapeutics (ChEBI), analytes (PRO), assay types (OBI), protocol application roles (OBI), etc.

- Finally, in some cases a database resource may not be interested in capturing all of the details for each subprocess in a protocol application in a structured way. For example, the database resource may only want to parse into the table structure selected process inputs (the virus, the mouse strain, the lung specimen) and selected outputs (interferon gamma concentration and the t-test p-value results). In this case, the other entities necessary to describe the derivation of the outputs from the inputs described in the database record must be described in a text document, similar to the experiment description provided in the methods section of the paper.

21.5   Ontology-based Data Mining

The ontology-based approach to knowledge representations offers many significant opportunities for new approaches to data mining that go beyond the simple search for patterns in the primary data by integrating information incorporated in the structure of the ontology representation. We term these approaches "meta-mining" because they represent the mining and analysis of integrated knowledge sets derived from multiple, often disparate, sources. Meta-mining approaches can be used for a wide range of different data mining activities, including indexing and retrieval of data and information, mapping among ontologies, data integration, data exchange, semantic interoperability, data selection and aggregation, decision support, natural language processing applications, and knowledge discovery (Rubin et al., 2008; Bodenreider, 2008). For example, Cook et al. (2007) have described the use of ontological representations to infer causal chains and feedback loops within the network of entities and reactions in biological pathway representation. O'Connor et al. (2009) have described the use of ontological representation of clinical trials information to support temporal reasoning in electronic clinical trials management systems. Here we focus on two specific examples. In the first example of meta-mining, the role of ontologies in the description of the experiment meta-data for the identification and use of related data sets will be discussed. In the second example of meta-mining, the use of GO-based biological process gene annotation for the interpretation of gene expression microarray results (Lee et al., 2006) and protein interaction network structure (Luo et al., 2007) will be discussed.

21.5.1　Metadata mining

The goal of establishing experiment data archives like ArrayExpress and ImmPort is to allow the reuse of data derived from previous experimentation in the interpretation of new experiment results. Indeed, scientists currently do this in an informal, subjective way in the discussion sections of their journal articles where they interpret their experiment result in the context of the current state of scientific knowledge in the relevant biological discipline. One of the goals of meta-mining is to approach this integrative interpretation in an objective, computational manner. For example, let's imagine that you have recently completed a gene expression microarray experiment in which you have determined gene expression levels in a series of samples from pancreatic specimens of rats immunized with insulin to induce type 1 diabetes through an autoimmune mechanism, in the presence and absence of treatment with the immunosuppressive drug cyclosporin. While gene expression microarrays have revolutionized the way we do gene expression analysis, by allowing the simultaneous assessment of all genes in the organisms genome, they are still hampered by the presence of natural biological and experimental variability, resulting in relatively high false positive and false negative rates. One approach for addressing these inaccuracies is to compare your data with related data sets under the assumption that any discoveries made with independent, related data sets are likely to be real and relevant. So how does one determine which data sets are 'related' in a comprehensive, objective way. This is where ontology-based representation of experiment meta-data can play a valuable role.

A simple approach would be to look for data sets derived from identical experiment designs, but the chances that there are sufficient numbers of these data sets in the public domain tends to be relatively small. And so, it would be beneficial to extend what we would consider to be "related" data sets to include those that are "similar" in experiment design. In this case we have used microarrays as the assay methodology for quantifying mRNA transcript levels. However, other types of assay methodologies used for assessing transcript levels, like RT-PCR, SAGE, and even northern blotting, would provide similar data that could be useful for meta-mining. Using an ontology like OBI to describe assay types would allow for this kind of definition of similarity assay. The pancreas specimen used for microarray assessment is a type of endocrine organ. We might be interested in incorporating data derived from experiments using other types of endocrine organs, e.g., thyroid gland, adrenal gland, ovaries, etc. Using organ terms derived from the Foundational Model of Anatomy for the annotation of the specimen derived from the biomaterial transformation step would allow this kind of inference about "similarity' to be made. The rat is used as an experiment animal model because it has a similar anatomy and physiology as humans, as do other mammalian species. Related species could be identified using a species taxonomy, like the NCBI taxonomy, as the basis for organism annotation. Type 1 diabetes is an autoimmune disease of endocrine glands, as are Graves' disease, Hashimoto's thyroiditis, Addison's disease; experiments that investigate these types of diseases are also likely to be helpful in the interpretation of your results. Indeed, we might also want to include any autoimmune disease (e.g., lupus, multiple sclerosis, etc.) for this purpose. Use of an ontology like the Disease Ontology would facilitate the identification of experiments based on these kinds of relationships. Finally, the use of an ontology like ChEBI for the identification of other immunosuppressive compounds could further extend the meta-mining analysis.

Thus, the use of ontology-based knowledge representation to define the qualities (disease state, immunosuppressive) of the input entities (rat, cyclosporin) playing specific roles (therapeutic) in the material transformation that gives rise to the specimen output (pancreas), which serves as the input to the assay (gene expression microarrays) that results in the measurement of mRNA transcript levels as output allows one to extend the analysis for associations between the dependent and independent variables in a range of related experiments through this ontology-driven meta-mining approach.

21.5.2   Knowledge integration

The second example of meta-mining that takes advantage of the knowledge incorporated into the semantic structure of the ontology comes from the well-established approach of using the GO (Ashburner et al., 2000; Harris et al., 2004, Diehl et al., 2007) annotation of gene products to analyze experiment data sets. The classic example of this comes from the use of the GO in the interpretation of gene groups derived from gene expression microarray data clustering. A common goal of microarray data analysis is to group genes together based on similarity in their expression pattern across different experiment conditions, under the assumption that genes whose expression correlates with the experiment condition being investigated are likely to be involved in a relevant underlying biological process. For examples, genes whose expression pattern correlates with the cancer phenotype of the specimens might be expected to be involved in cell proliferation control. The GO consortium has performed two activities of relevance for this use case. The GO developed by the consortium includes three comprehensive term hierarchies for biological processes, molecular functions and cellular components. The consortium has then curated the scientific literature and annotated gene products with GO terms from this ontology to generate a knowledgebase of biology.

Several groups have developed approaches for utilizing GO annotation as a means for identifying relevant biological processes associated with gene expression clusters derived from microarray data by assessing whether specific GO annotations are over-represented in the gene cluster (e.g., Lee et al., 2006 and http://geneontology.org/GO.tools.shtml). The CLASSIFI algorithm (http://patheurie1.swmed.edu/pathdb/classifi.html) not only assesses the co-clustering of the primary GO annotations for genes in a cluster, but also captures the parent terms from the GO hierarchy for this assessment (Lee et al., 2006).

The analysis of data sets from B cells stimulated with a panel of ligands illustrates how the semantic structure of the GO hierarchy allowed the discovery of important biological processes that would not have been readily apparent from the use of a flat vocabulary for gene function annotation (Lee et al., 2006). B lymphocytes were isolated from mouse spleen and treated with a variety of different ligands to simulate natural environmental stimuli. RNA isolated from these specimens was evaluated by gene expression microarray to measure the expression levels of a large cross section of genes in the mouse genome. The ~2500 genes that were differentially expressed were grouped together into 19 gene clusters based on their expression pattern in response to three important B cell stimuli—anti-CD40, lipopolysaccharide (LPS) and anti-Ig. The CLASSIFI algorithm was then used to determine if any categories of genes were overrepresented in any of the gene clusters based on an analysis of their GO annotations. For example, gene cluster #18 includes genes that were upregulated in response to anti-Ig but not anti-CD40 or LPS, and contains 191 of the 2490 genes in the entire data set, seven of the ten genes annotated with the GO term "monovalent inorganic cation transport," and 24 of the 122 genes annotated with the GO term "transporter activity." In the latter case, the probability that this degree of co-clustering would have occurred by chance is $\sim 9 \times 10^{-6}$. The result of this data mining exercise was the hypothesis that stimulation of B lymphocytes through their antigen receptor using anti-Ig results in the activation of a set of specific transporter processes involving receptor endocytosis and intracellular vesicle trafficking to facilitate antigen processing and presentation. Subsequent experimental studies confirmed this hypothesis.

The important point is that these data mining results would not have been possible without the semantic structure of the GO, which was used to infer relationships between the genes in the gene clusters based on prior knowledge of the interrelationships of biological processes. Figure 21.5 shows the hierarchical structure of a small portion of the GO biological process branch focused on transporter processes. The genes found in cluster #18 that gave rise to the cluster classification of "transporter activity" are listed next to the specific GO term with which they are annotated. At most only three genes were annotated with a given GO term. However, many genes are annotated with terms that are closely related within this small region of the GO biological process hierarchy. By incorporating the GO hierarchy in the analysis, CLASSIFI allowed the discovery of these relationships, which would not have been possible with a flat vocabulary.

21.6 Concluding Remarks

In order to take maximum advantage of the primary data and interpretive knowledge derived from the research enterprise it has become increasingly important to agree on standard approaches to represent this information in a consistent and useful format. Many international consortia have been working toward the establishment of standards related to what kind of information should be captured, how the information should be described and how the information should be captured in database resources. The combined effect is that experimental data from biomedical investigation is becoming increasingly accessible to reuse and reanalysis and thus is playing increasingly important roles in the discovery of new knowledge of the workings of biological systems through improved approaches to data mining.

Acknowledgment

This work was supported by the U.S. National Institutes of Health-N01AI40041, N01AI40076, U54RR023468, and U54HG004028.

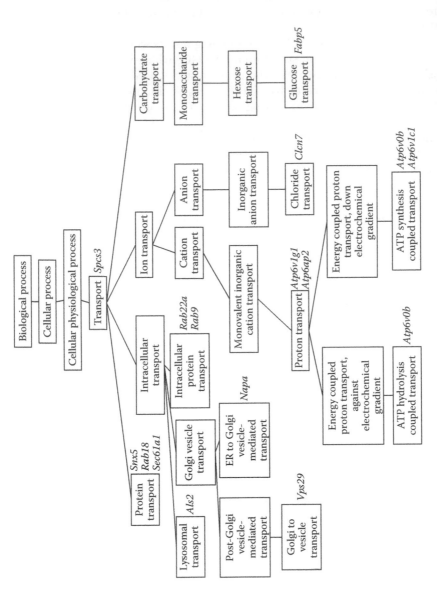

FIGURE 21.5: Gene Ontology hierarchy in the mining of gene expression microarray data. A piece of the GO biological process hierarchy that includes the cellular transport terms is displayed with GO terms listed in the boxes. Genes that have been found in the gene cluster #18 (Lee et al., 2006) and are annotated with the specific GO term are listed in *italic* to the right of the GO term box. See text for details.

References

Aranguren ME, Antezana E, Kuiper M, Stevens R. Ontology Design Patterns for bio-ontologies: a case study on the Cell Cycle Ontology. *BMC Bioinform.* 2008, 9(Suppl 5):S1.

Ashburner M, Ball CA, Blake JA, Botstein D, Butler H, Cherry JM, Davis AP et al. Gene ontology: tool for the unification of biology. The Gene Ontology Consortium. *Nat Genet.* 2000, 25(1):25–9.

Barrett T, Troup DB, Wilhite SE, Ledoux P, Rudnev D, Evangelista C, Kim IF, Soboleva A, Tomashevsky M, Edgar R. NCBI GEO: mining tens of millions of expression profiles—database and tools update. *Nucleic Acids Res.* 2007, 35(Database issue):D760–65.

Bodenreider O. Biomedical ontologies in action: role in knowledge management, data integration and decision support. *Yearb Med Inform.* 2008, 67–79.

Bornberg-Bauer E, Paton NW. Conceptual data modeling for bioinformatics. *Brief Bioinform.* 2002, 3:166–80.

Brazma A, Hingamp P, Quackenbush J, Sherlock G, Spellman P, Stoeckert C, Aach J et al. Minimum information about a microarray experiment (MIAME)-toward standards for microarray data. *Nat Genet.* 2001, 29(4):365–71.

Conenello GM, Zamarin D, Perrone LA, Tumpey T, Palese P. A single mutation in the PB1-F2 of H5N1 (HK/97) and 1918 influenza A viruses contributes to increased virulence. *PLoS Pathog.* 2007, 3(10):1414–21.

Cook DL, Wiley JC, Gennari JH. Chalkboard: ontology-based pathway modeling and qualitative inference of disease mechanisms. *Pac Symp Biocomput.* 2007, 16–27.

Demeter J, Beauheim C, Gollub J, Hernandez-Boussard T, Jin H, Maier D, Matese JC et al. The Stanford Microarray Database: implementation of new analysis tools and open source release of software. *Nucleic Acids Res.* 2007, 35(Database issue):D766–70.

Diehl AD, Lee JA, Scheuermann RH, Blake JA. Ontology development for biological systems: immunology. *Bioinformatics.* 2007, 23(7):913–15.

Fridsma DB, Evans J, Hastak S, Mead CN. The BRIDG project: a technical report. *J Am Med Inform Assoc.* 2008, 15(2):130–37.

Greene JM, Collins F, Lefkowitz EJ, Roos D, Scheuermann RH, Sobral B, Stevens R, White O, Di Francesco V. National Institute of Allergy and Infectious Diseases bioinformatics resource centers: new assets for pathogen informatics. *Infect Immun.* 2007, 75(7):3212–19.

Harris MA, Clark J, Ireland A, Lomax J, Ashburner M, Foulger R, Eilbeck K et al. Gene Ontology Consortium. The Gene Ontology (GO) database and informatics resource. *Nucleic Acids Res.* 2004, 32(Database issue): D258–61.

Hucka M, Bolouri H, Finney A, Sauro H, Doyle JHK, Arkin A, Bornstein B et al. The systems biology markup language (SBML): a medium for representation and exchange of biochemical network models. *Bioinformatics.* 2003, 19:524–31.

Jones AR, Miller M, Aebersold R, Apweiler R, Ball CA, Brazma A, Degreef J et al. The functional genomics experiment model (FuGE): an extensible framework for standards in functional genomics. *Nat Biotechnol.* 2007, 25(10):1127–33.

Laibe C, Le Novere N. MIRIAM Resources: tools to generate and resolve robust cross-references in systems biology. *BMC Systems Biol.* 2007, 1:58.

Lee JA, Sinkovits RS, Mock D, Rab EL, Cai J, Yang P, Saunders B, Hsueh RC, Choi S, Subramaniam S, Scheuermann RH. Alliance for cellular signaling. Components of the antigen processing and presentation pathway revealed by gene expression microarray analysis following B cell antigen receptor (BCR) stimulation. *BMC Bioinform.* 2006, 7:237.

Lee JA, Spidlen J, Atwater S, Boyce K, Cai J, Crosbie N, Dalphin M et al. MIFlowCyt: the minimum information about a flow cytometry experiment. *Cytometry: Part A.* 2008, 73(10):926–30.

Le Novère N, Finney A, Hucka M, Bhalla US, Campagne F, Collado-Vides J, Crampin EJ et al. Minimum information requested in the annotation of biochemical models (MIRIAM). *Nat Biotechnol.* 2005, 23(12):1509–15.

Lloyd C, Halstead M, Nielsen P. CellML: its future, present and past. *Prog Biophy Mol Biol.* 2004, 85:433–50.

Luciano J. PAX of mind for pathway researchers. *Drug Discovery Today.* 2005, 10(13):937–42.

Luo F, Yang Y, Chen CF, Chang R, Zhou J, Scheuermann RH. Modular organization of protein interaction networks. *Bioinformatics.* 2007, 23(2): 207–14.

O'Connor MJ, Shankar RD, Parrish DB, Das AK. Knowledge-data integration for temporal reasoning in a clinical trial system. *Int J Med Inform.* 2009, 78:577–85.

Parkinson H, Kapushesky M, Shojatalab M, Abeygunawardena N, Coulson R, Farne A, Holloway E et al. ArrayExpress—a public database of microarray experiments and gene expression profiles. *Nucleic Acids Res.* 2007, 35(Database issue):D747–50.

Qian Y, Tchuvatkina D., Spidlen J, Wilkinson P, Garparetto M, Jones AR, Manion FJ, Scheurmann RH, Sekaly RP, and Brinkmann RR. FuGEFlow: data model and markup language for flow cytometry. *BMC Bioinform.* 2009, submitted.

Ramakrishnan R, Gehrke J. *Database Management Systems*, 3rd Edition. McGraw-Hill Co., New York, 2003.

Rubin DL, Shah NH, Noy NF. Biomedical ontologies: a functional perspective. *Brief Bioinform.* 2008, 9(1):75–90.

Smith B, Ceusters W, Klagges B, Köhler J, Kumar A, Lomax J, Mungall C, Neuhaus F, Rector AL, Rosse C. Relations in biomedical ontologies. *Genome Biol.* 2005, 6(5):R46.

Smith B, Ashburner M, Rosse C, Bard J, Bug W, Ceusters W, Goldberg LJ et al. The OBO Foundry: coordinated evolution of ontologies to support biomedical data integration. *Nat Biotechnol.* 2007, 25(11):1251–55.

Squires B, Macken C, Garcia-Sastre A, Godbole S, Noronha J, Hunt V, Chang R, Larsen CN, Klem E, Biersack K, Scheuermann RH. BioHealthBase: informatics support in the elucidation of influenza virus host pathogen interactions and virulence. *Nucleic Acids Res.* 2008, 36(Database issue): D497–503.

Taylor CF, Field D, Sansone SA, Aerts J, Apweiler R, Ashburner M, Ball CA et al. Promoting coherent minimum reporting guidelines for biological and biomedical investigations: the MIBBI project. *Nat Biotechnol.* 2008, 26(8):889–96.

Chapter 22

Redescription Mining and Applications in Bioinformatics

Naren Ramakrishnan

Virginia Tech

Mohammed J. Zaki

Rensselaer Polytechnic Institute

22.1 Introduction

Our ability to interrogate the cell and computationally assimilate its answers is improving at a dramatic pace. The transformation of biology into a data-driven science is hence continuing unabated, as we become engulfed in ever-larger quantities of information about genes, proteins, pathways, and even entire processes. For instance, the study of even a focused aspect of cellular activity, such as gene action, now benefits from multiple high-throughput data acquisition technologies such as microarrays [4], genome-wide deletion screens [7], and RNAi assays [14–16]. Consequently, analysis and mining techniques, especially those that provide data reduction down to manageable quantities, have become a mainstay of computational biology and bioinformatics. From simple clustering of gene expression profiles [10], researchers have begun uncovering networks of concerted (regulatory) activity [20,26], reconstructing the dynamics of cellular processes [9,23], and even generating system-wide perspectives on complex diseases such as cancer [25].

The successes at being able to rapidly curate, analyze, and mine biological data obscure a serious problem, namely an overload of vocabularies now available for describing biological entities. For our purposes, a vocabulary is any way to carve up a domain of interest and posit distinctions and equivalences. While one biologist might study stress-responsive genes from the perspective of their transcriptional levels, another might assess downstream effects such as the proteins the genes encode, whereas still others might investigate the phenotypes of deletion mutants. All of these vocabularies offer alternative and mostly complementary (sometimes, contradictory) ways to organize information and each provides a different perspective into the problem being studied. To further knowledge discovery, biologists need tools to help uniformly reason across vocabularies, integrate multiple forms of characterizing datasets, and situate knowledge gained from one study in terms of others.

The need to bridge diverse biological vocabularies is more than a problem of data reconciliation, it is paramount to providing high-level problem solving functions for the biologist. As a motivating context, consider a biologist desiring to identify a set of *C. elegans* genes to knock-down (via RNAi) in order to confer improved desiccation tolerance in the nematode. Assume the biologist would like to decompose this problem via two lines of reasoning. First, proceeding along a stress response argument, the investigator would like to identify genes that serve as master controls (transcription factors) whose knock-down will cause significant change in downstream gene expression of other genes, leading to a modulation in the desiccation response (positive or negative), culminating in a disruption or delay of any shutdown processes. Second, following a phenotypical argument, efforts would be directed at identifying key physiological indicators of tolerance and adaptation, restate these indicators in terms of pathways that must be activated (or inactivated), and identify genes central to these objectives. To support such lines of reasoning, and integrate their answers, the biologist needs to be able to relate diverse data domains using a uniform analytical methodology. Redescription mining is such an approach. Redescriptions empower the biologist to define his own vocabularies, relate descriptors across them uniformly, and relationally compose sequences of redescriptions to realize complex functions. We will show how redescriptions are not specific to any data acquisition technology, domain of interest, or problem solving scenario. Instead, they can be configured to support a range of analytical functions that straddle vocabularies.

22.2 Reasoning About Sets Using Redescriptions

As the term indicates, to redescribe something is to describe anew or to express the same concept in a different vocabulary. The input to

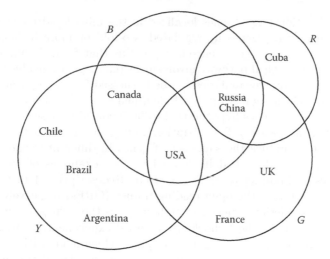

FIGURE 22.1: Example input to redescription mining. The expression $B-Y$ can be redescribed into $R \cap G$.

redescription mining is a set of objects and a collection of subsets defined over this set. It is easiest to first illustrate redescription mining using an everyday, non-biological, example; consider, therefore, the set of ten countries shown in Figure 22.1 and its four subsets, each of which denotes a meaningful grouping of countries according to some intensional definition. For instance, the circles (from right, counterclockwise) refer to the sets "permanent members of the UN security council," "countries with a history of communism," "countries with land area >3,000,000 square miles," and "popular tourist destinations in the Americas (North and South)." We will refer to such sets as *descriptors*. A redescription is a shift-of-vocabulary and the goal of redescription mining is to identify subsets that can be defined in at least two ways using the given descriptors. An example redescription for this dataset is then: "Countries with land area >3,000,000 square miles outside of the Americas" are the same as "Permanent members of the UN security council who have a history of communism." This redescription defines the set {Russia, China}, once by a set intersection of political indicators $(R \cap G)$, and again by a set difference involving geographical descriptors $(B - Y)$. Notice that neither the set of objects to be redescribed nor the ways in which descriptor expressions should be constructed is input to the algorithm. The underlying premise of redescription analysis is that sets that can indeed be defined in (at least) two ways are likely to exhibit concerted behavior and are, hence, interesting.

What makes redescription mining pertinent to biology is that the domain of discourse, i.e., the descriptors, is defined by the biologist. For instance, descriptors of genes in a given organism can be organized into vocabularies such as cellular location (e.g., "genes localized in the mitochondrion"), transcriptional activity (e.g., "genes up-regulated two-fold or more in heat stress"), protein function (e.g., "genes encoding proteins that form the immunoglobin complex"), or biological pathway involvement (e.g., "genes involved in glucose biosynthesis"). More vocabularies can be harnessed from computational studies (e.g., "genes forming module #124 identified by the Segal et al. algorithm") or literature (e.g., "genes hypothesized to be involved in desiccation response in the Potts et al. paper"). Redescription mining then constructs set-theoretic expressions that induce the same set of genes in different vocabularies. See Figure 22.4, to be described in detail later, for examples of redescriptions from studies on budding yeast *S. cerevisiae*. Redescription 1 in Figure 22.4, for instance, restates "the open reading frames (ORFs) negatively expressed in the histone depletion experiment (6 hours)" as those "negatively expressed two-fold or more in the heat shock (10 minutes) experiment." Notice that this is an approximate, albeit strong, redescription which holds with Jaccard's coefficient (the ratio of the size of the overlap to the size of the union) 0.78. These ORFs comprise functions related to metabolism, catalytic activity, and their action is localized in the cytoplasm. The Pearson coefficients for these ORFs in the histone depletion experiments match very strongly, showcasing the use of redescription in identifying a concerted set of ORFs. Similarly, it is easy to conceptualize redescription scenarios where the descriptors are defined over proteins, processes, or other domains.

In fact, the importance of "descriptors" to encode domain specific sets in biology and using them as a starting point for biological data mining has been recognized by other researchers. Segal et al. [25] focus on predefined sets of genes and this work defines descriptors based on the results of clustering, on expression in specific cell types, and membership in certain functional categories or pathways. The molecular signatures database (MSigDB) [29] supporting the gene set enrichment analysis (GSEA) algorithm is another resource that defines gene sets based on pathways, annotations, and similar information. There are many more such methods but essentially all of them are interested in casting interpretations over predefined, biologically meaningful, sets. Redescription mining views these databases as the primary resource for mining and reveals interdependencies within them.

22.3   Theory and Algorithms

Formally, the inputs to redescription mining are the universal set of objects (e.g., genes) $G = \{g_1, g_2, \ldots, g_n\}$, and a set $D = \{d_1, d_2, \ldots, d_m\}$ of proper subsets (the descriptors) of G. This information can be summarized in a $n \times m$ binary *dataset matrix* whose rows represent genes, columns represents the descriptors, and the (i, j) entry is 1 if object g_i is a member of descriptor d_j, and 0 otherwise. Typically, descriptors are organized into vocabularies, each of which provides a covering of G. An expression bias (more on this below) dictates allowable set-theoretic constructs involving the descriptors. This setting is similar to one studied by Pu and Mendelzon [21] but there the goal is to find one most concise description of a given set of objects using the vocabulary. The goal in redescription mining is to find equivalence relationships of the form $E \Leftrightarrow F$ that hold at or above a given Jaccard's coefficient θ (i.e., $|E \cap F|/|E \cup F| \geq \theta$), where E and F are expressions in the specified bias comprising the descriptors D. The key property of a redescription, like most data mining patterns, is that it must be falsifiable in *some* interpretation (dataset). Notice that this rules out tautologies, such as $d_i - (d_i - d_j) \Leftrightarrow d_i \cap d_j$, which are true in *all* datasets.

Redescription mining exhibits traits of many other data mining problems such as conceptual clustering, constructive induction, and Boolean formula discovery. It is a form of conceptual clustering [11,17] because the mined clusters are required to have not just one meaningful description, but two. It is a form of constructive induction since the features important for learning must be automatically constructed from the given vocabulary of descriptors. Finally, since redescriptions are equivalence relationships, the problem of mining redescriptions can be viewed as (unsupervised) Boolean formula discovery [6].

22.3.1 Structure theory of redescriptions

In [19], a structure theory of redescriptions is presented that yields both impossibility and strong possibility results. For instance, if the dataset matrix is a truth table, i.e., the number of genes n is 2^m, where m is the number of descriptors, then there can be no redescriptions. This is because the number of subsets of genes (2^n) coincides with the number of possible Boolean formulas over m variables (2^{2^m}). Each Boolean formula is then in an equivalence class by itself and induces a different subset of objects from other formulas. On the other hand, if the dataset is less than a truth table (i.e., missing one or more rows), then the "islands" of Boolean formulas begin to merge, with each missing row reducing the number of equivalence classes by a factor of two. In such a case, *all* Boolean formulas have redescriptions! This dichotomy law is rather disconcerting but holds only under the assumption that the expression bias is general enough to induce all subsets of genes. If we algorithmically restrict the bias, e.g., to conjunctions, or length-limited constructs, then it is not obvious which subsets afford redescriptions, leading to a non-trivial data mining problem. As a result, all algorithms for mining redescriptions focus on a specific bias and mine expressions within that bias.

22.3.2 Algorithms for mining redescriptions

The CARTwheels algorithm [24] mines redescriptions between length-limited disjunctions of conjunctions. The CHARM-L algorithm [33] mines redescriptions between conjunctions (with no restrictions on length). A recently developed extension to CHARM-L (BLOSOM [34]) provides a way to systematically mine complex Boolean expressions in various biases, e.g., conjunctions, disjunctions, CNF, or DNF forms. We highlight the main features of all these algorithms in this section.

22.3.2.1   CARTwheels

CARTwheels mines redescriptions by exploiting two important properties of binary decision trees [22]. First, if the nodes in such a tree correspond to Boolean membership variables of the given descriptors, then we can interpret paths to represent set intersections, differences, or complements; unions of paths would correspond to disjunctions. Second, a partition of paths in the tree corresponds to a partition of objects. These two properties are employed in CARTwheels which grows two trees in opposite directions so that they are joined at the leaves. Essentially, one tree exposes a partition of objects via its choice of subsets and the other tree tries to grow to match this partition using a different choice of subsets. If partition correspondence is established, then paths that join can be read off as redescriptions. CARTwheels explores the space of possible tree matchings via an alternation process (see Figure 22.2) whereby trees are repeatedly regrown to match the partitions exposed by the other tree. Notice the implicit restriction of bias to disjunctions of one to three clauses, each involving one to two descriptors (in negated or non-negated form). By suitably configuring this alternation, we can guarantee, with non-zero probability, that any redescription existing in the dataset would be found. Exploration policies must balance the potential of identifying unseen regions of descriptor space against redundancy from re-finding already mined redescriptions.

22.3.2.2   CHARM-L

CHARM-L, employing the conjunctions bias, adopts a different approach and exploits connections between Boolean formulas and closed sets, a concept popular in the association mining community [1,2] A closed set is a set of genes together with a set of descriptors such that the conjunction of the given descriptors induces the given set of genes, and no subset of the given descriptors induces the same set of genes. In other words, the gene set and descriptor set are maximal w.r.t. each other and we cannot reduce either of these sets without losing any elements of the other. The closed sets form a lossless representation of the underlying dataset matrix in that the only redescribable sets are the closed sets. Additionally, the redescriptions of a closed set are precisely the non-maximal versions of the given set. CHARM-L further focuses on only the minimal variants, called *minimal generators*. The problem of redescription mining then reduces to mining closed sets and relating their minimal generators to form redescriptions (see Figure 22.3). However, datasets for redescription analysis, when studied in the association mining framework, are

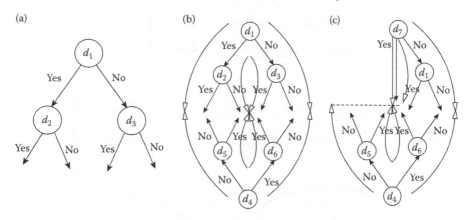

FIGURE 22.2: Mining redescriptions using the CARTwheels algorithm. The alternation begins with a tree (first frame) defining set-theoretic expressions to be matched. The bottom tree is then grown to match the top tree (second frame), which is then fixed, and the top tree is regrown (third frame). Arrow indicate the matching paths. Redescriptions corresponding to matching paths at every stage are read off and subjected to evaluation by Jaccard's coefficient. For instance, in the second frame, the matching paths give rise to three redescriptions: $d_1 \cap d_2 \Leftrightarrow G - d_4 - d_5$ from paths on the left, $G - d_1 - d_3 \Leftrightarrow d_4 - d_6$ from paths on the right, and $(d_1 - d_2) \cup (d_3 - d_1) \Leftrightarrow (d_4 \cap d_6) \cup (d_5 - d_4)$ from paths in the middle.

very dense. Since a gene participates in either a descriptor or its negation, the underlying dataset matrix is exactly 50% dense (or sparse). We hence, cannot rely merely on support pruning as a way to curtail the complexity of data mining. CHARM-L's solution is a constraint-based approach so that the lattice of closed sets is selectively computed around genes (or descriptors) of interest.

22.3.2.3 BLOSOM

A generalization of CHARM-L is now being developed in the BLOSOM data mining framework [34]. BLOSOM is a framework for mining closed Boolean expressions of all forms, and defines closure operators for specific families of expressions such as conjunctions, disjunctions, CNF, and DNF forms. It focuses on mining the minimal Boolean expressions that characterize a set of objects (e.g., genes). The main data mining engine is based on a new approach to mine disjunctions (OR-clauses) instead of conjunctions. A number of effective pruning techniques are utilized to effectively search for all the possible frequent Boolean expressions in normal form.

Besides their choice of biases, CARTwheels and CHARM-L/BLOSOM approach redescription mining from alternative viewpoints, namely exploratory search versus enumeration and pruning. In this chapter, we show the application of all these algorithms for biological studies.

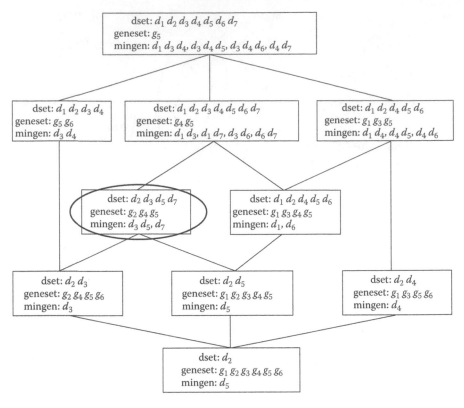

FIGURE 22.3: Mining redescriptions using the CHARM-L algorithm. Each node in the lattice denotes a closed set, comprising genes, descriptors (abbreviated as dsets), and their minimal generators. The only redescribable sets are the closed sets; redescriptions for these are obtained by relating their minimal generators. For instance, $d_3 \cap d_5 \Leftrightarrow d_7$ is a redescription because both $d_3 \cap d_5$ and d_7 are minimal generators of the closed set circled in the lattice.

22.4   Applications of Redescription Mining

22.4.1   Redescriptions for *Saccharomyces cerevisiae*

We have applied redescriptions to studying descriptors defined over the yeast genome [24,27], resulting in considerable biological insight. The biologist first narrows down on a reference set of a few hundred genes and then defines descriptors over this reference set. In [27] we defined the reference set to be the set of 210 "high-expressors"—genes exhibiting more than five-fold change in *some* time point (not necessarily all or even a majority) across the yeast desiccation and rehydration time course. The descriptors are drawn from a variety of sources. One vocabulary denotes expression levels in specific microarray measurements taken from Gasch et al. [12] and Wyrick et al. [32]. For instance, "genes negatively expressed two-fold or below in the 15 minute time point of the 1 M sorbitol experiment" is a descriptor in this vocabulary. A second type of vocabulary asserts membership of genes in targeted taxonomic categories of the gene ontology (biological processes (GO BIO), cellular components (GO CEL) or molecular functions (GO MOL)). A third class of descriptors is based on clustering time course datasets using a k-means clustering algorithm [28] and using the clusters as descriptors.

The redescriptions presented here, although approximate, have been whetted at a p-value significance level of .0001. Essentially, we characterize the distribution of set size overlaps for targeted descriptor cardinalities and reason about the possibility of obtaining the specified Jaccard's coefficient purely by chance.

Figure 22.4 depicts some statistically significant redescriptions obtained by CARTwheels, illustrating the diversity of set constructions possible. Of these, redescription R1 has been discussed earlier. R2 relates a k-means cluster to a set difference of two *related* GO cellular component categories. While the eight ORFs in R2 appear to be part of different response pathways, five of these eight ORFs are similarly regulated according to the work of Segal et al. [26]; these genes relate to the cellular hyperorganization and membrane dynamics in the regulation network.

R3 is a triangle of redescription relationships involving three different experimental comparisons, with 10 ORFs being implicated in all three expressions. From a biological standpoint, this is a very interesting result—the common genes indicate concerted participation across stress conditions; whereas the genes participating in, say, two of the descriptors, but not the third, suggest a careful diversification of functionality. six of the 10 ORFs are related to cell growth and maintenance. Five of the 10 ORFs have binding motifs related to the DNA binding protein REB1. The importance of phosphate and ribosomes appears to be salient in this redescription. It is important to note that the circularity of R3 is not directly mined by CARTwheels, but inferred post-hoc from a linear chain.

The theme in R4 is ribosome assembly/biogenesis and RNA processing. R4 is a linear chain comprising two redescriptions, and uses a GO descriptor as an intermediary between two expression-based descriptors. It is also interesting that this redescription involves a set of 45 ORFs!

R5 is an even longer chain involving 41 ORFs that are common to all descriptors. Notice the rather complicated set construct involving a disjunction of a conjunction and a difference, involving three different GO biological categories. Incidentally, this is the most complicated set expression representable in a two-level tree. Although R3, R4, and R5 are linear chains, CARTwheels is not a story telling algorithm since it cannot find such relationships between user-supplied descriptors. The examples shown here are snapshots from the continuous alternation of CARTwheels.

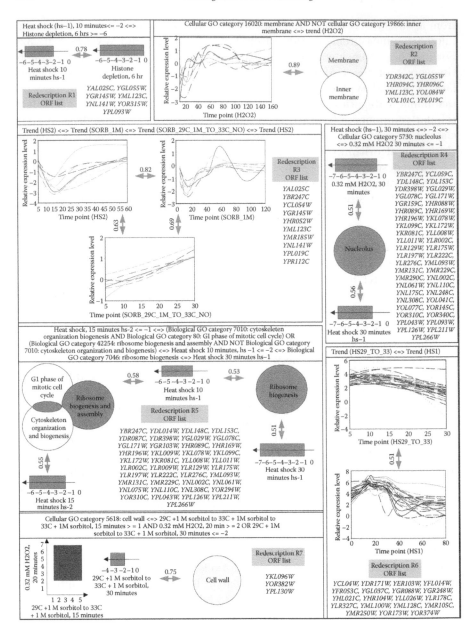

FIGURE 22.4: Seven (approximate) redescriptions mined from gene expression studies on *Saccharomyces cerevisiae*. Each box gives a readable statement of the redescription, presents it in graphical form, and identifies the ORFs conforming to the redescription. The Jaccard's coefficient is displayed over the redescription arrow. Notice that some redescriptions (e.g., R7) involve few ORFs, whereas others such as R5 involve larger numbers.

R6 is a relationship between two k-means clusters, between heat shock stresses. The ORFs participating in R6 demonstrate a clear focus on sugar or sugar phosphate metabolism.

R7 is a redescription relating a disjunction of descriptors to a GO cellular component category. It is also an interesting example of constructive induction, since a rectangular region is mined in a 2D space involving two different experimental comparisons.

The capabilities of the CHARM-L are best illustrated through an interactive scenario where, given constraints, the algorithm reasons about the conditions under which two given descriptors would be the equivalent. In one scenario described in [33], a biologist is exploring descriptors around his favorite gene—YOR374W, an ORF in *S. cerevisiae* that encodes an NAD-dependent aldehyde dehydrogenase (an enzyme—EC 1.2.1.3—that catalyzes the conversion of an aldehyde and NAD+ to a carboxylic acid and NADH), which has been determined to be very highly expressed in time point 20 minutes of the Gasch heat shock condition (more than five-fold). YOR374W (Ald4p) is important from the perspective of metabolism as it provide a means to generate reduced cofactor (NADH) for fueling electron transport and oxidative phosphorylation (ATP synthesis). The biologist is particularly interested in relating two descriptors that YOR374W participates in. One of them is descriptor d184 that denotes all ORFs that are expressed more than five-fold in the above time point; it contains 19 genes. Looking at the nearby time point (15 minutes) the biologist notices that the corresponding descriptor (d183) contains 21 genes, with 18 in common with d184. The Jaccard's coefficient between these descriptors is already high (0.857), but the biologist is curious to determine if there could be an exact redescription by using the GO vocabularies. CHARM-L uncovers the following redescription:

$$d183 - d388 - d460 - d515 \Leftrightarrow d184 - d309$$

In other words, to make d183 equivalent to d184, we need to subtract descriptors d388, d460, and d515 on the left (to remove three genes) and subtract descriptor d309 on the right (to remove one gene), bringing the commonality to 18, as desired. Here, d388 refers to the GO molecular function category: mannose transporter, d460 refers to the GO cellular component category: external protective structure, and d515 refers to the GO biological process category: fructose metabolism. d309, on the right side, incidentally happens to refer to genes whose molecular function, according to GO, is unknown. The implied message, from the above redescription, is that as we go from time point 15 minutes to time point 20 minutes, genes belonging to the above three categories drop out of the highly expressed ($\geq$five-fold) category.

22.4.2 Expanding redescriptions to uncover pathways

Figure 22.5 describes how we can uncover an entire pathway by integrating redescription analysis with domain theories. The source redescription in

Figure 22.5a states that the genes that are one-fold or more down-regulated during heat shock (30 min) can be restated as those that are between one- and five-fold down-regulated in the desiccation experiment. A conspicuous feature of this redescription (see Figure 22.5b) is the presence of three genes involved in sulfur metabolism: SAM1 (S-adenosylmethionine synthetase gene), encoding the protein that synthesizes the potential riboswitch ligand S-adenosylmethionine [8], SAH1 (S-adenosyl-L-homocysteine hydrolase gene), and YFR055W (cystathionine β-lyase gene). Riboswitch ligands such as SAM appear to serve as ancient master control molecules whose concentrations are monitored to ensure homeostasis of a much wider set of metabolic pathways [30,31], and indeed SAM has recently been implicated in G_1 cell cycle regulation [18].

We can procedurally uncover a pathway from the above redescription as follows. We discard genes encoding for ribosomal activity due to their consistent expression across the range of time courses in the dataset and study the remaining genes and their interacting partners. Sam1p was reported to interact with 13 other proteins [13], and the gene for one of these, URA7, is also present in the redescription. Using each of the genes (and their respective protein interactions culled from [5]), we systematically expand the given genes to form a network of interactions. We then use the primary microarray data to infer possible additional relationships (based on expression correlation). For example, MET30, encoding a cell cycle F-box protein and also involved in sulfur metabolism and protein ubiquination, can directly or indirectly be associated with TEF4 (translation elongation factor EF-1$_\gamma$ gene) and CLN2 (cyclin-dependent protein kinase regulator gene), both of which are present in the redescription. Note that MET30 itself was not present in the redescription. Finally, we improve the network further by incorporating genes from a pathway that show interactions with other genes in the redescription but not with one another (for example, the clustering of HIS4 and HIS1, ARO1 and ARO4, LYS14 and LYS12, and GAS1 and GAS3). The end result is depicted in Figure 22.5c. With the exception of MET30, SIP18, and CDC34, the transcription of each gene in the proposed network was either down-regulated or unchanged, suggesting a central role for sulfur metabolism in desiccation response. In view of the potential role of its protein in phospholipid binding SIP18 is shown close to other genes associated with lipid synthesis and binding (URA7, YBL085W), although it doesn't itself participate in the redescription. Studying this system further, Figure 22.5d illustrates the functions of genes involved in methyl group transfer and sulfur metabolism. Additional information can be obtained from comparison of Figures 22.5c and d; for example, note TPO2 (polyamine transport, Figure 22.5c) and role of SAM1 in polyamine synthesis (Figure 22.5d); in addition the connectivity of SER3 and ARO4 (serine biosynthesis, Figure 22.5c) and the role of serine in lipid biosynthesis (Figure 22.5d).

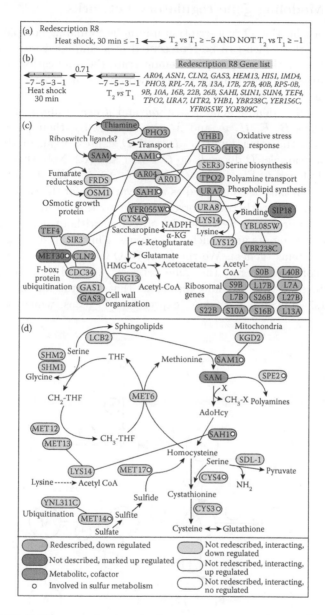

FIGURE 22.5: The use of redescriptions to uncover pathways. (a) Statement of redescription relating heat shock to desiccation experiment. (b) Graphical depiction of redescription and the genes identified. (c) Desiccation and heat shock lead to down-regulation of sets of genes with a central function in sulfur metabolism. (d) Functions of genes involved in methyl group transfer and sulfur metabolism.

22.4.3 Modeling gene regulatory networks

A final application of redescription mining is to finding complex gene regulatory networks, which can be represented in a simplified form, as Boolean networks [3]. For this purpose, we demonstrate the application of BLOSOM for redescription mining. Consider the network involving 16 genes, taken from [3], shown in Figure 22.6.

Here + and − denote gene *activation* and *deactivation*, respectively. For example, genes B, E, H, J, and M are expressed if their parents are not expressed. On the other hand G, L, and D express if all of their parents express. For example, D depends on $C, F, X1$ and $X2$. Note that F expresses if A does, but not L. Finally A, C, I, K, N, $X1$ and $X2$ do not depend on anyone, and can thus be considered as *input* variables for the Boolean network. We generated the truth table corresponding to the seven input genes but BLOSOM was provided the values for all genes, without explicit instruction about which are inputs and which are outputs. This yields a dataset with 128 rows and 16 items (genes). We then ran BLOSOM to discover the Boolean expression corresponding to this gene network; we used a minimum support of 100%, since we wish to find expressions that are true for the entire set of assignments. BLOSOM output 65 expressions in 0.36s, which hold true for

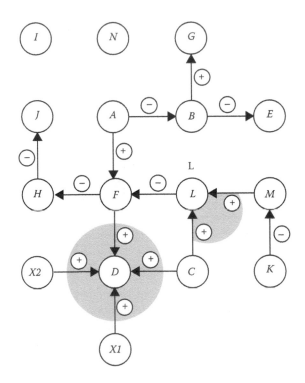

FIGURE 22.6: Gene network.

TABLE 22.1:   Boolean network expression.

$(D \mid (A\ B\ C\ E\ F\ G\ H\ J\ K\ L\ M\ X1\ X2))$ and
$(L \mid (C\ F\ H\ J\ K\ M))$ and
$((A\ B\ E\ G) \mid C \mid D \mid L \mid X1 \mid X2)$ and
$((A\ B\ E\ G) \mid (C\ L) \mid (F\ H\ J))$ and
$((F\ H\ J) \mid (A\ B\ C\ E\ G) \mid (A\ S\ E\ G\ K\ M))$

the entire dataset. After simplification these can be reduced to the equivalent expression, as shown in Table 22.1.

We verified that indeed this expression is true for all the rows in the dataset! It also allows us to reconstruct the Boolean gene network shown in Figure 22.6. For example, the first component of the expression in the first row $\overline{D} \mid (A\overline{B}CEF\overline{GH}JK\overline{LM}X1X2)$ can be converted into the implication $D \Rightarrow (A\overline{B}CEF\overline{GH}JK\overline{LM}X1X2)$, which means that D depends on the variables on the right hand side (RHS). If, at this point, we supply any partial knowledge about the input variables or of the maximum fan-out of the network, we could project the RHS only on those variables to obtain $(ACKX1X2)$, which happens to be precisely the relationship given in Figure 22.6. The second row tells us that L depends on the activation of C and inactivation of K, i.e., $\overline{K}$, if we restrict ourselves to the input variables. Note that C and $\overline{K}$ give the values for the remaining varibles. Note that other dependencies are also included in the mined expression. For example, we find that B and A always have opposite values, and so do B and E, and K and M. G and B always have the same values, and so on. Thus this example shows the power of BLOSOM in mining gene regulatory networks.

22.4.4   Cross-taxonomic and cross-genomic comparisons using redescriptions

Assume that we are provided with two families of functional annotations or ontologies, E and F, over the same space of objects (e.g., genes). The objective is to conduct an all-pairs redescription study relating categories or concepts between E and F. From the results of such a study, if $e_1 \in E$ is redescribed to $f_2 \in F$ with a very high Jaccard's coefficient, we could help impute annotations and properties typically associated with e_1 to f_2 (and vice versa). The results of such a study can then be used for funtional enrichment of unclassified genes, to analyze the structural consistencies (and inconsistencies) of different ontologies, and in general as an educational tool to communicate similarities and differences across taxonomies. Finally, when the ontologies apply to multiple organisms, we can study the extent to which redescriptions transfer across organisms and whether some organisms have more developed ontologies than others.

We conducted a cross-taxonomic GO comparison study using the GO BIO, GO CEL, and GO MOL assignments available for the *Arabidopsis thaliana*

TABLE 22.2: Summary of input GO categories for the six species considered.

	Arabidopsis	Fly	Human	Mouse	Worm	Yeast
Universal set size	13,572	8911	23,424	25,142	11,606	5731
Genes with BIO defined	6340	7424	18,068	18,193	9299	4711
Genes with CEL defined	3114	4131	16,002	17,362	5179	5713
Genes with MOL defined	12,817	7606	21,135	21,887	8975	5714
BIO categories involved	1043	2493	2837	2774	1361	1691
CEL categories involved	205	530	543	496	261	424
MOL categories involved	1212	2013	2516	2230	1062	1470

(arabidopsis), *Drosophila melanogaster* (fly), *Homo sapiens* (human), *Mus musculus* (mouse), *Caenorhabditis elegans* (worm), and *Saccharomyces cerevisiae* (yeast) genomes from the GO database website. The GO hierarchy information used for propogation of GO categories up the GO tree was also taken from the same website. For each organism, only those genes were considered that have at least one GO category, other than the categories for unknown GO BIO, GO MOL and, GO CEL, defined. The summary of the data used is provided in Table 22.2.

The experiment using the GO assignments for genes in each genome was performed as follows. The universal set of genes was defined as above. Within this universal set, the three families of GO categories were used individually as also in pairs to form input sets of descriptors. In all the runs, the descriptor family for which redescriptions were sought was used to build one-level trees. Two level trees were used for each pair of descriptor families used to construct derived descriptors. Thus, if a redescription was sought between the GO BIO categories on one side and combinations of GO CEL and GO MOL categories on the other, the study was done using a one-level tree for the GO BIO categories and upto a two-level tree for GO MOL and GO CEL categories. We also restricted all derived descriptors to involve just intersections and differences between descriptors. The support threshold was set at 3 to retain only the most significant redescriptions. The Jaccards coefficient threshold was set at 0.5.

Figure 22.7 shows a few examples of redescriptions mined using CARTwheels. Figure 22.7a shows a simple redescription between a GO BIO (51013) and GO CEL (8352) category. This redescription holds for the fly genome with Jaccard's coefficient of 1 and involves four genes. This type of a redescription can be easily used to relate functional enrichments across

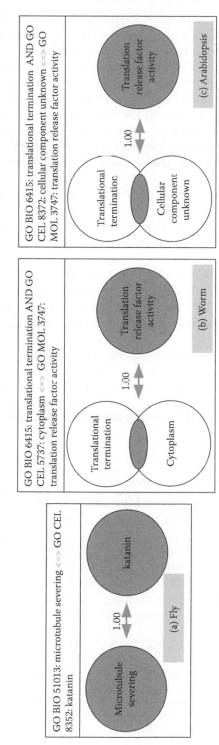

FIGURE 22.7: Examples of cross-taxonomic redescription: (a) A redescription between two functional categories for the fly genome. (b) A redescription involving an intersection between two categories for the worm genome. (c) A redescription involving the GO CEL category "celullar component unknown" for the arabidopsis genome. This redescription relates to the one in (b).

different taxonomies as described earlier. Figure 22.7b shows a redescription involving a derived descriptor formed by the intersection of a GO BIO and a GO CEL category which relates to a GO MOL category. This redescription holds for the worm genome with Jaccard's coefficient 1 and involves three genes. Figure 22.7c shows a redescription for the arabidopsis genome where the GO BIO and GO MOL involved are the same as in Figure 22.7b. The difference here is that GO CEL is assigned the GO category "cellular category unknown." This redescription also holds with a Jaccards coefficient of 1 and involve 11 genes. The pair of redescriptions found could potentially be used to better characterize the GO CEL categorization for the genes involved for the arabidopsis genome.

Table 22.3 summarizes the number of GO categories for which redescriptions are available and the number of redescriptions mined for each of the six species. In all cases, the use of derived descriptors (intersection and difference based) results in a significant increase in the number of categories involved in at least one redescription. Also, as is to be expected, a much higher number of redescriptions are found for genomes that have more categories involved with the genes (giving more descriptors to form derived descriptors with). Comparing Tables 22.2 and 22.3, we can conclude that a large proportion of the functional categories have redescriptions associated with them for all genomes.

Redescriptions found in the cross-taxonomic study described above can be used to validate and check the consistency of GO category assignments across different genomes. For this analysis, we conducted a pairwise comparison of redescriptions found for two different species and checked for overlap (same descriptors involved). Importantly, we did not require that the support or

TABLE 22.3: Summary of redescriptions obtained for the six species.

	Arabidopsis	Fly	Human	Mouse	Worm	Yeast
BIO categories	259	169	389	375	257	260
involved	(244)	(138)	(915)	(314)	(239)	(224)
CEL categories	50	146	102	94	70	149
involved	(40)	(92)	(71)	(70)	(58)	(103)
MOL categories	176	230	369	358	271	205
involved	(140)	(149)	(237)	(241)	(217)	(162)
BIO	6852	15,483	43,828	41,969	37,174	23,473
redescriptions	(469)	(324)	(589)	(622)	(713)	(513)
CEL	4971	12,567	22,046	15,531	11,280	43,293
redescriptions	(139)	(207)	(163)	(147)	(178)	(329)
MOL	9352	39,788	68,765	66,920	52,445	20,163
redescriptions	(408)	(363)	(582)	(581)	(711)	(388)

Note: Numbers in bracket indicate number of redescriptions with no derived descriptors.

TABLE 22.4:   Pairwise overlap between redescriptions obtained for the six species.

	Arabidopsis	Fly	Human	Mouse	Worm	Yeast
Arabidopsis	–	2744	4550	4383	4133	4824
		(48)	(129)	(120)	(124)	(59)
Fly	2744 (48)	–	13,306	11,198	8237	5561
			(123)	(94)	(96)	(117)
Human	4550 (129)	13,306	–	59,674	29,912	9871
		(123)		(475)	(291)	(116)
Mouse	4383 (120)	11,198	59,674	–	26,884	5278
		(94)	(475)		(282)	(91)
Worm	4133 (124)	8237	29,912	26,884	–	4555
		(96)	(291)	(282)		(86)
Yeast	4824 (59)	5561	9871	5278	4555	–
		(117)	(116)	(91)	(86)	

Note: The numbers in bracket indicate the number of distinct descriptors involved.

Jaccards coefficient be the same for the same redescription across a pair of species. The overlap observed is summarized in Table 22.4.

The species with large number of redescriptions (mouse and human) have high overlaps between them as also with other species. The fly and arabidopsis redescriptions show the minimum overlap. This is a result of the low number of descriptors available and redescriptions found for these species. As would be expected, arabidopsis and yeast which differ from the other four species most drastically show low amount of overlap.

Figure 22.8 shows three examples of redescriptions found to be common for various species for a Jaccards threshold of 1. Figure 22.8a shows a redescription common to the yeast and worm genome. This redescription involves an intersection between a GO MOL and a GO CEL category related to a GO BIO category. It involves 12 genes in the yeast genome and four genes in the worm genome. Figure 22.8b shows a redescription common to the worm and human genome. This redescription again involves an intersection with the GO CEL category "cellular category unknown" that is conserved across the two species. It involves four genes in the human genome and three genes in the worm genome. Figure 22.8c shows a redescription common to the human, mouse and worm genome. It involves the intersection between a GO MOL and GO CEL category related to a GO BIO category. This redescription involves 45 genes for human, 30 genes for mouse and 16 genes for the worm genome. Out of the redescription counts shown in Table 22.4, 920 redescriptions involving 15 categories were found to be constant across all six species. These 15 categories are listed in Table 22.5. All these categories lie quite high in the GO hierarchy and involve a lot of genes. Thus, there is no example of

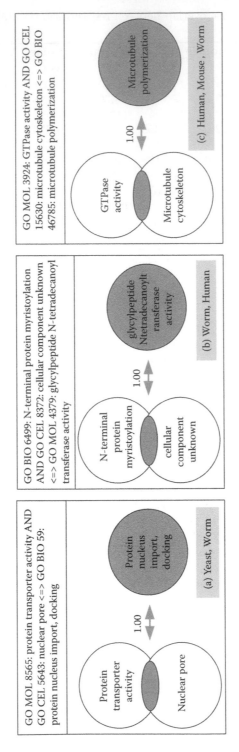

FIGURE 22.8: Examples of redescriptions that hold with Jaccard's coefficient for more that one species: (a) A redescription common to yeast and worm genomes. (b) A redescription common to the worm and human genomes. (c) A redescription common to the human, mouse, and worm genomes.

TABLE 22.5: GO categories for which redescriptions were found common to all six species.

GO category	Type	Description
GO:0003735	MOL	Structural constituent of ribosome
GO:0004672	MOL	Protein kinase activity
GO:0004674	MOL	Protein serine/threonine kinase activity
GO:0004812	MOL	tRNA ligase activity
GO:0005840	CEL	Ribosome
GO:0006413	BIO	Translational initiation
GO:0006418	BIO	tRNA aminoacylation for protein translation
GO:0006468	BIO	Protein amino acid phosphorylation
GO:0008452	MOL	RNA ligase activity
GO:0016310	BIO	Phosphorylation
GO:0016875	MOL	Ligase activity, forming carbon-oxygen bonds
GO:0016876	MOL	Ligase activity, forming aminoacyl-tRNA and related compounds
GO:0016886	MOL	Ligase activity, forming phosphoric ester bonds
GO:0043038	BIO	Amino acid activation
GO:0043039	BIO	tRNA aminoacylation

a redescription involving a very specific and precise functional category that could be found to be conserved across the six species.

22.5   Discussion

We hope to have shown here that redescription mining provides a domain-neutral way to cast complex data mining scenarios in terms of simpler primitives. This work makes possible to formulate and solve entirely new classes of research problems that are vital to biological knowledge discovery. The key to success in our approach is the use of domain-scientist-defined object sets (i.e., descriptors) as the starting point of analysis, ensuring relevance of mined results. As scientists are empowered to create their own vocabularies and descriptors and reason with them, there will be greater understanding of scientific datasets. Redescription mining promises to be an important tool in this endeavor.

Acknowledgments

This work was supported in part by NSF grants CNS-0615181, ITR-0428344, EMT-0829835, and CNS-0103708, and NIH Grant 1R01EB0080161-01A1. A significant amount of this work is the result of collaboration with many colleagues, including Deept Kumar, Richard F. Helm, Malcolm Potts, and Lizhuang Zhao. Deept Kumar helped gather the results presented in Section 22.4.4.

References

[1] R. Agrawal, T. Imielinski, and A.N. Swami. Mining association rules between sets of items in large databases. In *Proceedings of the ACM SIGMOD International Conference on Management of Data (SIGMOD' 93)*, Washington, DC, 207–216, 1993.

[2] R. Agrawal and R. Srikant. Fast algorithms for mining association rules in large databases. In *Proceedings of the 20th International Conference on Very Large Databases (VLDB' 94)*, Santiago de Chile, Chile, 487–499, 1994.

[3] T. Akutsu, S. Kuhara, O. Maruyama, and S. Miyano. Identification of dene regulatory networks by strategic gene disruptions and gene overexpressions. In *Proceedings of the Ninth Annual ACM-SIAM Symposium on Discrete Algorithms*, San Francisco, CA, 695–702, 1998.

[4] C.A. Ball, I.A. Awad, J. Demeter, J. Gollub, J.M. Hebert, T. Hernandez-Boussard, H. Jin, J.C. Matese, M. Nitzberg, F. Wymore, Z.K. Zachariah, P.O. Brown, and G. Sherlock. The Stanford microarray database accomodates additional microarray platforms and data formats. *Nucleic Acids Research*, 1(33):D580–D582, 2005.

[5] B.-J. Breitkreutz, C. Stark, and M. Tyers. The GRID: The general repository for interaction datasets. *Genome Biology*, 3(12), 2002.

[6] N.H. Bshouty. Exact learning Boolean functions via the monotone theory. *Information and Computation*, 123(1):146–153, 1995.

[7] A.E. Carpenter and D.M. Sabatini. Systematic genome-wide screens of gene function. *Nature Reviews Genetics*, 5(1):11–22, 2004.

[8] S.Y. Chan and D.R. Appling. Regulation of S-adenosylmethionine levels in *Saccharomyces cerevisiae*. *Journal of Biological Chemistry*, 278(44):43051–43059, 2003.

[9] U. de Lichtenberg, L.J. Jensen, S. Brunak, and P. Bork. Dynamic complex formation during the yeast cell cycle. *Science*, 307(5710):724–727, 2005.

[10] M.B. Eisen, P.T. Spellman, P.O. Brown, and D. Botstein. Cluster analysis and display of genome-wide expression patterns. *PNAS*, 95(25):14863–14868, 1998.

[11] D.H. Fisher. Knowledge acquisition via incremental conceptual clustering. *Machine Learning*, 2(2):139–172, 1987.

[12] A.P. Gasch, P.T. Spellman, C.M. Kao, O. Carmel-Harel, M.B. Eisen, G. Storz, D. Botstein, and P.O. Brown. Genomic expression programs in the response of yeast cells to environmental changes. *Molecular Biology of the Cell*, 11:4241–4257, 2000.

[13] A.C. Gavin, M. Bosche, R. Krause, P. Grandi, M. Marzioch, A. Bauer, J. Schultz, et al. Functional organization of the yeast proteome by systematic analysis of protein Complexes. *Nature,* 415(6868):141–147, 2002.

[14] K.C. Gunsalus and F. Piano. RNAi as a tool to study cell biology: Building the genome-phenome bridge. *Current Opinion in Cell Biology*, 17(1):3–8, 2005.

[15] M.A. Matzke and J.A. Birchler. RNAi-mediated pathways in the nucleus. *Nature Reviews Genetics*, 6(1):24–35, 2005.

[16] M.A. Matzke and A.J.M. Matzke. Planting the seeds of a new paradigm. *PLoS Biology*, 2(5):582–586, 2004.

[17] R.S. Michalski. Knowledge acquisition through conceptual clustering: A theoretical framework and algorithm for partitioning data into conjunctive concepts. *International Journal of Policy Analysis and Information Systems*, 4:219–243, 1980.

[18] M. Mizunuma, K. Miyamura, D. Hirata, H. Yokoyama, and T. Miyakawa. Involvement of S-adenosylmethionine in G1 cell cycle regulation in *Saccharomyces cerevisiae*. *PNAS*, 101(16):6086–6091, 2004.

[19] L. Parida and N. Ramakrishnan. Redescription mining: Structure theory and algorithms. In *Proceedings of the Twentieth National Conference on Artificial Intelligence (AAAI'05)*, Pittsburgh, PA, 837–844, 2005.

[20] Y. Pilpel, P. Sudarsanam, and G.M. Church. Identifying regulatory networks by combinatorial analysis of promoter elements. *Nature Genetics*, 29(2):153–159, 2001.

[21] K.Q. Pu and A.O. Mendelzon. Concise descriptions of subsets of structured sets. *ACM Transactions on Database Systems*, 30(1):211–248, 2005.

[22] J.R. Quinlan. Induction of decision trees. *Machine Learning*, 1(1):81–106, 1986.

[23] N. Ramakrishnan, M. Antoniotti, and B. Mishra. Reconstructing formal Temporal models of cellular events using the GO process ontology. In *Proceedings of the Eighth Annual Bio-Ontologies Meeting (ISMB'05 Satellite Workshop)*, Detroit, MI, 2005.

[24] N. Ramakrishnan, D. Kumar, B. Mishra, M. Potts, and R.F. Helm. Turning CARTwheels: An alternating algorithm for mining redescriptions. In *Proceedings of the Tenth ACM SIGKDD International Conference on Knowledge Discovery and Data Mining (KDD'04)*, Seattle, WA, 266–275, 2004.

[25] E. Segal, N. Friedman, D. Koller, and A. Regev. A module map showing Conditional activity of expression modules in cancer. *Nature Genetics*, 36(10):1090–1098, 2004.

[26] E Segal, M. Shapira, A. Regev, D. Pe'er, D. Botstein, D. Koller, and N. Friedman. Module networks: Identifying regulatory modules and their condition-specific regulators from gene expression data. *Nature Genetics*, 34(2):166–176, 2003.

[27] J. Singh, D. Kumar, N. Ramakrishnan, V. Singhal, J. Jervis, A. Desantis, J. Garst, S. Slaughter, M. Potts, and R.F. Helm. Transcriptional response of *Saccharomyces cerevisiae* to desiccation and rehydration. *Applied and Environmental Microbiology*, 71(12):8752–8763, 2005.

[28] A. Sturn, J. Quackenbush, and Z. Trajanoski. Genesis: Cluster analysis of microarray data. *Bioinformatics*, 18(1):207–208, 2002.

[29] A. Subramanian, P. Tamayo, V.K. Mootha, S. Mukherjee, B.L. Ebert, M.A. Gillette, A. Paulovich, S.L. Pomeroy, T.R. Golub, E.S. Lander, and J.P. Mesirov. Gene set enrichment analysis: A knowledge-based approach for interpreting genome-wide expression profiles. *PNAS*, 102(43):15545–15550, 2005.

[30] N. Sudarsan, J.E. Barrick, and R.R. Breaker. Metabolite-binding RNA domains are present in the genes of eukaryotes. *RNA*, 9:644–647, 2003.

[31] W.C. Winkler, A. Nahvi, N. Sudarsan, J.E. Barrick, and R.R. Breaker. An mRNA structure that controls gene expression by binding S-adenosylmethionine. *Nature Structural Biology*, 10:701–707, 2003.

[32] J.J. Wyrick, F.C. Holstege, E.G. Jennings, H.C. Causton, D. Shore, M. Grunstein, E.S. Lander, and R.A. Young. Chromosomal landscape of nucleosome-dependent gene expression and silencing in yeast. *Nature*, 402:418–421, 1999.

[33] M. Zaki and N. Ramakrishnan. Reasoning about sets using redescription mining. In *Proceedings of the Eleventh ACM SIGKDD International Conference on Knowledge Discovery and Data Mining (KDD'05)*, Chicago, IL, 364–373, 2005.

[34] L. Zhao, M. Zaki, and N. Ramakrishnan. BLOSOM: A framework for mining arbitrary Boolean expressions over attribute sets. In *Proceedings of the Twelfth ACM SIGKDD International Conference on Knowledge Discovery and Data Mining (KDD'2006)*, Philadelphia, PA, 827–832, 2006.

Part V

Genome Medicine Applications

Chapter 23

Data Mining Tools and Techniques for Identification of Biomarkers for Cancer

Mick Correll, Simon Beaulah, Robin Munro,
Jonathan Sheldon, and Yike Guo

InforSense Ltd.

Hai Hu

Windber Research Institute

23.1   Introduction

It has been nearly four decades since Richard Nixon famously launched the "war on cancer" in the United States. In that time, significant progress has been made on many fronts including major improvements in prevention, detection, and treatment. But despite billions of dollars and decades of research, cancer remains a leading cause of death worldwide. If current trends continue cancer could soon become the number one cause of death in the United States.

In the very same year that marks 40 year's since the signing of the National Cancer Act, another major milestone will also be reached; the ten year anniversary of the first release of the human genome. Today, genomics is increasingly finding a new role outside the realm of basic research in the hospitals, and in particular, the cancer treatment centers of the world. Rapid advances in technology have drastically changed the economics of high throughput genomic assays. High density and custom arrays measuring gene expression levels can now be used to differentiate clinically indistinguishable tumor subtypes, whole genome association studies are being performed on increasingly large and diverse populations to discover new genetic factors in disease, and on the horizon, ultra high throughput sequencing is offering a whole new host of possibilities for genomic analysis.

Biomarkers are biological markers for diseases, which are enabling major breakthroughs in the detection and treatment of diseases, particularly cancer. They improve our ability to detect the molecular changes associated with a tumor cell and enhance our ability to detect and identify tumors, select appropriate treatments, monitor the effectiveness of a treatment and determine prognosis. Well known examples, in breast cancer, include: HER2 over expression for molecular markers and breast density assessment for imaging biomarkers. Biomarkers can be identified through noninvasive approaches, for example, using urine and saliva as the specimens. Specimens obtained through minimally invasive approach for example blood samples and through invasive approaches for example solid body tissues, can also be used for biomarker studies. Medical images are often taken in a noninvasive manner but can also be through an invasive approach, for example by cutting an incision and probing the diseased area.

Historically, biomarkers were identified one at a time by using traditional scientific research techniques, such as molecular biology. In the postgenome era, a vast amount of data are generated using high-throughput or semi-high-throughput technologies, including chip-based technologies and mass spectrometry. In the biomedical informatics research field a large amount of clinical data and medical images are also collected or generated. Given the shear volume of the data, the traditional hypothesis-driven, single molecule-based research model can no longer be applied for efficient research. Data mining technologies are thus in demand to generate hypotheses which can then be followed by the traditional hypothesis-driven research approach.

One major problem of research using high-throughput technologies is that the noise level can be high. The noises can come from different sources, including false positives due to the size of the experiment, and the alteration of the environment in the in vitro experimental system compared to the original in vivo system. In statistical analysis, a false positive rate of 0.05 is often taken as the threshold to report a significant finding, in the sense that in the traditional way of doing experiments, if one molecule satisfies this threshold to be identified as a biomarker, then there is only a 5% chance that this biomarker is a false positive discovery. In a high-throughput experiment, 50,000 molecules may be assayed at one time, applying the statistical threshold, 2500 molecules will be reported as biomarkers by chance alone. The noise due to false positive not only needs to be controlled in in vitro experiments, but also needs to be controlled in in silico experiments and data mining of the large amount of data given the large number of tests.

In high-throughput experiments, the environment has to be optimized for all the molecules in study, in contrast to the optimization of a single molecule experimental environment in a traditional way of doing experiments. For example, in the high-throughput protein assays, proteins from different cell types and different organelles may be assayed for interaction, when in reality they may not reach the proximity of each other. Thus, in biomarker identification using high-throughput technologies, not only false positives need to be stringently controlled, but also multiple levels of validations need to be performed. This needs be done not only at the technical level (for example by doing the same experiments using different samples, using different experimental platforms, repeating experiments in a different lab), but also at multiple levels of biological systems including cell models and animal models. Only after going through more stringent false positive control and validation processes can biomarkers identified by high-throughput approaches be confirmed as real biomarkers.

This chapter will focus on the intersection of the global cancer research effort and the transformative technologies of genomics and image analysis. It is precisely at this intersection that many hope to find the long sought, epiphanic breakthrough.

23.2   Purpose and Promise of Biomarkers in Oncology

Biomarkers present a number of exciting possibilities for the medical field in general, and for cancer in particular, have the potential to impact almost every aspect of the disease management life cycle. For prevention, as our understanding of the genetic influences and risk factors behind cancer improves, individualized risk models can be developed and used to help guide life style choices and medical screening and intervention programs. For diagnosis, protein and other molecular markers can enable early detection through noninvasive procedures, improved imaging techniques can likewise provide valuable early detection capabilities and molecular pathology can provide increasingly precise and objective cancer classification. And finally, for treatment, pharmacogenomics and biomarkers have the potential to dramatically improve outcomes and reduce adverse events in cancer treatments.

23.2.1   Clinical uses of molecular biomarkers in cancer

23.2.1.1   Identification of high risk population

Biomarkers can be used to identify subjects with a high risk of developing a certain disease. For example, a subject with a mutation in the tumor suppressor gene breast cancer 1 (BRCA1) has a higher risk of developing breast cancer as well as ovarian and prostate cancers. High risk subjects are often monitored more closely than the general population with regard to routine screenings and some of them may even be given medication to prevent development of certain types of cancer such as breast cancer.

23.2.1.2   Diagnosis

A necessary first step in determining optimal treatment for cancer patients is an accurate diagnosis. Currently, cancer classification relies on a highly trained pathologist and a complex combination of clinical and histopathological data. Difficulties can arise however with atypical or morphologically indistinguishable tumors. Even in the best case, diagnosis in this fashion remains subjective and expensive. Molecular diagnostics have the potential to significantly improve the situation by providing an inexpensive, objective and precise mechanism of cancer classification. Today, the identification of the necessary molecular markers to enable these types of diagnostics is the subject of intense research.

23.2.1.3 Cancer classifications

Another area of cancer research where biomarkers hold great promise is in the development of improved cancer classification systems and the discovery of new classes of cancer from groups that are currently indistinguishable by traditional diagnostic methods. Finer grained and otherwise improved classification systems are invaluable in determining the proper course of treatment, as well as in the development of new treatments. For example, in breast cancer studies, the protein expression level of estrogen receptor (ER), progesterone receptor (PR), and HER2, as well as cytokeratines 5 and 6 have been used to classify five subclasses of breast cancers and some subclasses are more aggressive diseases than others.

23.2.1.4 Prognostic and predictive biomarkers

After an accurate diagnosis has been established, biomarkers have the potential to impact the next stage of the disease management life cycle, prognosis and treatment. Prognostic biomarkers are those that provide information about outcome independent of any particular treatment, while predictive biomarkers are those that can provide the potential response to a particular treatment. The potential to predict the efficacy of a drug treatment based on genetic variation, also known as pharmacogenomics, is currently the subject of much research and is believed by many to hold the potential for true "personalized medicine" where disease treatment can be tailored to the physiology of individual patients.

23.2.2 Examples of molecular biomarkers used in oncology today

Today there are a number of molecular biomarkers that are commonly used at various points in the disease treatment life cycle of cancer. One of the most well known examples is prostate-specific antigen (PSA). PSA is produced by the cells of the prostate and its levels can be measured by a simple blood test. Elevated levels of PSA can be an indication of prostate cancer and for this reason the PSA test is a commonly used screening measure for patients at high risk of prostate cancer. PSA is therefore, an example of diagnostic biomarker. In breast cancer, three commonly used molecular biomarkers are ER, PR, and human epidermal growth factor receptor 2 (HER2). All three are proteins that may or may not be present on the surface of tumor cells. All three are also targets of various drug treatments. ER and PR are biomarkers that are used to assess whether or not a patient should receive hormone therapy, while HER2 is a biomarker that is used to assess the possibility of treating a patient with herceptin. Each of the previous examples (PSA, HER2, ER, PR) are instances of protein biomarkers. An example of a gene expression biomarker is the OncotypeDX test developed by Genomic Health. OncotypeDX measures the gene expression levels of a panel of 21 genes. OncotypeDX provides prognostic information on disease recurrence as well as predictive information on the potential of chemotherapy.

23.2.3   Examples of image biomarkers used in oncology today

Probably the most well known imaging biomarker is the mammogram test for breast cancer. Widely adopted throughout the world this test enables breast tumors to be detected early and treatments applied at the most effective stage. In the United States alone, this test has been associated with an incredible 24% decrease in the death rate from breast cancer (even after taking patient age into account) from 1989 to 2003 (Smith, 2007). There is growing use of fluorodeoxyglucose positron emission tomography (FDG-PET) to assess treatment response in lymphoma patients with residual tumors after two sessions of chemotherapy. FDG-PET enables treatment intensity and type to be tailored to each patient. This improves the chances of a cure and the therapeutic value of the treatment. It also reduces safety side effects for patients with good survival prospects and enables an informed decision to be made for patients with a poor prognosis to use either an aggressive or palliative approach. In 1998, the Centers for Medicare and Medicaid Services (CMS) approved reimbursement for FDG-PET to assess solitary lung nodules. Small cell lung, lymphoma, melanoma, esophageal and recurrent colorectal cancers are now also covered as the value of PET in these disorders is well documented. Quantitative and qualitative assessment of breast density has also been widely studied (Kerlikowske et al., 2007) and shows that increases in breast density are associated with an increased risk of breast cancer and reduced density with a lower risk of breast cancer.

Due to the shortage of skilled imaging personnel and growing use of these techniques there is widespread research into and adoption of computer assisted image analysis. Advanced algorithms are used to identify features of interest in images from many different modalities. This approach can be used to draw the operators' attention to particular patients and areas of possible disease. Commercial vendors often refer to this as being similar to a spell checker; it assists the user but does not replace them. The FDA has approved a number of commercial computer based image analysis systems for the identification of lung nodules, a precursor to lung cancer. For example, Siemens produces its Syngo systems for assessment of lung nodules, breast and colon cancer.

23.3 Data Mining Tools and Techniques for Identification of Molecular Biomarkers in Cancer

In this the information age, usage of the term "data mining" has grown almost as rapidly as the tremendously large datasets on which it is often applied. We will begin this section then by defining the term itself and for our definition we will borrow from the synonymous term "knowledge discovery in databases (KDD)" which has been defined by Fayyad et al. (1996) as: "The nontrivial process of identifying valid, novel, potentially useful and ultimately understandable patterns in data." While there is some debate on whether KDD and data mining are truly synonymous, we find this definition to be the most succinct and accurate to date.

Data mining can be differentiated from other analytical methods, in that the emphasis is placed on discovery. While other methods seek to prove or disprove a specific hypothesis, data mining is hypothesis generating. As an example, let us consider data generated from a clinical trial of a new cancer treatment. In this hypothetical study, data is collected in order to assess disease progression for hundreds of patients that are receiving either the drug itself or a placebo. At the conclusion of the study, data analysis methods would be used to test whether there was a statistically significant difference in disease progression between the group that received the drug and the group that received placebo. In this case, the analysis is testing whether the data supports a specific hypothesis, namely that the drug is or is not effective. This type of analysis is not data mining. Now, consider that during that same clinical trial the genotypes of each participant were determined for a few hundred thousand markers across the genome. Using this additional data, researchers might perform a new analysis that looks for genetic markers that are correlating significantly with different rates of disease progression or effectiveness of therapy. The purpose of this analysis is not to test any particular hypothesis, but rather to discover novel features or patterns that might exist in the data, it is therefore hypothesis generating. This is data mining.

There are a number of different tools and techniques that fall under the umbrella of data mining. In terms of biomarker identification in cancer, these tools can be grouped into three primary categories: feature selection, clustering, and classification. Feature selection methods attempt to identify the most informative or useful features in a dataset. Clustering algorithms are used to discover underlying patterns in complex data. And the aim of classification is to develop models that can discriminate between two or more known classes. On the face of things, it would seem that feature selection methods alone would be sufficient for biomarker identification; biomarkers are after all the informative or useful features in a dataset. However, it is other tools like classification and clustering that turn a feature into a biomarker by putting it into a biological or clinical context, i.e. make it useful. For example, a feature selection algorithm may tell us that gene X is one of 200 interesting features in a dataset, but by developing a classification model that can use gene X to differentiate between two different tumor types we are able to identify gene X as a useful biomarker.

23.3.1 Curse of dimensionality

A defining factor of many modern genomics assays is their highly parallel nature. In a single assay researchers are routinely able to explore tens of thousands if not millions of different features. As an example, the Affymetrix Genome-Wide SNP Array 6.0 contains a staggering 1.8 million markers for genetic variation. With such a remarkable number of features, modern experiments pose a special problem for many data mining algorithms that was described by Richard Bellman (1961) as the curse of dimensionality.

The curse of dimensionality problem occurs in situations where the number of features (p) is greater than the number of samples (n). With their extraordinary number of features this problem is particularly pronounced in modern genomic experiments, where p is orders of magnitude greater than n. Even as technological improvements make it economically feasible for larger sample sizes, a scenario where n exceeds or even approaches p is highly unlikely, if anything the n–p gap will almost certainly continue to grow.

23.3.1.1 Impact on analysis of highly parallel genomic assays

The curse of dimensionality problem has important consequences for many data mining algorithms. In terms of classification algorithms, some types of classifiers, specifically direct linear discriminant analysis (DLDA) and direct quadratic discriminant analysis (DLQA), require that the number of samples (n) exceeds the number of features (p). DLDA and DLQA type classifiers operate by developing class discriminate functions. Obtaining these functions requires determining the inverse of the variance-covariance matrix that is estimated from the training samples. In cases where p is much greater than n however, this matrix is singular and the inverse cannot be defined. For other types of classifiers, having a p that is much greater than n leads to other problems such as degraded performance, or over-fitting, where a classifier is not effective on new samples outside the training set (Asyali et al., 2006).

There are a number of different strategies for effective data mining on highly dimensional datasets. One strategy is to transform or reduce the number of features of a dataset through the use of dimension reduction algorithms. Another strategy is to use a subset of tools that are capable of directly supporting data where $p \gg n$, or have built-in feature selection capabilities.

23.3.2 Dimension reduction in genomics

As their name indicates, the purpose of dimension reduction techniques is to reduce the number of features in a dataset. Dimension reduction has a number of important roles in biomarker identification. As discussed, problems such as the curse of dimensionality often necessitate closing the gap between n and p, however, reducing the number of features can also have an important cost savings effect in terms of memory and computational power required for model development and subsequent predictions. There are two distinct varieties of dimension reduction: feature selection and feature extraction. Feature selection attempts to identify the subset of features that have the best discriminating ability. Importantly, with feature selection, the selected features themselves are unaltered from the original dataset. Feature extraction on the other hand takes a very different approach. Feature extraction techniques use the entire dataset as input and produce a new reduced set of features, often some linear combination of the originals, which can then be used in subsequent analysis.

23.3.2.1 Feature selection

Feature selection is perhaps the most intuitive approach to dimension reduction. Given a large number of input features, we select a small subset of the most useful ones and use these for subsequent analysis. We will consider features selection techniques in two categories: univariate, which consider each variable separately; and multivariate, which take into account interactions and correlations between features.

23.3.2.1.1 Univariate feature selection

There is a wide variety of univariate feature selection strategies, most of which are conceptually very straight forward and computationally inexpensive. Both parametric and nonparametric significance testing algorithms, such as Student's T-test, ANOVA, and Wilcoxon signed-rank test, can be used to test the discriminating power of individual features. The resulting probability values from significance testing algorithms can be used to rank the feature list and a subset of the top ranked features can be selected.

Other techniques, such as information gain, are perhaps less intuitive. These methods function by examining each feature and measuring the entropy reduction in the target class distribution when that feature is used to partition the dataset. Another heuristic method that is commonly used relies on a signal-to-noise ratio calculation that can be used to rank the value of each feature in the set.

A major concern when using univariate feature selection methods is that they do not take into account correlations and interactions between features. With univariate methods there is therefore a risk that the subset of features that are selected, despite having strong individual differentiating capability, will be so highly correlated that they are redundant and therefore lacking the necessary discriminating power for downstream classification (Boulesteix et al., 2008).

23.3.2.1.2 Multivariate feature selection

Multivariate methods differ from their univariate cousins in that rather than examining each feature independently they take into account correlations between features and attempt to identify a combination of features that will have the strongest discriminating power. One multivariate approach that has been shown to be particularly useful in bioinformatics makes use of the genetic algorithm (GA). The GA is designed to mimic evolution and natural selection by employing operations such as recombination, mutation, and selection in order to perform sophisticated search and optimization in order to identify globally optimal feature sets. Other multivariate methods make use of algorithms such as random forest and random subspace in exploring the correlation and interactions among features. The major drawback of multivariate methods is their computational complexity and their tendency toward over-fitting (Liu et al., 2005).

23.3.2.2 Feature extraction

Feature extraction methods offer an alternative to feature selection for dimension reduction. Unlike feature selection, which attempts to identify a subset of dimensions from a dataset, feature extraction uses all dimensions as input, and then through a series of mathematical transformations, creates an entirely new dataset with a reduced number of dimensions. The most commonly used methods for feature extraction are principle component analysis (PCA) and partial least squares (PLS). Both PCA and PLS are widely used in data mining and can be very effective techniques for dimension reduction. However, feature extraction techniques are usually not suitable for biomarker identification due to the fact that the new features they create cannot be directly assayed and are difficult or impossible to interpret further.

23.3.2.3 Methods accepting highly dimensional data

As we have discussed, there are a number of techniques that are available that can reduce the number of dimensions in a dataset and help avoid some of the problems encountered when $p \gg n$. An alternative is to use a subset of methods that directly support highly dimensional data and therefore, make the dimension reduction step unnecessary. Two types of classifiers that have been used successfully for biomarker identification and directly support highly dimensional data are support vector machines (SVMs) and nearest shrunken centroid (NSC). Both of these classifiers have intrinsic feature selection capabilities and are covered in more detail later in the chapter.

23.3.3 Class discovery

The goal of class discovery is to find relevant subsets or structures in data that can be identified by a set of input features. Class discovery is also often referred to as clustering and in the parlance of machine learning, is an example of an unsupervised learning technique. This type of problem is common in data mining and we find a number of problems of this form in the context of biomarkers for cancer. The most common example of a class discovery problem in cancer research is for developing and refining cancer classification systems. Cancer classification plays an important role in guiding treatment decisions as well as in the development of new treatments. Traditionally classification systems were developed based on the histopathological features of tumors. Molecular biomarkers have the potential to significantly improve traditional classification systems by providing a more sensitive, systematic and more objective mechanism for detecting cancer subtypes.

There are a number of clustering algorithms that can be used for class discovery in cancer, here we cover self-organizing maps (SOMs) and hierarchical clustering (HAC).

23.3.3.1 Self-organizing map (SOM)

SOMs are a technique that has been shown to be well suited for exploratory analysis of genomic data and particularly for gene expression analysis. SOMs are constructed by first defining a set of nodes with a simple topology and a distance function between them. Nodes are then iteratively mapped onto a k-dimensional "gene expression" space and adjusted, or moved, based on the calculation of a distance formula to data points that are selected at random. A large number of iterations (20,000–50,000) are used and in this way the nodes migrate from the initial topology to a new pattern that fits the input data. The number and topology of the initial nodes can be varied to explore how different structures in the data emerge. SOMs are computationally simple, fast to create and scale well for large datasets (Tamayo et al., 1999).

23.3.3.2 Hierarchical clustering (HAC)

HAC is a very common technique for analyzing multidimensional molecular data. The result of an HAC analysis is a dendrogram covering the entire dataset, where the branch length between the clusters represents the degree of similarity. The HAC algorithm starts by defining every element as a singular cluster. The algorithm then proceeds through successive rounds of calculating the pair-wise similarity between clusters using a distance function. At the end of each round the graph is updated with the new clusters. There are a number of methods that can be used for calculating the pair-wise distance between clusters, such as Euclidean distance, or Pearson's correlation coefficient.

23.3.4 Class prediction in cancer

The goal of class prediction is to develop a mathematical model that for a given set of input features from an unknown subject is able to predict which group or class the subject belongs in. Class prediction, or classification, is a supervised learning technique. For biomarkers in cancer, there are a number of problems that take this form. Diagnosis is an excellent example of a class prediction problem. Given a set of input parameters (patient symptoms, histopathology, morphology) a physician attempts to correctly predict where a tumor belongs in a complex classification schema. Clinical decision support is another area where classification problems can be found. In determining an optimal course of treatment, a physician has access to a number of input features about the patient, the status and progression of their disease, as well as data about the use and applicability of a range of treatment methods available. With all the input data, the physician then attempts to predict the treatment that will result in the most favorable outcome. In both of the preceding examples, as in many others similar scenarios, biomarkers and classification techniques have the potential to provide valuable assistance to physicians in making these predictions.

23.3.4.1 Supervised learning introduction

The basic methodology of supervised learning is quite straight forward. We start with two sets of data, a training set and a test set, both of whose members have known class membership. Next, we use the training set to build our classification model. The specifics of how the model is built will depend on the classification technique that is being used. Finally, once we've developed our model, we apply it to our test set and examine the results in order to determine its effectiveness. Once we are satisfied that the model has the desired sensitivity and specificity we can start predicting the class of unknown elements.

23.3.4.2 Review of methods

There are a number of classification techniques that have been applied successfully to cancer research. In this section we will cover the following techniques: SVMs, NSC, advanced neural networks (ANNs), and classification and regression trees also known as decision trees. Different classifiers will approach the same problem with very different strategies and methods. Therefore, the choice of classifier can have a significant impact on the quality and properties of the model. It is also recommended to build and evaluate models from a number of different classification methods.

23.3.4.2.1 Support vector machine (SVM)

The support vector machine (SVM) algorithm has been shown to be one of the most powerful supervised learning techniques in bioinformatics. SVMs have a number of attractive features making them the classification method of choice for a variety of problems in the biological and clinical domains. The theory behind SVM is that a binary classifier can be created by constructing a hyperplane in the input space to separate the two different groups.

In real-world examples however, constructing such a hyperplane is often not possible. SVMs overcome this problem by mapping the input data onto a higher dimensional space, called the feature space and construct the separating hyperplane there. SVMs make use of a function, called the kernel function, for mapping the input data onto the higher-dimensional space and constructing the hyperplane. Different kernel methods can be used and choice of the kernel method can have important consequences in the classifier's performance. From the resulting hyperplane in an SVM classifier, one can measure the relative contributing weight from each individual feature, in this way SVMS can be seen as having built-in feature selection capabilities (Brown et al., 2000; Ramaswamy et al., 2001).

23.3.4.2.2 Nearest shrunken centroid (NSC)

NSC is another commonly used and highly effective classification method, particularly for the area of gene expression analysis. NSC is also sometimes referred to as prediction analysis for microarray (PAM) from the software package that implements the method. NSC is a derivative of the nearest centroid classification method. Nearest centroid works by calculating the standardized centroid of each class, which is the mean expression value for each gene in the class, divided by the standard deviation for that gene in the same class. For predicting the class of an unknown sample, the centroid of the unknown is calculated and compared to the centroids from each class. The predicted class is selected based on the smallest distance between the two centroids. The NSC method functions in a similar way, with the added step of first "shrinking" the class centroids. Shrinking is accomplished by moving each centroid toward zero by a defined threshold value. If a centroid hits or crosses zero, the value is set to zero. In the end, all centroids with the value of zero are elminated, and classification proceeds using the nearest centroid method and using the reduced set of "shrunken" centroids. The shrinking step has been shown to make the classifier more accurate, as well as performing a built-in feature selection function (Tibshirani et al., 2002).

23.3.4.2.3 Advanced neural network (ANN)

ANNs are another commonly used classification technique. Neural networks are statistical learning models that are composed of layers of interconnected nodes, in a structure that is roughly analogous to the networks of neurons found in the brain. Neural networks function by propagating data from a set of input nodes, through one or more processing layers of hidden nodes, to a set of output nodes that represent the prediction classes. Each of the nodes in the network can be a linear or nonlinear model. Neural networks have been successfully applied to a number of different data mining problems, including those relevant to biomarkers and cancer research. A significant disadvantage of neural networks is that the methodology behind their decisions is nonintuitive. Besides, special attention should be paid to the over-fitting problem as an advanced neural network (ANN) can be composed of many linear and nonlinear models and thus the total number of parameters in an ANN can be very high.

23.3.4.2.4 Decision trees

Decision trees are widely used classification tools and a method of choice for a number of types of data mining problems. There are different implementations and strategies for constructing decision tree. For example, the C4.5 algorithm developed by Ross Quinlan (Quinlan 1993) is based on examining the entropy reduction in the target class distribution for each of the input attributes, if that attribute were used to partitions the dataset. Decision trees can be powerful classifiers that also have the distinct advantage of revealing their methodology (partitioning rules) in an intuitive form.

23.4   Introduction to Image Analysis Techniques

The past 20 years have seen the introduction and widespread adoption of a wide range of sophisticated imaging technologies, which bring new and improving views on human physiology and anatomy. Over that time computerized tomography (CT), positron emission tomography (PET), single photon emission computerized tomography (SPECT), ultrasonography (US), and magnetic resonance imaging (MRI) have transformed from concept to routine use in clinical practice. The importance of medical imaging as a discipline can be clearly seen in a quote from the editors of the *New England Journal of Medicine* (2000) which named medical imaging as one of 11 "developments that changed the face of clinical medicine" during the last millennium.

Diagnostic imaging has transformed screening for disease, reducing costs and improving clinical outcomes. For example, Mammograms have been associated with a 24% decrease in the death rate from breast cancer in the United States between 1989 and 2003.

Although the diagnosis and treatment of many diseases has been improved by advanced medical imaging, it is cancer treatment and diagnosis that has benefited most from these technologies. CT, PET, SPECT, US, and MRI have opened the door to more targeted radiation therapies and minimally invasive diagnostic and therapeutic procedures. In addition to providing a more complete picture of disease diagnosis, these technologies also enable clinicians to more effectively track treatment effectiveness and disease recurrence.

23.4.1 Image registration

A central concept in image analysis and computer vision is "image registration," which is the process of transforming or aligning multiple images onto a single coordinate system. In the context of cancer research image registration is used in a number of ways. For example, images from different time points can be aligned to look for the appearance of a tumor, or to measure tumor growth over a period of time and multiple image modalities (MRI, CT, PET) can be overlaid to provide enhanced resolution. For each of the above example, image registration is the essential first step.

The process of image registration consists of four steps: feature detection, feature matching, transform model estimation and image re-sampling and transformation (Zitova and Flusser, 2003).

- **Feature detection** is the process of identifying distinctive elements in the images that can be used as reference points. Examples include distinctive objects such as edges, closed boundary regions, corners, line intersections, edges, and contours. In early image registration system, feature identification was a manual process, however it is increasingly being handled in an automated fashion.

- **Feature matching** is the process of establishing the links between an identified feature in one image and the corresponding feature in another image. Various feature descriptors and measures of similarity can be used, along with spatial relationships between the features.

- **Transform mapping functions** are used, once the corresponding features have been identified in the images, to calculate the transformations that will be necessary to overlay the images onto a common coordinate system. The mapping function parameters are calculated using the established feature correspondence.

- **Image resampling and transformation** is where the new image is transformed using the previously defined mapping function and mapped onto the common coordinate system. Appropriate interpolation techniques, such as nearest-neighbor, are chosen based on the required precision.

Each registration step has its own typical issues and problems. The first step is selecting what kind of feature is appropriate for a given task. Ideally the features should be distinctive objects that are easily identifiable in reference and sensed images, even when scenes are slightly different or objects are occluded. For feature matching, issues can arise through image degradation or inaccurate feature detection. The feature descriptors should be immune to any predicted degradations, distinguish between different features and be stable enough not to be influenced by small changes in noise and features. The matching algorithm should be robust and efficient even when faced with invariants. Single features without corresponding counterparts in the other image should not affect its performance. Wherever possible the selection of mapping functions should be based on a priori knowledge of the acquisition process and possible image degradations. If no a priori knowledge is available, the model should focus on handling all possible degradations by being as flexible and general as possible. The most important decision is determining which differences should be removed by registration as it is vital not to remove the changes of interest if change detection is the aim of the procedure. Lastly, the selection of resampling technique is a balance between computational complexity and required accuracy of the interpolation. Some applications may require more precise methods, but nearest-neighbor or bilinear interpolation are appropriate in most instances.

23.4.2   Image analysis

As previously mentioned medical imaging is most widely and successfully used in oncology. However, this has created some major challenges, not least of which is a global shortage of trained radiologists and histopathologists. This creates a bottleneck that delays patient diagnosis and treatment. Also, due to the inherent variability of living organisms, even the most detailed scans can be difficult to interpret. Identifying tumors and lesions is challenging and can be prone to inconsistencies, operator fatigue and data overload. Detection of early stage cancer tumors is very difficult as the lesions are so small they can easily be missed. The combined effects of personnel shortages and task complexity can result in delayed detection, diagnosis and treatment.

To combat these problems there is growing adoption of computer assisted image analysis, often referred to as computer aided detection (CAD). This can be seen in analysis of the medical imaging modalities mentioned earlier and also in analysis of pre-clinical pathology data in drug discovery and in tissue micro-arrays and cell based assay analysis. The key advantage of computer based image analysis is the speed, accuracy and consistency of results compared to human operators. However, computer based systems are not replacing human operators but are used to automatically analyze data and highlight key images for review.

23.4.3   Software tools

23.4.3.1   Definiens

Definiens advanced technology for extracting intelligence from images was developed by Gerd Binnig, the 1986 Nobel Laureate for Physics, and his team to emulate human cognitive processes. The technology looks at pictures in the same way as humans do; they refer to this as Cognition Network Technology®. Instead of examining a given image pixel by pixel, Definiens' segmentation and classification processes recognize groups of pixels as objects, picking out shapes, colors and textures. The technology examines objects in relationship to each other and can understand scale, overlapping objects and the relationship of two-dimensional images to three-dimensional shapes.

Using these factors, the technology identifies individual objects within an image and makes precise and detailed measurements. It can handle the inherent variability in living organisms and can even recognize anomalies, such as an abnormal tumor cell.

23.4.3.2 SlidePath

SlidePath was founded in 2003 by researchers in the School of Biotechnology, Dublin City University, with particular expertise in Digital Microscopy. This approach involves the rapid digitization of tissue biopsies at appropriate diagnostic resolution, using specialized hardware. Digital slides are then distributed to remote workstations for review either on the intranet or internet using specialized software solutions. SlidePath provide two key products, Digital Slidebox, for the management of digital slide content and Distiller, a web enabled, adaptive database system for the easy management of large, complex life sciences datasets. In the information management domain, SlidePath have been involved in implementing a growing number of Digital Slide/clinical data management solutions including:

- Prostate cancer tissuebank management system for distributed hospital network. Central biobank data repository for the storage and retrieval of clinical, sample, pathological, and *omic data, with the aim of compiling and sharing cancer research data accrued from the Trinity Centre for Health Informatics and the Dublin Molecular Medicine Centre's (DMMC) Prostate Research Consortium and six affiliated hospitals.

- Framework 6 Marie Curie TOK: provision of Distiller TMA workflow management solution for disparate research partners in order to readily share and access images and data for analysis.

- Breast cancer research portal: online Distiller clinical/research data management system. It provides a means for disparate research groups to house their datasets within the one structure and to manage access for intercommunication and collaboration. The portal currently houses a number of datasets containing clinical, epidemiological and experimental data.

23.4.3.3 Nanozoomer

Developed by Hamamatsu Photonics, K.K. of Japan, the NanoZoomer Digital Pathology (NDP) is a system for scanning tissue and cellular glass slides and converting them to high-resolution digital slides. Through PCs connected to the internet or an intranet environment, digital slides can be reviewed at any position and desired magnification using the mouse or keyboard with the same quality and speed as if one were operating a conventional microscope. The interpretation and quantification of disease state by examining a tissue section depends on the knowledge and skill of an individual pathologist and is therefore inherently affected by a certain level of subjectivity. Virtual microscopy can help improve on this in a variety of ways such as easy consultation (without any need to physically transport the slides) and it also opens the road to introducing software-assisted methods for certain tasks which will become important in future.

23.4.4   Model building for image biomarkers

The output of image analysis software, such as those mentioned in Section 23.4.3, is a large amount quantitative data about the images which have been analyzed often representing volumetric assessments of features of interest in the images. This data can then be used in case and control comparisons to feed the types of data mining models mentioned earlier in the chapter. As with other data, various feature selection, clustering and classification tools can be used to highlight the best tools for diagnosis and treatment selection. Section 23.4.5 highlights some examples of image biomarkers.

23.4.4.1   Linking to patient information

For individual patient diagnosis, image interpreters do not need to interact with patient data about family histories, treatments or current conditions. The images are assessed and the diagnosis logged in the database. However, when developing biomarker models, patient data is vital to support the selection of patient cohorts of disease and non-disease patients. New tools are enabling clinicians to "slice and dice" patient data, from a disease specific perspective, to select groups of patients to compare. These groups then need to drive the retrieval of quantitative image data from image storage systems. The patient ID is critical to this and shows why laboratory SOP are so vital. These data retrievals are frequently a long, drawn out process for Principle Investigators who are dependent on IT and database personnel to retrieve the data. However, there are a growing number of systems supporting a self service model to access this type of integrated data (Beaulah et al., 2008).

23.4.5   Identification of image biomarkers

23.4.5.1   Breast cancer screening

The medical community is still seeking a cure for breast cancer, with new targeted treatments such as herceptin leading the way. To be eligible to receive herceptin, patients must have breast cancer that over express HER2. Unfortunately, measuring HER2 over-expression, particularly using immunohistochemically equivocal cases, continues to be problematic. Manual assessment of HER2 requires comparison with positive controls and analysis of membrane staining. Although great improvement in standardization of these procedures have been made, inaccurate assessments are very common. Hall et al. (2008) describes a pathologist assisted, computer-based continuous scoring approach for increasing the precision and reproducibility of assessing imaged breast tissue specimens. In the study computer-assisted analysis on HER2 IHC is compared with manual scoring and fluorescence in situ hybridization results on a test set of 99 digitally imaged breast cancer cases enriched with equivocally scored (2+) cases. A computer-aided diagnostic approach was developed using a membrane isolation algorithm and quantitative use of positive immunostaining controls. By incorporating internal positive controls into feature analysis a greater area under the curve (AUC) in ROC analysis was achieved than feature analysis without positive controls. Evaluation of HER2 immunostaining that utilized membrane pixels, controls and percent area stained showed significantly greater AUC than manual scoring and significantly less false positive rate when used to evaluate immunohistochemically equivocal cases.

Use of predictive models in association with mammograms is also an active area of research. One group compared linear genetic programming with back-propagated neural networks (Sheta et al., 2005) and found genetic programming to be statistically more accurate. Another used PCA, independent component analysis and fuzzy logic to assess mammograms delivering an 84% accuracy in abnormality detection and 74% accuracy in diagnosis (Abdel-Qader and Abu-Amara, 2008).

It is worth noting that there has been a high early adoption rate of digital mammograms and this appears to have also increased the number of patient recall rates and false positives found during use of the system. It has been postulated that as the sensitivity of such a method increases and is able to identify in situ tumors and more calcium deposits than before, that a "better safe than sorry approach" has been taken by many clinics using new CAD systems. Clinical practice is still evaluating what the image analysis of digital mammograms can capture, how best to interpret the results and what the pros and cons of the method are. An interesting article by Fenton et al. (2007) finds that CAD is associated with reduced accuracy of interpretation.

23.4.5.2 Tissue microarray (TMA)

Lack of availability of tissue for research purposes is a common problem because pathologists are taking samples for diagnostic purposes, not gathering material for future research. There are also strict guidelines concerning the use of this material so when tissue is available it needs to be used as effectively as possible. To combat this tissue microarrays (MAs) provide a mechanism for highly effective use of available tissue samples. TMAs enable tissue from multiple patients or blocks to be seen on the same slide by a method of relocating tissue from conventional histology paraffin blocks. The technique has been widely adopted following a publication by Kononen et al. (1998).

TMAs, much like cDNA microarrays, support detailed quantitative analysis. However, they present some specific challenges which require advanced analysis software. Image analysis systems for tissue microarray (TMA) must be able to select a region of interest from a tissue disc and then compare and normalize expression levels across any other disc. A variety of automated systems are now available for analyzing TMAs including BLISS, Chromavision ACIS, Biogenex, Applied Imaging.

There is a growing body of work in the use of TMAs for analysis of squamous cell carcinomas of the head and neck (HNSCC), which is well summarized in Radhakrishnan et al. (2008) along with a more detailed description of TMA. The paper highlights a series of molecular markers, including EGFR, ERBB, EGFR1, and Ajkt, which have been identified as markers of amplification and expression prevalence and prognosis. Various data mining techniques can be used to analyze the TMA data including, multi-variant Cox regression to identify prognostic biomarkers and HAC and survival analysis using Kaplan-Meier survival plots to identify prognostically significant clusters and biomarkers and their impact on the outcome.

23.4.5.3 Positron emission tomography (PET)

PET produces a three-dimensional image or map of functional processes in the body. It detects pairs of gamma rays emitted indirectly by a positron-emitting radionuclide (tracer), which is introduced into the body on a biologically active molecule. A 3-dimensional representation of tracer concentration in the body is then produced by computer analysis. This construction is often done with the aid of a CT X-ray scan performed using the same machine. FDG PET is the most common type of PET scan and uses fluorine-18 (F-18) fluorodeoxyglucose (FDG), an analog of glucose, to provide an image of tissue metabolic activity for regional glucose uptake. Although use of this tracer results in the most common type of PET scan, other tracer molecules are used in PET to image the tissue concentration of many other types of molecules of interest.

The brain, liver, and most cancers have a high glucose uptake and consequently show up strongly in FDG PET results. Due to this, FDG-PET is commonly used for diagnosis, staging and monitoring treatment of cancers, particularly in Hodgkin's disease, non-Hodgkin's lymphoma and lung cancer. FDG PET scans in oncology make up over 90% of all PET scans performed each year. The technique is also useful for many other solid tumors and is particularly good at identifying tumor metastasis and recurrence following surgery to remove a primary tumor of high activity.

Recent work by Tian et al. (2007) used a probe with dual specificity to target CCND1 mRNA and IGF1R in combination with radiohybridization PET imaging to detect rapidly dividing cancer cells in breast tumors. The study in mice showed an eight fold increase in PET intensity in the center of the breast cancer xenografts. They postulated that using this technique it would be possible to differentiate between malignant and benign lumps without biopsy.

23.5 Validation and Evaluation

23.5.1 Error rate

Empirically the error rate is the ratio of errors in new cases examined. Statistically, the true error rate is defined as the error rate of the classifier or predictive model on such a large number of new cases that they converge with the actual population distribution. However, in the real-world sample numbers are always limited and usually small. Extrapolating from empirical error rates calculated from small sample results to the true error rate is an important factor in understanding the value of a potential biomarker.

The effectiveness of a model or classifier is enhanced by separating the training dataset from a test set of data and then assessing its results in both. Within the training set the error rate is also referred to as the re-substitution or reclassification error rate. For most types of classifiers and models, the training set error rate is not a good indicator of quality because they are optimistically biased. This can happen because the model has been over-fitted for the sample dataset. In reality the true error rather is frequently much higher than for the training set. For a more accurate estimation of error rate any potential sample set should be divided randomly into two groups; one group to train the model and one group to test it. A common allocation of cases is two third for the training set and one third for the test set. This provides enough data for the model or classifier to be constructed while allowing sufficient cases to check the results.

23.5.2 Sensitivity/specificity

It is vital to understand the effectiveness of a particular biomarker before it can be widely applied. To understand this, detailed comparisons must be made to assess the distribution of true positives, true negatives, false positives and false negatives. Sensitivity and specificity describe how well the test discriminates between patients with and without disease. Sensitivity is the proportion of patients with disease who test positive and specificity is the proportion of patients without disease who test negative.

Selecting the threshold at which a test is considered to be disease or abnormal is critical. It is at this stage that the balance between the sensitivity and specificity of a diagnostic test needs to be explored. To illustrate this, Figure 23.1 shows an idealized representation of disease and nondisease groups organized by their diagnostic test value. As with most tests the results are not 100% accurate in dividing disease from normal because the distributions overlap. Within this grey area the test cannot distinguish between the two states. A threshold is selected (the black line) to indicate that above this value the test is abnormal. It is this threshold that dictates the number of true positive, true negatives, false positives and false negatives. Different thresholds

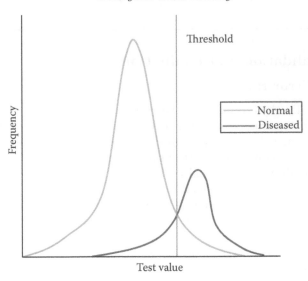

FIGURE 23.1: This graph shows where a threshold might be selected to specify differences between normal and diseased state. The tradeoff in sensitivity can easily be seen in the overlap between the two distributions where the test cannot distinguish between normal and diseased, misclassifying the tails of the distributions when they fall on the wrong side of the threshold.

may be used for different clinical situations to minimize one of the erroneous types of test results. The sensitivity of the test can be improved by making the threshold for a positive test less strict and specificity can be improved by making the threshold for a positive test more strict. However, there is always a tradeoff between sensitivity and specificity.

23.5.3   Cross-validation and bootstrapping

Cross-validation and bootstrapping techniques are widely used in assessment of potential biomarkers. In cross-validation data is divided into k subsets of roughly equal size. Predictive models are run over the data k times, each time leaving out one of the subsets from training, but using only the omitted subset to compute the error criteria. If k equals the sample size, this is called "leave-one-out" cross-validation. The leave-one-out cross-validation is an almost unbiased estimator of the true error rate of a classifier, but does have some disadvantages. Although leave-one-out is nearly unbiased, its variance can be high for small samples.

Bootstrapping works better than cross-validation in many cases. In the simplest form of bootstrapping, instead of repeatedly analyzing subsets of the data, subsamples of the data are repeatedly analyzed. Each subsample is a random sample with replacement from the full sample. Sampling with replacement means that the training samples are drawn from the dataset and placed back after they are used, so their repeated use is allowed. The most popular bootstrapping methods are known as e0 and 0.632. A variation, the 0.632+ bootstrap, has the advantage of performing well even when there is severe over-fitting. However, it has been found that bootstrapping can be excessively optimistic for some other methodologies such as empirical decision trees.

23.5.4 Receiver operating characteristic (ROC) curves

A receiver operating characteristic (ROC) curve is a plot of the true positive rate against the false positive rate for the range of possible thresholds of a diagnostic test.

A ROC curve, shown in Figure 23.2, is a tool to assess the comparative effects of sensitivity and specificity. An accurate test will closely follow the left and top border. A less accurate test will follow the 45-degree angle. Consequently the AUC can be used to measure the accuracy of a test. Values can be between 0.5 and 1, with 1 being a perfect test and 0.5 being worthless. To be considered good or excellent the test needs to be above 0.80.

Two methods are commonly used to calculate the AUC: a nonparametric method based on constructing trapeziods under the curve as an approximation

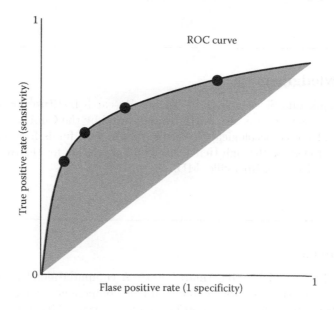

FIGURE 23.2: This graph shows an example of a ROC curve. The area under the curve is used to assess the accuracy of the test. An ideal test will have an area of 1, i.e., the entire graph area would be under the curve. In this case the blue area plus the lower rectangle enclosed by the line $x = y$, has an area of 0.8, indicating a reasonable test accuracy.

of area and a parametric method using a maximum likelihood estimator to fit a smooth curve to the data points. These methods provide an estimate of area and standard area and can be used to compare patient populations and different tests. For researchers in this area the selection of the right algorithm is very serious business.

23.6 Summary

In this chapter, we give a systematic overview of data mining technologies and an overview of molecular and image biomarkers and their clinical and research applications. Examples were given, how to use these technologies for identification and validation of biomarkers. In this context, identification of biomarkers involve the management and mining of a large amount of data and different data mining technologies have been applied to different types of data, including -omics molecular study data and medical image data that are shown as examples in this chapter. While this chapter is focusing on cancer studies, the same technologies and principals can be applied to the study of other diseases.

Acknowledgment

We thank InforSense and the Windber Research Institute for supporting this work. This work was also partially supported by the Clinical Breast Care Project and the Gynecological Disease Program with funds from the US Department of Defense through Henry Jackson Foundation for the Advancement of Military Medicine, Rockville, MD.

References

Abdel-Qader, I. and F. Abu-Amara. 2008. A computer-aided diagnosis system for breast cancer using independent component analysis and fuzzy classifier. *Modeling and Simulation in Engineering*, 2008, 1–9.

Asyali, M. H., D. Colak, O. Demirkaya, and M. S. Inan. 2006. Gene expression profile classification: A review. *Current Bioinformatics*, 1:55–73.

Bellman, R.E. 1961. *Adaptive Control Processes*. Princeton University Press, Princeton, NJ.

Beaulah, S. A., M. Correll, R. E. J. Munro, and J. G. Sheldon. 2008. Addressing informatics challenges in translational research with workflow technology. *Drug Discovery Today,* 2008 sept; 13(17–18): 771–7.

Boulesteix, A.-L., C. Strobl, T. Augustin, and M. Daumer. 2008. Evaluating microarray-based classifiers: An overview. *Cancer Informatics,* 6:77–97.

Brown, M. P. S, W. N. Grundy, D. Lin, N. Cristianini, C. W. Sugnet, T. S. Furey, M. Ares, Jr., and D. Haussler. 2000. Knowledge-based analysis of microarray gene expression data by using support vector machines. *Proceedings of the National Academy of Sciences USA,* 97(1):262–267.

Editors. 2000. Looking back on the millennium in medicine. *New England Journal of Medicine,* 342(1):42–49.

Fayyad, U. M., G. Piatetsky-Shapiro, and P. Smyth. 1996. *From Data Mining to Knowledge Discovery: An overview.* In: Advances in Knowledge Discovery, and Data Mining, eds. U. Fayyad, G. Piatetsky-Shapiro, P. Smyth, and R. Uthurasamy, Menlo Park, California, AAAI Press, 1–30.

Fenton, J. J., S. H. Taplin, P. A. Carney, L. Abraham, E. A. Sickles, C. D'Orsi, E. A. Berns, G. Cutter, R. E. Hendrick, W. E. Barlow, and J. G. Elmore. 2007. Influence of computer-aided detection on performance of screening mammography. *New England Journal of Medicine,* 356(14):1399–1409.

Hall, B. H., M. Ianosi-Irimie, P. Javidian, W. Chen, S. Ganesan, and D.J. Foran. 2008. Computer-assisted assessment of the Human Epidermal Growth Factor Receptor 2 immunohistochemical assay in imaged histologic sections using a membrane isolation algorithm and quantitative analysis of positive control. *BMC Medical Imaging,* 8:11. DOI:10.1186/1471-2342-8-11.

Kerlikowske K., L. Ichikawa, D. L. Miglioretti, D. S. Buist, P. M. Vacek, R. Smith-Bindman, B. Yankaskas, P. A. Carney, and R. Ballard-Barbash. 2007. Longitudinal measurement of clinical mammographic breast density to improve estimation of breast cancer risk. *Journal of the National Cancer Institute,* 99(5):386–395.

Kononen J, Bubendorf L, Kallioniemi A, Bärlund M, Schraml P, Leighton S, Torhorst J, Mihatsch MJ, Sauter G, and Kallioniemi O.P. 1998. Tissue microarrays for high-throughput molecular profiling of tumor specimens. *Nature Medicine,* 4(7):844–847.

Liu J. J., G. Cutler, W. Li, Z. Pan, S. Peng, T. Hoey, L. Chen, and X. B. Ling. 2005. Multiclass cancer classification and biomarker discovery using GA-based algorithms. *Bioinformatics.* 21(11):2691–2697.

Quinlan, J. R. 1993. C4.5: Programs for Machine Learning. Morgan Kaufmann Publishers.

Radhakrishnan, R. 2008. Tissue microarray—a high-throughput molecular analysis in head and neck cancer. *Journal of Oral Pathology and Medicine,* 37:166–176.

Ramaswamy S., P. Tamayo, R. Rifkin, S. Mukherjee, C.H. Yeang, M. Angelo, C. Ladd, M. Reich, E. Latulippe, J.P. Mesirov, T. Poggio, W. Gerald, M. Loda, E.S. Lander, and T.R. Golub. 2001. Multiclass cancer diagnosis using tumor gene expression signatures. *Proceedings of the National Academy of Sciences USA,* 98, 15149–15154.

Sheta, W, N. Eltonsy, G. Tourassi, and A. Elmaghraby. 2005. Automated detection of breast cancer from screening mammograms using genetic programming. *International Journal of Intelligent Computing and Information Sciences,* 5(1):309–318.

Smith R. A. 2007. The evolving role of MRI in the detection and evaluation of breast cancer. *New England Journal of Medicine,* 356:1362–1364.

Tamayo P., D. Slonim, J. Mesirov, Q. Zhu, S. Kitareewan, E. Dmitrovsky, E. S. Lander, and T. Golub. 1999. Interpreting patterns of gene expression with self-organizing maps: Methods and application to hematopoietic differentiation. *Proceedings of the National Academy of Sciences USA,* 96(6):2907–2912.

Tian X, M. R. Aruva, K. Zhang, N. Shanthly, C. A. Cardi, M. L. Thakur and E. Wickstrom. 2007. PET imaging of CCND1 mRNA in human MCF7 estrogen receptor-positive breast cancer xenografts with oncogene-specific [64Cu]chelator-peptide nucleic acid-IGF1 analog radiohybridization probes. *Journal of Nuclear Medicine,* 48(10):1699–1707.

Tibshirani, R., T. Hastie, B. Narasimhan, and G. Chu. 2002. Diagnosis of multiple cancer types by shrunken centroids of gene expression. *Proceedings of the National Academy of Sciences USA,* 99(10):6567–6572.

Zitová, B. and J. Flusser. 2003.Image registration methods: A survey. *Image and Vision Computing,* 21:977–1000.

Chapter 24

Cancer Biomarker Prioritization: Assessing the in vivo Impact of in vitro Models by in silico Mining of Microarray Database, Literature, and Gene Annotation

Chia-Ju Lee, Zan Huang, Hongmei Jiang,
John Crispino, and Simon Lin

Northwestern University

24.1 Introduction

An ultimate goal of cancer research is to identify the underlying molecular mechanisms of oncogenesis and then to translate these discoveries into developing valid biomarkers that are useful for cancer patient care or new drug development.

24.1.1 Use of cancer biomarker

The use of cancer biomarkers in patient care may encompass the whole continuum of cancer management, i.e., cancer prevention, early detection, diagnosis, and treatment [1]. A few remarkable strides have been made. For example, genetic testing for deleterious BRCA1 and BRCA2 mutations identifies women with markedly increased risks of ovarian and breast cancer; expression of the estrogen receptor is used to identify women who are likely to benefit from antiestrogen therapy.

Biomarkers can help to identify new targets for drug development as well as to serve as an indicator for evidence of safety and efficacy at each step of the drug development process [1]. The Food and Drug Administration has approved several new drugs that are designed to target a specific molecule. For example, Gefitinib, a tyrosin kinase inhibitor, blocking the activity of the epidermal growth factor receptor (EGFR), has been approved for the treatment of nonsmall cell lung cancer [2]. A good diagnostic biomarker could greatly accelerate new drug development by shortening clinical trials, identifying responsive patients and revealing toxic side effects. For example, a clinical trial using BCR–ABL (break point cluster region–Abelson) as an endpoint to compare treatments for chronic myelogenous leukemia (CML) would be reduced in time from several years to 12 months [3].

24.1.2 Development of cancer biomarker

The development of cancer biomarker is a long and challenging process, which is implied in the definitions of biomarker made by working groups from the National Institutes of Health and the Food and Drug Administration, respectively (Table 24.1).

Two primary challenges to develop cancer biomarkers are the discovery of candidate markers and the validation of those candidates for specific uses. The discovery process depends on the technologies available to investigate the complex physiology and biochemistry of health and disease in order to identify differences that can be detected consistently. The validation process is more arduous and costly, often involving several stages of confirming measurement performance and evaluating clinical validity and utility [4,5].

Recent development of high throughput "omics" technologies, i.e., genomics, transcriptomics, and proteomics, have revolutionized the way in which the researchers address biological problems. These technologies, with their capabilities of examining a large number of potential markers at once, are promising tools for cancer biomarker discovery. In this chapter, we focus specifically on the use of gene expression microarray technology in biomarker discovery because it is the most commonly available technology and many software tools have been developed for analysis and interpretation of microarray data.

24.2 Using Microarray in Biomarker Discovery

The goal of using microarray in biomarker discovery is to identify changes in gene expression that can be linked to a disease state or a response to

TABLE 24.1: Biomarker definitions.

Glossary	Definition	Source
Biomarker	A characteristic that is objectively measured and evaluated as an indicator of normal biologic processes, pathogenic processes, or pharmacologic responses to a therapeutic intervention.	Biomarker Definitions Working Group, NIH, 2001 [6]
Valid biomarker	A biomarker that is measured in an analytical test system with well-established performance characteristics and for which there is an *established scientific framework or body of evidence* that elucidates the physiologic, toxicologic, pharmacologic, or clinical significance of the test results.	Pharmacogenomics Guidance, FDA, 2005 [7]
Known valid biomarker	A biomarker that is measured in an analytical test system with well-established performance characteristics and for which there is *widespread agreement* in the medical or scientific community about the physiological, toxicological, pharmacological, or clinical significance of the results.	
Probable valid biomarker	A biomarker that is measured in an analytical test system with well-established performance characteristics and for which there is a *scientific framework or body of evidence that appears* to elucidate the physiological, toxicological, pharmacologic, or clinical significance of the results.	

a medical intervention. A multitude of DNA microarray-based studies have demonstrated the ability of gene expression signatures to define cancer subtypes, recurrence of disease and response to specific therapies.

In this section, we introduce a cancer biomarker prioritization model of assessing the in vivo impact of in vitro models by in silico mining of microarray database, literature, and gene annotation.

24.2.1 Cancer biomarker prioritization

A conceptual framework of cancer biomarker prioritization is presented in
Figure 24.1. The major components of the framework comprises the widely
used mouse model and DNA microarray assay, along with data mining tools to
discover the biological and clinical meaning hidden in microarray expression
data.

24.2.1.1 Mouse model

Mouse models of cancer have been widely used in understanding the genetic, molecular, and biological basis of human cancer. Biomarker discovery
through the use of mouse model has the advantages that are critical for successful biomarker discovery. First, inbred mouse models of cancer eliminate
sources of environmental and genetic variability. Second, to collect samples
from highly matched cases and controls under identical conditions further
reduces the variability [8].

24.2.1.2 Cross-species analysis of gene expression changes

Cross-species analysis of microarray gene expression profiles in human and
in animal models might yield important insights into pathological molecular
mechanisms. The analysis is also used to assess whether experimental animal models recapitulate critical aspects of human disease. The comparison

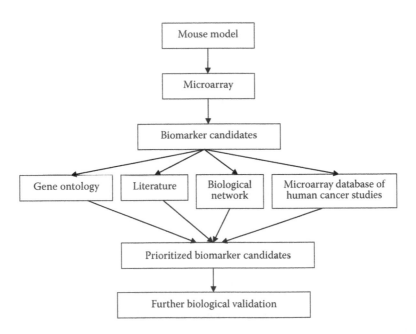

FIGURE 24.1: A conceptual framework of cancer biomarker prioritization.

of transcriptional profiles across different species makes use of homologous/orthologous gene database, e.g., HomoloGene, to map probes across different microarray platforms [9].

24.2.1.3 Functional interpretation of microarray data into biological knowledge

Microarray studies typically generate long lists of significant genes based on differential expression or coexpression. To help with the functional interpretation of such a gene list, a number of analysis tools have been developed. The common goal of these tools is to find patterns in the gene list to infer the biological meaning of the observed expression profile. Two critical strategies have been developed to achieve the goal. The first one is to define a biological concept by a gene set. The second one is to perform enrichment tests to assess the overlap significance between the gene list acquired from the microarray experiment and the gene set predefined to represent a biological concept or functional term. The methodological issue of analyzing gene expression data in terms of gene sets have been vigorously studied in recent years and interested readers are referred to the reviews by Troyanskaya, Dopazo and Nam [10–12] for details.

24.2.2 Data mining tools for functional annotation of microarray gene expression data

There are a number of data mining tools currently available that make use of gene enrichment analysis to provide functional interpretation for microarray gene expression data. Those software suites that integrate a variety of applications and data sources within a package are easier to be used by biologists. A conceptual framework that presents the major components of a microarray data mining software suite is shown in Figure 24.2. The powerful and easy-to-use data mining tool systematically compares the user's own query gene list with the predefined gene sets deposited in a database. The predefined gene sets are derived from published microarray data or prior knowledge stored in annotated databases or published literature. The enrichment analysis tool produces all of the lists that have a statistically significant overlap with the user's data. The visualization tool helps to present as well as explore the analysis result.

24.2.2.1 Gene set database

The usefulness of a microarray data mining software heavily relies on the scope of the gene sets included in the database. Theoretically, gene sets can be generated from any sources of biological knowledge that can be represented as a set of gene lists, each specifying a group of genes that share common biological meaning. Currently, plenty of gene sets have been created (Table 24.2), mainly from two types of information sources.

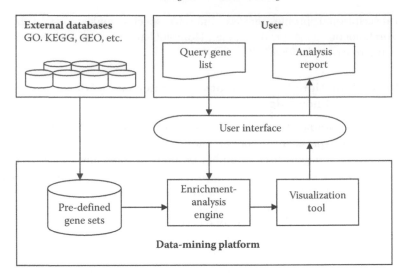

FIGURE 24.2: Conceptual framework of a microarry data mining platform.

Published microarray data, either in literature or public repositories such as GEO, are the major sources for compiling gene sets. For example, numerous cancer gene expression modules, genes sets in response to genetic and chemical perturbations, and tissue/phenotype-specific gene signatures were derived from mountainous microarray data published. A more specific collection of gene sets derived from Connectivity Map can be used to identify drugs that are maybe capable of reversing a disease signature.

Gene sets can also be created from any source of gene annotation. The most widely used ones are functional annotation from Gene Ontology (GO), pathway annotation from KEGG and Biocarta, regulatory annotation from TRANSFAC, CisRed, and miRBase, and protein-to-protein interaction from HPRD and Reactome.

A new type of gene sets definition was emerged along with the advance of text mining technologies. For examples, gene modules associated with specific biomedical terms such as diseases and drugs were derived based on their cooccurrences in sentences.

24.2.2.2 Current tools

A number of microarray data mining tools have been developed with different statistical methods for enrichment analysis, various types of gene sets compiled in the database, available either as a web-based online tool or as a downloadable desktop application. Two comprehensive surveys of currently available tools are provided [11,12]. We focus on four tools, Babelomics (http://www.babelomics.org), Oncomine (http://www.compendiabio.com), L2L (http://depts.washington.edu/l2l), and GSEA (http://www.broad.

TABLE 24.2: Types and sources of biological information for defining gene sets.

Type of data source	Biological properties	Babelomics [13]	Oncomine [14]	L2L [15]	GSEA [16]
Published microarray data	Cancer gene expression module	X	X	X	X
	Tissue/phenotype specific gene expression profiles	X	X	X	X
	Genetic and chemical perturbation		X		X
	Drug treatment signature		Connectivity map		
Annotation database	Biological process, molecular function, cellular component	GO	GO	GO	GO
	Regulatory elements	TRANSFAC, CisRed, miRBase	TRANSFAC, BROAD, picTar	microrna.org	TRANSFAC, BROAD
	Metabolic pathway	KEGG	KEGG		KEGG
	Signaling pathways	Biocarta	Biocarta		Biocarta
	Protein–protein interaction		HPRD	Reactome	HPRD
	Protein domains and families	Interpro	Interpro		
	Chromosome arm and cytoband		NCBI MAP Viewer		
	Orthologs		Inparanoid		
	Nuclear protein complexes		PIN		
	Keywords with functional meaning	SwissProt			
Literature	Gene modules associated with biomedical entities: diseases, chemicals	Text mining derived			HUGO, Unigene

Note: X indicates that gene sets are derived from published microarray data either in literature or in public repositories.

mit.edu/gsea), which are characterized by their comprehensiveness of biological topics covered by the gene set database. The coverage of each tool is summarized in Table 24.2.

24.3 A Case Study of GATA2

GATA2 is an essential transcription factor that regulates multiple aspects of hematopoiesis. It is required for hematopoietic stem cell proliferation. Mutations in GATA2 are associated with blast crisis phase of CML while overexpression of GATA2 is a hallmark of acute megakaryoblastic leukemia in children with Down syndrome (DS-AMKL), a malignancy that is defined by the combination of trisomic chromosome 21 and a GATA1 mutation. However, the mechanism by which GATA2 controls hematopoietic stem cell proliferation and its dysregulation contributes to malignant hematopoiesis is unknown.

In this study, we use G1ME cells as a model to test its function in proliferation. G1ME cells are GATA1-null megakaryocytic cells derived from mouse embryo stem cells and expresses high levels of GATA2. We knocked down GATA2 through retrovirus expressing a GATA2 specific hairpin RNA in G1ME cells. We discovered that GATA2 down regulation caused cell cycle arrest and cessation of proliferation in G1ME cells. To detect potential GATA2 target genes that may control cell cycle progression, we further performed a genome wide microarray analysis comparing GATA2 knockdown cells to control vector transduced cells. We identified 191 probes, representing 146 genes, whose signal changed more than 1.5-fold with a p value $<.01$ (after FDR adjustment). Of these 146 genes, 39 genes were down regulated and 107 were up-regulated in shGATA2-transduced cells.

We are especially interested in up-regulated genes because of the potential repressor role of GATA2. A list of the 107 up-regulated genes is shown in Table 24.3. All these genes already passed stringent statistical criteria. We will use Oncomine to demonstrate how to prioritize them for further investigation. Although GO analysis revealed that these genes fell into multiple classes, including cell cycle regulators, cell proliferation, transcription factors, and markers of terminal differentiation of myeloid cells, erythrocytes, and megakaryocytes, we are more interested to find out how to translate these in vivo finding from manipulating mouse cell lines into clinical relevance.

Because many of the gene symbols are conserved between mouse and human, we initially took a very simple approach to directly cut and paste the gene symbols in Table 24.3 to Oncomine for analysis. A refined approach may use HomoloGene for this cross-species conversion step.

To test whether "GATA2 signature" was reflected in human disease, we performed an Oncomine analysis, which crossed the GATA2 signature to differential expression profiles (Figure 24.3). As of the writing of this book

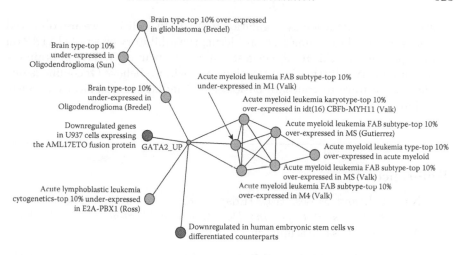

FIGURE 24.3: Oncomine analysis of the GATA2 signature. A molecular concept map of the GATA2 signature (node in the middle, labeled "GATA2_UP") and selected significantly linked concepts. Each node represents a molecular concept, i.e., a set of biologically related genes. The size of the node is proportional to the number of genes in the concept. Each edge signifies a statistically significant association ($p < 1e - 6$).

TABLE 24.3:   Up-regulated genes in the GATA2 experiment.

Tnfaip8l2, Pscdbp, Grcc10, Tgif1, Gab1, Ap3s1, Mc5r, Calml4, 5730593F17Rik, Ndufa6, Ndrg1, Psenen, Greb1, Scand1, Cox6a1, H2afy, Irak3, Atp6v0e, Arpc1a, Mgst2, H2-T23, Kctd12, Bloc1s1, Ubl5, Arid3a, Rin3, 2310005E10Rik, Rab38, Kcnn4, Dap, Emp3, Prdx5, Evi2a, Hes6, Gcnt2, Il16, Gsto1, Sh3bgrl3, Gmfg, Fcgr2b, Ifi205, Ccpg1, Tnfaip8, B2m, B230342M21Rik, Rhog, Sft2d2, Gabarap, Samsn1, Sipa1l1, Tgfbr2, Cd47, Slc12a6, BC028528, Fes, Tpcn1, Tapt1, Rab31, Adssl1, Mylc2b, Tmem50a, Tap2, Psmb9, Hbp1, Cdkn1a, Ctse, Cd244, Cst3, Ifitm1, Fut8, Gyg, St8sia4, Cd93, Glipr1, Gpx1, Stard10, Sat1, Dynlt1, Hhex, Clec4e, P2ry14, Myo1f, Cerk, 5830405N20Rik, Psmb8, 9830134C10Rik, Srgn, Tyrobp, Prtn3, Tspo, Mylc2pl, Rnf130, Selplg, 3110001A13Rik, Gdpd3, Gpr171, Lgals1, Hp, Alox5ap, Cd44, Fcgr3, Cmtm7, Ccl9, Kit, Cebpa, Cpa3, Mpo

chapter, Oncomine has a collection of 28,880 microarray profiles from 41 types of cancers. We found significant overlap between GATA2 signature and differential genes over-expressed in acute myeloid leukemia (AML) M4 and M5. Especially, one significant hit of GATA2 had a differential gene profile in acute promyeloid leukemia patients treated with retinoic acid (RA). Twelve genes that were up-regulated by GATA2 knockdown were down-regulated by RA treatment, of which Kit and Hhex down-regulation are well correlated to differentiation induced by RA. These results suggest that GATA2 may also contribute to proliferation dysregulation in AML. In addition, we also observed strong overlap in brain tumors. Considering the inhibitory effect of GATA2 on neuron progenitors, this result suggests an involvement of GATA2 in brain tumors that deserves further investigation. As such, we believe Oncomine is one of the critical tools to translate basic science findings into clinical relevance, especially in oncology.

References

[1] Nass, S.J. and Moses, H.L. (Eds). *Cancer Biomarkers: The Promises and Challenges of Improving Detection and Treatment.* The National Academies Press, Washington DC, 2007.

[2] Dalton, W.S. and Friend, S.H. Cancer biomarkers—an invitation to the table. *Science*, 2006, 312(5777), 1165–68.

[3] Hartwell, L., et al. Cancer biomarkers: a systems approach. *Nat Biotechnol*, 2006, 24(8), 905–8.

[4] Goodsaid, F. and Frueh, F.W. Implementing the U.S. FDA guidance on pharmacogenomic data submissions. *Environ Mol Mutagen*, 2007, 48(5), 354–58.

[5] Ratner, M. FDA pharmacogenomics guidance sends clear message to industry. *Nat Rev Drug Discov*, 2005, 4(5), 359.

[6] Biomarkers Definition Working Group. Biomarkers and surrogate endpoints: preferred definitions and conceptual framework. *Clin Pharmacol Ther*, 2001, 69(3), 89–95.

[7] Food and Drug Administration. U.S. Department of Health and Human Services, Guidance for industry—Pharmacogenomic data submissions. 2005.

[8] Kelly-Spratt, K.S., et al. A mouse model repository for cancer biomarker discovery. *J Proteome Res*, 2008, 7(8), 3613–18.

[9] Kuhn, A., Luthi-Carter, R., and Delorenzi, M. Cross-species and cross-platform gene expression studies with the Bioconductor-compliant R package 'annotationTools'. *BMC Bioinform*, 2008, 9:26, doi:10.1186/1471-2105-9-26.

[10] Troyanskaya, O.G. Putting microarrays in a context: integrated analysis of diverse biological data. *Brief Bioinform*, 2005, 6(1), p. 34–43.

[11] Dopazo, J. Functional interpretation of microarray experiments. *Omics*, 2006, 10(3), 398–410.

[12] Nam, D. and Kim, S.Y. Gene-set approach for expression pattern analysis. *Brief Bioinform*, 2008, 189–97.

[13] Al-Shahrour, F., et al. Babelomics: advanced functional profiling of transcriptomics, proteomics and genomics experiments. *Nucleic Acids Res*, 2008, 36(Web Server issue), W341–46.

[14] Rhodes, D.R., et al. Molecular concepts analysis links tumors, pathways, mechanisms, and drugs. *Neoplasia*, 2007, 9(5), 443–54.

[15] Newman, J.C. and Weiner, A.M. L2L: a simple tool for discovering the hidden significance in microarray expression data. *Genome Biol*, 2005, 6(9), R81.1–18.

[16] Subramanian, A., et al. Gene set enrichment analysis: A knowledge-based approach for interpreting genome-wide expression profiles. *PNAS*, 2005, 102(43), 15545–50.

Chapter 25

Biomarker Discovery by Mining Glycomic and Lipidomic Data

Haixu Tang, Mehmet Dalkilic, and Yehia Mechref

Indiana University

25.1 Introduction

The completion of the Human Genome Project (HGP) has catalyzed the most influencing paradigm shift in life sciences since the emergence of molecular biology in 1950s (symbolized by Crick and Watson's discovery of the double helical structure of the DNA molecule). Unlike the conventional biological research that focuses on a small number of biomolecules (e.g., a single gene or protein), HGP attempted to determine the whole class of biomolecules (i.e., the DNA sequences) inside a cell, which is now often referred to as the *genomics* (or *omics* in general) approach. The success of HGP can be traced to two key elements: (i) the advancement of a core biochemical technology—the DNA sequencing; and (ii) the development of the cyber-infrastructure for both sharing and data analyses which have facilitated the efficient collaboration among multiple laboratories across the world.

The HGP model has been applied to the studies of the other types of biomolecules. Proteins carry out most biological functions in the cell, and have (naturally) become the next research target of the omics approach. Proteomics aims at the elucidation of the whole set of proteins inside a cell (i.e., the *proteome*) and their dynamic changes under different physiological conditions [1]. Sensitivity appears the essential issue when analyzing proteins (or other types of molecules discussed below), because there is no easy way to replicate proteins as can be done for DNA molecules. Thus proteomic analysis has to be conducted on a small amount of material. Mass spectroscopy (MS) becomes the technology of choice in proteomics due to its high sensitivity and throughput. Among various MS instrumentations, liquid chromatogram coupled with tandem MS (LC/MS/MS), which can provide valuable information for peptide sequence and relative abundances, is the most commonly used platform for large scale proteomics projects. Nowadays, a single shotgun proteomics* experiment can generate >100 M (M, megabyte) of raw data within just a few hours. On the one hand, proteomics data management—automatic storing, sharing and analyzing these data—becomes a major challenge in proteomics; on the other hand, the large amount of proteomics data provides an unprecedented amount of data, and, as stated previously, holds promises through the application of novel data mining algorithms in discovering, for example, proteins related to essential biological functions and novel biomarkers for disease prognosis.

Although DNA and proteins have garnered most of the attention (in terms of scientific inquiry), other classes of biomolecules are no less important. Glycans (also known as carbohydrates or sugars) and lipids are the other two classes of biomolecules that have been well-studied in biochemistry for their roles as the structural molecules (e.g., as the main structural components for the cellular and intracellular membranes) and in the cellular metabolisms (e.g., as the main energy carriers and providers). The recent advancement in the research of these molecules (i.e., the *glycobiology* and *lipidology*, respectively), however, was concentrated on their relatively "new" roles as *signaling* molecules. Oligosaccharides and glycoconjugates (e.g., the glycoproteins and glycolipids) participate in many cellular processes, such as development, inflammation, and cell–cell recognitions, and have been well-known for decades in higher organisms [2]. The number of genetic disorders attributed to the dysregulation of glycan synthesis or mal-functioning of glycans keeps growing [3]. Similarly, many lipids were also revealed to participate in the intra- and intercellular signaling processes, including the deterrence and defense against pathogens. Studies show lipid functions are relevant to the diseases in which the role of lipids is either revealed (e.g., the diabetes) or being investigated (e.g., the Alzheimer's disease). Glycan and lipid research are even more challenging than proteins and nucleic acid research because of three reasons. First, unlike linear polymers, nucleic acids and proteins, glycans and lipids are nonlinear polymers that contain multiple linkage types (e.g., the single or double bonds) and often branching structures. Therefore, there

---

* Shotgun proteomics, also called bottom-up proteomics, is a popular strategy for MS-based proteome analysis, in which protein mixtures are first digested by using trypsin and subsequently analyzed by LC/MS/MS instrument.

is a lack of high throughput technologies to rapidly characterize their primary structures. Second, inside the cell, glycans and lipids are both synthesized through a template-free and step-wise fashion. The functional glycans (oligosaccharides) are assembled from monosaccharides by a complex machinery involving hundreds of proteins (encoded by ∼2% genes in human genome [4]). Similarly, the natural lipids are synthesized by many different enzymes using rich metabolites as substrates [5]. It is hard to study the functions of glycans and lipids using the well developed methodologies in molecular genetics. Finally, functional glycans and lipids often exist in high complexity but low abundance inside the cell, which demands the advancement of bioanalytical technologies.

In recent years, because of its high sensitivity and throughput, MS, the core technology used in proteomics, has also become the core bioanalytical technology for the analysis of glycans [6–9] and lipids [10,11], and has resulted two new multidisciplinary fields, *glycomics* and *lipidomics*, which aims at the full characterization and quantification of the functional oligosaccharides and lipids inside a cell (i.e., the *glycome* and *lipidome*), respectively. In addition to MS, array-based techniques have also been developed for large scale glycan and lipid analysis, akin to the DNA or protein arrays used in genomics and proteomics. In these arrays, ligands (e.g., glycans for the glycoarray [12–15] or lipids for lipid microarrays [16,17]) are (covalently or noncovalently) immobilized onto a microchip which is then used for a binding assay to screen interactive molecules (typically proteins) and/or quantitatively measure their interaction strengths. Although preliminary, these techniques provide complementary information as the MS-based glycomic methodologies (see Section 25.4 for details).

It is worth noting that not only the content but also the organization of glycomics and lipidomics research resembles that of the genomics. Large multiinvestigator consortia were formed in several continents, and collaborate (share technologies and data) actively. The consortium for functional glycomics (CFG, http://www.functionalglycomics.org) is one of the primary groups.

Glycomics and lipidomics research has been driven by the expectation of the discovery of novel biomarkers for disease prognosis usually from blood samples [18–21]. It is crucial to validate an experimental platform before applying it to the search of biomarkers. The validation should involve the accuracy and reproducibility of the whole procedure, including the sample preparation, retrieval of the molecule-of-interests, the instrumentation and the experimental conditions. Because most of these experiments are conducted on a large number of samples over time, the recording of the details of each experiment (referred to as the *metadata* in bioinformatics) is important, which can be later used as the bases for the validation and the high dimensional data analysis (e.g., the outlier detection).

We have so far concentrated on the commonalities between glycomics and lipidomics. Indeed, they both focus on biomolecules synthesized in similar fashions and also share the same core technology (MS). However, glycans and lipids each possess certain specific properties that require distinct approaches to these two areas. For instance, glycans are water-soluble compounds, whereas the lipids are fat-soluble, though not necessarily water-soluble. This distinction necessitates different experimental protocols to be employed in their respective analyses. These divergent protocols pose different computational challenges for data analyses, in particular for quantitative analysis. Another distinction resides in the structural characterization of glycans and lipids. Glycans are composed of several types of building blocks (*monosaccharides*). In contrast the common lipids have one or more fatty acid chains of varying lengths connected by a head group (e.g., the phosphate for the phospholipids). As a result, both glycans and lipids contain numerous *isobars* (molecules with different chemical formula but the same mass) and *isomers* (molecules with the same chemical formula but different configurations). Their resolution is a great challenge since they are essentially indistinguishable using MS. However, the emphasis of their resolution is quite different in glycomics and lipidomics. On the one hand, it is known that isomeric glycans may have different functions. Hence, a main goal of glycomics is to develop techniques to characterize the exact structure (including the branching topology and the linkages) of glycans, either through non-MS-based separation technologies or through the analysis of MS/MS or MS^n spectra (see Section 25.2 for details). On the other hand, the main function related structural feature of lipids appears to be the head group and the saturation of the fatty acid chains (i.e., the number of double bounds). Therefore, the main goal in lipidomics is to quantify different classes of lipids (with different head groups) and the saturated and unsaturated lipids in each class, which usually do not involve the resolution of isobars or isomers.

As in the areas of proteomics and genomics, bioinformatics approaches play an essential role in acquiring, analyzing and mining the data in glycomics and lipidomics [22–25]. The bioinformatics for glycomics and lipidomics has three major objectives: (1) the development of databases and data sharing strategies for the collection of structures of glycans and lipids, associated with the corresponding raw and annotated spectroscopic data; (2) algorithm and software development for the high throughput technologies used in glycomics and lipidomics, including the MS and the array-based techniques; and (3) the development of statistical and data mining methods for the discovery of novel biomarkers by using large scale glycomic and lipidomic data and integrating them with additional data recourses from genomics and proteomics. We will organize this chapter as following. In Sections 25.2 and 25.3, we will discuss the main technologies (with a focus on the MS-based ones) used in glycomics and lipidomics, respectively, and the computational challenges facing these areas. In Section 25.4, we will propose a computational framework that incorporates the prior biology knowledge by a systems biology methodology to mine the massive glycomic and lipidomic data.

25.2   Glycomics: Profiling Functional Carbohydrates Inside the Cell

The common monosaccharides found in higher animal glycans are listed in Table 25.1 [6]. Several others can be found in lower animals, bacteria and plants [2]. The carbon atoms in monosaccharides are numbered following the organic chemistry nomenclature, such that the hemiacetal carbon is referred to as C1 (Figure 25.1a). Two monosaccharides react and form a glycosidic bond between the C1 group of one monosaccharide and the alcohol group of the other while releasing a water molecule. Depending on which alcohol group participates in the reaction, there are four different types of glycosidic bonds, i.e., 1–2, 1–3, 1–4, and 1–6 bonds (often abbreviated as 2, 3, 4, and 6, respectively). A monosaccharide can also be connected with more than one monosaccharide at a time and form branching structures (Figure 25.1b). Branching structures can be represented by a tree (Figure 25.1c), in which each monosaccharide is represented by a symbol (see Table 25.1 for the list of symbols) and each glycosidic bond is represented by an edge.

Glycosylation of a protein is accomplished through linkage to Asn residues (designated as N-glycosylation) or to Ser/Thr residues (designated as O-glycosylation). All N-linked glycans share a common core structure (called the "pentamer") consisting of two N-acetylglucosamine (GlcNAc) residues and three mannose residues. Additional monosaccharides, mainly GlcNAc, mannose, galactose and sialic acid, can be further linked to this core structure to form diverse branching glycan structures. O-linked glycans do not have a common core structure and have higher sequence diversity. It is common that several glycan structures attach to the same glycosylation sites. This process is commonly referred to as the "microheterogeneity" [26].

The high throughput glycomic approaches roughly fall into two categories. The first class of approaches (referred to as the glycan profiling) studies

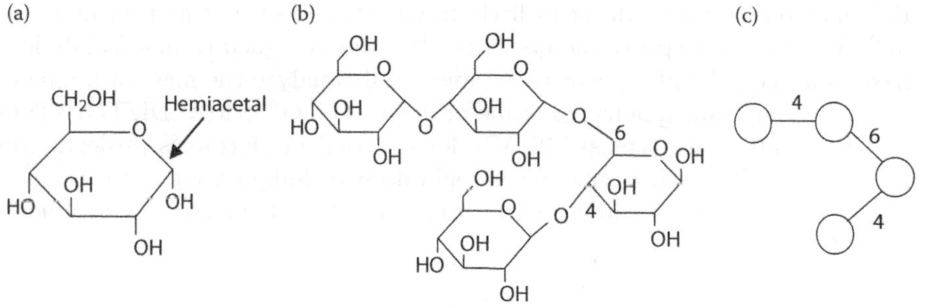

FIGURE 25.1: The structure of glycans. (a) The cyclic structure of glucose. (b) A tetraglucose, consisting of four glucoses with a branching. (c) The tree representation of the tetraglucose.

TABLE 25.1: The common monosaccharides in animal glycans.

Name (abbreviation)	Examples	Mass	Symbol
Hexose (Hex)	Glucose (Glc), mannose (Man), galactose (Gal)	162.05	○
HexA	Glucuronic acid (GlcA), Iduronic acid (IdcA)	203.08	⬭
HexNAc	N-acetylglucosamine (GlcNAc), N-acetylgalactosamine (GalNAc)	176.03	□
Xylose (Xyl)	–	132.04	▼
Fucose (Fuc)	–	146.06	▲
N-acetyl neuraminic acid (NeuAc)	–	291.10	◇

the released oligosaccharides in an attempt to elucidate their glycan structures tandem or even higher MS and to quantify different glycan species (by measuring the ion intensities in MS) [27–31]. The second class of approaches (referred to as the glycoproteomics analysis) studies the glycoproteins using conventional shotgun proteomics protocols. Glycoproteins can be identified by MS/MS analysis of those nonglycosylated tryptic peptides [8], whereas the MS/MS analysis of glycosylated tryptic peptides can be used to characterize and profile *microheterogeneity* of the site-specific glycosylations [32].

MS instruments used in the analysis of glycans differ in two key steps: ionization and mass analysis [7]. There are two main ionization techniques: electrospray ionization (ESI) and matrix-assisted laser desorption-ionization (MALDI), and three major techniques for mass analysis: ion-trap (IT), time-of-flight (TOF), and Fourier-transform ion cyclotron resonance (FTICR). Due to the large range of mass-to-charge ratio that can be covered by the TOF-MS, it has become the common mass analyzer (usually coupled with MALDI) for glycan profiling. Both ESI and MALDI are low energy ("soft") ionization methods, thus resulting in little fragmentation on the ionized molecule [33]. Tandem mass spectrometers (MS/MS) were designed to first isolate ionized molecules, then fragment them, and finally analyze the mass of fragment ions. ESI-ion trap, quadrupole-time-of-flight (Q-TOF), MALDI/TOF/TOF and ESI-FTICR are typical MS/MS devices used in glycomics projects. MS are often used together with other separation techniques, e.g., ion mobility spectroscopy (IMS), liquid chromatography (LC), and capillary electrophoresis (CE), in glycomics.

25.2.1 Bioinformatics resources and tools for mass spectroscopy (MS) data analysis in lipidomics

Table 25.2 summarizes the current databases for known glycan structures and software tools for the automatic identification and quantification

TABLE 25.2: Carbohydrate-related databases and software tools for glycan profiling using mass spectrometry.

Name	Developer	Resources/functionalities	Availability
EuroCarbDB	DKFZ, Germany	~ Entries of glycans; spectroscopic data associated to glycans; advanced search tools available.	http://www.eurocarbdb.org
CFG glycan database	CFG, USA	~ 7500 entries of glycan structures; advanced search tools available; both biological (e.g., N- and O-linked) and synthetic glycans.	http://www.functionalglycomics.org
CCSD	CCRC, USA	~ 50,000 entries of glycan structures from the literatures published before 1995; data were manually curated; accessed by the CarbBank program; continuous maintenance stopped.	http://www.boc.chem.uu.nl/sugabase/carbbank.html
BCSDB	Russia	~ 9000 bacterial glycan structures; advanced search tools available.	http://www.glyco.ac.ru/bcsdb
KEGG Glycan	KEGG, Japan	~ 12,000 entries of glycan structures; links to glycan synthetic pathways in KEGG	http://www.genome.ad.jp/kegg/glycan/
SimGlycan	Premier Biosoft	Matching experimental MS/MS spectra with probable fragment ions of glycans within a build-in database (with ~8000 entries); commercial available software.	http://www.premierbiosoft.com/glycan/features/glycoproteins.html
Glyco-workbench	Imperial college, London	Semiautomatic interpretation and annotation of glycan mass spectra; tools for assisting drawing glycan structures from scratch.	http://www.dkfz-heidelberg.de/spec/EUROCarbDB/GlycoWorkbench/

(Continued)

TABLE 25.2: (Continued)

Name	Developer	Resources/functionalities	Availability
Glycomod	SIB, Europe	Compute putative monosaccharide compositions for an experimentally determined mass.	http://www.expasy.org/tools/glycomod/
GlycoPepID	University of Kansas, USA	Matching experimental MS/MS spectra with the fragment ions of glycopeptides in the database (GlycoPep DB).	http://hexose.chem.ku.edu/
GlyPID	NCGG, USA	Mapping site-specific glycosylations in proteins using MS full scan and MS/MS spectra; report microheterogeneities associated to each glycosylation site.	http://ncgg.indiana.edu
Cartoonist	PARC, USA	Automatic drawing assignment of glycan structures (N- and O-linked) for experimentally determined m/z values.	Available from the author
OSCAR	UNH, USA	Assemble oligosaccharide topology (branching and linkage) from MS^n ion fragmentation pathways.	Available from the authors

of glycans using MS. The CarbBank CCSD database was the first curated database of glycan structures and remains one of the largest to date. However, its maintenance has stopped since 1996. Currently, the glycan structures can be accessed by three major databases: the CFG database maintained by the Consortium for functional glycomics (CFG), the KEGG glycan database maintained by Kyoto Encyclopedia of Genes and Genomes (KEGG), and the EuroCarbDB maintained by the German Cancer Research Center (DKFZ). There is active data sharing and collaboration among the development teams of these databases. The integration of the glycan structure data into a central database is planned.

Many computational tools have been developed for glycan identification and quantification using MS. Cartoonist and Cartoonist2 [34,35] are tools for automatic annotation of mass spectra for profiling N-linked and O-linked glycans, respectively. Glyco-workbench [36] is a suite of programs assisting the annotation of the fragment ions in the MS/MS spectra and semi-automatic identification of glycans. SimGlycan is the first commercial software tool toward a fully automatic identification of glycan using MS/MS spectra. It adopted a database searching approach (as previously described in [37–39]) that compares the MS/MS spectra with the list of probable fragment ions of glycans within a build-in database. A similar database searching approach was also proposed for the identification of intact glycopeptides [40]. Finally, the software solution to characterize glycan structure from the fragmentation pathways in MS^n experiment was also developed [41].

25.2.2 Computational challenges in glycomics

As we discussed above, the identification of glycan structural isomers remains among the most difficult problems in glycomics. It has been observed that high energy collision induced dissociation (CID), resulting from MS/MS, contains significant cross-ring ions providing valuable information (e.g., the characteristic ions occurring in some glycan isomers, but not in all, or the distinct intensity patterns of cross-ring ions for different glycan isomers) for distinguishing the structural glycan isomers [42]. MALDI/TOF/TOF is the commonly used technology to acquire the CID-MS/MS spectra of glycans, and several software tools have been developed to analyze these spectra for isomer characterization [43]. Continuous development of the probabilistic models and the scoring functions (as used for the peptide identification in proteomics) is required for better prediction of the glycan isomer structures from the observed patterns of fragment ions in their MS/MS spectra.

Sometimes, MS/MS spectra do not contain sufficient information to distinguish a glycan from all of its isomers, and it is desirable to identify its exact (nonambiguous) chemical structure. In this case, MS^n based on ion-trap-MS is needed [44]. Assisted by appropriate software tools [41], the complete chemical structure of a glycan can be characterized from its fragmentation pathway under MS^n in a semiautomatic fashion. Although this methodology is reliable and robust, it is time consuming and with only limited applications in high throughput glycomics projects. Thus, developing the software that directly interfaces the MS instrument and fully automates the procedure of ion-trapping, fragmentation and data analysis is currently needed.

Technically, the glycan isomers should have already been separated before dissociation analysis described below. It is, however, not trivial to separate them in a high throughput experiment. MS is not the right choice since isomers have the identical mass. Other bioanalytical technologies, including the condensed phase separation like LC and electrophoresis and the gas phase separation like the ion mobility spectrometry, have been successful in some cases [45–47]. Another interesting approach is to directly fragment a mixture of (a few) isomeric ions trapped in the mass spectrometer, and then to identify each of the structures of these isomers based on the pattern of the fragment ions. It has been shown that, using novel dissociation methods such as photo-dissociation [48], the characteristic fragment ions corresponding to each of these isomeric ions, were all observed providing sufficient information to predict the glycan isomers in the sample.

Recently, the glycoproteomics approach (the direct analysis of glycoprotein mixtures) has attracted much attention, and several new approaches are being proposed that utilize both MS full scan and MS/MS analysis of glycopeptides in an attempt to simultaneously identify the glycosylation sites and the microheterogeneities associated to each site. A recently published algorithm, GlycoX [20], aims at identifying the N-glycosylation sites and associated microheterogeneities within known proteins by mining putative glycopeptide ions captured in FT-MS scans. This algorithm, however, did not incorporate useful information from MS/MS spectra, and thus may report spurious glycopeptides. Other pertinent algorithms [49,50] do not assume any prior knowledge of protein contents in the sample, and attempt to characterize the glycoproteins, the glycosylation sites and the microheterogeneities simultaneously by integrating information from MS and MS/MS spectra. All the above algorithms suffer from the missing data problem in a single LC/MS/MS analysis, i.e., a majority of the predicted glycopeptide ions are not selected for the MS/MS analysis due to the duty cycle of MS. As a result, many of these ions are not real glycopeptides, but are simply artifacts by matching the expected glycopeptide mass by chance. To overcome this problem, software-guided targeted analysis of the glycoprotein samples is desirable.

Like proteins, many glycans with similar structures often carry out the same biological functions. Hence, it is an essential bioinformatics problem to elucidate the common substructures (known as the *glycan motifs*) among a given set of known glycans to study glycan-protein interactions and to discover biomarkers (see [51,52] for an excellent review of this area). Various machine learning techniques have been applied to this problem, ranging from the kernel methods (e.g., the support vector machines, SVMs [53], and the tree kernels [54]) to probabilistic graphical models (e.g., the hidden Markov models, HMMs [55,56]). Although considerable progress has been made in this area, this problem remains largely open because the lack of benchmarking data to test these proposed methods. In the near future, with the availability of glycan profiling data from disease samples, glycan motif finding will become one of the core computational problems in glycomics.

25.3 Lipidomics: Global Analysis of Lipid Derivatives

Fatty acids are the common building blocks of most bioactive lipids. Traditionally, lipids are classified into eight categories according to the chemical structures of their head groups linked to the fatty acid chains [57]: fatty acids (FA), glycerolipids (GL), glycerophospholipids (GP), sphingolipids (SP), sterol (ST), prenol (PR), saccharolipids (SL), and polyketides (PK). Of them, phospholipids are the most abundant lipid class present in eukaryotic cells [58].

Three MS methods have been employed in lipidomics [10], including the gas chromatography coupled electron impact ionization (GC-EI-MS), the liquid chromatography coupled electrospray ionization (LC-ESI-MS) and less commonly used matrix assisted laser desorption/ionisation (MALDI-MS). GC-EI-MS generates a reproducible pattern of fragment ions that can be compared with a library of spectra of previously characterized lipids to identify the lipids in the sample. The identification becomes quite challenging when the lipid mixture is quite complex, and contains some isomeric species (e.g., the unsaturated fatty acids with the same length). ESI and MALDI are "soft" ionization methods that can detect nonvolatile intact lipids. LC-ESI-MS is the most popular MS platform in lipidomics because of the easy coupling of two complementary separation techniques (LC and MS) through the liquid phase elusion. MS/MS is often employed in ESI-MS to obtain a pattern of characteristic fragment ions (i.e., the *fingerprint*) for the identification of lipid species [59]. When analyzing complex lipid mixtures, the sensitivity of ESI-MS can increase by applying nano-ESI, an electrospray process with reduced flow rates [60]. Compared with GC-EI and LC-ESI, MALDI-MS has not been widely used in lipidomics, mainly because the choice of MALDI material has an undue high impact on the lipid quantification [61]. However, MALDI imaging MS (MALDI-IMS) was recently implemented as a powerful high throughput tool to monitor the time and spatial distribution of lipids in various tissues [62,63].

The infusion of a complex lipid mixture into the mass spectrometer is the most efficient way for lipid analysis using ESI-MS. Combining it with effective separation techniques for predetermined lipid classes, the *shotgun lipidomics* approach was proposed to the analyses of directly extracted lipids from biological samples [58]. Afterward, appropriate characteristic ions can be used to identify and quantify different lipid classes. It has been shown that this

approach can identify and quantify five classes (including over 450 phospholipid species) in mammalian cells [58,59], which corresponds to roughly 95% of
the total cellular lipidome. Despite its high efficiency, shotgun lipidomics may
suffer large variations in lipid quantification because of the ion suppression
during the ionization procedures. Among many coeluting lipid species, some
classes of lipids (e.g., some phospholipids [64,65]) are more easily ionized than
the others, especially when using positive ion mode [64].

25.3.1 Bioinformatics resources and tools for MS data analysis in lipidomics

Table 25.3 summarizes the current databases for lipids and software tools
for the automatic identification and quantification of lipids using MS. Currently, the biological functional lipids are collected by three major lipid
databases around the world: the LIPID MAPS database maintained by the
LIPID metabolites and pathways strategy (MAPS) consortium [24], the LDB
located at the University of Limerick, Ireland, and the LipidBank maintained
by Japanese Conference on the Biochemistry of lipids (JCBL) [66]. Large consortia such as LIPID MAPS, the European Lipidomics Initiative and JCBL
adopt high throughput lipidomics methodologies in an attempt to identify all
lipid signaling molecules. The resulting data from these projects are deposited
to the corresponding lipid databases within a short period of time after being
generated.

To automatically analyze the MS data from large scale lipidomic projects,
software tools have been developed for lipid identification and quantification.
The main goal of lipidomics is two-fold: to determine the total abundance for
each major lipid class, and to quantify the distribution of saturated and non-
saturated lipids in each class. It is a relatively easy task to distinguish a lipid
class from another based on their respective fragment ions, since upon CID,
a characteristic fragment ion (corresponding to the loss of the head group)
is usually observed. For the most abundant class, phospholipids, a fragment
ion of m/z 184 is produced corresponding to the protonated phosphocholine
group. These characteristic ions can be used to classify the precursor ions. Several software tools (e.g., the Lipid Navigator [67], Lipid MS predictor [24] and
LIPIDQA [68]) have been developed to automate this process. This approach
can also be extended to the quantification of lipid classes by comparing the total intensities of the identified ions with those of the spiked internal standards,
as implemented in LIMSA [69–71].

Although the relative quantification of lipids in different samples is quite
successful, predicting the absolute lipid abundances remains a challenge, even
when internal standards are applied. This is mainly due to the strong ion
suppression that may happen for certain lipid species during ionization. To
overcome this hurdle, computational models are needed to predict the suppression effects for different lipid species based on their chemical properties.
A related concept of peptide *detectability* was proposed in proteomics [72,73],

TABLE 25.3: Lipid-related databases and software tools for lipid identification and quantification using mass spectrometry.

Name	Developer	Resources	Web address
LipidBank	JCBL, Japan	~ 6000 entries of lipid species; structural formula, thermodynamic properties, biological activities, metabolism and spectroscopic data; advanced search tools available.	http://lipidbank.jp
LDB	University of Limerick, Ireland	LIPIDAT: ~20,000 of lipid records of their thermodynamic properties; LIPIDAG: ~1600 phase diagrams related to lipid miscibility; LSD: ~13,000 lipid structures.	http://www.caffreylabs.ul.ie
LIPID MAPS	LIPID MAPS Consortium, USA	LMSD: ~10,000 lipid species with links to NCBI Substance ID (SID); LMPD: lipid related protein sequences; tools for advanced search and pathway analysis available.	http://www.lipidmaps.org
SOFA	FRCNF, Germany	Lipids related to plant oils.	http://www.bagkf.de/sofa
NPLC	National Plant Lipid Cooperative	~ 18,000 references on plant acyl-lipids.	https://www.msu.edu/user/ohlrogge/nplc.html

(Continued)

TABLE 25.3: (Continued)

Name	Developer	Resources	Web address
Lipid Search/Lipid Navigator	JCBL, Japan	Matching raw MS data with theoretical m/z values of precursor and fragment ions; for phospholipids, fatty acids, glycerolipids and their metabolites; selective mass surveys; quantification after correcting relative ionization efficiency by internal standards; support positive and negative ion modes.	http://lipidsearch.jp
LIMSA	University of Helsinki, Finland	Automatic quantification of lipids using mass spectrometry.	http://www.helsinki.fi/science/lipids/software.html
LIPID DB	VTT, Finland	Matching MS data with characteristic ions for a scaffold of theoretically possible lipids; support positive and negative ion modes. for glycerophospholipids, sphingolipids, glycerolipids, and sterol esters.	
LIPID MS predictor	LIPID MAPS Consortium	Matching a m/z value of interest with "high-probabilities" product ions from a list of candidate lipids; for glycerolipids, cardiolipins and glycerophospholipids.	http://www.lipidmaps.org
LipidQA	Washington University, USA	Matching fragment ions with the theoretical spectrum of standard lipids in the library; automatic accommodation of internal standards and calibration curves for quantifying complex lipid mixture.	Available from the authors

and has led to a reasonable improvement for protein quantification. The similar algorithm may be applied here to improve absolute lipid quantification.

25.4   Toward a Systems Biology Approach to Biomarker Discovery in Glycomics and Lipidomics

Glycan or lipid profiling provides tools for the discovery new potential *features* (i.e., the identity or quantity of a class of glycans or lipids), which are significantly different in disease and healthy samples, as candidate biomarkers. The combination of these features using multidimensional statistical methods (such as the principal component analysis, PCA) can enhance the resolution power of individual features [10,30,74]. A similar approach has been taken for biomarker discovery using transcriptomic [75,76] and proteomic data [77]. However, the mRNAs and proteins are synthesized independently in a template-based fashion; hence, the inherent correlations among genes are low unless they are "functionally" related. Furthermore, employing standard machine learning techniques can be applied since each gene/protein can be viewed as an individual feature. Glycans and lipids, on the other hand, are synthesized in a series of reactions catalyzed by enzymes, in which some are used as substrates for the others. As a result, there are strong correlations among the abundances of different glycan/lipid species in this stream-like flow. Note these correlations are not necessarily related to their biological functions. For example, if a glycan G is related to certain disease (and abundant in the disease samples), it is likely that many glycans $G_1, G_2, G_3, \ldots G_n$ in the pathway leading to the final synthesis of G are also abundant in the disease sample, although they are not necessarily directly related to disease process. One may argue that even though these features are not biologically relevant to the disease per se, as long as they are different between disease and normal samples, all these glycans are worth examining. Incorporating the inherent correlations among features may significantly improve statistical power of biomarker discovery. A novel systems biology approach, called *network marker* approach, is proposed recently for the analysis of gene expression data, which attempts to identify the subnetwork of biomolecules (e.g., protein interaction subnetworks) that show significantly different abundances in disease and normal samples [78]. Since many synthetic pathways of lipids and glycomics have been well studied, this novel approach is readily applied to the mining of biomarkers in glycomics and lipidomics. It also provides a straightforward way to integrate the glycan/lipid profiling data and the expression data of the enzymes involved in these pathways, which are represented as the nodes (glycans/lipids) and the edges (enzymes), respectively. We note, however, there are two computational challenges in implementing this approach. First, due to the difficulty characterizing the structural isomers (see Sections 25.2 and 25.3), while glycan/lipid

(a)   (b)
Known pathways   Observed glycan levels, conditions, under disease, and control

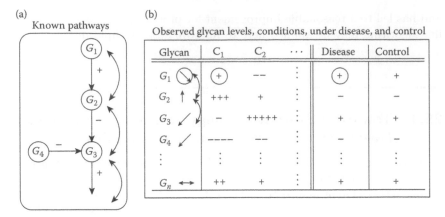

FIGURE 25.2: A data mining approach to discovering association among glycans and disease.

profiling often provides the abundances of a group of molecules, it does not work for specific molecules. In this case, how to combine the pathway information and the profiling data is nontrivial. Second, some synthetic pathways are not completely known. It is an open problem how to handle the missing edges (reactions) in the pathway so that the network markers and the missing reactions can be simultaneously characterized.

In Figure 25.2 we present a high-level data mining approach to assessing and understanding glycans and disease. In panel A, we have the known metabolic glycomic pathways. In panel B we have the relative abundances of the glycans associated with conditions and two classes, disease and control. In this case, we are interested in comparing the simulated metabolic levels with the observed (vertically) and cross classifying this with the condition features. The double headed arrows represent the relative ratios of pairs of species in the pathway. This, together with the conditions can be used to classify disease and control states.

25.5 Conclusions

Bioinformatics research in glycomics and lipidomics has so far focused on automating the analysis of high throughput data. With the rapid growth of the size of data in these fields, we anticipate the shift of research interests from data analysis to data mining, especially for the purpose of disease biomarker discovery. In particular, sophisticated data mining algorithms are greatly needed to integrate glycomic/lipidomic data with other data sources (e.g., the gene expression and proteomics data) as well as with prior biological knowledge.

Acknowledgment

This study was funded by NSF grant #DBI-0642897 to HT, and was also supported by the National Center for Glycomics and Glycoproteomics funded by NIH-National Center for Research Resources (NCRR grant #RR018942-02).

References

[1] Aebersold, R. and M. Mann. Mass spectrometry-based proteomics. *Nature*, 2003, 422(6928), 198–207.

[2] Varki, A., et al. *Essentials of Glycobiology*. Cold Spring Harbor, New York: Cold Spring Harbor Laboratory Press, 1999.

[3] Freeze, H.H. and M. Aebi. Altered glycan structures: the molecular basis of congenital disorders of glycosylation. *Curr. Opin. Struct. Biol.*, 2005, 15(5), 490–498.

[4] Bertozzi, C.R. and L.L. Kiessling. Chemical glycobiology. *Science*, 2001, 291, 2357–2364.

[5] Buhman, K.K., H.C. Chen, and R.V. Farese, Jr. The enzymes of neutral lipid synthesis. *J. Biol. Chem.*, 2001, 276(44), 40369–40372.

[6] Dell, A. and H.R. Morris. Glycoprotein structure determination by mass spectrometry. *Science,* 2001, 291(5512), 2351–2356.

[7] Zaia, J. Mass spectrometry of oligosaccharides. *Mass Spectrom. Rev.*, 2004, 23(3), 161–227.

[8] Novotny, M.V. and Y. Mechref. New hyphenated methodologies in high-sensitivity glycoprotein analysis. *J. Sep. Sci.*, 2005, 28(15), 1956–1968.

[9] Mechref, Y. and M.V. Novotny. Miniaturized separation techniques in glycomic investigations. *J. Chromatogr. B*, 2006, 841(1–2), 65–78.

[10] Roberts, L.D., et al. A matter of fat: an introduction to lipidomic profiling methods. *J. Chromatogr. B*, 2008, 174–181.

[11] Isaac, G., et al. New mass-spectrometry-based strategies for lipids. *Genet. Eng. (NY)*, 2007, 28, 129–157.

[12] Stevens, J., et al. Glycan microarray technologies: tools to survey host specificity of influenza viruses. *Nat. Rev. Microbiol.*, 2006, 4, 857–864.

[13] Dyukova, V.I., et al. Hydrogel glycan microarrays. *Anal. Biochem.*, 2005, 347(1), 94–105.

[14] Feizi, T., et al. Carbohydrate microarrays—a new set of technologies at the frontiers of glycomics. *Curr. Opin. Struct. Biol.*, 2003, 13, 637–645.

[15] Xia, B., et al. Versatile fluorescent derivatization of glycans for glycomic analysis. *Nat. Methods*, 2005, 2(11), 845–850.

[16] Feng, L. Probing lipid-protein interactions using lipid microarrays. *Prost. & Other Lipid Mediators*, 2005, 77(1–4), 158–167.

[17] Kanter, J.L., et al. Lipid microarrays identify key mediators of autoimmune brain inflammation. *Nat. Med.*, 2006, 12(1), 138–143.

[18] Miyamoto, S. Clinical applications of glycomic approaches for the detection of cancer and other diseases. *Curr. Opin. Mol. Ther.*, 2006, 8(6), 507–513.

[19] Wiest, M.M. and S.M. Watkins. Biomarker discovery using high-dimensional lipid analysis. *Curr. Opin. Lipidol.*, 2007, 18(2), 181–186.

[20] Gross, R.W. and X. Han. Unlocking the complexity of lipids: using lipidomics to identify disease mechanisms, biomarkers and treatment efficacy. *Future Lipidol.*, 2006, 1(5), 539–547.

[21] Wuhrer, M. Glycosylation profiling in clinical proteomics: heading for glycan biomarkers. *Expert Rev. Proteomics*, 2007, 4(2), 135–136.

[22] von der Lieth, C.-W., T. Lütteke, and M. Frank. The role of informatics in glycobiology research with special emphasis on automatic interpretation of MS spectra. *Biochim. Biophys. Acta (BBA)*, St. Louis, MO, USA. 2006, 1760(4), 568–577.

[23] Perez, S. and B. Mulloy. Prospects for glycoinformatics. *Curr. Opin. Struct. Biol.*, 2005, 15(5), 517–524.

[24] Fahy, E., et al. Bioinformatics for lipidomics. In *Methods in Enzymology.* Academic Press, 2007, 247–273.

[25] Dwek, R.A. Glycobiology: toward understanding the function of sugars. *Chem. Rev.*, 1996, 96(2), 683–720.

[26] Mechref, Y. and M.V. Novotny. Structural investigations of glycoconjugates at high sensitivity. *Chem. Rev.*, 2002, 102(2), 321–369.

[27] Nishimura, S.-I., et al. High-throughput protein glycomics: combined use of chemoselective glycoblotting and MALDI-TOF/TOF mass spectrometry. *Angew Chem. Int. Ed. Engl.*, 2004, 44, 91–96.

[28] Ressom, H.W., et al. Analysis of MALDI-TOF mass spectrometry data for discovery of peptide and glycan biomarkers of hepatocellular carcinoma. *J. Proteome Res.*, 2008, 7(2), 603–610.

[29] Kyselova, Z., et al. Breast cancer diagnosis and prognosis through quantitative measurements of serum glycan profiles. *Clin. Chem.*, 2008, 54(7), 1166–1175.

[30] Kyselova, Z., et al. Alterations in the serum glycome due to metastatic prostate cancer. *J. Proteome Res.*, 2007, 6(5), 1822–1832.

[31] Jang-Lee, J., et al. Glycomic profiling of cells and tissues by mass spectrometry: fingerprinting and sequencing methodologies. *Methods Enzymol.*, 2006, 415, 59–86.

[32] Ethier, M., D. Figeys, and H. Perreault. N-glycosylation analysis using the StrOligo algorithm. *Methods Mol. Biol.*, 2006, 328, 187–197.

[33] Mechref, Y., N.V. Novotny, and C. Krishnan. Structural characterization of oligosaccharides using MALDI-TOF/TOF tandem mass spectrometry. *Anal. Chem.*, 2003, 75(18), 4895–4903.

[34] Goldberg, D., et al. Automatic determination of O-glycan structure from fragmentation pectra. *J. Proteome Res.*, 2006, 5(6), 1429–1434.

[35] Goldberg, D., et al. Automatic annotation of matrix-assisted laser desorption/ionization N-glycan spectra. *Proteomics*, 2005, 5(4), 865–875.

[36] Ceroni, A., et al. GlycoWorkbench: a tool for the computer-assisted annotation of mass spectra of glycans. *J. Proteome Res.*, 2008, 7(4), 1650–1659.

[37] Ethier, M., et al. Global and site-specific detection of human integrin $\alpha 5\beta 1$ glycosylation using tandem mass spectrometry and the StrOligo algorithm. *Rapid Commun. Mass Spectrom.*, 2005, 19(5), 721–727.

[38] Lohmann, K.K. and C.-W. von der Lieth. GlycoFragment and GlycoSearchMS: web tools to support the interpretation of mass spectra of complex carbohydrates. *Nucl. Acids Res.*, 2004, 32(suppl 2), W261–W266.

[39] Ethier, M., et al. Application of the StrOligo algorithm for the automated structure assignment of complex N-linked glycans from glycoproteins using tandem mass spectrometry. *Rapid Commun. Mass Spectrom.*, 2003, 17(24), 2713–2720.

[40] Irungu, J., et al. Simplification of mass spectral analysis of acidic glycopeptides using GlycoPep ID. *Anal. Chem.*, 2007, 79(8), 3065–3074.

[41] Lapadula, A.J., et al. Congruent strategies for carbohydrate sequencing. 3. OSCAR: an algorithm for assigning oligosaccharide topology from MSn data. *Anal. Chem.*, 2005, 77(19), 6271–6279.

[42] Mechref, Y., M.V. Novotny, and C. Krishnan. Structural characterization of oligosaccharides using Maldi-TOF/TOF tandem mass spectrometry. *Anal. Chem.*, 2003, 75(18), 4895–4903.

[43] Tang, H., Y. Mechref, and M.V. Novotny. Automated interpretation of MS/MS spectra of oligosaccharides. *Bioinformatics,* 2005, 21(suppl 1), i431–i439.

[44] Ashline, D.J., et al. Carbohydrate structural isomers analyzed by sequential mass spectrometry. *Anal. Chem.*, 2007, 79(10), 3830–3842.

[45] Takegawa, Y., et al. Separation of isomeric 2-aminopyridine derivatized N-glycans and N-glycopeptides of human serum immunoglobulin G by using a zwitterionic type of hydrophilic-interaction chromatography. *J. Chromatogr. A*, 2006, 1113(1–2), 177–181.

[46] Zhuang, Z., et al. Electrophoretic analysis of N-glycans on microfluidic devices. *Anal. Chem.*, 2007, 79(18), 7170–7175.

[47] Isailovic, D., et al. Profiling of human serum glycans associated with liver cancer and cirrhosis by IMS/MS. *J. Proteome Res.*, 2008, 7(3), 1109–1117.

[48] Devakumar, A., et al. Laser-induced photofragmentation of neutral and acidic glycans inside an ion-trap mass spectrometer. *Rapid Commun. Mass Spectrom.*, 2007, 21(8), 1452–1460.

[49] Yin, W., et al. A computational approach for the identification of site-specific protein glycosylations through ion-trap mass spectrometry. In *Proceedings of RECOMB Satellite Conferences on: Systems Biology and Computational Proteomics*. The third RECOMB satellite meeting on Proteomics, Lecture Notes in Bioinformatics, Springer, 2007, 4532(96–107).

[50] Goldberg, D., et al. Automated N-glycopeptide identification using a combination of single- and tandem-MS. *J. Proteome Res.*, 2007, 6(10), 3995–4005.

[51] Mamitsuka, H. Informatic innovations in glycobiology: relevance to drug discovery. *Drug Discov. Today*, 2008, 13, 118–23.

[52] Aoki-Kinoshita, K.F. An introduction to bioinformatics for glycomics research. *PLoS Computation. Biol.*, 2008, 4(5), e1000075.

[53] Hizukuri, Y., et al. Extraction of leukemia specific glycan motifs in humans by computational glycomics. *Carbohydr. Res.*, 2005, 340(14), 2270–2278.

[54] Yamanishi, Y., F. Bach, and J.-P. Vert. Glycan classification with tree kernels. *Bioinformatics*, 2007, 23(10), 1211–1216.

[55] Aoki, K.F., et al. Application of a new probabilistic model for recognizing complex patterns in glycans. *Bioinformatics*, 2004, 20(suppl 1), i6–i14.

[56] Aoki-Kinoshita, K.F., et al. ProfilePSTMM: capturing tree-structure motifs in carbohydrate sugar chains. *Bioinformatics*, 2006, 22(14), e25–e34.

[57] Fahy, E., et al. A comprehensive classification system for lipids. *J. Lipid Res.*, 2005, 46(5), 839–862.

[58] Han, X. and R.W. Gross. Shotgun lipidomics: multidimensional MS analysis of cellular lipidomes. *Expert Rev. Proteomics*, 2005, 2(2), 253–264.

[59] Milne, S., et al. Lipidomics: an analysis of cellular lipids by ESI-MS. *Methods*, 2006, 39(2), 92–103.

[60] Brugger, B., et al. Quantitative analysis of biological membrane lipids at the low picomole level by nano-electrospray ionization tandem mass spectrometry. *Proc. Natl. Acad. Sci. USA*, 1997, 94(6), 2339–2344.

[61] Knochenmuss, R. Ion formation mechanisms in UV-MALDI. *The Analysts*, 2006, 131, 966–986.

[62] Cornett, D.S., et al. MALDI imaging mass spectrometry: molecular snapshots of biochemical systems. *Nat. Methods*, 2007, 4(10), 828–833.

[63] Walch, A., et al. MALDI imaging mass spectrometry for direct tissue analysis: a new frontier for molecular histology. *Histochem. Cell Biol.*, 2008, 130(3), 421–434.

[64] Wenk, M.R., et al. Phosphoinositide profiling in complex lipid mixtures using electrospray ionization mass spectrometry. *Nat. Biotech.*, 2003, 21(7), 813–817.

[65] Petkovic, M., et al. Detection of individual phospholipids in lipid mixtures by matrix-assisted laser desorption/ionization time-of-flight mass spectrometry: Phosphatidylcholine prevents the detection of further species. *Anal. Biochem.*, 2001, 289(2), 202–216.

[66] Watanabe, K., E. Yasugi, and M. Oshima. How to search the glycolipid data in LIPIDBANK for Web: the newly developed lipid database. *Japan Trend Glycosci. Glycotechnol.*, 2000, 12, 175–184.

[67] Houjou, T., et al. A shotgun tandem mass spectrometric analysis of phospholipids with normal-phase and/or reverse-phase liquid chromatography/electrospray ionization mass spectrometry. *Rapid Commun. Mass Spectrom.*, 2005, 19(5), 654–666.

[68] Song, H., et al. Algorithm for processing raw mass spectrometric data to identify and quantitate complex lipid molecular species in mixtures by data-dependent scanning and fragment ion database searching. *J. Am. Soc. Mass Spectrom.*, 2007, 18(10), 1848–1858.

[69] Haimi, P., et al. Software tools for analysis of mass spectrometric lipidome data. *Anal. Chem.*, 2006, 78(24), 8324–8331.

[70] Yetukuri, L., et al. Informatics and computational strategies for the study of lipids. *Mol. BioSyst.*, 2008, 4, 121–127.

[71] Yetukuri, L., et al. Bioinformatics strategies for lipidomics analysis: characterization of obesity related hepatic steatosis. *BMC Syst Biol.*, 2007, 1(12), 12–26.

[72] Tang, H., et al. A computational approach toward label-free protein quantification using predicted peptide detectability. *Bioinformatics*, 2006, 22(14), e481–e488.

[73] Lu, P., et al. Absolute protein expression profiling estimates the relative contributions of transcriptional and translational regulation. *Nat. Biotech.*, 2007, 25(1), 117–124.

[74] Ressom, H.W., et al. Analysis of MALDI-TOF mass spectrometry data for detection of glycan biomarkers. *Pac. Symp. Biocomput.*, 2008, 216–227.

[75] Welsh, J.B., et al. Analysis of gene expression profiles in normal and neoplastic ovarian tissue samples identifies candidate molecular markers of epithelial ovarian cancer. *Proc. Natl. Acad. Sci. USA*, 2001, 98(3), 1176–1181.

[76] Ramaswamy, S., et al. Multiclass cancer diagnosis using tumor gene expression signatures. *Proc. Natl. Acad. Sci. USA*, 2001, 98(26), 15149–15154.

[77] Zhang, Z., et al. Three biomarkers identified from serum proteomic analysis for the detection of early stage ovarian cancer. *Cancer Res.*, 2004, 64(16), 5882–5890.

[78] Chuang, H.-Y., et al. Network-based classification of breast cancer metastasis. *Mol. Syst. Biol.*, 2007, 3, 140–149.

Chapter 26

Data Mining Chemical Structures and Biological Data

Glenn J. Myatt

Myatt & Johnson, Inc.

Paul E. Blower

Ohio State University

26.1 Introduction

Technologies such as combinatorial chemistry, high throughput screening (HTS), and high throughput genomics/proteomics have increased the speed at which new chemicals are made and tested within various biological systems. This has led to a significant increase in the amount of chemicals and related biological data available to researchers. The associated biological data includes biological activity data, toxicity, and differential gene/protein expression on cellular response to compound treatment. The ability to data mine these large chemical databases alongside the biological information is critical to the discovery of new therapeutic agents and to providing insight into disease mechanisms. Table 26.1 illustrates the availability of a number of large chemical databases in the public domain.

Data mining chemical information has been applied to a number of areas including:

- Chemogenomics: chemogenomics involves the use of genomics to measure the system-wide effect of a compound on a biological system, either single cells or whole organisms. It combines high-throughput genomics or proteomics profiling with data mining chemical information to study the response of a biological system to chemical compounds (Bredel and Jacoby, 2004).

- Selecting screening sets: the choice of compounds to screen impacts the quality of any resulting findings. To enhance the probability of identifying novel lead compounds, chemical libraries are often designed to be diverse, that is, containing a wide variety of chemical scaffolds. In addition to diversity considerations, compounds in the screening sets are also chosen to include properties known to be important to marketed drugs. These property profiles are often identified through data mining pharmaceutical databases. Large chemical databases can also be screened in silico using computational models that have been generated from empirical in vivo or in vitro screening data to predict biological response.

- Identifying lead series: the testing of hundreds of thousands of chemicals using HTS is routine in the pharmaceutical industry. Data mining the chemical structures in combination with the resulting biological activity data is commonly used to identify chemical attributes, such as chemical scaffolds, that can be used in identifying the next batch of chemicals to screen (Blower et al., 2004).

- Optimizing lead series: the iterative process of optimizing chemical lead series generates data over thousands of candidate drug compounds for many different properties that include efficacy and transport properties.

TABLE 26.1: Selected sources of chemical data available in the public domain.

Name	Description	Number of compounds	URL
PubChem	Database of chemicals and related biological information, developed by the National Center for Biotechnology Information (NCBI)	19,248,988*	pubchem.ncbi.nlm.nih. gov
ZINC	Database of available compounds for virtual screening from the University of California, San Francisco	12,865,326*	zinc.docking.org
E-Molecules	Database of available chemicals with supplier information	~9,900,000*	www.emolecules.com
DSSTOX	Chemical structure files associated with toxicity data from the US EPA.	12,950**	www.epa.gov/NCCT/ dsstox

* Count in July 2008.
** Number of chemical structures deposited in PubChem in July 2008 with a DSSTOX label.

This data must be simultaneously optimized to identify the most promising chemicals to take forward.

- Predictive ADMET properties: early detection of problematic issues in chemicals under consideration is a critical issue. Over the years, information on why compounds fail has been generated as a by-product of different research programs. This information related to ADMET characteristics (absorption, distribution, metabolism, elimination, and toxicity) must be collected, integrated, and cleaned before being data mined to understand the chemical characteristics of those chemicals that failed.

Data mining chemical information directly is not possible because of the format of the data. The data is not in a traditional table that is directly amenable to data analysis or data mining methods. The data comprises of the atoms and bonds of the chemical structure. As a result of the many specific issues associated with analyzing and data mining chemical structure data, a field often referred to as chemoinformatics (or cheminformatics) has been established (Gasteiger and Engel, 2003; Leach and Gillet, 2007). In fact, the use of chemical information within standard relational database management systems often requires the use of specialized chemical cartridges to handle the chemical structures. A large portion of the field of chemoinformatics deals with data mining chemical structures in combination with biological data. In addition, chemoinformatics also covers the analysis of reactions, patents, and chemicals in 3D, which are not covered in this chapter. Chemical data is rarely analyzed in isolation from other data such as biological activity data or genomics information. The processing of this related data is not discussed in detail in this chapter.

The first step in any analysis is to read the chemical structures, using a description of its connectivity. Once read in, a certain amount of processing of the chemical structures is initially required to take care of issues such as the use of different representations for the same chemicals or chemical fragments.

Generally, any data mining exercise requires a table containing rows (observations) and columns (variables) describing the data. Three types of variables are usually required: (1) a *primary key*, (2) *independent variables*, and (3) a *response* variable. A primary key is a unique identification for an individual chemical structure and is valuable for referring to a specific chemical and for integrating data from different sources. Independent variables or *descriptors* are variables that describe attributes thought to be important characteristics of the chemicals. In unsupervised data mining approaches, such as clustering, only descriptors are used to perform data mining. In supervised approaches, such as building predictive models, both chemical descriptors and a response (often related to biological activity) are needed. The generation of these chemical descriptors is a major focus of chemoinformatics. Selection of appropriate chemical descriptors is also important, since there are thousands of ways of describing chemicals and only a handful may be important to the problem being addressed. A critical task is to match the chemical descriptors to the problem at hand, since it is in these terms that the results are presented.

Having generated the necessary data table, an analysis of the chemical information is accomplished by a combination of searching, visualization, and analytics. Searching chemical databases is essential for selecting chemicals to analyze and qualifying results since information relevant to the analysis is usually found in multiple sources. Viewing the individual chemicals, groups of chemicals, and their relationship to other data, such as biological activity data, is essential to interpreting the results. Grouping chemicals, finding relationships between chemical attributes and biological data, and building prediction models are routine operations when data mining chemical information.

The following chapter reviews the steps and issues involved in data mining chemical information. It covers standard ways of representing chemical information, along with commonly used approaches to the generation of a primary key. The initial process of cleaning, annotating, and normalizing chemical structures is reviewed, followed by a discussion of a number of different ways to generate chemical descriptors. The three core decision-support operations of searching, visualization, and analytics are described. In recent years, chemoinformatics methods have been applied to the analysis of chemogenomics data, and this chapter is concluded with an overview of this important new field.

26.2   Representing Chemical Structures

26.2.1   Overview

Any analysis or searching of chemical structures requires the atoms, bonds, and the connections between them, since it usually involves tracing paths through the chemical structure. The first step is reading the chemical structures in from a file or a database. Figure 26.1 is an example of a computer representation of a chemical structure. Each atom and bond is numbered and listed, shown as atoms 1–29, and bonds 1–30. Information concerning the atoms is shown, in this case the atom type. Atom 1 is a carbon; atom 2 is a carbon; and so on. Other information related to an atom, such as the charge may also be saved with each atom. The atom numbers correspond to the annotation in the structure drawing. The bonds are listed similarly, with bond 1 a single bond, bond 2 a single bond, bond 3 a single bond, bond 4 a single dashed bond, and so on. How the bonds connect the atoms to create the specific chemical structure is also noted. Bond 1 connects atoms 2–4; bond 2 connects atoms 14–15; and so on. This electronic representation of the atoms, bonds, and connectivity is usually referred to as a *connection table*. There are a number of common formats for standardizing the connection table information, including the MDL® MOLFILE/SDF format, SMILES, Chemical Markup Language or CML, PATRAN, Wisswesser Line Notation, and Sybyl® Line Notation (SLN). The two most widely used formats are the MDL MOLFILE/SDF Files and SMILES strings.

26.2.2   MDL® MOLFILE and structure data file (SDF) format

An MDL® MOLFILE is an electronic representation of a connection table for a chemical structure. A detailed description of this format can be found at the web site: http://www.mdl.com/downloads/public/ctfile/ctfile.pdf. The file format consists of five parts, (1) a header block, (2) a count block, (3) an atoms block, and (4) a bonds block, and (5) a properties block:

- Header block: the first three lines of the file are dedicated to header information. The first line may contain the name of the chemical structure. In the example in Figure 26.2, the name is shown: chloramphenicol.

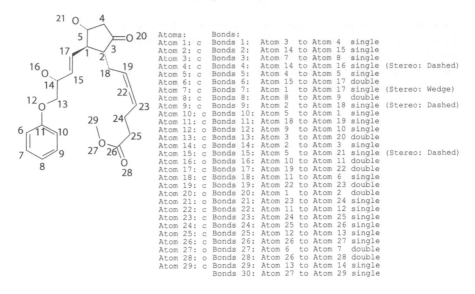

```
Atoms:        Bonds:
Atom 1: c     Bonds 1:   Atom 3  to Atom 4  single
Atom 2: c     Bonds 2:   Atom 14 to Atom 15 single
Atom 3: c     Bonds 3:   Atom 7  to Atom 8  single
Atom 4: c     Bonds 4:   Atom 14 to Atom 16 single (Stereo: Dashed)
Atom 5: c     Bonds 5:   Atom 4  to Atom 5  single
Atom 6: c     Bonds 6:   Atom 15 to Atom 17 double
Atom 7: c     Bonds 7:   Atom 1  to Atom 17 single (Stereo: Wedge)
Atom 8: c     Bonds 8:   Atom 8  to Atom 9  double
Atom 9: c     Bonds 9:   Atom 2  to Atom 18 single (Stereo: Dashed)
Atom 10: c    Bonds 10:  Atom 5  to Atom 1  single
Atom 11: c    Bonds 11:  Atom 18 to Atom 19 single
Atom 12: o    Bonds 12:  Atom 9  to Atom 10 single
Atom 13: c    Bonds 13:  Atom 3  to Atom 20 double
Atom 14: c    Bonds 14:  Atom 2  to Atom 3  single
Atom 15: c    Bonds 15:  Atom 5  to Atom 21 single (Stereo: Dashed)
Atom 16: o    Bonds 16:  Atom 10 to Atom 11 double
Atom 17: c    Bonds 17:  Atom 19 to Atom 22 double
Atom 18: c    Bonds 18:  Atom 11 to Atom 6  single
Atom 19: c    Bonds 19:  Atom 22 to Atom 23 double
Atom 20: o    Bonds 20:  Atom 1  to Atom 2  double
Atom 21: o    Bonds 21:  Atom 23 to Atom 24 single
Atom 22: c    Bonds 22:  Atom 11 to Atom 12 single
Atom 23: c    Bonds 23:  Atom 24 to Atom 25 single
Atom 24: c    Bonds 24:  Atom 25 to Atom 26 single
Atom 25: c    Bonds 25:  Atom 12 to Atom 13 single
Atom 26: c    Bonds 26:  Atom 26 to Atom 27 single
Atom 27: o    Bonds 27:  Atom 6  to Atom 7  double
Atom 28: o    Bonds 28:  Atom 26 to Atom 28 double
Atom 29: c    Bonds 29:  Atom 13 to Atom 14 single
              Bonds 30:  Atom 27 to Atom 29 single
```

FIGURE 26.1: An example for how a chemical structure can be described electronically.

The second line is an optional line and usually contains miscellaneous program and usage data. The third line is again optional and is available for adding comments.

- Count block: The fourth line carries information concerning the number of atoms and bonds in the chemical structure, as well as some additional information. The line is formatted such that characters 1–3 are the number of atoms, characters 4–6 are the number of bonds, and characters 13–15 represent a flag as to whether the chemical is chiral or not (0 is not chiral and 1 is chiral). In the example in Figure 26.2 there are 20 atoms and 20 bonds in the chemical structure.

- Atoms block: following the count line is information on the atoms in the chemical structure, with each line corresponding to a single atom. The number of lines is determined by the number of atoms in the count block. Each of the atoms is implicitly numbered from one. The different fields are summarized in Table 26.2. In the example in Figure 26.2, the x-coordinate for the first atoms is 3.8985, the y-coordinate is -7.9333, the z-coordinate is 0.0000, and the atom is C or carbon. One of the advantages of this type of format is that the chemical structure can be redrawn exactly as it was drawn originally, since it retains the geometrical atom coordinates. The other fields are not so widely used or are present for compatibility with older versions of the MOLFILE format.

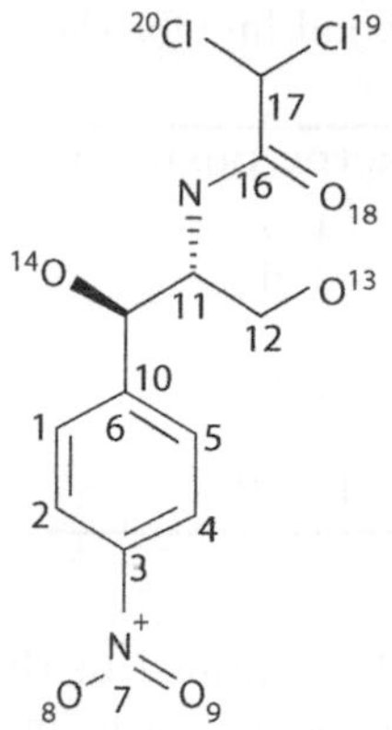

```
Chloaramphenicol
 -ISIS-  07150814262D
 20  20  0  0  0  0  0  0  0  0999 V2000
    3.8985   -7.9333    0.0000 C   0  0  0  0  0  0  0  0  0  0  0  0
    3.8932   -8.7656    0.0000 C   0  0  0  0  0  0  0  0  0  0  0  0
    4.6024   -9.1709    0.0000 C   0  0  0  0  0  0  0  0  0  0  0  0
    5.3173   -8.7632    0.0000 C   0  0  0  0  0  0  0  0  0  0  0  0
    5.3186   -7.9368    0.0000 C   0  0  0  0  0  0  0  0  0  0  0  0
    4.6089   -7.5262    0.0000 C   0  0  0  0  0  0  0  0  0  0  0  0
    4.5958   -9.9958    0.0000 N   0  3  0  0  0  0  0  0  0  0  0  0
    3.8792  -10.4042    0.0000 O   0  5  0  0  0  0  0  0  0  0  0  0
    5.3083  -10.4083    0.0000 O   0  0  0  0  0  0  0  0  0  0  0  0
    4.6042   -6.7000    0.0000 C   0  0  2  0  0  0  0  0  0  0  0  0
    5.3167   -6.2833    0.0000 C   0  0  1  0  0  0  0  0  0  0  0  0
    6.0292   -6.6958    0.0000 C   0  0  0  0  0  0  0  0  0  0  0  0
    6.7417   -6.2792    0.0000 O   0  0  0  0  0  0  0  0  0  0  0  0
    3.8875   -6.2875    0.0000 O   0  0  0  0  0  0  0  0  0  0  0  0
    5.3125   -5.4583    0.0000 N   0  0  0  0  0  0  0  0  0  0  0  0
    6.0250   -5.0417    0.0000 C   0  0  0  0  0  0  0  0  0  0  0  0
    6.0208   -4.2167    0.0000 C   0  0  3  0  0  0  0  0  0  0  0  0
    6.7375   -5.4542    0.0000 O   0  0  0  0  0  0  0  0  0  0  0  0
    6.7333   -3.8000    0.0000 Cl  0  0  0  0  0  0  0  0  0  0  0  0
    5.3042   -3.8042    0.0000 Cl  0  0  0  0  0  0  0  0  0  0  0  0
  6 10  1  0  0  0  0
  2  3  1  0  0  0  0
 10 11  1  0  0  0  0
  5  6  2  0  0  0  0
 11 12  1  0  0  0  0
  6  1  1  0  0  0  0
 12 13  1  0  0  0  0
  1  2  2  0  0  0  0
 10 14  1  1  0  0  0
  3  7  1  0  0  0  0
 11 15  1  6  0  0  0
  3  4  2  0  0  0  0
 15 16  1  0  0  0  0
  7  8  1  0  0  0  0
 16 17  1  0  0  0  0
 16 18  2  0  0  0  0
  7  9  2  0  0  0  0
 17 19  1  0  0  0  0
  4  5  1  0  0  0  0
 17 20  1  0  0  0  0
M  CHG  2   7   1   8  -1
M  END
```

FIGURE 26.2: An example of an MDL MOLFILE format.

TABLE 26.2: Summary of key information in an atom line for the MDL MOLFILE file format.

Information	Character position
x-coordinate	1–10
y-coordinate	11–20
z-coordinate	21–30
Atom type	32–34

TABLE 26.3: Summary of key information in an bond line for the MDL MOLFILE file format.

Information	Character position
From atom	1–3
To atom	4–6
Bond type (1 is single, 2 is double, and 3 is triple)	7–9
Stereo (0 is no stereo, 1 is down stereo, 6 is up stereo, 4 is either, 3 is cis or trans)	10–12

- Bonds block: corresponding to the number of bonds identified in the count block, the bonds block describes a specific bond on each line. Table 26.3 summarizes the fields. In the example in Figure 26.2, the first bond is from atom number 6 to atom number 10 and is a single bond. The bond on the ninth line is a single bond from atom 10 to atom 14 with a stereo indicator set to up.

- Properties block: the properties block contains information including charge, radical, and isotope data. Each line contains the property information and the properties block is terminated with the line "M END." In the example in Figure 26.2, the second to last line "M CHG 2 7 1 8 -1" describes charges on two atoms: atom 7 has a positive charge and atom 8 a negative charge. For more information see http://www.mdl.com/downloads/public/ctfile/ctfile.pdf.

An structure data file (SDF) format allows for multiple chemical structures within a single file, as well as data associated with each chemical structure. Any number of chemical structures can be included in the file. A line containing the text "$$$$" is used to separate each chemical structure record. A MOLFILE is used to represent an individual chemical structure followed by a data block containing information on the associated data. In the example in Figure 26.3, a portion of a SDF record is shown. Up to the line "M END" is the MOLFILE detailing the first chemical structure. Following this through to the "$$$$" record delimiter is the information on the associated data. The first data entry is for the field "PUBCHEM_COMPOUND_CID" and has a value 241, the second data entry is for the field "PUB-CHEM_CACTVS_HBOND_ACCEPTOR" and has a value 0, and so on.

26.2.3 Simplified molecular input line entry system (SMILES) format

SMILES is another format for representing chemical structures and stands for Simplified Molecular Input Line Entry System. The chemical structure is represented as a single string which can be conveniently included within a field of a table. A computer program is required to redraw a structure

FIGURE 26.3: An example of an SDF file.

diagram from the SMILES string. Since it does not retain the coordinates of
the atoms, it will typically be redrawn differently than originally depicted.
A detailed description of the SMILES format can be found at the web site:
http://www.daylight.com/dayhtml/doc/theory/theory.smiles.html.

Individual atoms in the connection table are represented by their atom
symbol, such as C, Br, N, and so on. Additional information, such as charges
is represented by a square bracket around the atom symbol followed by the
information. For example, [N+] represents a nitrogen with a single positive
charge. Bonds are represented by a single character: "−" for single bonds, "="
for double bonds, and "#" for triple bonds. For example C-C represents a
single bond between two carbons and C#N represents a triple bond between
a carbon and a nitrogen. It is not necessary to use single bonds explicitly,
for example, C-C and CC both represent a single bond between two carbon
atoms.

Any time a path through a chemical structure splits, a branch is created.
Using the chemical structure in Figure 26.4 to illustrate, starting from the
carbon marked with an "*", there is a path from the carbon to the adjacent carbon to the nitrogen. This path is represented as a SMILES string
CCN. A branch point is created at the second carbon, by a double bond to
an oxygen. Paths are represented within brackets. For example CC(=O)N
represents the SMILES string for the four atom fragment described. The
full SMILES string for the chemical structure is shown below the chemical
structure.

A cycle is formed where a continuous path can be traced through the chemical structure that starts and ends on the same atom. A cycle is represented
in the SMILES format by designating an atom on which the cycle starts and
ends, numbering the atom, and then closing the cycle using this designated
atom number. For example, a six-membered ring with alternating single and
double bonds can be represented by the SMILES string C1=CC=CC=C1.

The SMILES format can handle disconnected chemical structures, iso-
topes, olefinic and tetrahedral stereochemistry. For more information see:
http://www.daylight.com/dayhtml/doc/theory/theory.smiles.html.

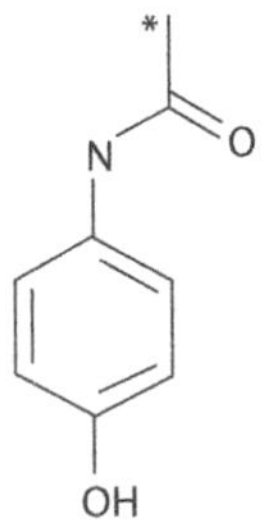

CC(=O) NC1=CC=C (C=C1) O

FIGURE 26.4: SMILES string for the chemical structure shown.

26.3   Unique Identifier

A unique identification for a chemical structure is important for a number of reasons. It provides a convenient way of communicating experimental observations, and provides a way of rapidly determining whether the dataset has duplicate entries for the same compound. This is particularly important when combining datasets from different sources. In many data mining applications it is also important to have access to the chemical structure in the form it was tested as well as in a form better suited for analysis. For example, a chemical may be tested in a salt form, but the salt may be removed prior to generating a computational model. Both the tested form and the model-ready structure should have a different id, since they are different chemical structures.

There are many different methods for generating a unique identification for a single chemical structure:

- Common names: these are convenient particularly when discussing the results with others; however, the names are not necessarily unique.

- Systematic names: these are chemical names where there is no ambiguity as to chemical composition. There may be multiple names that describe the same chemical structure; however, the application of additional rules will ensure a unique name. The systematic rules for naming organic chemicals have been developed by International Union of Pure and Applied Chemistry (IUPAC) along with the additional rules that ensure a unique name for any chemical structure, irrespective of how it is drawn. The disadvantage of this approach is that the names are usually long, making them difficult for communicating the results efficiently.

- Canonical SMILES: like systematic names, different SMILES strings can represent the same chemical structure; however, rules can be applied to ensure only a unique SMILES string is generated for a particular chemical structure. The advantage of this approach is that the connection table information is retained.

- Canonical numbers: algorithms have been developed to generate a unique number for an individual chemical structure. One example is the Morgan algorithm created in the 1960s. Generally, a unique number is initially assigned to each atom in the chemical structure. This number reflects the atom type, and its connection to all other atoms in the chemical structure. Atoms with the same number are symmetrical. A final number is calculated through a function that uses all the individual atoms' scores.

- Registered ids: organizations, such as pharmaceutical companies and vendors of chemical structure databases, rely on registration systems to ensure they have a systematic method of determining unique chemical structures. A registration system generates a unique identifier for a chemical structure that has not been previously seen. The systems generally operate by searching for the candidate chemical structure in the database of previously assigned chemicals. If the candidate is in the database, it is given the previously assigned id; otherwise a new id is assigned.

- InChI: the InChI id describes a particular chemical structure as a unique string. One of the major advantages of the InChI format is that it is layered. One layer may cover information such as the chemical structure's connectivity; another may cover its stereochemistry. This makes it flexible when used for searching.

Figure 26.5 illustrates a common name, an IUPAC name, a canonical SMILES, an InChI id, and a registered id (Leadscope ID) for the chemical structure shown.

26.4 Cleaning, Annotating, and Normalizing

26.4.1 Overview

Once a chemical structure has been parsed, it is usual to perform a series of operations to clean, annotate, and normalize specific atoms and bonds in

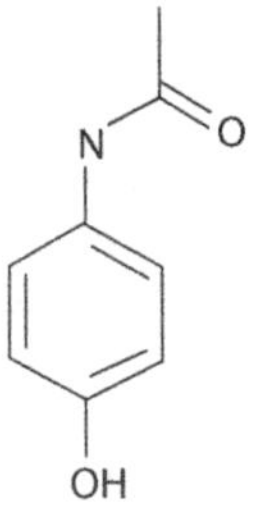

Common name: Tylenol
IUPAC name: N-(4–hydroxypheny) acetamide
Canonical SMILES: CC (=O) NC1=CC=C (C=C1) O
InChI: InChI=1/C8H9NO2/c1-6 (10) 9-7-2-4-8(11) 5-3-7/h2-5, 11H, 1H3, (H,9,10)/fh9H
Leadscope ID: LS–32

FIGURE 26.5: Examples of different names or ids for a particular chemical structure.

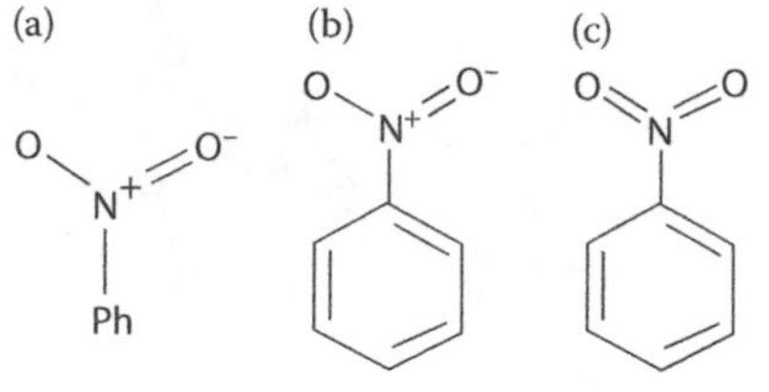

(a) (b) (c)

FIGURE 26.6: Three ways of drawing the same chemical structure.

the chemical structure. These steps are necessary to ensure any searching will be performed correctly and to generate consistent chemical descriptors.

26.4.2 Representation

There are multiple ways to draw the same chemical structure. For example, in Figure 26.6 the same chemical structure has been drawn in three different ways. Structure (a) uses the symbol Ph to represent the benzene ring. The Ph will need to be converted to a full connection table containing the individual atoms and bonds for analysis. In structures (b) and (c), the benzene ring has been drawn, yet the same nitro functional group has been drawn using different drawing conventions. Both (b) and (c) are the same structure and to ensure consistency in any analysis they should be either converted to a consistent representation or the atoms and bonds should be represented by a common normalized chemical structure.

26.4.3 Hydrogens

The number of hydrogens attached to each atom is rarely included in the connection table for a chemical structure. The number of hydrogens is usually calculated for each of the atoms using the atom type along with the number and type of adjacent bonds and charge information, if present.

26.4.4 Cycles and chains

It is usual to annotate atoms and bonds as either cyclic or acyclic. Algorithms to perceive rings are usually performed to achieve this. A single ring can be traced by exploring all paths starting from an atom. Where a path start and ends on the same atom, a ring is formed. Standard graph searching algorithms, such as a depth first search, can be used to identify paths through the chemical structure. Heuristics are often applied to ensure the search is carried out efficiently, for example eliminating all exterior chains or starting from atoms with three or more connections (except in the situation where there is only a single ring in the chemical structure). In many situations finding all possible cycles is not needed and it is usual to identify a subset referred to

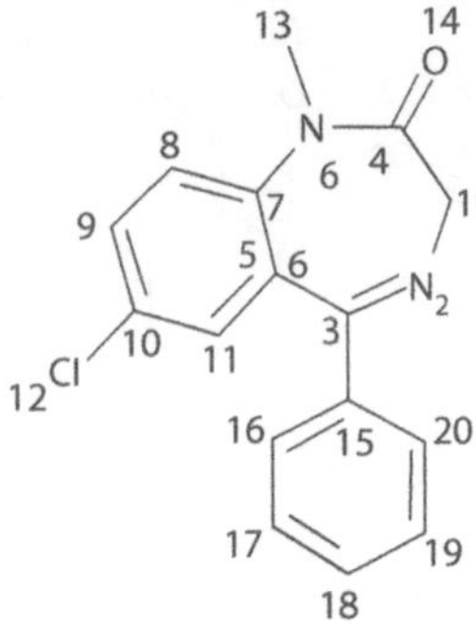

Cyclic atoms: 10, 11, 5, 7, 8, 9, 6, 4, 1, 2, 3, 15, 20, 19, 18, 17, 16
Cyclic bonds: 10–11, 11–5, 6,–7, 7–8, 8–9, 9–10, 7–6, 6–4, 4–1, 1–2,
2–3, 3–6, 15–20, 20–19. 19–18. 18–17. 17–16, 16–15

Acyclic atoms: 12, 13, 14
Acyclic atoms: 12–10, 13–6, 4–14, 3 –15

Ring (1) atoms: 10, 11, 5, 7, 8, 9
Ring (1) bonds: 10–11, 11–5, 5–7, 7 –8, 8–9, 9–10

Ring (2) atoms: 5, 3, 2, 1, 4, 6, 7
Ring (2) bonds: 5–3, 3–2, 2–1, 1–4, 4–6, 6–7, 5–7

Ring (3) atoms: 15, 20, 19, 18, 17, 16, 15
Ring (3) bonds: 15–20, 20–19, 19–18, 18 –17, 17–16, 16–15

FIGURE 26.7: Example of the identification of rings.

as the smallest set of smallest rings (SSSR). Where all atoms and bonds of
a larger ring are part of rings of smaller size, the larger ring is usually discarded. The analysis of rings allows atoms and bonds to be flagged as cyclic or
acyclic. Information on the ring sizes is also associated with the cyclic atoms
and bonds. Figure 26.7 illustrate the atoms and bonds identified as cycles, as
well as the specific rings identified. It should be noted that the cycle which is
the envelope to Ring(1) and Ring(2) is not shown as a separate ring since the
SSSR approach was used. Also bond 5–7 is part of both a 6- and a 7-membered
ring, and bond 3–15 is acyclic despite atoms 3 and 15 being assigned as cyclic.

26.4.5   Aromatic atoms and bonds

By convention aromatic rings are not drawn explicitly; however, they
should be identified since they exhibit properties different from other types of
bonds. One approach is to use the Hückel $4n+2$ rule to determine if rings are
aromatic or not. An alternative approach is to use a series of pre-defined molecular templates that define aromatic rings. Where there is a match between
these aromatic rules and the chemical structures, those atoms and bonds are
assigned as aromatic. Figure 26.8 illustrates a chemical structure with those
atoms and bonds that are flagged as aromatic.

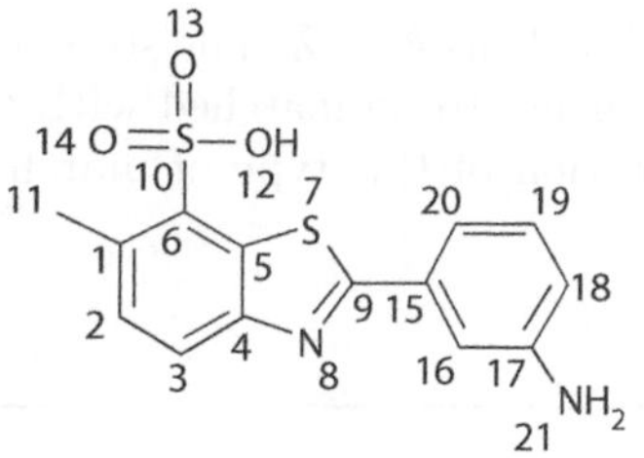

Aromatic atoms: 1, 2, 3, 4, 5, 6, 7, 8, 9, 15, 16, 17, 18, 19, 20
Aromatic bonds: 1–2, 2–3, 3–4, 4–5, 5–6, 6–1, 4–8, 8–9, 9–7,
7–5, 15–16, 16–17, 17–18, 18–19, 19–20, 20–15

FIGURE 26.8: Aromatic atoms and bonds.

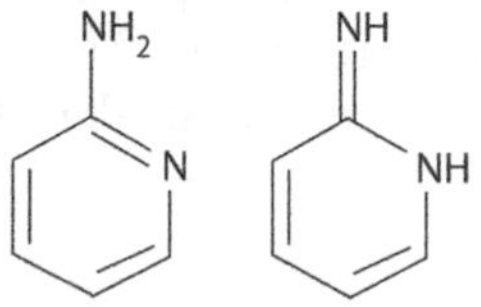

FIGURE 26.9: Amine–imine tautomers.

26.4.6 Tautomers

There exist situations where distinct chemical structures can easily convert from one to the other as a result of tautomerism. Figure 26.9 illustrates two chemicals that would be considered different tautomers. As a result of the ease of conversion between these two forms, it is important to identify different tautomers where a hydrogen or a charge is mobile, resulting in a change to the chemical structure. One approach to identifying these systems is to construct a dictionary of small molecular templates that are considered tautomers, and where these templates match parts of a chemical structure, the matching atoms and/or bonds are flagged as tautomers.

26.4.7 Stereochemistry

Stereochemistry most often refers to the spatial arrangement of atoms about specific carbon atoms and double bonds. Carbon atoms with four different attachment (looking at all atoms and bonds attached, not just those adjacent) are considered chiral centers. The three dimensional spatial arrangement is important since two chemical structures with the same connection table may be different stereoisomers. Two dimensional structure drawings usually represent stereochemistry using wedged and dashed bonds. One way to unambiguously identify stereo configurations within a computer program is to automatically apply the Cahn–Ingold–Prelog rules for assigning stereocenters as either R or S and double bonds as E or Z. The stereochemistry pre-processing enables chemicals or fragments to be matched with the same stereochemical configuration. The application of this type of matching is discussed later in the chapter.

26.5 Generation of Descriptors

26.5.1 Overview

To enable data mining of chemical structures, new columns of data are generated to describe the whole chemical structure and portions of the chemical structure. There are many different approaches, including counts of different types of atoms or bonds, properties that describe various attributes of the whole molecule, and fingerprints that identify the presence or absence of specific molecular fragments.

26.5.2 Counts

A simple approach to calculating chemical descriptors is to count the number of specific atoms, bonds, or fragments contained in the structure. They can be useful descriptors when predicting certain biological activities, especially when used in combination with other descriptors. Number of rotatable bonds and counts of different pharmacophore atom types are commonly used. A rotatable bond is defined as a single, acyclic, nonterminal bond that is not on the extremities of the molecule. Counts of the number of rotatable bonds reflect the overall flexibility of a molecule. A pharmacophore atom is a generalized atom type used to summarize specific types of interactions between a drug bound to a protein at a molecular level. These atom types are usually defined using molecular fragment that describe the environment around the atom. Common pharmacophore atom types include: hydrogen bond donors, hydrogen bond acceptors, aromatic rings, positive charge centers, negative charge centers, and hydrophobic centers.

26.5.3 Properties

Properties of the whole molecule are often useful descriptors for chemicals. Some commonly calculated properties include:

- Molecular weight: generally, drugs have a molecular weight below 500. Molecular weight is a calculation made by simply summing the individual isotopic masses for all atoms in the chemical structure.

- Hydrophobicity: this property is related to the drug likeness of a compound in terms of its ADME characteristics and its ability to bind to

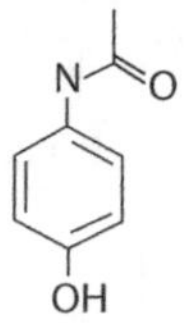

Name	Molecular weight	XLogP	H-Bond donor	H-Bond acceptor	Rotatable bond count	Topological polar surface area
Tylenol	151.16	0.4	2	2	2	49.3

FIGURE 26.10: Example of a series of properties calculated for a single compound.

a target receptor. It is commonly estimated by the property LogP or the partition coefficient between octanol and water. There are a number of computational approaches to calculating LogP, for example cLogP, aLogP, and XlogP. Generally, a score is assigned to a series of predefined atom types or molecular fragments, which are matched to the target molecule. The final LogP estimate is generated by summing the individual scores for all atoms or fragments present.

- Polar surface area: this property reflects the surface area over the polar atoms and is related to the drug's ability to permeate a cell. This property can also be estimated through the use of fragment scores.

The example in Figure 26.10 shows of a series of descriptors generated for a single molecule: molecular weight, XlogP, number of hydrogen bond acceptors, number of hydrogen bond donors, number of rotatable bonds, and topological polar surface area. In addition, there are many properties of the whole molecule that can be calculated and potentially used as descriptors, such as topological indices, molar refractivity, and Kappa shape indices (see Gasteiger and Engel, 2003; and Leach and Gillet, 2007 for more details.)

26.5.4 Fingerprints

The presence of a specific molecular fragment is often used as a descriptor for data mining purposes. A dummy variable is generated for a given fragment, where a value of "1" is used to indicate the fragment exists in the chemical structure, and "0" is used to indicate its absence. In addition, a variable may be generated that contains a count of the number of times the feature appears in the molecule. There are a number of ways to generate fingerprints, including:

- Predefined substructures: this approach uses a dictionary of molecular fragments constructed before any analysis. When a dictionary fragment matches a portion of the chemical structure a "1" is recorded

	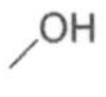 Alcohol	Carbonyl	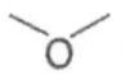 Ether	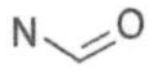 Carboxamide	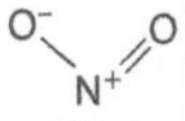 Nitro	 Benzene
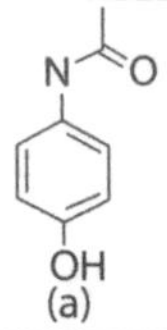 (a)	1	1	0	1	0	1
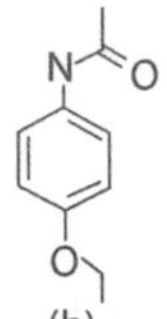 (b)	0	1	1	1	0	1
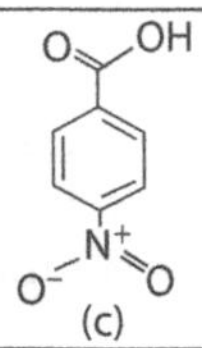 (c)	1	1	0	0	1	1

FIGURE 26.11: Generated fingerprint dummy variables using a dictionary of structural fragments shown along the top.

for that feature, otherwise a "0" is entered. In Figure 26.11, six pre-defined structural features are shown along the top of the table as an example: alcohol, carbonyl, ether, carboxamide, nitro, and benzene. For structure (a) alcohol, carbonyl, carboxamide, and benzene fragments match shown as a "1," whereas ether and nitro do not match. The presence or absence of the structural features for chemical structure (b) and (c) is also shown. There are a number of fragment collections that are used, including the MDL® MACCS public keys (166 fragments) and the Leadscope® fingerprints (over 27,000 molecular fragments) (Blower et al., 2004).

- Generated (supervised): fingerprints are often used to generate prediction models, in which variables showing the presence or absence of features are used in the model to make activity predictions. One approach is to generate molecular fragments on-the-fly (rather than rely on a fixed dictionary of features), where the fragments characterize chemicals with either unusually high or low levels of biological activity. For example, in the macrostructure assembly approach, a set of pre-defined structural features are combined in different ways to form new fragments (Blower et al., 2004). These new fragments are then matched against the underlying set to determine whether the resulting new fragment maps onto chemical structures with unusually high or low levels of biological activity. The most differentiating fragments are retained. This process is computationally intensive and requires the application of multiple heuristics to make the algorithm run fast.

- Generated (unsupervised): fingerprints can also be generated dynamically without the need for a fragment dictionary and without a response variable to guide the generation of the keys. For example, paths of various lengths between pairs of atoms can be generated. Atoms may also be defined as generalized atom types. In the PowerMV program, one of the types of fingerprints is referred to as pharmacophores, where all paths of length 2-7 bonds between pairs of pharmacophore atoms are generated (Liu et al., 2005).

26.6 Searching Chemical Data

26.6.1 Overview

The ability to search the contents of a chemical structure database is critical. For a single compound look-up where the name or id is known, a simple text search would return the chemical structure if it is in the database. However, the name may not be known and only the chemical structure is available. In this situation, an exact structure search should be performed using the available structure as a query. For situations where classes of related chemical structures are required, searches such as a substructure search or a similarity search should be used.

In order to perform a structure search, a query describing the connection table of the chemical structure of interest should be defined. There are three primary ways for specifying a structure query. A structure query could be drawn in a chemical structure editor through a graphical user interface. The atoms and bonds are drawn graphically, as well as any necessary restrictions on the atoms or bonds. Alternatively, where a structure query has been committed to file, usually from a chemical structure editor—for example using the MDL MOLFILE format, it can be uploaded to initiate a query. Finally, in the same way a text search is initiated, the connection table for the query can be typed in using a format such as a SMILES string.

26.6.2 Exact

Having specified a structure query, an exact structure search will compare the connection table of the query to each structure in the database to determine whether there is a match. An exhaustive match of every atom and bond in the query to every atom and bond in the database would result in a slow

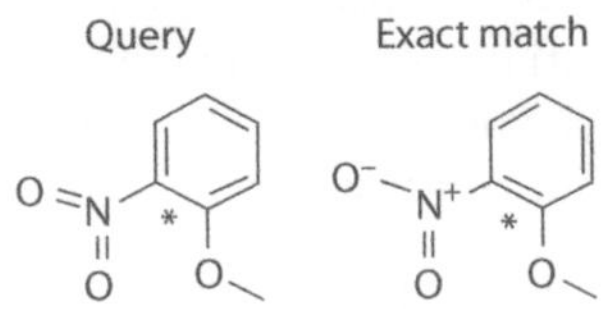

FIGURE 26.12: Results of an exact match query.

search. The search algorithms are highly optimized to ensure the search is executed quickly. In addition, the pre-processing described earlier of is essential (applied to both the query and the database structures) to ensure accurate recall. In Figure 26.12, the query structure shown is used to perform an exact match search. The resulting match from the database is also shown. In this case the atoms and bonds of the query do not match exactly onto the matched structure. The bond in the query marked with an "*" is a single bond, which matches onto a double bond in the matched structure. This is because the ring's bonds are considered to be aromatic (as a result of the aromatic preprocessing described earlier) and so the aromatic bonds in the query matches the aromatic bonds in the structure from the database. In addition, a different convention has been used to draw the nitro group (O=N=O) in the query from the nitro group in the database structure. Other issues that should be addressed when searching an exact match include matching tautomers, stereochemical matching, and matching components within multi-component chemicals in the database.

26.6.3 Substructure

There are a number of ways of performing more generalized chemical structure searches. Substructure search is one of the most popular where, as in the case of the exact match search, a structure query is initially specified. When a substructure search is then executed, a search of the database chemical structures is performed to identify those chemical structures containing the query chemical fragment. Again, like the exact match search, substructure searching needs to address issues concerning aromaticity, tautomers, and stereochemistry to ensure appropriate recall. In addition to the connection table for the query, restrictions on the atoms and bonds can also be specified. Common restrictions include whether an atom or bond is part of a ring, the size of the ring the atom or bond is part of, whether particular atoms are open to further substitution, and so on. Atoms and bonds can also be defined as a member of a group, such as any one of the atoms Cl, Br, I, or F. Extensions to the MDL® MOLFILE and the SMILES (called SMARTS) formats accommodate these additional restrictions. In the example shown in Figure 26.13, a query connection table is drawn. Along with the chemical structure, a restriction is placed on one of the bonds as shown. The restriction states that the

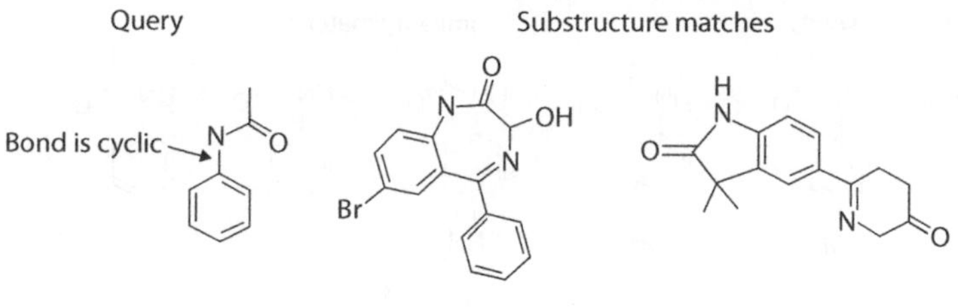

FIGURE 26.13: Example of a substructure search.

specific bond should only match a cyclic bond in a matched chemical structure from the database. The figure identified two matched chemical structures from the database. The highlighted atoms and bonds indicate where the query matches.

26.6.4 Similarity

A third popular structure search is called a similarity search. Like substructure search, similarity search returns multiple structures that are, for the most part, different from the chemical structure specified in the query. Unlike substructure search, similarity searching may return structures with different atoms and bonds as well as chemical structures smaller than the query. A typical method for performing a similarity search is to use fingerprint descriptors for both the query structure and the database chemical structures. A similarity score is generated for each database structure, based on a comparison of the fingerprints for the two structures (query and database). Initially, the individual values for the fingerprints are compared to determine the number of common and different values. Four counts are calculated:

$Count_{11}$: this is a count of the number of variables where the value in the query is 1 and the value in the database structure is 1.

$Count_{10}$: this is a count of the number of variables where the value in the query is 1 and the value in the database structure is 0.

$Count_{01}$: this is a count of the number of variables where the value in the query is 0 and the value in the database structure is 1.

$Count_{00}$: this is a count of the number of variables where the value in the query is 0 and the value in the database structure is 0.

Using the example in Figure 26.11 where chemical structure (a) is the query and (b) is the database structure, $count_{11}$ is 3, $count_{10}$ is 1, $count_{01}$ is 1, and $count_{00}$ is 1. The similarity score is most often calculated using the *Tanimoto* similarity coefficient: S_{Tanimoto}:

$$S_{\text{Tanimoto}} = \frac{count_{11}}{count_{10} + count_{01} + count_{11}}$$

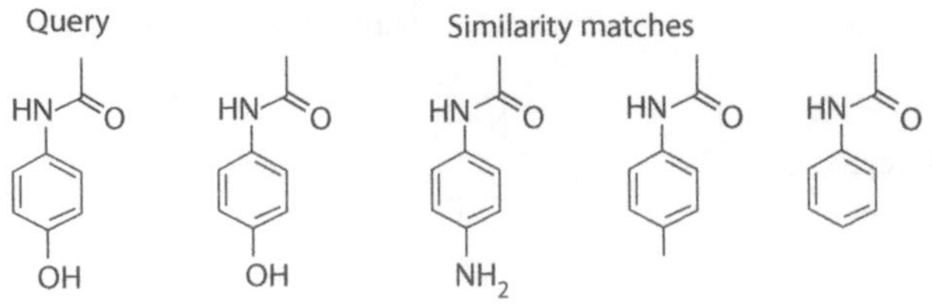

FIGURE 26.14: Results from a similarity search.

It should be noted that $count_{00}$ is not used with the Tanimoto similarity coefficient.

In the example in Figure 26.11:

$$S_{\text{Tanimoto}}(a, b) = \frac{3}{1 + 1 + 3} = 0.6$$

In Figure 26.14 a query chemical structure is shown along with the results of a similarity search.

26.7   Visualization Methods

The ability to visualize sets of chemical structures is critical to inspecting the underlying information, as well as results from the analysis. The two most commonly used formats for browsing individual chemical structures and any related data are the structure grid and the chemical spreadsheet, illustrated in Figures 26.15 and 26.16. The structure grid is useful when the primary operation is to browse through sets of chemicals with limited data, whereas the chemical spreadsheet is well-suited to browsing structure and related data simultaneously.

Where a set of chemical structures is characterized by a common molecular fragment, an R-group analysis is a succinct method for summarizing the chemical and related data. The core fragment is shown, along with points on the core where there are examples involving the substitution at those point (R-groups). The different combinations of R-groups are then summarized in a table. Where the table summarizes all substitution points, each row relates to a specific chemical structure; however, when less than the total number of substituents are displayed in the table, each row now relates to one or more chemical structures. The number of chemical structures in each row is often provided. Data, such as biological activity, can also be placed alongside each row. Where a row is a single chemical structure the actual data would be presented, and where a row is a set of chemical structure, the data would be a

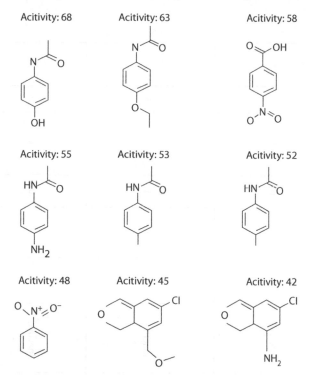

FIGURE 26.15: Viewing chemical structures in a grid.

summary of the subset's chemicals, such as the mean. Figure 26.17 illustrates an R-group analysis for a set of 13 compounds sharing a common core.

Interactive data visualizations approaches, such as histograms and scatterplots, are often used to view individual data points (measured or calculated) along with summaries of the data. In Figure 26.18, a histogram shows the frequency distribution of activity data (pIC_{50}) and a scatterplot shows the relationship between the properties XlogP and biological activity. Data brushing is an interactive technique which allows a user to highlight a subset of the data in multiple graphs. In this example, a specific group of chemical structures sharing a common feature is highlighted using dark shading on the histogram and the scatterplot. Summary tables and graphs are also used to summarize collections of chemicals structures, along with any relationships to biological data. For example, the Leadscope software summarizes sets of chemical structures by chemical features, using histograms to represent the number of chemical structures containing a particular feature (Blower et al., 2004). In addition, the relationship of the chemical groups to any associated biological activity is highlighted using different colors to represent groups with unusually high or low levels of biological activity.

Structure	Name	Mol wt	XLogP	HBA	HBD	RB	PSA	Mec2-1 strain	Alcohol	Carbonyl	Ether	Carboxamide	Nitro	Benzene
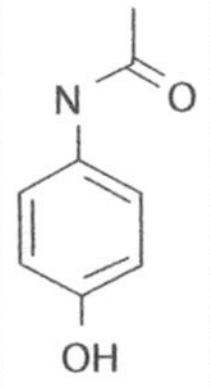	Acetaminophen	151.16	0.4	2	2	2	49.3	Inactive	1	1	0	1	0	1
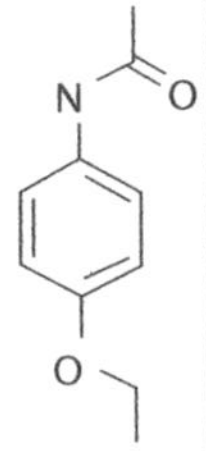	Phenacetin	179.22	1.6	1	2	4	38.3	Inactive	0	1	1	1	0	1
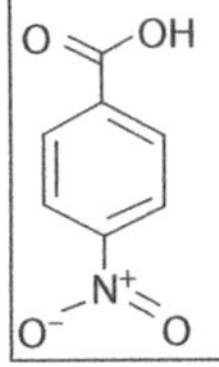	P-Nitrobenzoic acid	167.12	1.5	1	4	2	80.4	Inactive	1	1	0	0	1	1

FIGURE 26.16: Viewing chemical structures and related data in a spreadsheet.

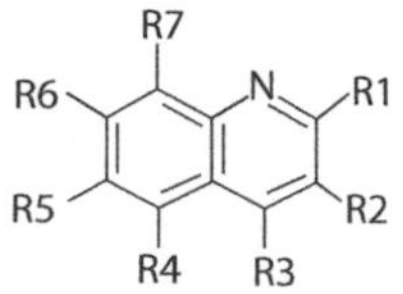

R1	R2	R3	R4	R5	R6	R7	Count	P-IC50
H	H	H	Chloride	H	Iodide	Alcohol	1	1.26
H	H	H	H	H	H	Alcohol	1	1.18
H	H	H	H	H	H	Methyl	1	1.16
H	H	H	H	H	Methyl	H	1	1.13
H	H	H	H	Methyl	H	H	1	1.09
H	H	H	Bromide	H	Bromide	H	1	1.05
H	H	H	H	Nitro	H	H	1	0.95
Carboxylate	H	Alcohol	H	H	H	H	1	0.84
Chloride	H	H	H	H	H	H	1	0.83
H	H	H	H	H	H	Chloride	1	0.79
H	Alcohol	H	H	H	H	H	1	0.68
H	H	Alcohol	H	H	H	H	1	0.65
H	H	Primary amine	H	H	H	H	1	0.64

FIGURE 26.17: R-group analysis.

26.8 Analytics

26.8.1 Overview

Numerous analytical approaches are routinely used to sift through large volumes of chemical and related biological information, including the generated chemical descriptors. These approaches attempt to identify interesting patterns and trends in the data that must be translated into decisions concerning how to move projects forward. As one example, data mining is used to sift through HTS data, simultaneously analyzing hundreds of thousands of chemical structures and associated biological activity data, to identify lead series that will be the focus for lead optimization. Additional chemicals will be selected for testing, based on the attributes identified, in order to enhance activity as well as other drug like properties. In addition to identifying patterns and trends in the data, building predictive models is a major focus of data mining chemical information. These computational models are used to screen actual or hypothetical chemicals in silico, prioritizing those with desired characteristics.

26.8.2 Feature assessment

Whether the objective of the analysis is to identify structure–activity relationships (SAR), or to build models with a prioritized list of descriptor

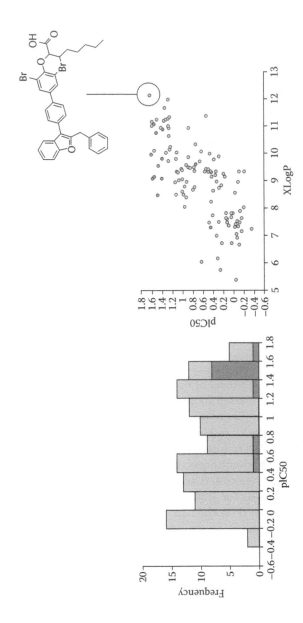

FIGURE 26.18: A set of chemical structures summarized using histogram and scatterplots of the data, incorporating data brushing to highlight a set of structures.

variables, it is important to rank descriptors. When a set of variables has been generated using fingerprints, a number of simple steps are routinely used. First, where the variables contain little or no information, such as when all or the majority of values are "1"s or "0"s, these variables can be removed. When building models, any correlation between the descriptor variables to be used in the model should be avoided. Methods such as principal component analysis or matrices of scatterplots are used to prioritize the properties. In addition, those variables with strong relationships to the biological response data may need to be prioritized. For example a z-score calculates the number of standard deviations the mean activity of the subset containing a structural feature ($\bar{x}_1$) is away from the mean activity of the entire set ($\bar{x}_0$), where n_1 is the number of chemical structures in the subset, n_0 is the number of chemical structures in the entire set, and s_0^2 is the sample variance for the full set (Blower et al., 2004):

$$z = (\bar{x}_1 - \bar{x}_0)\sqrt{\frac{n_1 n_0}{s_0^2(n_0 - n_1)}}$$

26.8.3   Clustering

Grouping sets of chemical structures helps to quickly understand the set and is commonly used to divide the structures into structurally homogenous groups (Barnard and Downs, 1992). Any generated descriptors such as calculated properties or fingerprints may be used to describe the individual chemical structures. The distance between the chemical structures is required for clustering and uses selected descriptors. There are many methods to calculate distances between observations, based on the selected descriptors. Two commonly used distance methods used when grouping chemical structures are the Tanimoto distance and the *Euclidean* distance.

The Tanimoto distance is used when fingerprints characterize the chemical structures. As described in the section on similarity searching, the Tanimoto distance uses counts of the number of common "1"s and "0"s, as well as counts of the number of "1"/"0" or "0"/"1" combinations. The Tanimoto distance is calculated as:

$$D_{\text{Tanimoto}} = 1 - \frac{count_{11}}{count_{10} + count_{01} + count_{11}}$$

Using the example in Figure 26.11, the distance between (a) and (b) would be 0.4.

The Euclidean distance is often used when numeric properties (calculated or measured) are used to characterize the chemical structures, such as log P, molecular weight, and so on. The following formula calculates a Euclidean distance between two observations (p and q), measured over n variables:

$$D_{\text{Euclidean}}(p, q) = \sum_{i=1}^{n} \sqrt{(p_i - q_i)^2}$$

One of the more popular methods for clustering sets of chemical structures is the agglomerative hierarchical clustering approach. There are three initial choices prior to clustering: (1) the variables to use in describing the chemical structures, (2) the distance method (Tanimoto, Euclidean, and so on), and (3) a linkage method. A linkage method is used since throughout the process of hierarchical clustering, two groups of chemical structures will be compared and a distance needs to be computed between the two groups. Three alternative linkage methods include single (the shortest distance), complete (the farthest distance), and average (the average distance between all pairs of chemical structures between the two groups).

Having made these selections, a distance matrix is initially calculated. This matrix is a summary of the distances between all combinations of chemical structures. A process of generating a hierarchy of chemical structures is now initiated. First, the pair of structures with the smallest distance is selected as the initial group to cluster. From now on, the two structures in this group are no longer considered individually. A distance matrix is recomputed by removing the two individual chemical structures, and computing the distances between this new group and the remaining observations (using the selected linkage methods). Again, the shortest distance is selected, merging either two chemicals structures or the new group with another chemical structure. A new distance matrix is calculated, and the process continues by continually combining individual structures and groups until there is one single group containing all chemical structures. The hierarchical structure of the grouping can be seen visually as a dendrogram. An example is shown in Figure 26.19, where a set of five chemicals are shown (for illustration purposes). The right side of the dendrogram relates to the individual chemical structures, with the five chemical structures shown. The points at which two observations and/or groups are joined are shown with the vertical lines. For example, the first two structures are joined at a distance of approximately 0.4.

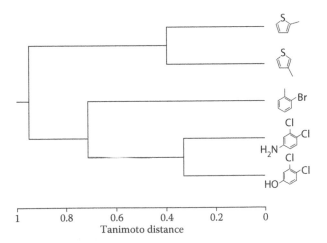

FIGURE 26.19: Clustering a set of chemical structures.

A dataset can be divided into groups by specifying a cut-off distance. In Figure 26.20, two cut-off distances have been applied to the same dendrogram. In the dendrogram to the left, a cut-off distance of around 0.6 was set. Where the cut-off line crosses a horizontal line of the dendrogram, a cluster is formed from all the chemical structures that are children at that point (to the right). In this example, the cut-off line creates three groups. On the dendrogram on the right, a cut-off distance line was set to just less than 0.4, creating more groups. In general, setting a lower cut-off distance will result in fewer numbers of chemical structures in each group, but each group will contain more structurally similar structures.

In addition to agglomerative hierarchical clustering, a number of other clustering approaches are applied to problems of grouping chemical structures including:

- K-means: this approach groups a set of chemical structures into a fixed number of clusters (k). The number of clusters must be specified prior to performing any clustering and the results do not reflect any hierarchical relationships within the data; however, the approach can operate on larger numbers of chemical structures than hierarchical agglomerative clustering. Initially, the approach assigns a chemical structure to each of the k groups. All other chemical structures in the set are then assigned to the group with the smallest distance to the already assigned structure. Once all structures have been assigned to a group, a centroid for each group (representing the middle of the cluster) is calculated and the structures are re-assigned to the group with the smallest distance to the centroid. The process of recalculating the centroid and reassigning the structures is repeated until an optimal solution is identified, for example, when there is no change in the assignment of the structures.

- Jarvis Patrick: this is a nonhierarchical approach that can be used to cluster large datasets quickly. No value is set for the number of clusters; however two parameters are required: the number of neighbors to examine (J) and the number of neighbors in common (C). Initially, for each structure a list of the J most similar structures are identified. These lists are then processed further to generate the cluster. A cluster is formed when two or more lists have at least C structures in common.

Clustering is used in many situations, including being used to understand screening results, partitioning datasets into structurally homogeneous subsets for modeling, and picking chemical structures from individual clusters for selection of a diverse subset.

26.8.4 Decision trees

The generation of decision trees (also referred to as recursive partitioning) is one approach to sifting through sets of chemical structures to identify subsets with unusually high or low levels of biological activity (Hawkins et al.,

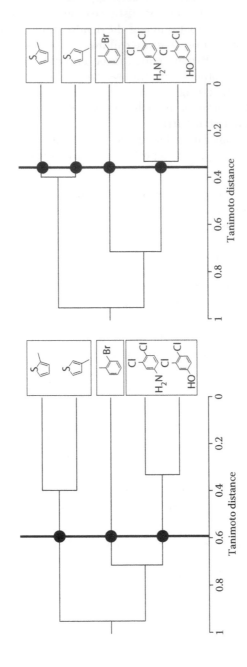

FIGURE 26.20: Partitioning a set of chemical structures into a series of groups based on different cut-off distances.

1997). These subsets are characterized by the presence (or absence) of combinations of structural features. The process of generating a decision tree starts with the entire set of chemical structures. Initially, the specific descriptor that results in the most biologically interesting subset, as measured by some statistic, is identified. The set is divided into two subsets: those containing the feature and those without. There are multiple approaches to determining the most interesting descriptors. Having partitioned the set, the process begins again on each of the two recently identified subsets to again identify the most interesting descriptor in each set. The process of partitioning the subsets continues until some terminating criteria are met, such as the size of the subset falls below a specific number of chemical structures.

In Figure 26.21, a set of chemical structures has been characterized by a series of pharmacophore fingerprints (Liu et al., 2005). The response variable is a specific biological activity, and the mean activity is shown in each node in

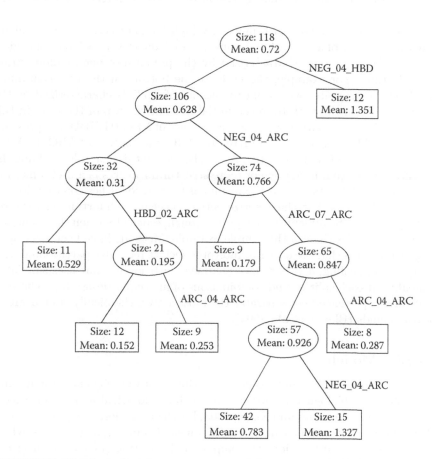

FIGURE 26.21: A decision tree generated from a set of chemical structure with biological activity data.

the decision tree. The initial node at the top of the tree is the entire dataset, containing 118 observations with a mean biological activity of 0.702. After having considered all descriptors, the original set is divided into two subsets: a set of 12 and 106 structures. The descriptor NEG_04_HBD is selected as the most biologically interesting, with those chemical structures containing this feature divided into a subset of 12, and those remaining placed into the node containing 106 chemical structures. As can be seen from the mean activities, the subset containing the feature has a considerably higher mean activity, compared to the set without the feature. These two nodes are considered further to attempt to identify further interesting subsets. In the case of the node containing 106 chemical structures, the descriptor NEG_04_ARC is determined to divide the 106 structures to form the most interesting groups. There is no further division of the set containing 12 structures since any further splitting violates one or more of the terminating conditions. The decision tree continues to grow until the terminating condition stops all further division of the branches.

The nodes in the tree with unusually high levels of activity represent interesting classes of chemical structures, based on the associated biological activity. The nodes are characterized by the presence of one or more structural features. For example, the node at the bottom of the tree containing 15 chemical structures with a mean activity of 1.327 is characterized by the features used on the path from the initial node at the top of the tree. In this example, the set is characterized by the absence of NEG_04_HBD, the presence of NEG_04_ARC, the presence of ARC_07_ARC, the absence of ARC_04_ARC, and the presence of NEG_06_ARC. As can be seen in this example, dividing the set into subsets containing one or more structural features is a useful method for analyzing datasets such as the results of HTS. It identifies combinations of structural features that characterize active molecules. Unfortunately, the resulting trees can be difficult to navigate through, especially when they generate many more nodes than in this simple example. Methods have been developed to generate trees that are easier to interpret such as recursive partitioning with simulated annealing, a variation of recursive partitioning that uses simulated annealing at each split to find combinations of molecular descriptors whose simultaneous presence best separates the most active, chemically similar group of compounds (Blower et al., 2004).

26.8.5 Modeling

A prediction model can be generated using a dataset of chemical structures and associated biological response data, such as biological activity or a measured toxicity (Gasteiger and Engel, 2003; Leach and Gillet, 2007). These models enable estimates to be calculated for untested compounds (in silico) which in turn can be used to prioritize compounds for testing (in vivo or in vitro). This approach is commonly referred to as quantitative structure–activity relationships or QSAR. Generally, models built from structurally homogeneous structures can be used to predict the biological response of an untested chemical structure, providing the untested compound is within the same structural class from which the model was built. In certain situations, global models are built from diverse sets of chemical structures; however, prior to calculating an estimate from the model, a check should be performed to verify that the untested chemical structure is within the *applicability domain* of the training set used to build the model.

A dataset is usually prepared as a table containing multiple descriptor variables to characterize the chemical structures, along with the biological response data. It takes a great deal of effort to prepare the dataset prior to building the model, including prioritizing which descriptor variables to use and potentially transforming the biological response data. Also prior to building a prediction model, portions of the dataset are usually set aside in order to test the model's accuracy. This avoids the use of the actual data used to build the model (training set) for testing purposes, as this may result in an over-optimistic assessment of the model's accuracy. Methods such as R^2 (coefficient of determination) for regression models or concordance/error rate/sensitivity/specificity metrics for binary classification models are used to assess the overall quality of any models built.

Having selected the dataset, as well as a small set of descriptors to use, a number of approaches can be used to build models. Improvements to the model's performance can be achieved by building models with different combinations of descriptor variables, as well as fine-tuning any model parameters. The following are some of the methods used in chemoinformatics to build prediction models:

- Linear regression: this approach will generate a model when the biological response data is a continuous variable. It will model the linear relationship between one or more descriptor variables and a single response variable. The relationship is described as a weighted sum of the descriptor variables. Linear models are easy to understand as they generate a straightforward equation; however, the approach is limited to linear relationships.

- Logistic regression: this method can be used when the biological response data is a binary categorical variable, that is, contains two values: positive ("1") and negative ("0"). Like linear regression, it will generate a simple and easy to interpret formula from one or more descriptor variables.

- Neural networks: a neural network is comprised of interconnected and weighted nodes that join the input descriptor variables to the output response variable. In the learning phase, the network learns from examples by adjusting the network's internal weights. Once a neural network has been trained, untested data can be fed into the network and an estimate for the biological response calculated. Neural networks are a flexible approach to modeling data, and are capable of modeling non-linear relationships; however, they are considered a black box since it is not possible to easily understand the reasons why the estimate was calculated.

- Naive Bayes: this approach will calculate estimates for categorical response data, where the descriptors are also categorical. It makes use of Bayes theorem using the assumption that the descriptor variables are all independent, which is not usually the case. This assumption enables the method to operate efficiently and has been particularly effective when used over large datasets.

- k-nearest neighbors: k-nearest neighbors or kNN is an approach that can be used to build both classification and regression models. An estimate for an untested chemical structure is made by identifying the k most similar chemical structures in the training set. This similarity is often based on the Tanimoto coefficient using structural fingerprints. The average activity of the identified structures is used when the biological response is continuous or the mode value when the biological response is categorical and assigned to the untested compound. The actual number of nearest neighbors to use (k) is optimized prior to using the model.

- Random forests: decision trees can also be used to generate predictions since they classify all the chemical structures into one of the terminal nodes in the tree. Untested compounds can be assigned to a terminal node in the tree using the rules associated with the branch points in the tree. For example in Figure 26.21, an untested compound that has the feature NEG_04_HBD would be assigned to the node shown as having 12 members and an average activity of 1.351. The prediction for an untested compound is either the assigned node's average value (for regression models) or the mode value (for classification models). A random forest is a collection of decision tree models, where each decision tree used a different set of descriptors to build the model. Estimates for the untested compounds are calculated from each model, and a final consensus score is generated from all the results.

- Support vector machines: support vector machines or SVMs attempt to model categorical response variables by identifying a hyperplane that separates the different response classes using the descriptor variables. They are a useful approach; however the models are difficult to interpret.

Once a model has been built, it can be used to calculate estimates for untested compounds. These chemical structures should be prepared in the same manner as the chemical structures used to build the model, that is using the same calculated descriptors and the same set of mathematical transformations. After being satisfied that the untested compounds are within the applicability domain of the model, an estimate for the biological response data can be calculated. This estimate can then be used to prioritize research directions. For example, the model results may help to prioritize additional chemicals to acquire and test. Alternatively, models built from safety data may help to guide which chemicals to avoid or prioritize *in vivo or in vitro* safety testing options, focusing on those highlighted as potential issues by the toxicity *in silico* model results.

26.9 Chemogenomics

Chemogenomics involves the use of genomics to measure the system-wide effect of a compound on a biological system, either single cells or whole organisms (Bredel and Jacoby, 2004). It combines high-throughput genomics or proteomics profiling with chemoinformatic and statistical analysis to study the response of a biological system to chemical compounds. Chemogenomics also investigates the consequences of differential gene/protein expression on cellular response to compound treatment which is measured by phenotypic readouts in a high-throughput assay. The use of chemoinformatics allows one to deal with numerous compounds and compound classes simultaneously. This has proven valuable where the analysis of the effect of single compounds is insufficient to draw inferences on biological mechanisms.

A model system for chemogenomic analysis is the panel of 60 diverse human cancer cell lines (the NCI-60) used by the National Cancer Institute (NCI) to screen >100,000 chemical compounds and natural product extracts for anticancer activity since 1990 (Weinstein, 2006). Dose-response curves generated by the screening process provide 50% growth inhibitory (GI_{50}) values for each compound-cell line pair. Screening data for approximately 43,000 nonproprietary compounds are publicly available (http://dtp.nci.nih.gov). To take advantage of the pharmacological profiling, the NCI-60 panel has been the subject of numerous genomic, proteomic, and other "-omic" studies (Weinstein, 2006). The collection of datasets related to the NCI-60 provides an unparalleled public resource for integrated chemogenomic studies aimed at elucidating molecular targets, identifying biomarkers for personalization of therapy, and understanding mechanisms of chemosensitivity and -resistance.

In 1997, Weinstein et al. laid out a conceptual framework for large-scale integration of databases that reveals connections between potencies of cytotoxic compounds and cellular characteristics (Weinstein et al., 1997). This approach involves the three databases shown in Figure 26.22. Database [A] contains the activity patterns of tested compounds over the NCI-60 cell lines, [S] contains molecular structural features of the compounds, and each row in database [T] contains the pattern (across the NCI-60) of a measured cellular characteristic that may modulate cellular response to compound dosing. Databases [A] and

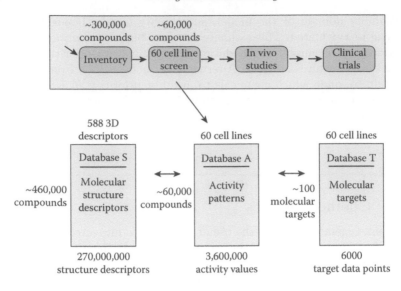

FIGURE 26.22: Simplified schematic overview of an information-intensive approach to cancer drug discovery and molecular pharmacology at the NCI. Each row of the activity (A) database represents the pattern of activity of a particular compound across the 60 cell lines. As described in the text, the (A) database can be related to a structure (S) database containing 2D or 3D chemical structure characteristics of the compounds and a target (T) database containing information on possible molecular targets or modulators of activity within the cells. (Reprinted from Weinstein, J.N., Myers, T.G., O'Connor, P.M., Friend, S.H., Fornace, A.J. Jr, Kohn, K.W., Fojo, J. et al., *Science*, 275, 343–349, 1997; statistics are 1997 data; With permission.)

[T] were integrated by calculating the correlation between the activity pattern of a compound and the characteristic pattern of a target. Each row of [A] and [T] was normalized and the inner product of the two matrices was calculated, producing a matrix of Pearson correlation coefficients. Each row in the correlation matrix corresponds to a compound, each column to a target, and entry (i, j) gives the correlation between the activity pattern of compound C_i and the characteristic pattern of target T_j.

Pursuing a suggestion of Weinstein et al. (1997), Blower et al. (2002) developed statistical techniques for selecting genes with characteristic expression patterns, and applying structure-based data mining techniques to identify substructural compound classes correlated with gene expression. This semiempirical method projects genomic information from cells through compound activity patterns to structural classes of potential drugs.

In a series of studies of the NCI-60 datasets (Wallqvist et al., 2003), Wallqvist and co-workers developed data mining tools to study associations between chemosensitivity patterns and gene expression, molecular target measurements and mutational status over the NCI-60 cell lines. Compounds are clustered using a self-organizing map (SOM). Each node is assigned a fingerprint which encodes the growth inhibition pattern for compounds within the node, and a similarity score is calculated between each node fingerprint and the fingerprints for measurements of mRNA expression, molecular targets and mutational status. For each gene or molecular target, the similarity scores are mapped onto the SOM using a coloring scheme which gives a visual image of the correlations. Numerous examples were found where a correlation suggested a biochemical basis for cellular cytotoxicity.

Huang and coworkers conducted a series of studies (Huang et al., 2005) that use informatics approaches to associate compounds or compound classes with relevant gene families, followed by experimental validation. These studies—focusing on transporters such as ABCB1, encoding the multidrug resistance protein MDR1, and SLC7A11, encoding the transporter subunit of the heterodimeric amino acid transport system—exemplify the utility of chemogenomics in combination with rapid in vitro testing of drug candidates in selected cell lines expressing the target gene.

Giaever et al. (2004) developed a heterozygous yeast deletion library for drug target identification based on the hypothesis that reducing gene dosage of drug targets from two copies in a wild-type strain to one copy after heterozygous deletion will increase drug sensitivity. Chemical compounds were tested against pooled cultures of approximately 6000 heterozygous yeast deletion strains, and DNA from the culture was hybridized to oligo-microarrays to identify the strains that are sensitive and inhibited by the compounds. The authors characterized gene products interacting with anticancer, antifugal, and anticholesterol agents. This technique can accelerate the drug discovery process by revealing unknown drug targets, increase mechanistic understanding, and aid in prioritizing compounds for development.

Using cell lines that were most resistant or sensitive to specific drugs, Potti and coworkers developed gene expression signatures that predict sensitivity to individual chemotherapeutic drugs (Potti et al., 2006). In many cases, the signatures can predict clinical response to drug treatment, and individual drug signatures could be combined to predict response to multidrug regimens.

Lamb et al. (2006) developed the *connectivity map*, a database of drug-gene signatures for linking drugs, genes and diseases. The authors profiled 164 bioactive small molecules including many FDA approved drugs in four of the NCI-60 cell lines. Following compound treatment, mRNA expression levels were measured giving a database of reference profiles. Each reference profile comprises a rank ordered list of genes ordered by differential expression relative to the batch control. The database can be searched by comparing a query signature of up- and down-regulated genes with reference signatures.

References

Barnard, J.M. and Downs, G.M. 1992. Clustering of chemical structures on the basis of two-dimensional similarity measures. *J. Chem. Inf. Comput. Sci.*, 32:644–49.

Bredel, M. and Jacoby, E. 2004. Chemogenomics: an emerging strategy for rapid target and drug discovery. *Nat. Rev. Genet.*, 5:262–75.

Blower, P.E., Cross, K.P., Fligner, M.A., Myatt, G.J., Verducci, J.S., and Yang, C. 2004. Systematic analysis of large screening sets in drug discovery. *Curr. Drug Discov. Technol.*, 42:393–404.

Blower, P.E., Yang, C., Fligner, M.A., Verducci, J.S., Yu, L., Richman, S., and Weinstein, J.N. 2002. Pharmacogenomic analysis: correlating molecular substructure classes with microarray gene expression data. *Pharmacogenomics J.*, 2:259–71.

Gasteiger, J. and Engel T. 2003. *Chemoinformatics: A Textbook*. Wiley-VCH, Weinheim.

Giaever, G., Flaherty, P., Kumm, J., Proctor, M., Nislow, C., Jaramillo, D.F., Chu, A.M., Jordan, M.I., Arkin, A.P., and Davis, R.W. 2004. Chemogenomic profiling: identifying the functional interactions of small molecules in yeast. *Proc. Natl. Acad. Sci. USA*, 101:793–98.

Hawkins, D.M., Young, S., and Rusinko, A. 1997. Analysis of large structure-activity data set using recursive partitioning. *Quant. Strct.-Act. Relat.*, 16:296–302.

Huang, Y., Blower, P.E., Yang, C., Barbacioru, C., Dai, Z., Zhang, Y., Xiao, J.J., Chan, K.K., and Sadée, W. 2005. Correlating gene expression with chemical scaffolds of cytotoxic agents: ellipticines as substrates and inhibitors of MDR1. *Pharmacogenomics J.*, 5:112–25.

Leach, A.R. and Gillet, V.J. 2007. *An Introduction to Chemoinformatics*. Springer, Dordrecht, The Netherlands.

Lamb, J., Crawford, E.D., Peck, D., Modell, J.W., Blat, I.C., Wrobel, M.J., Lerner, J. et al. 2006. The Connectivity Map: using gene-expression signatures to connect small molecules, genes, and disease. *Science*, 313:1929–35.

Liu, K., Feng, J., and Young, S.S. 2005. PowerMV: a software environment for molecular viewing, descriptor generation, data analysis and hit evaluation. *J. Chem. Inf. Model.*, 45:515–22.

Potti, A., Dressman, H.K., Bild, A., Riedel, R.F., Chan, G., Sayer, R., Cragun, J. et al. 2006. Genomic signatures to guide the use of chemotherapeutics. *Nat. Med.*, 12:1294–300.

Wallqvist, A., Rabow, A.A., Shoemaker, R.H., Sausville, E.A., and Covell, D.G. 2003. Linking the growth inhibition response from the National Cancer Institute's anticancer screen to gene expression levels and other molecular target data. *Bioinformatics.* 19:2212–24

Weinstein, J.N. 2006. Spotlight on molecular profiling: "Integromic" analysis of the NCI-60 cancer cell lines. *Mol. Cancer Ther.*, 5:2601–5.

Weinstein, J.N., Myers, T.G., O'Connor, P.M., Friend, S.H., Fornace, A.J. Jr., Kohn, K.W., Fojo, T. et al. 1997. An information-intensive approach to the molecular pharmacology of cancer. *Science*, 275:343–49.

Index

Printed and bound by CPI Group (UK) Ltd, Croydon, CR0 4YY

21/10/2024

01777103-0017